HYPERBOLE Symposium 2017

HYdropower plants PERformance and flexiBle Operation towards Lean integration of new renewable Energies

Journal of Physics: Conference Series Volume 813

Porto, Portugal
2 – 3 February 2017

ISBN: 978-1-5108-3899-4
ISSN: 1742-6588

Printed from e-media with permission by:

Curran Associates, Inc.
57 Morehouse Lane
Red Hook, NY 12571

Some format issues inherent in the e-media version may also appear in this print version.

Printed by Curran Associates, Inc. (2017)

For permission requests, please contact the Institute of Physics
at the address below.

Institute of Physics
Dirac House, Temple Back
Bristol BS1 6BE UK

Phone: 44 1 17 929 7481
Fax: 44 1 17 920 0979

techtracking@iop.org

Additional copies of this publication are available from:

Curran Associates, Inc.
57 Morehouse Lane
Red Hook, NY 12571 USA
Phone: 845-758-0400
Fax: 845-758-2633
Email: curran@proceedings.com
Web: www.proceedings.com

Table of contents

Volume 813

HYPERBOLE Symposium 2017 (HYdropower plants PERformance and flexiBle Operation towards Lean integration of new renewable Energies)

2–3 February 2017, Porto, Portugal

Accepted papers received: 15 February 2017
Published online: 4 April 2017

Preface

Papers

Electric Power System

Energy Regulation

Hydro Units

IAHRWG

Preface to the book of proceedings

The HYPERBOLE Conference was held in Porto on February 2 -3, 2017 within the framework of the HYPERBOLE research project, granted by the European Commission (ERC/FP7- ENERGY-2013-1-Grant 608532). This research project aimed to enhance **HY**dropower plants **PER**formance and flexi**B**le **O**peration towards **L**ean integration of new renewable **E**nergies. To achieve this, the following three fold objectives were defined: *i)* understanding the root causes for the unit operating range limitations of hydropower plants, *ii)* modeling and simulating the hydropower plants dynamics over their full range of operation and, then, *iii)* assessing the impact of the enhanced operation of hydropower plant on the development of so-called New Renewable Energy Resources. The project methodology was based on a system modeling relying on multidisciplinary approach considering hydraulic, mechanical, electrical and, finally, control engineering aspects. Both experimental and numerical simulations tools were used to investigate the aforementioned aspects for selected case studies: i.e. a Francis hydroelectric unit of 440 MW, a reference reversible pump-turbine and the Portuguese Transmission System as a whole.

The conference brought together 123 energy specialists from hydraulic, power systems and energy markets to setup a forum to share research advances and technical achievements in the aforementioned topics. In addition, the conference was a unique opportunity to disseminate the results of EU-funded FP7 HYPERBOLE project.

During the conference opening ceremony, Prof. François Avellan, EPFL, coordinator and scientific representative of the HYPERBOLE research project presented the general background of the project. Then, Prof. António Sá da Costa, chair of the Portuguese Renewable Energy Association (APREN) presented the status of renewable electricity in Portugal. Finally, Mr. Décio Silva, Elec. Eng. MSc., from EDP generation, presented the Venda Nova III Repowering Project, to introduce the technical tour organized for the afternoon following the conference.

Then, during the plenary session, the leaders of the project work packages gave an overview of the HYPERBOLE research project. Prof. François Avellan introduced the project, while Dr. Thomas Aschenbrenner, VOITH Hydro, and Mr. Pierre-Yves Lowys, presented the hydraulic and the mechanical challenges, respectively. Then, Mr. Johan Hell, ANDRITZ, presented the concepts of high flexible hydropower generation for future grids and Prof. Carlos Moreira, INESTEC, the role of pumped storage power plants in electric power systems. Finally, Prof. Eduard Egusquiza, UPC, presented the on-site monitoring of the Francis hydroelectric unit case study.

For the closing ceremony, Mr. Luciano dos Santos, GE Renewable Energy Hydro, gave an outlook about the digital disruption of energy. Then, Mr. Manuel Oliveira, EDP Generation, presented as project manager, the details of the Venda Nova III Repowering Project.

Finally, Prof. François Avellan addresses the concluding remarks of the Conference and thanks the attendees, the local organizers, Dr. Bernardo Silva, Prof. Carlos Moreira and Prof. João Peças Lopes, the partners of the HYPERBOLE research project and the European Commission for their supports.

It is worth to mention that, to increase further the public impact of the HYPERBOLE Conference among the scientific community, the 7th Meeting of the IAHR Working Group on Cavitation and Dynamic Problems was organized in Porto on February 1, 2017, 1 day just before the HYPERBOLE Conference. The authors wanted to thank Prof. Romeo Susan-Resiga, leader of the IAHR Working Group, and Dr. Sebastian Muntean both from the Romanian Academy Timişoara.

Lausanne, February 13, 2017 Porto, February 13, 2017

Prof. François Avellan, Eng. Dr. Dr. Bernardo Silva

Prof. Carlos Moreira

Peer review statement

All papers published in this volume of *Journal of Physics: Conference Series* have been peer reviewed through processes administered by the proceedings Editors. Reviews were conducted by expert referees to the professional and scientific standards expected of a proceedings journal published by IOP Publishing.

Assessment of power step performances of variable speed pump-turbine unit by means of hydro-electrical system simulation

A Béguin[1], C Nicolet[1], J Hell[2], C Moreira[3]

[1] Power Vision Engineering Sàrl, Ch. Des champs courbes 1, CH 1024 Ecublens, Switzerland
[2] ANDRITZ HYDRO GmbH, Eibesbrunnergasse 20, A-1120 Vienna, Austria
[3] Centre for Power and Energy Systems, INESC TEC – INESC Technology and Science, Porto, Portugal

Abstract. The paper explores the improvement in ancillary services that variable speed technologies can provide for the case of an existing pumped storage power plant of 2x210 MVA which conversion from fixed speed to variable speed is investigated with a focus on the power step performances of the units. First two motor-generator variable speed technologies are introduced, namely the Doubly Fed Induction Machine (DFIM) and the Full Scale Frequency Converter (FSFC). Then a detailed numerical simulation model of the investigated power plant used to simulate power steps response and comprising the waterways, the pump-turbine unit, the motor-generator, the grid connection and the control systems is presented. Hydroelectric system time domain simulations are performed in order to determine the shortest response time achievable, taking into account the constraints from the maximum penstock pressure and from the rotational speed limits. It is shown that the maximum instantaneous power step response up and down depends on the hydro-mechanical characteristics of the pump-turbine unit and of the motor-generator speed limits. As a results, for the investigated test case, the FSFC solution offer the best power step response performances.

1. Introduction

The integration of the constantly growing capacity of New Renewable Energies, NRE, mainly composed by wind and solar energies, is a challenging task as far as the power network stability is concerned, due to the intermittent nature of these energy sources [1]. Beside storage and substitution production capabilities, pumped storage power plants can significantly contribute to improve the stability of power network by providing ancillary services, taking advantage of their production flexibility, fast response time and large energy storage capability [2], [3]. The capability of pumped-storage power plant to achieve very fast power injection or absorption assuming a conversion from fixed speed to variable speed technology (see [7]) was investigated in the framework of the HYPERBOLE European Research Project. In the framework of hydro unit refurbishment, such conversion may further increase the unit flexibility thus enable a better integration of NRE. This capability was investigated by means of detailed hydroelectric system time domain simulation to evaluate pump-turbine unit active power step response up and down. The response time of fixed speed units is limited by the maximal rate of opening and closing of the pump-turbine guide vanes, which is optimized to comply with the maximum allowable penstock pressure and the maximum unit over speed during transient regimes. The variable speed motor-generator solutions not only enable efficient off-design operation and pump input control of pump-turbine units, but can also considerably improve active power response time, by decoupling grid frequency from the unit rotational speed, and taking advantage of the fly-wheel effect of the mechanical rotating train. This paper assesses the performances of variable speed pump-turbine units regarding the response time at the grid connection, using numerical simulations of the entire hydro-electrical system of an existing pumped storage plant which is assumed to be converted from fix speed to variable speed. This paper aims at highlighting the performances and limitations of power step performances obtained with two different variable speed technologies, namely the Doubly Fed Induction Machine (DFIM) ([4],[5]) and the Full Size Frequency Converter (FSFC) ([6],[7]). This paper first introduces both variable speed technologies and related control strategies. Then, the 2x210 MW pumped storage power plant case study and the related simulation load cases are presented. Finally simulation results are presented to compare the power step performances achieved with DFIM and FSFC solutions for the considered test case.

2. Variable speed technologies and the control strategy

Figure 1 presents the hydro-electrical model of the existing 2x210 MW pumped-storage power plant case study, modelled with the simulation software SIMSEN, with the hydro-mechanical model (upper part) and electro-mechanical model (lower part). The hydro-mechanical model takes into account the 2 individual penstocks, the 2 reversible Francis pump-turbines with related governors, the downstream surge tanks and the tailrace tunnel. The electro-mechanical layout is shown for both technologies, DFIM (lower left) and FSFC (lower right). For DFIM, the electrical machine is a Wound Rotor Induction Generator with an open terminal rotor, which is fed from a power frequency converter, while for FSFC, the electrical machine is a classical synchronous machine with the stator directly connected to the power frequency converter itself connected to the grid. DFIM solution has one advantage over FSFC which is that the rated power of the frequency converter is only a fraction (typically 10%) of the rated power of the unit. However, for DFIM, the speed range is limited to a given range around the rated speed, typically ≤ 10%, mainly due to thermal limitation of the rotor of the induction machine, while for FSFC, the speed range is not limited by the technology itself.

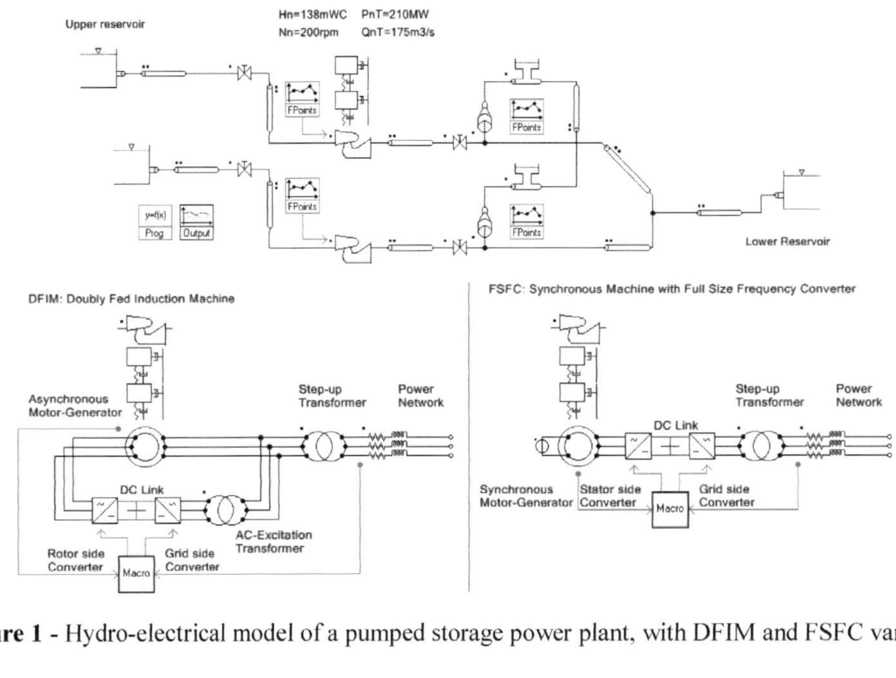

Figure 1 - Hydro-electrical model of a pumped storage power plant, with DFIM and FSFC variants.

For both technologies, the power converter is used to control the active and reactive power that is injected into the grid while the speed of the unit is controlled by the turbine speed governor, acting on the guide vane opening. In pump mode, the pump characteristic imposes the speed for a given power and head, while the guide vane opening is set to an optimal value for maximum pump efficiency. This control strategy allows to use the capability of a power converter to change the operating point very quickly, typically in the order of hundreds of millisecond or smaller [4], [5]. The turbine speed governor response time is much slower due to the hydraulic and mechanical constraints, but the rotational speed is allowed to vary within a given range, thanks to the decoupling between the grid frequency and the units rotational speed. The response time of the turbine speed governor is limited by the opening and closing time of the guide vanes which are set to comply with maximum allowable penstock pressure and unit rotational speed during in transient regimes. The variable speed technologies allow for fast power performance of the converter by exploiting the so called "fly–wheel" effect the rotating inertias of the unit. However, the fly-wheel effect is limited by the allowed speed range of the motor-generator and also by the maximum torque that the turbine can deliver which depends on its torque characteristic. As a consequence, in some cases, the ideal response time that can be expect from a power converter and is restricted and results in a slower power set point ramp.

3. Simulations model and load cases

The SIMSEN simulation of Figure 1 comprises high order models of the hydraulic system, taking into account for penstock water hammer effect, downstream surge tanks mass oscillations and pump-turbine 4 quadrant characteristics including the unstable S-shape characteristic, and the PID turbine governor with guide vane rate limiter. The rated apparent power of the unit is 210MVA, under a voltage of 15kV and the frequency of the electrical grid is 50Hz. The nominal rotational speed is 200rpm. The limitations for the speed deviation from rated speed, for DFIM, has been set to +-7%, which is a typical value. For FSFC, there is no lower limit for speed but upper limit corresponds to maximum allowable over-speed of the unit. Simulation are performed considering three different active power set point decrease of respectively 0.1, 0.2 or 0.4 pu, from nominal power and 100s latter an active power set point increase back to the nominal power. For each power set point decrease and increase, the minimal duration of the set point transition is determined by successive trial, starting with an ideal instantaneous transition and progressively increasing the ramp time until the resulting transients are stable and respect the constraints on rotational speed and penstock's pressures. The constraints on pressures are in fact ensured by the closing time of the guide vane, which has been determined beforehand also by simulations on specific dimensioning load cases. In variable speed technologies, the speed of the mechanical shaft is usually set to optimize the pump-turbine efficiency and hence depends on the operating point. In this study, it is chosen to let the speed set point to the nominal speed in order to benefit from the maximal speed range for both over-speed and under-speed deviations.

4. Simulation results

Table 1 summarizes the response time achieved by the pump-turbine unit of interest, for both DFIM and FSFC technology, for the three different magnitudes of power step and for power set point decrease and increase. The simulation results are shown in figures 2a and 2b for the DFIM and FSFC technology respectively. In figure 2, the speed of the unit and the guide vane opening are shown, as well as the active power at the grid connection point. Power decrease results in a speed rise that is later restored to the nominal speed set point by the action of the turbine speed governor. Similarly, for power increase, a speed decrease first happens followed by the recovery of the nominal speed set point.

Table 1- Response time of power set point decrease and increase, for DFIM and FSFC

Set point initial value (MW)	Set point final value (MW)	(pu)	Power step magnitude (MW)	(pu)	ramp time decrease (s)		ramp time increase (s)	
					DFIM	FSFC	DFIM	FSFC
210	189	0.9	21	0.1	< 0.1	< 0.1	< 0.1	< 0.1
210	168	0.8	42	0.2	7	< 0.1	7	4
210	130	0.6	80	0.4	14	< 0.1	14	11

From table 1 it could be noticed that for power decrease, the response time are improved significantly when FSFC is used as compared to DFIM, because the upper speed limitation is higher for FSFC than for DFIM. Hence, for FSFC, the power decrease considered here, up to -0.4pu, can be performed instantaneously. A power set point ramp is considered instantaneous when it is in the order of hundreds of millisecond or below, when this is the power converter that imposes this minimal ramp time, which depends on its current controller's dynamic and on its maximum output voltage. For DFIM, only small steps (~ -0.1 pu) can be performed instantaneously. Larger step decrease in DFIM mode require to ramp the power set point change to comply with the upper limit of rotational speed, which is approximately linearly dependant on the power step size, as it can be observed in table for DFIM for the step magnitude -0.2 and -0.4 pu. For DFIM, it can be seen that the response time for a given step size is similar for the power decrease as for the power increase for the three load case considered, mainly because the speed range is symmetrical.

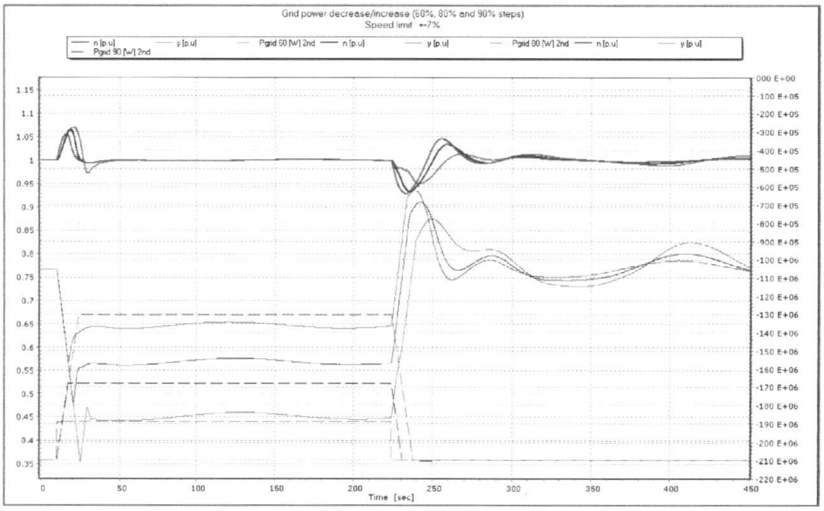

a)

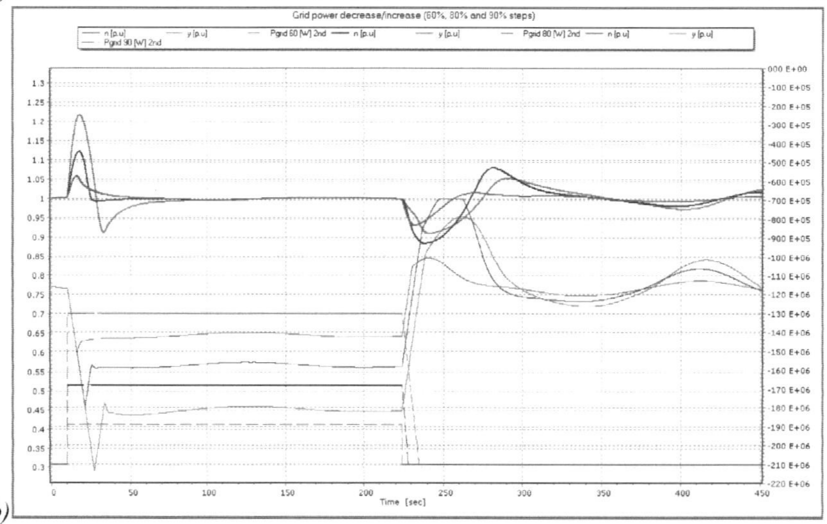

b)

Figure 2 - Simulated transient responses of power set point decrease/increase, for DFIM (a) FSFC (b), unit's speed, guide vane opening ad converter output power

For power set point increase, the situation is still better with FSFC than for DFIM but not as good as for power step decrease. Indeed the improvement in the response time of FSFC compared to DFIM is about 2/3 faster. Although there is no lower limit for the speed in FSFC mode, the fly-wheel effect cannot be fully exploited in power increase because of the maximum torque that the turbine can deliver. Indeed, for a given constant power at the grid, the lower the speed decreases, the higher the torque to apply in order to recover the speed set point. Hence, the maximum torque that the turbine can deliver is quickly reached, which can be clearly observed in figure 2, where the guide vane opening reaches the maximum value and stays saturated for a few seconds. Imposing a quicker power ramp would lead to unstable behaviour of the pump-turbine unit, as the speed governor will not be able to recover the speed set point. The stronger oscillations observed on the speed transient for power set point increase than the one observed for power set point decrease are due to this limited accelerating torque.

5. Conclusions

This paper presents simulation results of a hydro-electrical model of a 2x210MW variable speed pumped-storage power plant, considering in two different variable speed technologies: DFIM and FSFC. The simulated load cases are changes of the active power set point of the generating unit. The shortest ramping time of the set point changes are tight to resulting transient responses on speed and penstock pressures respect the hydro-mechanical constraints. It could be highlighted that using FSFC can globally improve the response time of the power plant compared to DFIM, in particular for power decrease. It was also highlighted that despite the converters fast response time, the speed range, the turbine torque characteristic as well as the penstock pressure limitation restricts the improvement of response time offered by variable speed technologies. For the considered case study, the maximum power set point change that can be performed quasi instantaneously is 0.1 pu. For larger step magnitude, a response time of about 10s is a was obtained for a set point change of magnitude ±0.4pu, which is still an improvement compared to a fixed speed technology. However, it should be mentioned that higher active power steps have been found for other case studies, with about 0.2 pu achieved within less than 0.2s. see [4], [5]. Therefore, it is important to mention that the maximum active power step that can be achieved with variable speed technology strongly depends on the hydro-mechanical characteristics of the pumped storage power plant, and thus has to be carefully evaluated on a case by case basis at early stage of a project using detailed simulation model.

6. References

[1] Martins, N., et al., "The integration of large amounts of renewable energy in the Portuguese Power System", SHF : «Pumped storage Powerplants», Lyon, nov. 2011.

[2] Fisher, R. K., et al., "A Comparison of Advanced Pumped Storage Equipment Drivers in the US and Europe", Louisville, USA, Hydrovision 2012.

[3] Koritarov, V., "Modeling and Analysis of Value of Pumped Storage Hydro", NHA Annual Conference 2013, Washington, DC, USA. http://www.dis.anl.gov/psh

[4] Pannatier, Y., Kawkabani, B., Nicolet, C., Simond, J.-J., Schwery, A., Allenbach, P., *Investigation of control strategies for variable speed pump-turbine units by using a simplified model of the converters*, IEEE Transactions on Industrial Electronics, Volume: 57, Issue: 9, 2010, pp: 3039-3049.

[5] Kuwabara, T., Shibuya, A., Furuta, H., Kita, E., Mitsuhashi, K., "Design and Dynamic Response Characteristics of 400 MW Adjustable Speed Pumped Storage Unit for Ohkawachi Power Station", IEEE Transactions on Energy Conversion, vol. 11, issue 2, pp. 376-384, June 1996

[6] Hell J., Egretzberger M., Lechner A., Schürhuber R., Vaillant Y., "Full size converter solutions for pumped storage plants – a promising new technology", Hydro2012, Euskalduna Congress Centre Bilbao, Spain, 29-31 October 2012.

[7] Schlunegger, H., Thöni, A., "100MW full-site converter in the Grimsel 2 pumped-storage plant", Innsbruck, Hydro 2013

Acknowledgments

The research leading to the results published in this paper is part of the HYPERBOLE research project, granted by the European Commission (ERC/FP7- ENERGY-2013-1-Grant 608532) [32]. The authors would like to thank Iberdrola Generación, S.A.U. in Madrid for their collaboration in the modelling of the prototype power plant test case considered in this paper.

IOP Conf. Series: Journal of Physics: Conf. Series **813** (2017) 012002 doi:10.1088/1742-6596/813/1/012002

Role of pump hydro in electric power systems

R Bessa[1], C Moreira[1], B Silva[1], J Filipe and N Fulgêncio[1]

[1] INESC TEC – INESC Technology and Science, Porto, Portugal

E-mail: carlos.moreira@inesctec.pt

Abstract. This paper provides an overview of the expected role that variable speed hydro power plants can have in future electric power systems characterized by a massive integration of highly variable sources. Therefore, it is discussed the development of a methodology for optimising the operation of hydropower plants under increasing contribution from new renewable energy sources, addressing the participation of a hydropower plant with variable speed pumping in reserve markets. Complementarily, it is also discussed the active role variable speed generators can have in the provision of advanced frequency regulation services.

1. Introduction

Energy storage devices in general and pumped storage power (PSP) plants in particular play an important role for handling new renewable energy (NRE) sources (such as solar and wind) variability and uncertainty[1]-[3]. In what concerns associated capacity, PSP plants are the sole technology that nowadays can effectively face the challenges of variability resulting from a large share of NRE integration. In a large number of existing studies addressing energy storage alternatives for future electric power systems, PSP is found to be the most mature, efficient and cost-effective technology [4]-[5]. For example, the available PSP capacity in Germany (currently at 7.6 GW) is projected to increase by up to 60% until 2020 [6]. Other studies have shown similar significant potential in the United States [7]. A whole-system approach of potential storage solutions in Great Britain has demonstrated that beyond NRE accommodation, large-scale storage schemes entail a number of additional system-wide benefits, such as reduction of required investment in generation, transmission and distribution assets [8]. Currently, exist 270 PSP stations around the world (fully operational and under construction) with a combined capacity of 120 GW. The vast majority of existing hydro power plants employs a conventional setup based on reversible single-stage Francis pump-turbines [9]-[10]. An important drawback of the existing schemes is the prominent use of fixed-speed pump-turbines. Although the traditional fixed-speed design has demonstrated good performance during many decades, there are significant limitations that are becoming apparent as power system requirements are changing while facing the increasing penetration of NRE. The conventional design's main limitations are related to the inability to adjust the pumping input power and the fact that the unit cannot operate over the whole turbine operating range. These limitations significantly reduce the ability for PSP plants to have a more important contribution in order to balance the rapidly-varying output of NRE and to provide ancillary services, especially when pump operating mode is in cause.

The intrinsic variability and uncertainty of NRE create several challenges in power system operation and planning. In every instant, generation must follow load variations in order to maintain the generation-load balance. The highly variable nature of NRE (e.g., rapid generation ramps) represents a

This work was developed in the framework of the HYPERBOLE project – HYdropower plants PERformance and flexiBle Operation towards Lean integration of new renewable Energies – grant agreement number 608532, financed by European funds through the 7th RTD Framework Programme.

challenge, in particular for systems without hydropower, since it introduces variations in the generation-side that can only be smoothed within the physical constraints of the conventional power plants (e.g., ramping up and down, minimum generation limits). In general, the available ramping rates of flexible generation units and fast-starting units (e.g., hydropower) are used for accommodating this variability. If these new type of solutions are not adopted, variability and uncertainty of NRE could lead to situations with high operational cost. For instance, the curtailment of renewable generation during low load periods and the start-up of expensive fast-starting units lead to a cost increase. Moreover, even with perfect forecast for the next hours/days, it is necessary to have in the system flexible generation units for accommodating the generation ramps.

Aiming to provide additional flexibility in electric power systems with high share of NRE, variable-speed hydro machines are a technological option that was firstly deployed in Japan and later in Europe. Nowadays, two main technological solutions can be identified: the use of variable speed technology through the installation of a doubly-fed induction machine (DFIM) or the use of a synchronous generator connected to the grid through a full size frequency converter (FSFC). Worldwide, there are a total of 36 variable-speed PSP plants projects: 17 projects with a total installed capacity of 3.6 GW are operational and 19 projects with a total capacity of 4.6 GW are under construction [10]. In Europe, the main variable-speed operating units are installed in Goldisthal (Germany) with 1060 MW being half of its capacity equipped with variable speed technology, Grimsel II (Switzerland) with 400 MW comprising 100 MW variable speed units equipped with FSFC, Avče (Slovenia) with 185 MW, Nant de Drance and Linth-Limmern with respectively 900 MW and 1000 MW (Switzerland) and Venda Nova III (Portugal) with 736 MW (under final construction phase).

The increasing penetration levels of NRE into electric power grids, together with its specific energy converters technological characteristics, led system operators (SO) to define very restrict rules (grid codes) for allowing increasing NRE penetrations. In general, present grid codes require different generation sources to withstand several disturbances and to support network stability through the provision of ancillary services similarly to those provided by conventional synchronous units. Recalling for international grid code requirements for wind power integration into electric power systems, they can be generally organized in five main categories: (a) fault ride-through requirements, (b) active and reactive power responses following disturbances, (c) extended variation range for the voltage–frequency, (d) synthetic inertia, active power control or frequency regulation support, and (e) reactive power control or voltage regulation capability. Taking the envelop of these requirements and the specific case of hydro power technology, the controllability offered by variable speed units makes possible to provide enhance flexibility for improving global system operation by providing important contributions with respect to system requirements.

2. Participation of Variable Speed PSP in the Electricity Market

Compared to conventional fixed-speed PSP, which in general participate in electrical energy sessions (i.e., day-ahead, intraday) and replacement reserve (RR), variable-speed PSP presents a flexibility band that makes it an important asset for frequency restoration reserve (FRR). This reserve category is generally remunerated with a capacity (€/MW) and electrical energy price (€/MWh), which makes it attractive from the financial point of view. Therefore, it is necessary to develop advanced optimization algorithms for the identification of the best opportunities for the participation of variable-speed PSP units in reserve markets, as well as to deal with wind power variability and uncertainty.

As a first step, a survey of the electricity market rules was conducted to study the international rules for FRR and RR. It was identified that the current fast dynamic automatic generation control (AGC) in place at several U.S.A. system operators might be beneficial for PSP units and it is important to study the economic gains of this new AGC signal. Moreover, most of the markets have a capacity price for the FRR (see [11] for a complete overview). Afterwards, an optimization problem for the operation of hydropower plants under increasing contribution from new renewable energy, addressing the participation of a hydropower plant with variable speed PSP in the FRR and RR markets, was developed.

The optimization problem is formulated for determining the reserve capacity bids for each hour of the next day.

A second optimization problem is also formulated, and covers a future scenario where wind power plants participate in the electricity market. In this case, an electrical utility, with hydropower and wind power in its portfolio, combines both generation sources to mitigate wind power forecast errors and try to maximize the income from selling FRR capacity [12]. A stochastic optimization problem, including wind power forecast uncertainty, is developed to compute the market bids (energy and reserve) of this virtual power plant (VPP) for the next day.

2.1. *Individual Participation of Hydropower Plants*

The accurate economic evaluation of variable-speed PSP in the electricity market requires a mathematical model for its flexibility band, suitable to be integrated in the formulation of optimization problems [13]-[14]. This mathematical model was embedded in a chain of optimization models depicted in Figure 1. The framework is divided in the following blocks:

- *Data analytics*: bootstrapping and forecasting algorithms employed to generate long-term and short-term input data related to market prices (electrical energy and reserve) and natural water inflows.
- *Medium-term optimization*: takes into account the seasonality and variability of the natural water inflows combined with market prices to find the optimal allocation of the water resource for one year, considering weekly periods (i.e., 52 weeks) (adapted from [16]). In order words, this module defines the initial and final levels of the water reservoirs for each week.
- *Short-term optimization*: aims to maximize the revenue of the system for a period of one week, considering two different sources of income, the day-ahead electrical energy market and the FRR market (which is also dispatched for the next day).
- *Evaluation module*: evaluates the performance of the bidding strategy, where the markets offers determined by the optimization problem are applied to real market conditions. The real prices are used to evaluate the economic gains, as well as realized natural water inflows.

The final outcome is the market revenue obtained by the hydropower plants from the participation in the electrical energy and reserve market sessions. The results are presented in a complement paper [17] and showed that by having a variable speed pump unit the system revenue increased by 12% in comparison with a fixed speed.

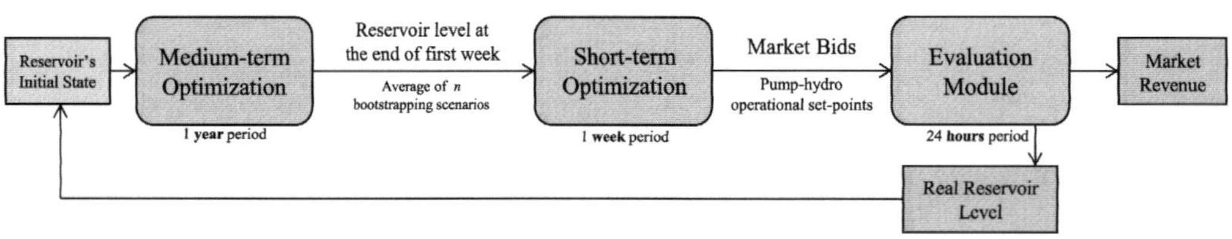

Figure 1. Optimization framework of a variable speed hydropower plant participating in the electricity market.

2.2. *Hydropower Plants Combined with Wind Power Plants*

In order to analyze the economic gains of combining variable speed hydropower and wind power plants, the framework depicted in Figure 2 was implemented.

In this framework, the PSP flexibility (i.e., between min and max values) is divide into three categories: (i) capacity margin to handle wind power variability; (ii) capacity margin to mitigate wind power uncertainty (forecast errors); (iii) upward and downward FRR capacity band around the offer presented to the day-ahead electrical energy market.

Using a machine learning algorithm (gradient boosting tree), a probabilistic forecast is generated for the maximum difference between min and max wind power generation levels within a hourly period (captures variability) and for the wind power generation level (captures uncertainty). The first probabilistic forecasts is an input to define the variability capacity margin, which guarantees a flat generation profile of the hydro-wind VPP (as depicted in Figure 2). The second probabilistic forecast is used to define the capacity margin to mitigate wind power forecast errors, which results in a minimization of the imbalances costs from the wind power plant.

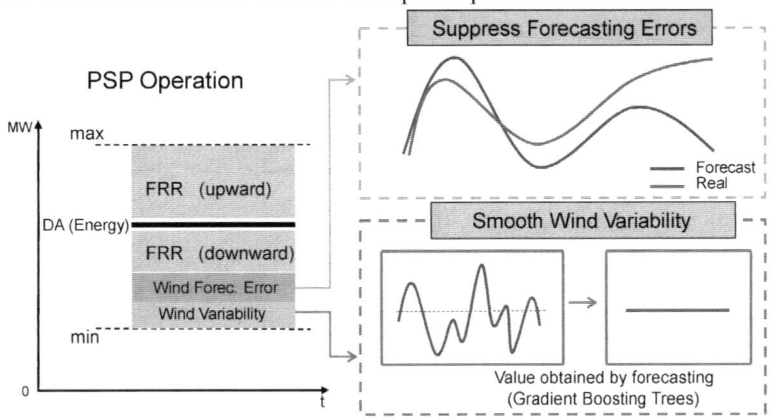

Figure 2. Joint optimization strategy for the combination of a variable speed hydropower and wind power plants.

After estimating these capacity margins, an optimization problem is formulated to determine the optimal offers for electrical energy and FRR capacity from the VPP. An evaluation module was also developed to estimate the economic gains considering the realized prices and natural water inflows. Figure 3 illustrates the hydro and wind power operating curves for a scenario where the FRR price is greater than the electrical energy prices.

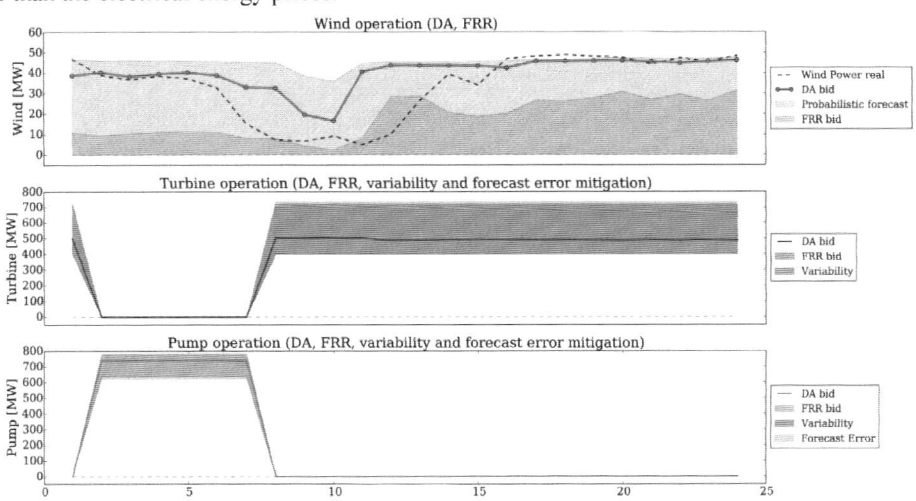

Figure 3. Operating curves for scenario where FRR price is greater than electrical energy prices.

Obtained results showed that there is an added value, both technical and economically, on coordinating the wind farm with a PSP unit with variable speed pumping capabilities. From an economic point of view, the added remuneration comes from the provision of FRR by the wind farm without the need of operating throttled (the Portuguese TSO mandates that 2/3 of the FRR band must be for upward regulation). Furthermore, the PSP unit is able to smooth the wind power variability and mitigate forecasting errors that, otherwise, would lead to market penalties.

3. Provision of Frequency Regulation Services from Variable Speed Hydro Units

Electrical power systems and their control are facing a continuous transformation process as a result of large scale integration of NRE. Nowadays, the grid control services are largely based on large-scale power plants with classical synchronous generators. In the future grid scenario of up to 100% shares of renewables, the active contribution of inverter-based generators to the grid stability and to the security of power supply becomes absolutely necessary. In this view, the flexibility of hydropower is, in one hand, an expected key contribution, and in the other hand an additional difficulty, as in many cases the desired enhanced flexibility range is itself achieved with inverter based generators. Therefore, new services are required like fast frequency containment reserve (FCR) or provision of synthetic inertia in order to meet stable network operation conditions in case of critical disturbances.

The provision of this type of system services (FCR and the provision of synthetic inertia) in variable speed hydro units can largely benefit from the controllability that is offered form inverter-based generators used in both FSFC and DFIM technologies. In this technologies, the use of inverter-based generator allows that in both the turbine and pump modes, the active power output of the machine can be directly controlled by the power electronic interface. Consequently, the fast control action of the power electronic interface makes possible to achieve the desired modulation of the active power output of the unit through the use of a local frequency controller, as it is depicted in Figure 4 and Figure 5. Regarding the turbine operation mode, when a power set-point change is requested to the PSP plant, (for example, resulting from the action of the frequency controller as a result of the speed-droop) it can be directly used at the machine power converters in order to control the active power output. At the same time, the power set-point can be used at the "speed optimizer" in order to define the best efficiency speed reference at which the machine should operate. As a result of a change in the machine's power output, the unit's speed varies as well, being then regulated by the speed governor. Regarding the pump operating mode, it is well known that gate opening control only slightly influences the efficiency of the PSP plant. Therefore, the control strategy of the unit consists on the active power control through the VSC converters, being the speed defined through the characteristic curves of the unit.

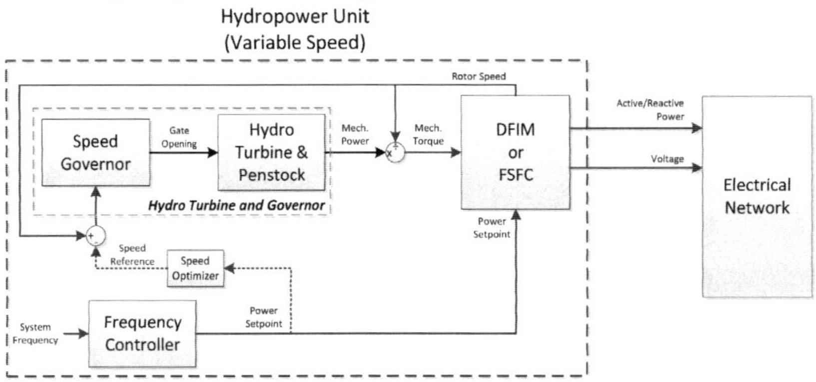

Figure 4. Functional block diagram of power generation and control system for a variable speed PSP plant.

IOP Conf. Series: Journal of Physics: Conf. Series **813** (2017) 012002 doi:10.1088/1742-6596/813/1/012002

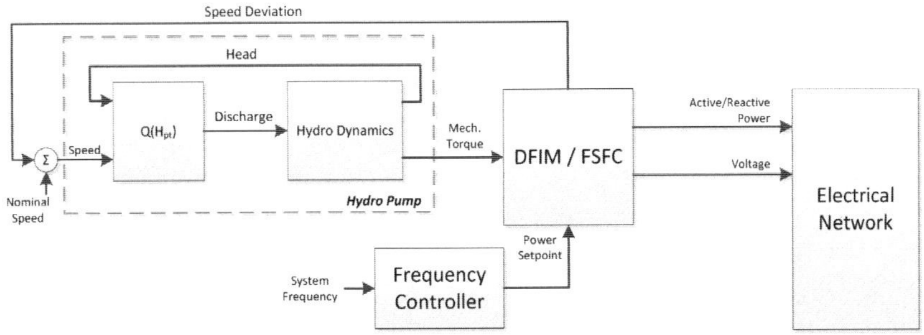

Figure 5. Functional block diagram of a variable speed hydro unit in pumping mode.

In both pumping and turbine operating modes, the frequency is responsible for modulating the unit active power response with respect to grid frequency changes. Therefore, it can incorporate the conventional active power/frequency speed-droop based controlled for the provision of FCR as well as dedicated controllers for the provision of synthetic inertia, for example by modulating the output power proportionally to the rate of change of the grid frequency. As an example, Figure 6 illustrates the response of a variable speed unit in comparison to the response of a classic synchronous-based hydro generator.

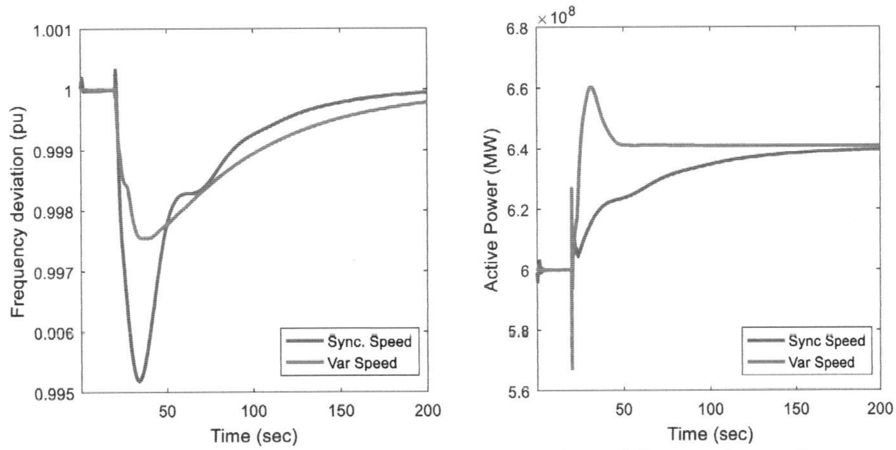

Figure 6. Functional block diagram of a variable speed hydro unit in pumping mode.

4. Final Remarks

Hydropower constitutes nowadays a mature technology that can actively contribute to leverage a sound integration of NRE. The controllability and flexibility characteristics of variable speed generators are of utmost importance regarding the need of providing advanced system support functions in scenarios with massive integration of NRE. In fact, the large scale integration of NRE introduces new challenges associated to the variability and uncertainty of these sources that require specific mitigation strategies. Energy storage capability is one possible mitigation strategy that can be conveniently exploited from hydropower making it by far one of the most effective resources to provide the necessary flexibility and controllability for managing future electricity supply/demand challenges. At the same time, the controllability provided by inverter-based generators applied in variable speed hydro units provides increased performance with respect to provided enhanced frequency regulation services for securing grid operation in future electric power systems.

References

[1] M. A. Hozouri, A. Abbaspour, M. Fotuhi-Firuzabad, and M. Moeini-Aghtaie, "On the use of pumped storage for wind energy maximization in transmission-constrained power systems," IEEE Transactions on Power Systems, vol. 30, no. 2, pp. 1017-1025, March 2015.

[2] F. Díaz-González, A. Sumper, O. Gomis-Bellmunt, and R. Villafáfila-Robles, "A review of energy storage technologies for wind power applications," Renewable and Sustainable Energy Reviews, vol. 16, pp. 2154-2171, May 2012.

[3] K. Hedegaard, and P. Meibom, "Wind power impacts and electricity storage – a time scale perspective," Renewable Energy, vol. 37, pp. 318-324, January 2012.

[4] B. Nyamdash, E. Denny, and M. O'Malley, "The viability of balancing wind generation with large scale energy storage," Energy Policy, vol. 38, pp. 7200-7208, November 2010.

[5] A. Ibrahim, J. Ilinca, and J. Perron, "Energy storage systems - characteristics and comparisons," Renewable and Sustainable Energy Reviews, vol. Volume 12, no. 5, pp. 1221-1250, June 2008.

[6] Bjarne Steffen, "Prospects for pumped-hydro storage in Germany," Energy Policy, vol. 45, pp. 420-429, June 2012.

[7] R. B. J. Chi-Jen Yang, "Opportunities and barriers to pumped-hydro energy storage in the United States," Renewable and Sustainable Energy Reviews, vol. 15, no. 1, pp. 839-844, January 2011.

[8] D. Pudjianto, M. Aunedi, P. Djapic, and G. Strbac, "Whole-Systems Assessment of the Value of Energy Storage in Low-Carbon Electricity Systems," IEEE Transactions on Smart Grid, vol. 5, no. 2, pp. 1098-1109, March 2014.

[9] J. P. Deane, B. P. Ó. Gallachóir, and E. J. McKeogh, "Techno-economic review of existing and new pumped hydro energy storage plant," Renewable and Sustainable Energy Reviews, vol. 14, no. 4, pp. 1293-1302, May 2010.

[10] National Hydropower Association. (2014). Challenges and opportunities for new pumped storage development. [Online]. Available: http://www.hydro.org/wp-content/uploads/2014/01/NHA_PumpedStorage_071212b12.pdf

[11] R. J. Bessa, "Methodologies for the Participation of an Electric Vehicles' Aggregator in the Electricity Markets," Faculdade de Engenharia da Universidade do Porto (FEUP), July 2013.

[12] R. Louvet, A. Neto, L. Billet, J.c. Passelergue, "Smart dispatch of variable-speed Pump Storage Plants to facilitate the insertion of intermittent generation," Cigré Session 2016, Paris, France, Aug. 2016.

[13] M. Chazarra, J. I. Perez-Díaz, and J. Garcia-González, "Optimal operation of variable speed pumped storage hydropower plants participating in secondary regulation reserve markets," in Proc. of the 11th International Conference on the European Energy Market (EEM14), Krakow, Poland, May 2014.

[14] J. I. Pérez-Díaz, M. Chazarra, J. García-González, G. Cavazzini, and A. Stoppato, "Trends and challenges in the operation of pumped storage hydropower plants," Renewable and Sustainable Energy Reviews, vol. 44, pp. 767-784, Apr. 2015.

[15] J. M. Filipe, R. J. Bessa, B. Silva, C. Moreira, "Optimization of the variable speed pump storage participation in frequency restoration reserve market," in Proc. 13th European Energy Market Conference (EEM 2016), Porto, Portugal, 6-9 June 2016.

[16] J. C. Sousa, V. T. Mendes, and J. T. Saraiva, "Estimation of the remuneration of hydro plants in a market environment using an iterative under-relaxation approach," in Proc. of the IEEE PowerTech Conference 09, Bucharest, Romania, Jun. 2009.

[17] J. M. Filipe, R. J. Bessa, B. Silva, C. Moreira, "On the profitability of variable speed pump storage power in frequency restoration reserve," in Proc. of the HYPERBOLE Symposium, Porto, Portugal, 2-3 February 2017.

Application of Power Electronics on Hydropower Generation

Johann Hell

ANDRITZ HYDRO GmbH, Vienna, Austria
E-mail: Johann.Hell@andritz.com

Abstract. The developments in power electronics are offering new opportunities in operation of hydro power generating units. The applied load in pump and turbine operation cannot be changed easily. By using of frequency converters, the speed of the units can be changed in a defined range, without losing much efficiency. An additional benefit of such kind of concept is the improved transient performance of the entire system. In the presented paper the advantage of speed variable power generating system equipped with frequency converters are shown.

1. Classical concepts in Hydro power Generation

Electrical Power Systems are operating with a 3-phase alternating current (AC) system with 50Hz or 60Hz. The synchronous speed is defined by the system frequency and the number of poles of the motor/generator. In most of the generation applications the prime mover is direct coupled with the generator. For very high or very low rotational speed of the turbine a gear box will be used. So the size of the generator can be reduced, but on the other hand the efficiency will be reduced and maintenance costs of the unit are increased. In general, AC generating units have to provide active power for balancing of production and consumption and reactive power for voltage stability.

1.1. Electric Machine concepts

Depending on the application different types of electrical machines are used in hydro power generation. For all machines the active power flow only can be controlled by the turbine.

The electrical machine concepts are also different in the amount of provided Inertia. This can be important when special requirements from the Network Code (NC) must be fulfilled.

1.1.1. Synchronous machines

This type of machine in used in 99% of all the power generating systems. In the classical concept the reactive power can be controlled by variation of the excitation voltage. The excitation current can be transferred by slip-rings to the rotor or brushless exciters can be used.

For transient system stabilization a Power System Stabilizer (PSS) can be used. By measuring speed and electrical power, and additional damping torque will be provided for an active stabilization of the units. In most of the Network Codes this additional control function is mandatory.

For small units in special applications permanent excited machines can be used. Because of missing control of reactive power flow, the application of this option must be confirmed by the Transmission System Operator (TSO).

IOP Conf. Series: Journal of Physics: Conf. Series **813** (2017) 012003 doi:10.1088/1742-6596/813/1/012003

1.1.2. Asynchronous Machines

This type of machine is characterized by a very simple and robust design. On the other hand these units cannot supply reactive power to the grid. For that purpose additional electrical components (reactive power compensation) must be installed.

1.1.3. Asynchronous Machines with slip-rings

For special applications the rotor of the machine can be equipped with a 3 –phase winding system which is connected to 3 slip-rings. This type of machines, together with a converter in the rotor circuit, nowadays is used for variable speed hydro power generation. A detailed description of this system can be found in chapter 3.1 below.

1.2. Requirements for Grid Connection

For grid connection each generating units has to fulfill different requirements defined in the Network Code of the TSO.

All the parameters of the generating units (voltage and frequency variation) are defined on the network connection point – Point of Common Coupling (PCC). The main requirements are: voltage and frequency variation, reactive power capability, frequency regulation capability and transient stability. The last point is most critical, especially for units with low inertia (e.g. Bulb type Hydro units, or gas-turbine driven units). In Figure 1 below, the rotor-angle as a function of the inertia constant H following a 150ms grid disturbance is shown.

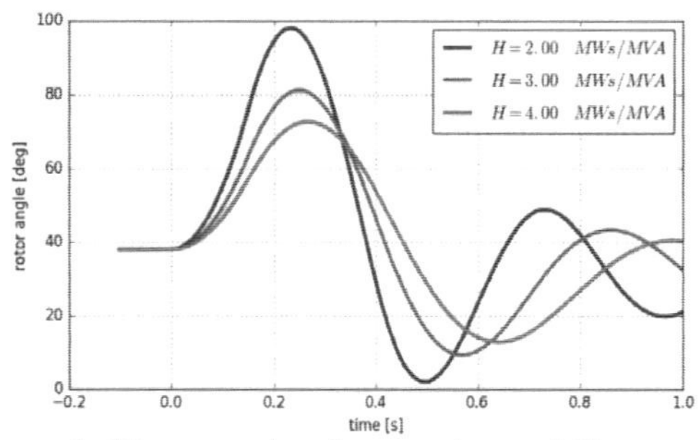

Figure 1: Diagrams showing transient stability on a grid fault with clearing time of 150ms and different Inertia H

The normalized energy of the rotating mass is expressed in the Inertia constant H measured in MWs/MVA.

Units with low inertia are not able to fulfill the transient stability requirements.

The maximal fault clearing time will be defined by the TSO and is in Central Europe 150ms. This time can be different in other regions of the world

2. Power Electronics

The ongoing development in power electronic switching devices enables now to design and build converters in the multi-Megawatt range. Devices like IGBT, IGCT and IEGT from different supplier are available for voltages up to 8000Volts and 4000Amps. The used converter topologies are different depending on supplier. All the offered solutions have an excellent transient performance and a high efficiency.

The principal diagram of a frequency converter is show in Figure 2. The converter itself cannot store energy, so the active power **P** on both ends must be the same at any time. The reactive power ($\mathbf{Q_1}$, $\mathbf{Q_2}$) can be controlled independently and can be different.

By means of a power electronic circuit a voltage phasor is generated which is used for control the power flow between the AC und the DC terminals (Figure 3). The very short response time of the voltage control allows very accurate and fast power control of the converter.

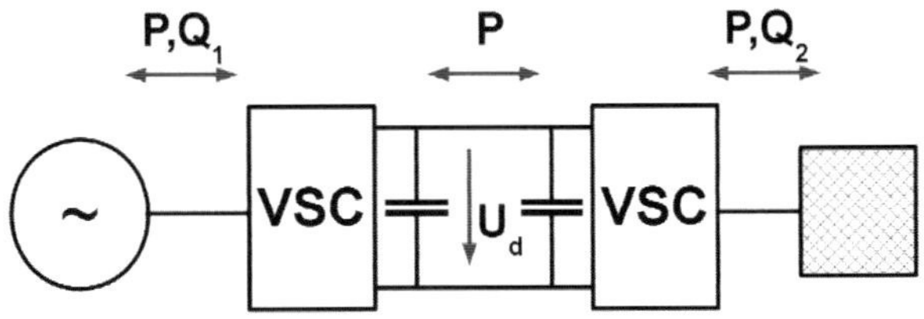

Figure 2: Principal diagram of a Full Size Frequency Converter

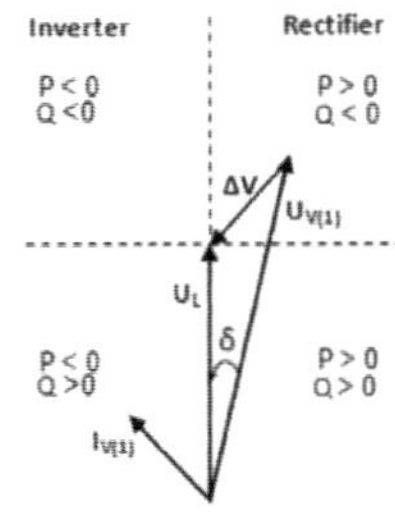

Figure 3: Voltage phasor used for power control

2.1. Classical Three-Level converter

This type of converter is mainly used in drive application for low voltage as well as in medium voltage applications. The shape of the phase voltage of such kind of converter is shown in Figure 4 on the left diagram.

2.2. Multi-level converter

This converter topology will also be used for High Voltage Direct Current (HVDC) power transmission. Hereby identical power electronic modules are connected in series so high output voltages (some 100kV's) can be achieved. The same modules can also be used in application for power generation for voltages up 20kV. For operation at frequencies lower than 15Hz, the output current must be reduced.

In Hydro Power generation this restriction does not limit the application of the converter, because of the speed operating range of a hydro turbine is small and the frequency can be defined by choosing the number of poles of the generator.

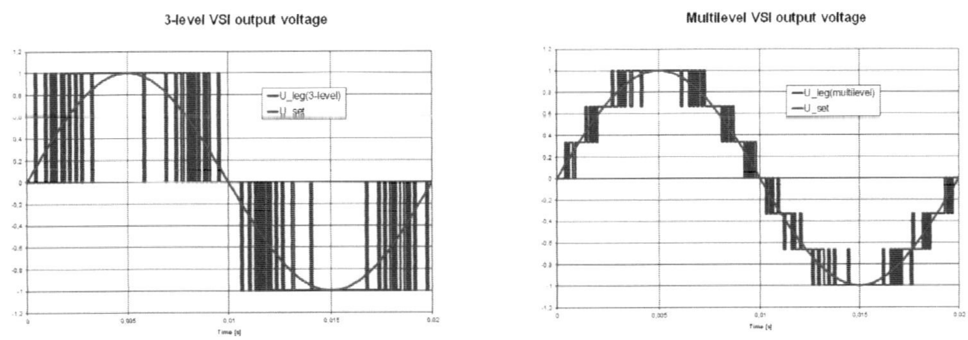

Figure 4: Shape of Phase-Voltage of 3-level (left) and Multi-level converter (right)

3. Variable speed concepts for Hydro Power Generation

Based on the machine and the converter technology different concepts for variable speed power generation systems are possible. In the following paragraphs, an example for a variable speed unit with a rated output of 420MVA is presented.

IOP Conf. Series: Journal of Physics: Conf. Series **813** (2017) 012003 doi:10.1088/1742-6596/813/1/012003

3.1. Double Fed Induction Machine

For the Double Fed Induction Machine concept an asynchronous machine with slip rings is used. The converter is connected between the rotor circuit and the grid for exchange of the "slip-power" (see Figure 5 below). It should be noted, that in generator mode and sub-synchronous operation a circular power flow is generated (left diagram). A detailed description of the electrical model is available in [5]. General transient power system aspects can be found in [6], [8] and [7].

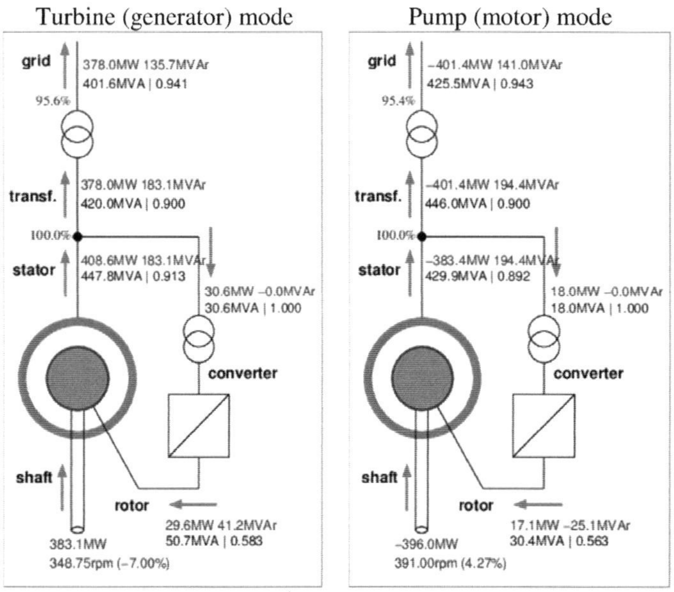

In the diagrams in Figure 5, the power flow from the mechanical shaft through the generator and converter to the grid for different load cases is shown.

The power flow through the converter depends on load flow and on the sign of slip.

In pumped storage applications the units are usually operating in sub-synchronous mode in turbine operation and in super-synchronous mode in pump operation.

Figure 5: Power flow diagram for Turbine operation (sub-synchronous) and Pump operation (super-synchronous) of an example case

This kind of system can operate only in a limited speed range. In the diagrams below (Figure 6) the efficiency for different loads in generator mode and in a defined speed range of +/- 7% is presented.

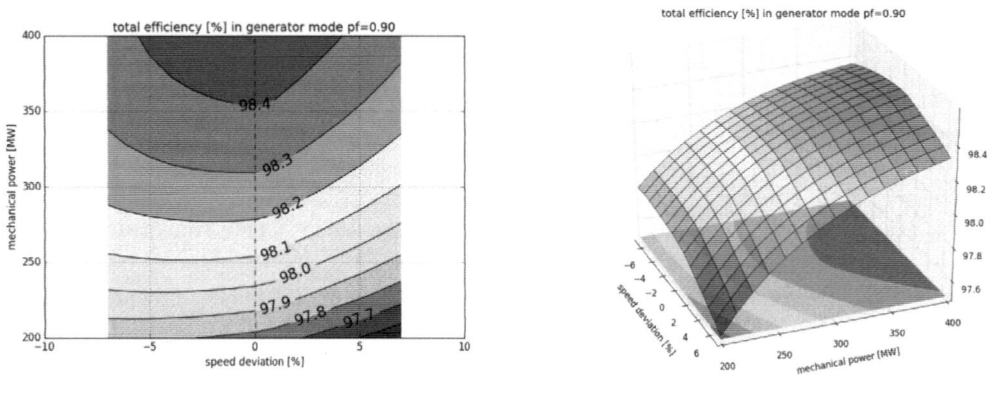

Figure 6: Efficiency for a DFIM for different load and speed

The total losses are calculated as sum of the losses of the involved components with different characteristics. Mechanical losses are depending only on speed, generator losses are more depending on load and some losses are constant.

3.2. Full Size Frequency Converter

For Full Size Frequency Converter application (see Figure 2) the diagrams for efficiency are shown in Figure 7. Due to different loss mechanisms the shape of the losses curves looks different. In general the losses are higher, compared to a DFIM. On the other hand this system offers more flexibility in terms of operating range and transient performance.

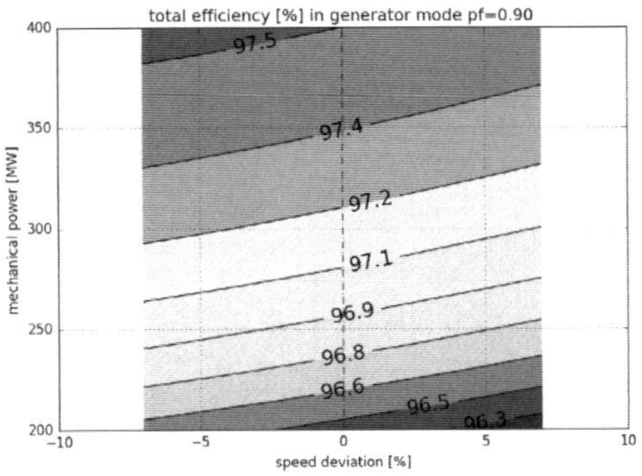

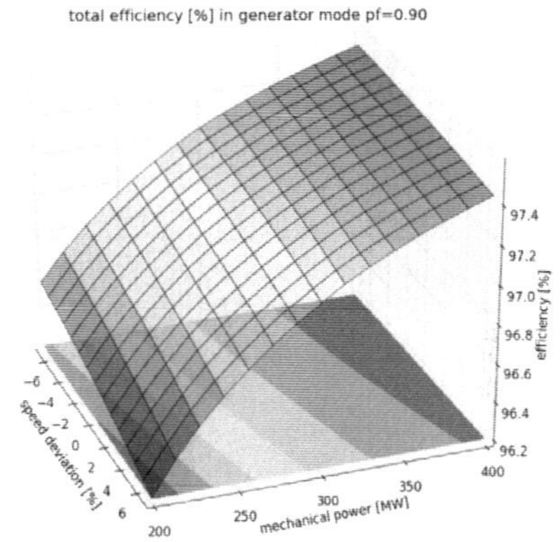

Figure 7: Efficiency for a Full Size Frequency Converter concept for different load and speed

4. Comparison & Outlook

In the table below the two proposed concepts are compared on different criteria's.

Table 1: Comparison of different concepts for variable speed power generation

	Double Fed Induction Machine	**Full Size Frequency Converter**
Electrical Machine	3-phase winding system in the rotor, connected to slip-rings.	Simple electric excited Synchronous machine
Speed range	Limited: about +/- 7% from synchr. speed in practical hydro application	Speed can defined with restriction to fit best to hydraulic conditions
Start/Stop	Additional equipment for pump-start needed	Pump-start in water with converter possible[1]
Grid Code Compliance: Fault Ride Through[2]	Can be fulfilled with additional equipment or oversized converter	Can be simple fulfilled
Reactive Power	Must be provided by machine and converter	Provided by converter only
Sizing of equipment	Generator and connection more complex, Converter small	Generator simple, Converter larger

[1] Operation of the unit in the full speed range (FSFC) is reducing the mechanical stresses of the hydraulic equipment during transitions. This is a main advantage when the units are operating with frequent start/stops.
[2] This is a mandatory requirement from network code and will be defined in detail by the related TSO. A basic definition is provided by entso-e [3]

Both of the proposed concepts are showing major improvements in hydro power generation by:

- Extension of operating range, and
- Improving the transient performance

The drawback of installation of additional power electronic equipment with higher losses will be compensated by higher flexibility which becomes more important in the future. Due to future developments in power electronics reduction in cost and size of the converters can be expected.

5. Bibliography

[1] J. Hell, Grid Frequency Response – Contribution of Hydro-Power for Grid Stabilization, HYDRO conference Cernobbio, 2014

[2] HYPERBOLE project EU FP7-ENERGY-2013-1 Grant 608532
web: https://hyperbole.epfl.ch/SitePages/Home.aspx

[3] ENTSO-E Network Code for Requirements for Grid Connection Applicable to all Generators, ENTSO-E, Brussels, Belgium, March 2013.

[4] M. Matos Fernandes, "Connection of the First Pump – Storage Variable Speed Machines in Portugal: TSO Main Requirements", *11th Int'l Workshop on Large-Scale Integration of Wind Power and on Transmission Networks for Offshore Wind Power Plants,* November 2012

[5] Argonne National Laboratory, Modelling Adjustable Speed Pumped Storage Hydro Units Employing Doubly-Fed Induction Machine, ANL/DIS-13/06, available in: http://www.dis.anl.gov/projects/psh/ANL_DIS-13_06_Modeling_AS_PSH.pdf

[6] P. Kundur, Power System Stability and Control, McGraw-Hill Education, 1994, ISBN: 9780070359581

[7] F. Rodriguez, P. Ledamsa , Simplified wind generator model for Transmission System Operator planning studies, 7th International Workshop on large scale integration of Wind Power and on Transmission Networks of Offshore Wind farms

[8] J.Machowski, J.W. Bialek Power System Dynamics – Stability and Control, John Wiley & Sons 2008

6. Acknowledgments

The research leading to the results published in this paper is part of the HYPERBOLE research project [2], granted by the European Commission (ERC/FP7-ENERGY2013-1-Grant 608532).

IOP Conf. Series: Journal of Physics: Conf. Series **813** (2017) 012004 doi:10.1088/1742-6596/813/1/012004

Operation ranges and dynamic capabilities of variable-speed pumped-storage hydropower

Thomas Mercier[1], Mathieu Olivier[2] and Emmanuel Dejaeger[1]

[1] Center for Research in Energy and Mechatronics, Université catholique de Louvain, 1348, Louvain-la-Neuve, Belgium

[2] Laboratoire de Mécanique des Fluides Numérique, Université Laval, G1V 0A6, Québec City, Canada

E-mail: thomas.mercier@uclouvain.be

Abstract. The development of renewable and intermittent power generation creates incentives for the development of both energy storage solutions and more flexible power generation assets. Pumped-storage hydropower (PSH) is the most established and mature energy storage technology, but recent developments in power electronics have created a renewed interest by providing PSH units with a variable-speed feature, thereby increasing their flexibility. This paper reviews technical considerations related to variable-speed PSH in link with the provision of primary frequency control, also referred to as frequency containment reserves (FCRs). Based on the detailed characteristics of a scale model pump-turbine, the variable-speed operation ranges in pump and turbine modes are precisely assessed and the implications for the provision of FCRs are highlighted. Modelling and control for power system studies are discussed, both for fixed- and variable-speed machines and simulation results are provided to illustrate the high dynamic capabilities of variable-speed PSH.

1. Introduction

Up to now, PSH has mainly consisted in large fixed-speed reversible pump-turbines driving or being driven by a grid-connected synchronous machine. The power in turbine mode is varied at the expense of efficiency by controlling the flow in the machine while the power is fixed in pump mode. The best efficiency point (BEP) corresponds to different speeds in pump and turbine modes, resulting in a loss in efficiency as the machine is operated at the synchronous speed.

In the last decades, developments in power electronics have enabled the supply of electrical machines with variable-frequency voltages, resulting in the possibility to vary the speed of PSH plants. Large machines, of more than 50 MW, use a doubly-fed induction machine (DFIM) with a power converter rated to only a few percent of the nominal power while smaller machines use a synchronous machine with a full size power converter [1, 2, 3].

This variable speed possibility can be used to always operate the hydraulic machine at its BEPs, as these are related to different speeds in pump and turbine modes, thereby increasing the revenues from price arbitrage on the energy markets. Sometimes the variable speed becomes a necessity in order for the pump mode to support high head variations and be able to operate between its stability and cavitation limits [1]. Besides, the increased head range improve the use of basins' capacity. Another option is to use of the variable speed to provide transmission system operators (TSOs) with ancillary services, in particular primary and secondary frequency control,

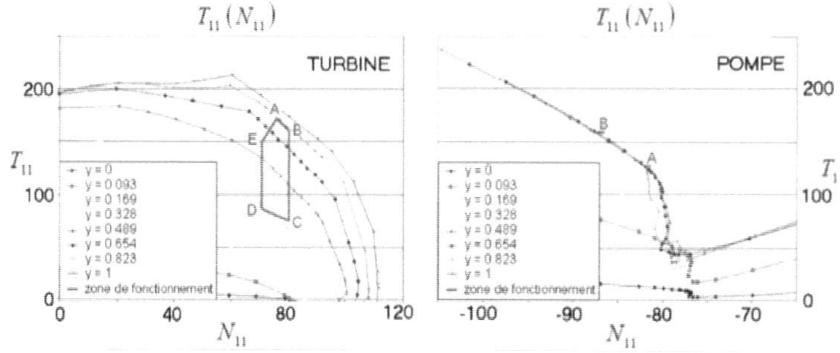

Figure 1: Characteristics and operation ranges of the studied pump-turbine, as found in [9].

both in pump and turbine modes. These services consist in adjusting power injections/offtakes in real-time, so as to keep the balance between electricity generation and consumption, which imposes high technical requirements on power plants, as these must have a range of power within which injections/offtakes can continuously and rapidly be varied.

Several authors have emphasized that a significant share of PSH revenues could be achieved through the provision of ancillary services [4, 5]. While these studies aim to give an overview of PSH, they fail to provide important technical details regarding the variable-speed operation of PSH to provide ancillary services. Economic calculus is often based on rather vague assumptions regarding the operation range of PSH [2], while it is also noticed that control aspects and dynamic capabilities are not well known. Besides, hydro turbine models and control schemes traditionally used in power system studies [6, 7, 8] have become inappropriate for variable-speed units.

This paper aims to clarify the technical capabilities of variable-speed PSH, and to put in light state of the art control schemes and transient modelling for power system studies.

2. Operation range of variable-speed PSH

The detailed characteristics of a Francis pump-turbine are given in [9]. These characteristics are obtained from laboratory measurements on a scale model by varying the head H, for different wicket gate positions y, while the rotational speed N is kept constant and measurements of discharge Q and torque T are made, which gives measurement vectors (y,N,H,Q,T). The measurements are then provided as charts giving the relations between unit speed N_{11}, unit flow Q_{11} and unit torque T_{11}. The operation ranges are also provided [9], and indicated on the T_{11}-N_{11} charts, as shown in Figure 1. In turbine mode, the limitation of the operation range is mainly due to several flow phenomena taking place in the hydraulic machine [10]. The A-B and C-D limits are defined by the manufacturer in order to avoid discharge ring swirl cavitation. Indeed, at high and low discharge, a swirling flow appears at the runner outlet, with eventually cavitation at its center, in which case it is referred to as vortex rope. The whirl dynamics compromises the machine's operation stability, since it is the main source of pressure fluctuations in the hydraulic installation. The B-C and D-E limits are set to respectively avoid leading edge pressure- and suction-side cavitation as it can lead to a severe erosion of the blades. Finally, the A-E limit is set by the rating of the electrical generator. In pump mode, the operation range is limited by lower and upper N_{11} bounds, i.e. higher and lower head values, to respectively prevent suction- and pressure-side cavitation on the runner blade [3, 10]. In addition, the operation range is limited by another upper N_{11} bound, corresponding to the pump's stability limit [3].

The power P_m associated to each measurement can be derived from the unit numbers using the following relation:

$$P_m = \frac{2\,\pi\,N\,T}{60} = \frac{\pi}{30}\,H^{3/2}\,D_{ref}^2\,N_{11}\,T_{11} \tag{1}$$

with D_{ref} the pump-turbine's reference diameter. Given N_{11}, Q_{11} and T_{11} numbers, e.g. those associated with the BEP or the operation limits, one has just to set two variables among D_{ref}, N, Q, H, T and P_m in order to define a pump-turbine, i.e. to derive its diameter and operation ranges in terms of head, flow and power. Given a nominal rotational speed of 675 RPM, and a maximum mechanical power of 2 MW in pump mode, one can derive the corresponding head and reference diameter, 93.04 and 1.2413 m respectively. It can be shown that at given head, the variable-speed feature provides the pump mode with a non-zero power range, while the turbine mode does not see its power range affected. The variable speed enables both modes to operate at lower heads. Besides, the extent of the power range varies with head, both in pump and turbine modes, and can precisely be quantified using 1 and the ranges of Figure 1, i.e. substituting in 1 the extreme values that can be taken by N_{11} and T_{11}. Using this methodology, it is found that given a head, the allowed power range is 24 and 53 % of the maximum allowed power at that head, respectively for the pump and the turbine modes. The maximum allowed power depends on the available head. The lower the available head, the lower the maximum allowed power. The fact that the power range varies with head is crucial for the provision of FCRs. Indeed, as the level of water varies with time, so does the available power range and the amount of FCRs that can be provided. Whether in pump or turbine mode, the amount of FCRs that can be provided by a pump-turbine should be computed based on the minimum operation head. Besides, variable-speed PSH must be operating rather far from the zero output to be able to provide FCRs, which implies frequent switches from pump to turbine, and vice versa. Energy market implications significantly impact the revenues of variable-speed PSH providing FCRs.

3. Transient modelling of PSH for power system studies

3.1. Hydraulic model

Various models can be used to study transients in hydropower plants. Although most of them are dedicated to hydraulic turbines, those models can be adapted/generalized to pumps and pump-turbines. They most often consist in the combination of a model for the water column, in the penstock, with the characteristics of the turbomachinery, which can be with or without guide vanes. The water column is generally considered either elastic or inelastic and is modelled, in the former case, through a first-order dynamic model. The characteristics of the hydraulic turbine is static, i.e. non-dynamic, meaning that it does not consist in differential equations. The characteristics can be represented through two classes of models. First, several models, linear or non-linear, have been developed for small to medium deviations around an operating point [6, 7, 8]. Those models target hydraulic machines operated at fixed or nearly fixed speed, i.e. those connected to the grid through a synchronous machine with no power electronics converter. In the second class of models, the turbine is considered through its detailed characteristics [9, 11, 12], i.e. interpolation curves joining various operation points, as in Figure 1. Those models are suited for the study of large variations in operating conditions, as variations in rotational speed. Modelling the water column and coupling it to the characteristics of the turbomachinery, as is done in [9, 11] but not in [12], enables to account for water hammer phenomena resulting from variations in guide vane position and/or rotational speed.

3.2. Control scheme

Reference [6] describes a governor for fixed-speed hydraulic machines, enabling the control of power thanks to the wicket-gate position Y. It implements a droop characteristics whose stability is ensured by a low transient droop. This scheme should be different for variable-speed machines. In pump mode, power should be controlled, thanks to the variable speed, while in turbine mode, power can be controlled through rotational speed and/or wicket gate position. In [13], power in turbine mode is controlled through Y while the variable speed enables to completely dampen power oscillations normally resulting from changes in Y. In [9], two control

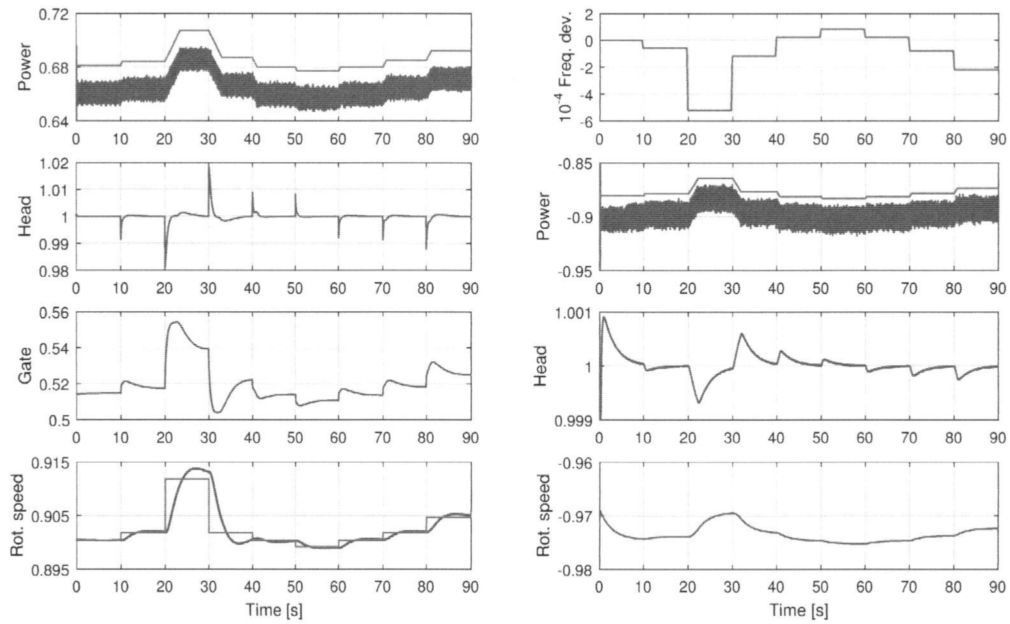

Figure 2: Transient operation of variable-speed PSH in turbine (left) and pump (right) modes for the provision of FCRs. All values are given in per unit. Red curves are reference signals.

strategies are proposed in turbine mode, which consist in (i) controlling power thanks to the wicket gate position, while speed control enables to optimize the efficiency and (ii) controlling power through the electromagnetic torque while varying the wicket gate position to optimize the efficiency through the rotational speed. It is shown that the second strategy, which goes one step further than [13] through efficiency optimisation, implies much faster dynamic response. This strategy is definitely the best suited for the provision of ancillary services. Whether it is in [9] or in [13], no information is given on the methodology followed for tuning the speed controller.

In order to design the control in speed for the turbine mode, transfer functions linking the guide vanes position to the rotational speed have to be established. An approximate transfer function is therefore needed for the turbine characteristics. Besides, this transfer function should be linear in order to be able to use linear control theory to design the speed controller.

3.3. Simulation results

A PSH case has been set up in *MATLAB/Simulink* in order to illustrate the dynamic capabilities brought by the variable speed. The hydraulic model consists of a rigid water column coupled to the pump-turbine characteristics of Figure 1. On the electrical side, a permanent magnet synchronous machine (PMSM) is fed by two-level voltage source inverters (VSIs) in back-to-back configuration with a DC link. The power converter is connected to an infinite medium-voltage bus through a short distribution line. The goal of the simulation is to observe the dynamics of the whole system while the PSH unit is providing FCRs with a control scheme developed as explained in the previous subsection. Simulation results are given in Figure 2. Based on requirements imposed by the Belgian TSO, an active power reference is derived from the 2013 frequency deviation measurements at ten-second time steps. FCRs are provided for an amount equal to 20 and 12.5 % of the pump rated power, respectively in turbine and pump modes. For both modes, the results show that power injections can effectively be varied to follow a power reference for the provision of FCRs. Mismatches between power injection and reference are around two percentage points and are explained by losses in the electrical machine, the VSIs and the distribution line. In turbine mode, speed control through guide vanes position is effective. Variations in guide vanes position induce water hammer effects. The latter are reduced in pump mode, where guide vanes are kept fixed and power is controlled through the VSIs.

4. Conclusion

Compared to their fixed-speed counterparts, variable-speed hydraulic machines are able to operate over wider head ranges. For a given head, the variable speed allows power variation in pumping, while a unique power is possible at fixed speed, but leaves the power range unchanged in turbine mode. Variable-speed PSH is thus able to provide FCRs both in pump and turbine modes. The operation ranges are limited to prevent undesired flow phenomena. Given a head, the studied pump-turbine's allowed power range spreads 24 % and 53 % of and from the maximum allowed power at the given head. In absolute values, these ranges vary with head, which impacts the amount of FCRs that can be provided. Besides, the continuous provision of FCRs requires PSH units to constantly be operating far from the zero output, which implies frequent switches from pump to turbine and vice versa. The energy market consequences should be taken into account when assessing the revenues of variable-speed PSH providing FCRs.

Modelling and control for power systems studies have been discussed, both for fixed- and variable-speed machines. A variable-speed PSH model has been set up based on the detailed characteristics of a Francis pump-turbine, a rigid water column, the dynamic model of a PMSM, a short impedance line and two two-level back-to-back VSIs. In pump as in turbine mode, power control has been achieved through the VSIs. The guide vanes position has been kept fixed in pump mode while its adjustment enabled speed control in turbine mode. Simulation results have shown the high dynamic performances of variable-speed PSH, in line with FCR requirements.

5. Acknowledgments

The paper was made in the frame of SmartWater, a project supported by the Walloon Region.

References

[1] Ardizzon G, Cavazzini G and Pavesi G 2014 *Renewable and Sustainable Energy Reviews* **31** 746–761
[2] Beevers D, Branchini L, Orlandini V, De Pascale A and Perez-Blanco H 2015 *Applied Energy* **137** 629–637
[3] Jürgen K, Helmut K and Manfred S 2013 *Energy and Power Engineering* **2013**
[4] Pérez-Díaz J I, Chazarra M, García-González J, Cavazzini G and Stoppato A 2015 *Renewable and Sustainable Energy Reviews* **44** 767–784
[5] Pinto J, de Sousa J and Neves M V 2011 *2011 8th International Conference on the European Energy Market (EEM)* (IEEE) pp 306–311
[6] Kundur P, Balu N J and Lauby M G 1994 *Power system stability and control* vol 7 (McGraw-hill New York)
[7] Munoz-Hernandez G A, Jones D I *et al.* 2012 *Modelling and controlling hydropower plants* (Springer Science & Business Media)
[8] De Jaeger E, Janssens N, Malfliet B and Van De Meulebroeke F 1994 *IEEE Transactions on Power Systems* **9** 1709–1715
[9] Pannatier Y 2010 *Optimisation des stratégies de réglage d'une installation de pompage-turbinage à vitesse variable* Ph.D. thesis École Polytechnique Fédérale de Lausanne
[10] Avellan F 2004 *6th International Conference on Hydraulic Machinery and Hydrodynamics, Timisoara, Romania, October* pp 21–22
[11] Chaudhry M H 1979 *Applied hydraulic transients* (Springer)
[12] Belhadji L, Bacha S, Munteanu I, Rumeau A and Roye D 2013 *IEEE Transactions on Power Systems* **28** 34–43
[13] Lung J K, Lu Y, Hung W L and Kao W S 2007 *IEEE Transactions on Energy Conversion* **22** 250–258

Integration of Variable Speed Pumped Hydro Storage in Automatic Generation Control Systems

N Fulgêncio[1], C Moreira[1] and B Silva[1]

[1] INESC TEC – INESC Technology and Science, Porto, Portugal

E-mail: {nuno.r.fulgencio, carlos.moreira, bsilva}@inesctec.pt

Abstract. Pumped storage power (PSP) plants are expected to be an important player in modern electrical power systems when dealing with increasing shares of new renewable energies (NRE) such as solar or wind power. The massive penetration of NRE and consequent replacement of conventional synchronous units will significantly affect the controllability of the system. In order to evaluate the capability of variable speed PSP plants participation in the frequency restoration reserve (FRR) provision, taking into account the expected performance in terms of improved ramp response capability, a comparison with conventional hydro units is presented. In order to address this issue, a three area test network was considered, as well as the corresponding automatic generation control (AGC) systems, being responsible for re-dispatching the generation units to re-establish power interchange between areas as well as the system nominal frequency. The main issue under analysis in this paper is related to the benefits of the fast response of variable speed PSP with respect to its capability of providing fast power balancing in a control area.

1. Introduction

The growing integration of new renewable energy (NRE) sources such as wind and solar power has driven electrical power systems to adapt to a new operational paradigm. Although their acknowledged advantages, generation from NRE induce high levels of uncertainty and variability in the system when compared to the conventional technologies (e.g., coal-fired thermal plants, combined and open cycle gas turbines as well as hydro power plants). To overcome these major drawbacks, energy storage devices have been referred as a possible solution that is able to leverage high levels of NRE deployment throughout worlds' electrical power systems [1]. Particularly, along with these are the pumped storage power (PSP) plants, which explore the possibility of using reversible units, able to either generate power or to pump water to an upstream reservoir. PSP plants have proved to be the most utilized, mature, efficient and cost-effective technology for large-scale energy storage, and may be playing and important role in handling the NRE integration issues in the future by providing storage capacity and advanced system support function in the form of fast ramps for power balancing purposes [2]. Thus, alongside a worldwide effort for energy decarbonisation, a commercial and technical interest for these (and other types of storage) units has been renewed.

The need for generation units' increased operational flexibility has driven the exploitation of new power generation solutions within the hydropower context by exploring variable-speed machines. The controllability provided by converter-connected units enables faster power-to-grid responses through the exploitation of higher ranges of turbines' speed and respective kinetic energy stored in its rotating masses, since the generator rotational speed is decoupled from the grid frequency.

However, the massive integration of NRE and subsequent replacement of conventional machines have significant impacts on the frequency control robustness either due to the enormous reduction of the system inertia as well as because of the reduction of the number of fully controllable units connected to the grid. In order to provide grid frequency control services, generation units are required to provide enough power reserves. As soon as a disturbance in the grid occurs, frequency containment reserve (FCR) is provided automatically and locally by some generation units, assuring excursions are contained and rapidly stabilized. Afterwards, the provision of frequency restoration reserve (FRR) which is activated centrally by the system operators through automatic generation control (AGC) systems is responsible for re-establishment of the nominal system frequency and the power interchanges between the so-called control areas.

In line with this control organization in electric power systems, this paper presents and discusses the role of a variable speed hydro PSP plant in the regulation mechanism supported by the AGC system. Taking into account the controllability characteristics of the variable speed units, it is demonstrated its effectiveness in the provision of fast responses for an improved power balance in a given control area in both the pumping and generation modes. The improved power balance capabilities that can be achieved with variable speed units are compared with those provided by a conventional synchronous-based machine. Besides the importance of validating the integration of such units in the AGC system, this work intends to highlight the advantages of using variable-speed units, interfaced with the grid through power converters, when dealing with FRR provision in order to re-establish frequency and inter-areas power flow to the pre-defined values.

2. Pumped storage power (PSP) plants

From a technological perspective, PSP plants store energy using the height difference between the headwater and tailwater reservoirs. The main operation principle of PSP is storing water during periods of low demand and releasing it during periods of high demand. These plants are typically highly efficient (>80%) and have been proving to be very beneficial in providing load balancing services to the electric power systems, while also allowing its promotors to obtain commercial profits from these temporal arbitrage energy management activities [3]. Despite the aforementioned benefits, an important drawback of the existing PSP schemes is the prominent use of fixed-speed pump-turbines. Although the traditional fixed-speed design has worked well for many decades, there are limitations to its performance which are becoming apparent as power system requirements are changing in view of the increasing penetration of NRE. In response to these limitations, variable-speed machines integration in hydropower plants is a new technology being now exploited. Recognizing that hydropower is the most mature and controllable renewable based energy source currently available, it is undertaking profound technically developments such that it becomes more effective to support the sound integration of NRE, based on its intrinsic high regulation and storage capabilities. Current advanced designs for PSP plants lay on the use of electronic converters that act as the interface with the grid as well as perform the machine's control, using technologies as the doubly-fed induction machines (DFIM) and, yet in a less extend, synchronous machines connected to the grid through full-scale frequency converters (FSFC). In all the cases, the use of power electronic interfaces allows for the pump turbine rotation speed to be controlled, offering improved flexibility in the amount of power that can be absorbed in pumping mode. Moreover, the variable speed technology enables to operate Francis pump-turbines in pumping mode with extended head operating range, offering the possibility to cover the entire head range of PSP subjected to large head variations. Additionally, and regarding the turbine operation mode, this enables a PSP plant to operate at higher efficiency over a larger portion of its operating range, also contributing for the enhancement of the integration of renewables. In the same perspective, the ability to provide frequency regulation services is also considerably enhanced due to smoother operation at partial load conditions and the ability for fast power adjustment [4]–[6].

To study the performance of a variable-speed PSP unit and its benefits when integrated in AGC systems, thus allowing its active participation in the FRR provision, a reduced-order model of a Francis turbine connected to a DFIM was used. In order to perform system stability studies, coping with limited computational effort, a reduced detail modelling approach was used. Since modelling process is out of the scope of this paper, detailed information about the PSP unit modelling deduction, structure and respective validation can be found in [7]. Generally, the adopted model tackles both operation modes – turbine and pumping – which are addressed separately. As it can be understood in the aforementioned reference, hydropower units allow the modelling decoupling of both the hydro-mechanical and the electro-mechanical parts. In the case of the turbine mode, the representation of the hydraulic system was based on the non-ideal turbine general expression given in [8], which relates gate opening and mechanical power variations around a given operating point. Contrary to what happens in the turbine mode, when the machine is working as a pump, the influence of the gate opening in the system operation is not significant. For that reason, an alternative model that explores an electrical analogy to relate hydraulic components was used. Details about the hydraulic model for pumping mode can also be found in [6]. A phasor modelling approach is used to represent the DFIM. The electric machine is represented by a transient model based on a voltage source being a transit reactance, being the mechanical part represented by a single-mass model (rigid shaft).

3. Automatic generation control (AGC)

Most synchronously-interconnected electric power systems are nowadays regulated by automatic systems that are responsible to maintain a stable and steady operation, including technical and economic considerations. In general, system operators (SO) have the responsibility for maintaining target values of grid frequency and scheduled tie-lines power flows, thus requiring the exploitation of FRR [9]. This type of action is automatically assured by each control area transmission system operators (TSO) through an AGC system, responsible to re-dispatch committed generation units within each load-frequency control (LFC) area.

For illustration purposes the block diagram of Figure 1 depicts the overall organization of an AGC system for a control area with two generation units.

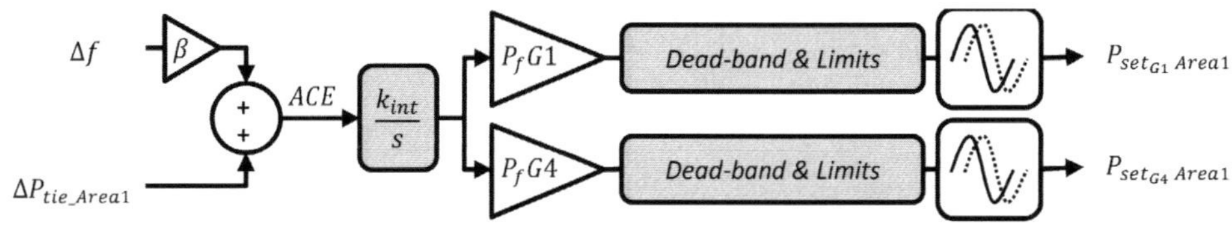

Figure 1. Area 1 detailed view of the control architecture.

For a given control area m, as the frequency deviation (Δf) and the power deviation in the tie-lines ($\Delta P_{tie\ Area_m}$) with respect to the pre-scheduled value changes, in the moments subsequent to a disturbance affecting the generation and load balance, the AGC is set to respond automatically by sending new active power set-points for the n generation units ($P_{set_{G_n}\ Area_m}$) under its control responsibility. Depending on the magnitude of the disturbance, the FRR needed from each generator is defined as a function of the area control error (ACE). This value adds in a linear combination of the Δf, which is affected by the stiffness ($\beta_m = \frac{1}{R_n} + D_m$, given in p.u./Hz, where R_n is the power speed droop of generator n and D_m is the load damping coefficient [10]), and the $\Delta P_{tie\ Area_m}$. The ACE is then fed to an integral controller (k_{int} gain in Figure 1) that computes the total amount of FRR required to cancel the ACE error and therefore assure frequency recovery to the nominal value as well as the re-establishment of the tie-line programmed flows. Based on the output of the integral controller, participation factors ($P_f G_n$) are used to define the power set-point to be assigned to each generator.

4. Test case definition

This section summarizes the simulation conditions used to study the behaviour and the benefits of including a hydro PSP plant in the AGC scheme, where it actively participates in the provision of FRR. The network and respective models were developed and implemented in the simulation platform MATLAB/Simulink®.

4.1. Network

A three-area network test system (depicted in Figure 2) was used in the performed simulation studies. This network comprises three main voltage levels: 21,6kV for the generation systems, being the transmission grid established at high and extra-high voltage levels of 230kV and 500kV respectively. It is assumed that generator G3 (located in control area 3) consists of a hydro PSP while all the other generators are referred to thermal units using synchronous machines. A brief summary of the network's

IOP Conf. Series: Journal of Physics: Conf. Series **813** (2017) 012005 doi:10.1088/1742-6596/813/1/012005

operational conditions can be consulted in Table 1 and Table 2. It was considered that power flowing between areas is defined to be positive when power is being extracted from a given area. Therefore, the tie-lines powers, in the same tables, are presented with the negative signal. The same occurs for the hydro PSP plant: positive active power means the unit is generating power while the negative signal indicates it is consuming power, while working in pumping mode. For the purpose of implementing the AGC system for each of the control areas, inter-area tie-line power flows are related according to the following equations (*a)* and *b)*:

$$a)\quad \begin{cases} \Delta P_{tie_{Area1}} = \Delta P_{tie12} - \Delta P_{tie31} \\ \Delta P_{tie_{Area2}} = \Delta P_{tie23} - \Delta P_{tie12} \\ \Delta P_{tie_{Area3}} = \Delta P_{tie31} - \Delta P_{tie23} \end{cases} \qquad b)\quad \begin{cases} \Delta P_{tie12} = P_{tie12} - P_{tie12_{ref}} \\ \Delta P_{tie23} = P_{tie23} - P_{tie23_{ref}} \\ \Delta P_{tie31} = P_{tie31} - P_{tie31_{ref}} \end{cases}$$

The power flow in the three transmission lines connecting area 1 and area 2 are used to define P_{tie12}, being its respective scheduled reference value $P_{tie12_{ref}}$. In the same way, the power flow in the four transmission lines connecting area 2 and area 3 are used to define P_{tie23} (its reference represented by $P_{tie23_{ref}}$). Areas 1 and 3 are not directly connected so P_{tie31} is only defined (and its reference value $P_{tie31_{ref}}$) to facilitate the mathematical comprehension. Its values are therefore considered to be null.

Table 1. Network power summary.

	Area 1		Area 2		Area 3	
Load Power (MW)	500		1200		1500	
	G1	**G4**	**G2a**	**G2b**	**G3**	**G5**
Rated power (MW)	1000	130	900	900	±750	900
Init. Power (Turbine) (MW)	616.5	82.5	572.5	572.5	600	567.5
Init. Power (Pumping) (MW)	830	117	814	814	-375	810

Table 2. Inter-area ref. power.

Tie-line reference power		
Turbine mode	Pumping mode	
-225	-472	$P_{tie12_{ref}}$
-250	-984	$P_{tie23_{ref}}$
0	0	$P_{tie31_{ref}}$

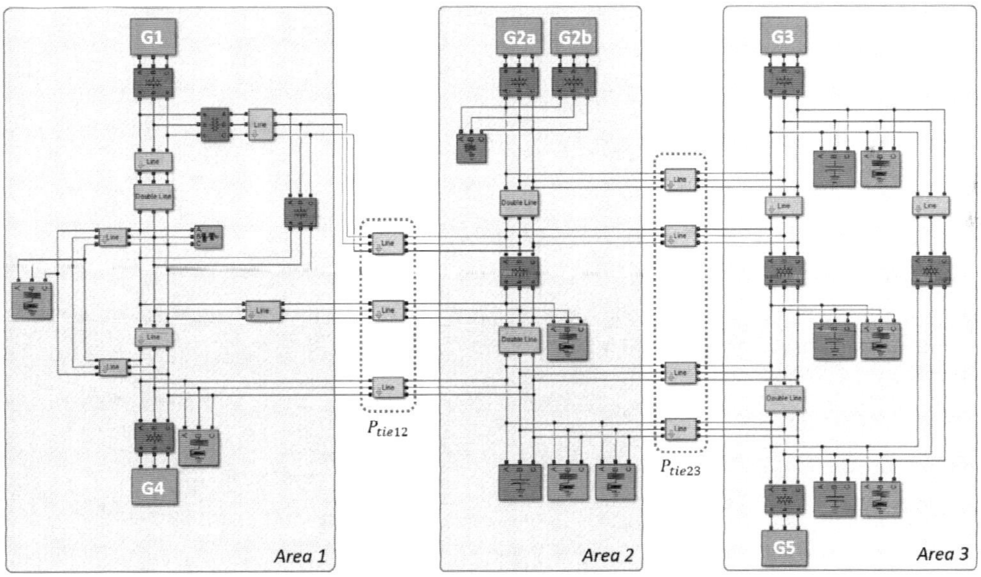

Figure 2. Three-areas network used in the simulation platform.

The main parameters used in the simulation studies regarding the AGC system are summarized in Table 3.

Table 3. AGC simulation parameters.

	Area 1		Area 2		Area 3	
Parameters	**G1**	**G4**	**G2a**	**G2b**	**G3**	**G5**
Droop (R)	0.05	0.05	0.05	0.05	0.05	0.05
Participation factor (P_f)	50%	50%	50%	50%	100%	0%
Damping (D)	0.05		0.05		0.06	
Integral gain (k_i)	-0.03		-0.03		-0.09	

4.2. Test cases

To study the response of the hydro PSP plant within the AGC scheme, three main tests were performed and are presented in this paper. A 50 MW load increase in Area 3 (where the hydro PSP plant is located) was simulated considering two the PSP plant to be operating according to the following conditions:

- **Turbine mode**: conventional synchronous-based hydro unit performance was compared to a variable-speed unit using a DFIM.
- **Pumping mode**: variable speed unit using DFIM and controlled by power converters.

The comparison, in turbine mode, between the fixed and variable speed units allow to demonstrate the advantages of using fast-responses units in the participation of FRR within the AGC scheme. Alongside, and once the reversible characteristic of the PSP plants is one of the main services these units can provide, contributing hence for NRE integration, the pumping mode is included.

5. Simulation results

Regarding turbine operation mode, the tie-lines power flow (Figure 3) re-establishment in the moments subsequent to the disturbance occurred in Area 3 is significantly faster when using a variable-speed PSP unit when compared to a PSP with a synchronous generator. Its controllability and improved power-frequency response characteristics leads to a faster tracking capability of the AGC control signal and consequently to a faster response when analysing the re-establishment of tie-line flows. To support this power-to-grid response, as it can be observed in Figure 4, the unit takes advantage of the wider range of available turbine speed. When the load increase occurs the power converters respond almost immediately (constrained by the ramping limits defined in order to impose a limit on the power converters dimension, set within the turbine speed deviation reference of ±7% of its nominal speed) and the speed deviation around the nominal value tends to be much larger. Additionally, the frequency recovery occurs also faster for the variable-speed unit due to the same reasons.

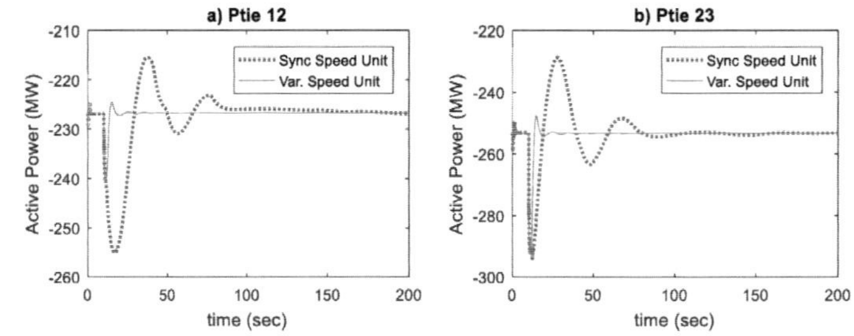

Figure 3. Turbine mode: Tie-lines power flow compared responses to 50MW load increase in Area 3, for synchronous and variable-speed hydro PSP plant.

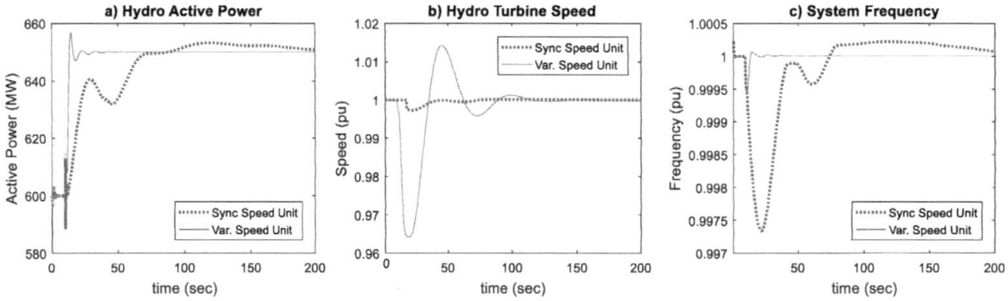

Figure 4. Turbine mode: Synchronous and variable speed hydro PSP plant response to 50MW load increase in Area 3.

Regarding pumping mode operation (Figure 5), the behaviour of the hydro unit and the impact of its response regarding FRR provision is similar to turbine mode operation. Both tie-line power flows are re-established considerably fast – within around 40 seconds – as occurred when working as a generator, due to its fast power-to-grid response.

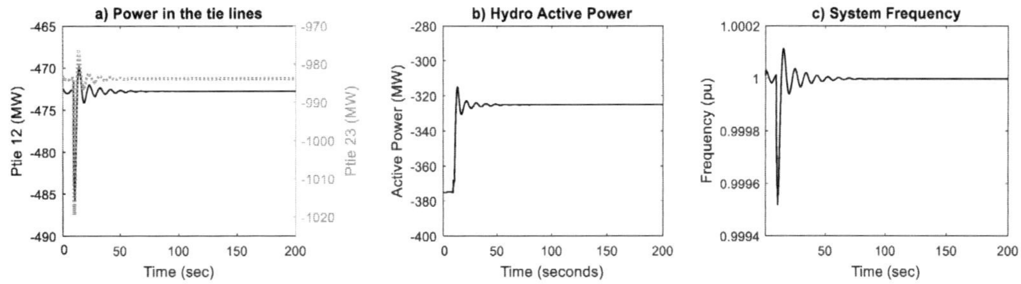

Figure 5. Pumping mode: Hydro PSP plant response to 50MW load increase in Area 3.

6. Conclusions

The need for new and better strategies that support the increasing growth of new renewable energies (NRE) is deploying a progressive interest in new forms of large-scale energy storage devices. Along with this, hydro power plants, due to its mature and highly efficient technologies are raising a renewed interest, especially when offering the possibility of working with reversible characteristics enabling pumping operation mode. Pumped storage power (PSP) plants are expected to play an important role when dealing with NRE integration by accommodating some of its uncertainty and variability. However, the replacement of conventional synchronous units for NRE – which normally rely on power-electronics connections to the grid – significantly decreases the system inertia and, hence, the robustness of frequency control. In order to understand how can PSP plants participate in the frequency restoration reserve (FRR) provision, a three-areas network controlled by an automatic generation control (AGC) system, responsible for re-dispatching units for FRR provision, was implemented and tested in this paper. Due to its electrical characteristics, these units have proven to be much faster, when compared to conventional hydro units – synchronous machines based – responding very much positively to disturbances and in the provision of FRR. Their power converters controlled connection, which allow a more flexible operation, have also proven to be rather advantageous when dealing with fast set-point changes, hence contributing to a better performance of the AGC and the overall system. Such conclusions demonstrated variable speed PSP ability to provide key system balancing functionalities in scenarios with high shares of NRE.

References

[1] Y. Cheng, M. Tabrizi, M. Sahni, A. Povedano, and D. Nichols, "Dynamic available AGC based approach for enhancing utility scale energy storage performance," *IEEE Trans. Smart Grid*, vol. 5, no. 2, pp. 1070–1078, 2014.

[2] C. K. Ekman and S. H. Jensen, "Prospects for large scale electricity storage in Denmark," *Energy Convers. Manag.*, vol. 51, no. 6, pp. 1140–1147, 2010.

[3] D. Pudjianto, M. Aunedi, P. Djapic, and G. Strbac, "Whole-Systems Assessment of the Value of Energy Storage in Low-Carbon Electricity Systems," *IEEE Trans. Smart Grid*, vol. 5, no. 2, pp. 1098–1109, Mar. 2014.

[4] C. Nicolet, B. Greiveldinger, J. J. Herou, B. Kawkabani, P. Allenbach, J.-J. J. Simond, and F. F. Avellan, "High-order modeling of hydraulic power plant in islanded power network," *IEEE Trans. Power Syst.*, vol. 22, no. 4, pp. 1870–1880, Nov. 2007.

[5] A. C. Padoan, B. Kawkabani, A. Schwery, C. Ramirez, C. Nicolet, J. J. Simond, and F. Avellan, "Dynamical behavior comparison between variable speed and synchronous machines with PSS," *IEEE Trans. Power Syst.*, vol. 25, no. 3, pp. 1555–1565, 2010.

[6] Y. Pannatier, B. Kawkabani, C. Nicolet, J. J. Simond, A. Schwery, and P. Allenbach, "Investigation of control strategies for variable-speed pump-turbine units by using a simplified model of the converters," *IEEE Trans. Ind. Electron.*, vol. 57, no. 9, pp. 3039–3049, 2010.

[7] C. Moreira, N. Fulgêncio, B. Silva, C. Nicolet, and A. Béguin, "Identification of Dynamic Simulation Models for Variable Speed Pumped Storage Power Plants," in *Hyperbole Conference*, 2017, pp. 1–6.
[8] P. Kundur, N. J. Balu, and M. G. Lauby, "Power system stability and control." p. 1176, 1994.
[9] ENTSO-E, "Continental Europe Operation Handbook." [Online]. Available: https://www.entsoe.eu/publications/system-operations-reports/operation-handbook/Pages/default.aspx.
[10] H. Bevrani and T. Hiyama, *Intelligent automatic generation control*. CRC Press, 2011.

IOP Conf. Series: Journal of Physics: Conf. Series **813** (2017) 012006 doi:10.1088/1742-6596/813/1/012006

Identification of Dynamic Simulation Models for Variable Speed Pumped Storage Power Plants

C Moreira[1], N Fulgêncio[1], B Silva[1], C Nicolet[2] and A Béguin[2]

[1] INESC TEC – INESC Technology and Science, Porto, Portugal
[2] Power Vision Engineering, Ecublens. Switzerland

Abstract. This paper addresses the identification of reduced order models for variable speed pump-turbine plants, including the representation of the dynamic behaviour of the main components: hydraulic system, turbine governors, electromechanical equipment and power converters. A methodology for the identification of appropriated reduced order models both for turbine and pump operating modes is presented and discussed. The methodological approach consists of three main steps: 1) detailed pumped-storage power plant modelling in SIMSEN; 2) reduced order models identification and 3) specification of test conditions for performance evaluation.

1. Introduction

Pumped Storage Power (PSP) plants are expected to play a key role for handling the operation of electrical power systems with increasing shares of New Renewable Energy (NRE) sources (such as solar and wind) which are characterized by high variability and uncertainty [1]. Currently there exist 270 PSP stations worldwide (a combined capacity of 120 GW) and the vast majority of these plants employ a conventional setup based on reversible single-stage Francis pump-turbines [2]. An important drawback of these existing schemes is the prominent use of fixed-speed pump-turbines. Although the traditional fixed-speed design has worked well for many decades, there are inherent limitations which significantly reduce PSP plants ability to contribute to the balancing of the rapidly-varying output of NRE and ancillary services provision, especially when operating in pumping mode.

In response to these limitations, variable-speed machines have been a technological option that mostly rests on two main technological solutions: variable speed technology through the a Doubly-Fed Induction Machine (DFIM) or a synchronous generator connected to the grid through a Full Size Frequency Converter (FSFC). In order to study the PSP plants integration and operation in electric power systems, dynamic and transient stability studies are required. In fact, a wide range of models with different levels of complexity, capable of representing different operational conditions, have been developed and are widely accepted for the characterization of the dynamic response of hydro power plants, though only when equipped with fixed-speed units [3]. For the specific case of the variable-speed PSP technology, the associated dynamic models capable of representing the main dynamic response, both in turbine and pump operation modes, are not widely available and are not properly validated. A few references [4]–[6] can be found in the literature addressing some control issues of the power electronic interfaces in variable speed PSP, though without providing a global model that can be used in dynamic stability studies of large electric power (and operating as a generator or as a pump). Therefore, this paper aims at providing new contributions within the scope of variable speed PSP plant dynamic modelling, both for pump and turbine operating modes. The proposed models are extensively validated in coordination with detailed simulation results that were obtained from SIMSEN, a specific software package for hydro units simulation [7]. Additionally, the approach for model validation relies on the real hydraulic configuration of an existing 2x210 MW PSP plant.

This work was developed in the framework of the HYPERBOLE project – HYdropower plants PERformance and flexiBle Operation towards Lean integration of new renewable Energies – grant agreement number 608532, financed by European funds through the 7th RTD Framework Programme.

2. System Modelling

This section provides an overview regarding the modelling approaches for PSP plants that are used in SIMSEN as well as the reduced order model that was entailed to represent the variable speed PSP, which was assumed to use a DFIM.

2.1. Hydro-mechanical and DFIM modelling in SIMSEN

The SIMSEN simulation package is used here as a reference software in order to develop and perform detailed simulation of a PSP plant. Next sections introduce the hydro-mechanical and the DFIM electrical systems models implemented in SIMSEN. Both hydro-mechanical and electrical models are combined in a unique model – the hydro-electrical model of the hydro power plant, [6].

2.1.1. Hydro-mechanical system: The SIMSEN model of the hydraulic system of a PSP plant includes the following elements: an upper reservoir with constant water level, two sub-horizontal square sections intake pipes, upstream gates, two vertical circular pressure shafts, two reversible Francis pump-turbines, linked to the rotating inertia of the turbine and of the generator rotor, downstream pipes, two downstream surge tanks, a downstream junction and tailrace tunnel, a lower reservoir with constant water level and the turbine governor. Regarding the SIMSEN hydraulic model, it is important to mention it takes into account: the water hammer phenomena in piping systems (pipe distributed head losses, water inertia and pipe elastic behaviour – fluid compressibility and wall deformation); the 4 quadrants transient behaviour of the pump-turbine including so-called S-shape unstable characteristics and the link with rotating inertias [6]; the surge tank mass oscillations phenomena between the lower surge tanks and the lower reservoirs taking into account variable cross section area of the surge tanks and asymmetric head losses at the inlet of the surge tank; the generating unit rotating train torsion dynamics with turbine and motor-generator rotor inertia linked through coupling shaft with given torsional stiffness.

2.1.2. DFIM electrical system: In the DFIM variable speed technology, a wounded rotor induction machine is fed by a back-to-back voltage source converter. Controlling the rotor current via the inverter control enables to achieve variable speed of the electrical machine with respect to the rated synchronous speed. The model developed in SIMSEN includes the representation of the induction machine with wound rotor, slip recovery cascade with rotor and grid side converters as well as the cascade transformer, and the unit step-up transformer, see [6].

2.2. Reduced order models for hydro-mechanical systems and DFIM

Regarding variable speed PSP plants modelling, turbine and pumping operation modes needs to be taken into account. Since these modes have distinct ways of control and operation, it was considered to develop different dynamic models for each case. Therefore, modelling the transitions from turbine to pump or vice-versa is out of the scope of this paper (however, it can be conveniently analysed using the full model in SIMSEN).

2.2.1. Turbine operation mode: IEEE standard models [3] for hydro turbines and its controls usually allow a direct separation between the turbine speed governor and the hydraulic system dynamics, which is the same approach that was also used in this work. The representation of the hydraulic system in the reduced modelling approach was based on the non-ideal turbine general expression given in [8], which relates gate opening and mechanical power variations around a given operating point. Due to the high dependence of the model parameters in relation to the machine characteristics and operational point, an approach relating the penstock coefficients to the ideal lossless model was used. This model constitutes the main approach that was used to represent the hydro dynamics for the turbine operation mode, once it is easy to handle in terms of simulation integration in power system studies.

2.2.2. Pump operation mode: Contrary to what happens in turbine mode, when the hydro unit is operating in the pump mode, there is no significant effect of the gate opening in the control of the system operation. Therefore, pump-turbines are typically operated with constant guide vane opening set for the best efficiency during pump operation. When variable speed machines are considered, the operation of the unit in pump mode is typically based on the control of the active power (which can be achieved by the control of the rotor side converters in DFIM-type units). Consequently, there is a direct effect on the unit speed and discharge, that must be taken into account in the model development. In order to model the different hydraulic components, a possible modelling approach is based on the use of an electrical analogy, where pressure is analogous to voltage and discharge is analogous to electric current [6]. This approach, applied to a simplified PSP system, is represented in Figure 1, being its mathematical formulation represented by the following equations: $H_0 = L_p \cdot \frac{\partial Q}{\partial t} + R_p \cdot Q + H_{pt}$ and $J \cdot \frac{\partial \omega}{\partial t} = T_{pt} + T_{el}$. In this sense, the hydraulic resistance R and the hydraulic inductance L correspond to energy losses and hydraulic inertia, respectively. The system comprises an upstream reservoir, a penstock and a pump-turbine linked to a rotating inertia then connected to a downstream reservoir. The upper reservoir is modelled with a constant pressure source through the reservoir gross head (H_0) and the penstock is modelled considering water inertia and head losses effects respectively modelled by hydraulic inductance L_p and hydraulic resistance R_p. The pump-turbine is modelled by a pressure source $H_{pt}(Q, \omega)$ (net head) driven by the associated 4-quadrant characteristic curve (knowing the pump-turbine discharge Q and rotating speed ω), while the guide vane opening is assumed to be constant. As the pump-turbine characteristic shows a typical S-shaped curve, it leads to numerical problems when interpolating between points of the curve (due to the multiple values problem). To overcome this numerical problem, the so-called Suter polar representation is used [6]. In this mathematical representation, the swing equation allows to derive the unit rotational speed, as far as the hydraulic torque T_{pt} is derived from the pump-turbine characteristics (as a function of the rotational speed and discharge), being T_{el} the electromagnetic torque from the DFIM and J the combined inertia of the system (hydraulic part and rotor of the DFIM).

2.2.3. DFIM model: A phasor modelling approach is used to represent the DFIM. The electric machine is represented by a transient model based on a voltage source being a transient reactance [9], being the mechanical part represented by a single-mass model (rigid shaft). Regarding the converters operation and control in the DFIM, a phasor model was also used, in line with the approach that is followed in SIMSEN.

3. Control Strategies for Variable Speed PSP

The control strategies used in variable speed PSP units are the same both for SIMSEN detailed modelling and the reduced order models. Regarding the turbine operation mode (see Figure 2), when a power set-point change is requested to the PSP plant, (for example, resulting from the action of the frequency controller) it can be directly used at the machine power converters to control the active power output. The power set-point can be also used at the "speed optimizer" to define the most efficient speed reference at which the machine should operate. In order to have better compatibility conditions between detailed simulation to be performed in SIMSEN and reduced order models it was adopted the same type of turbine governors in both cases. In the specific case of the variable speed unit, as a result of the improved performance it usually demonstrates, a PID-type governor with a feed-forward power-to-gate control action was adopted [6]. Its main block diagram is presented in Figure 3. The main parameters of the PID-type governor are: the unit permanent droop (R_p, (p.u. MW)/(p.u. rad.s^{-1}), the tachometer time constant (τ_ω), the proportional, integral and derivative constants (K_p, T_i and T_d, respectively) and the filter time constant T_{filter}. The output of the PID governor is limited in amplitude (gate position limits between 0 and 1.p.u.), but also in the gate opening/close speed limit rate (U_o, U_c). The main function of the turbine governor is to control the guide vane opening in response to turbine speed or power variations. By controlling the opening or closing of the guide vanes it performs a primary speed/load control. The normal approach to this type of governing systems consists in modelling the action of a servomotor, by eliminating the speed error feedback from the machine using a permanent and a transient droop feedback and/or a PID or a PI controller. Regarding the pump operation mode, and as already referred, gate opening control has only slightly influences on the efficiency of the PSP plant. Therefore, the control strategy of the unit consists on the active power control through the VSC converters, being the speed defined through the characteristic curves of the unit. A general overview about the corresponding block diagram is presented in Figure 4.

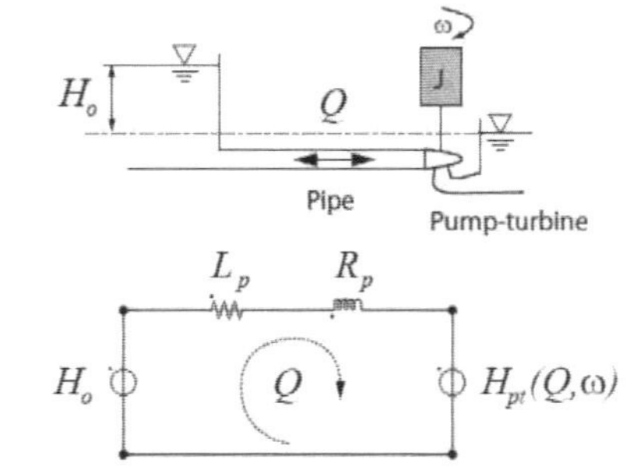

Figure 1. Reduced hydraulic model description used to represent pumping mode operation.

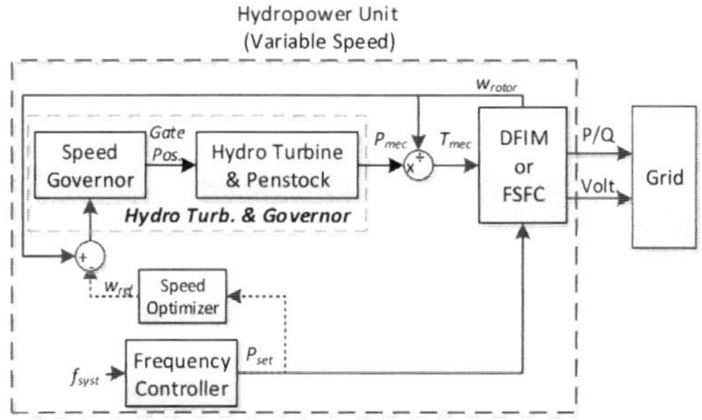

Figure 2. Functional block diagram for of a variable speed hydro unit – turbine operation mode

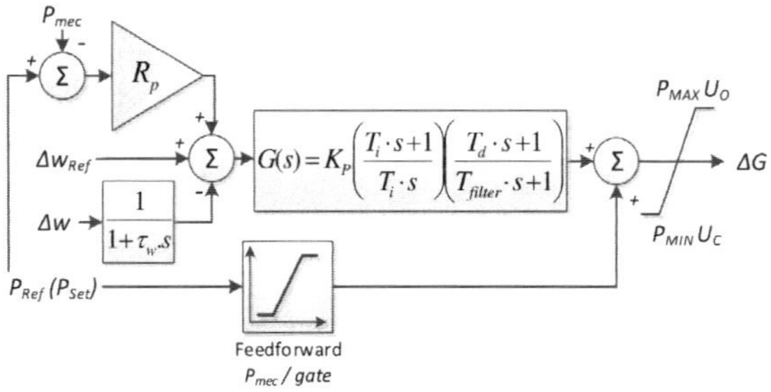

Figure 3. PID-type speed governor for variable speed PSP plant.

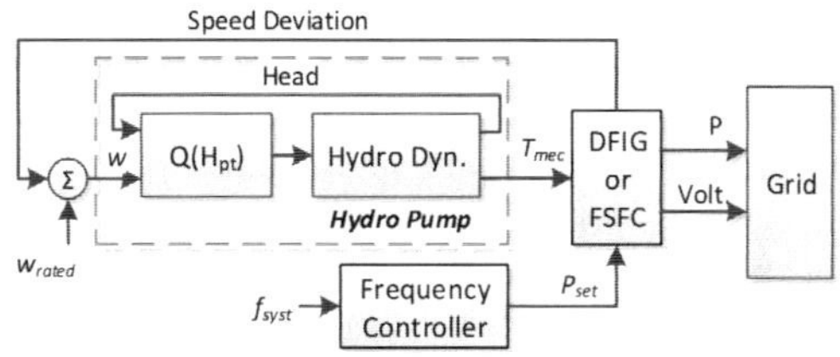

Figure 4. Functional block diagram for a variable speed hydro unit – pumping mode.

4. Test case and simulation results

To perform the derivation and validation of reduced-order models of PSP units, a detailed model was implemented in the dedicated simulation platform of SIMSEN. The highly detailed approach that was followed, considering the necessary hydraulic phenomena and its associated behaviour, was then compared with the developed models implemented exploiting the software MATLAB/Simulink®. Electric models and their parameters, the hydraulic governors and control strategies for voltage source convertors in variable speed units were kept equal between both approaches. The PSP plant considered for the test case is based on an existing 2x210MW power plant equipped with reversible Francis pump-turbines. The key characteristics of the hydroelectric scheme are summarized in Table 1 referring to the original fixed speed configuration with synchronous generators.

Table 1. Test case PSP plant original hydroelectric scheme characteristics.

Number of units	2	50 Hz	Frequency
Total installed capacity	421 MW	400 kV	Voltage at the connection point
Total discharge in pumping mode	350 m3/s	420 MVA	Total rated apparent power
Total discharge in turbine mode	266 m3/s	15 kV	Generator's stator line voltage
Maximum gross head	148 mWC	15	Generator's poles pair
Penstock length	260 m	200 rpm	Generator's rotational speed

The simulations were based on a single pump-turbine unit, the other being at rest. The validation procedure consisted then in studying the operation in turbine and pump operation modes, comparing the response of both detailed and reduced order models. The representativeness of the reduced order models was validated based on the pre and post disturbance values, as well as response times and associated magnitude.

4.1. Turbine operation mode

To perform the reduced-order models validation, the following test condition was specified, in the turbine operation mode: *unit operating at the nominal power conditions ($P_{elec} = 210\ MW$), followed* *to a step load reference reduction to* $P_{elec} = 130\ MW$ *and back to* $P_{elec} = 210\ MW$ *after a suitable time for unit stabilization.*

A critical issue of the DFIM operation is related with the unit speed deviation resulting from the load change. To preserve rotor converters' control capabilities, the unit needs to keep its speed within a given band. Therefore, a ±7% speed deviation range around the nominal value was considered, which was achieved by the introduction of ramp rates that limited the active power to the grid. To comply with this requirement, it was necessary to identify appropriated ramp rates to the prescribed load step (controlled through the active power control loop implemented in the rotor-side converters). For the specified test conditions, the ramp time was set to 14 s.

As it is depicted in Figure 5, a very good agreement between reduced-order and detailed models is achieved, being possible to state that a good degree of representativeness was obtained. As expected, some oscillations can be observed in the speed and gate opening in the detailed solution, which results from the detailed modelling of a downstream surge tank, not reproduced in the reduced order model.

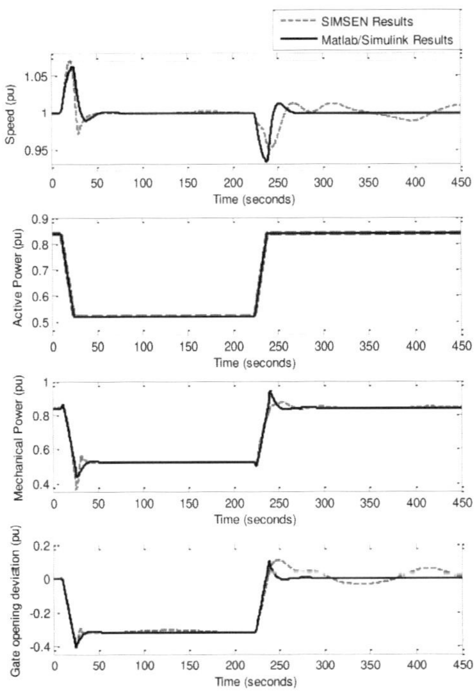

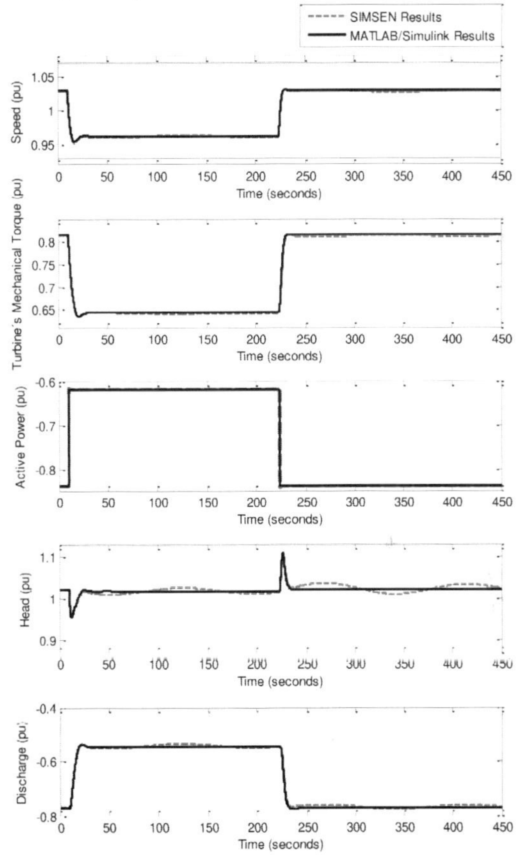

Figure 5. Turbine operation mode: comparison between the SIMSEN and the Matlab/Simulink® simulation results for a PSP plant with a DFIM.

Figure 6. Pump mode: comparison between the SIMSEN and the Matlab/Simulink® simulation results for a PSP plant with a DFIM.

4.2. Pump operation mode

The ability of proving power variations while operating in the pump mode is a distinctive feature of variable speed PSP plants. In line with the previous test case, the following test condition was considered: *unit operating at the nominal power conditions (*$P_{elec} = -210\ MW$*), followed by a step load reference reduction to* $P_{elec} = -155\ MW$ *and back to* $P_{elec} = 210\ MW$ *after a suitable time for unit stabilization.*

The simulation results are presented in Figure 6 and feature a very good agreement in terms of rotational speed and active power, even if the head and discharge fluctuations related to the downstream surge tank are not reproduced. It should be mentioned that due to short penstock length, the inertia effect of the penstock is dominant as compared to penstock elasticity effects, especially when the hydro unit controllability is concerned, while for long penstocks, the elasticity effect and the related water hammer phenomenon has to be considered, [3] and [4]. Nevertheless, from the power system stability perspectives, good simulation results can be obtained with reduced simulation model of the hydraulic system for pumping mode provided that the model includes penstock inertia and realistic pump characteristics.

5. Conclusions

Following the integration of new PSP technologies in electric power systems (along with the integration of NRE), specific studies are required to evaluate system stability with respect to different grid disturbances and considering different operational scenarios. Though dynamic simulation models of different sources have been defined for many decades, defining a well-established base for most of the available generation technologies, the specific case of the new PSP technology is not widely available. Particularly when considering new variable speed technologies, the associated dynamic models are generally not capable of representing the main dynamic response either in turbine and pump operation modes. In this sense, the work presented in this paper provides important contributions regarding the identification and validation of variable speed PSP plants dynamic models, both for pump and turbine operating modes. The proposed models are validated in several operating conditions, considering different load changes and relying on existing PSP data. Such approach provides a solid background on the extensive validation of the proposed models with respect to its future use in electric power system studies involving large disturbances requiring significant changes in the units' power output. The process described provides the necessary confidence for model exploitation in system studies where these type of units are expected to have a key role in important grid services, such as the participation in the frequency containing reserve or synthetic inertia to the system.

Acknowledgments

The authors would like to thank Mr. Johann Hell from ANDRITZ Hydro GmbH, Partner of the Hyperbole EU project, for his contribution in the definition of the DFIM solutions considered in this paper.

References

[1] M. A. Hozouri, A. Abbaspour, M. Fotuhi-Firuzabad, and M. Moeini-Aghtaie, "On the Use of Pumped Storage for Wind Energy Maximization in Transmission-Constrained Power Systems," *IEEE Trans. Power Syst.*, vol. 30, no. 2, pp. 1017–1025, Mar. 2015.

[2] J. P. Deane, B. P. Ó Gallachóir, and E. J. McKeogh, "Techno-economic review of existing and new pumped hydro energy storage plant," *Renew. Sustain. Energy Rev.*, vol. 14, no. 4, pp. 1293–1302, 2010.

[3] E. De Jaeger, N. Janssens, B. Malfliet, and F. Van De Meulebroeke, "Hydro turbine model for system dynamic studies," *IEEE Trans. Power Syst.*, vol. 9, no. 4, pp. 1709–1715, 1994.

[4] C. Nicolet, B. Greiveldinger, J. J. Herou, B. Kawkabani, P. Allenbach, J.-J. J. Simond, and F. F. Avellan, "High-order modeling of hydraulic power plant in islanded power network," *IEEE Trans. Power Syst.*, vol. 22, no. 4, pp. 1870–1880, Nov. 2007.

[5] A. C. Padoan, B. Kawkabani, A. Schwery, C. Ramirez, C. Nicolet, J. J. Simond, and F. Avellan, "Dynamical behavior comparison between variable speed and synchronous machines with PSS," *IEEE Trans. Power Syst.*, vol. 25, no. 3, pp. 1555–1565, 2010.

[6] Y. Pannatier, B. Kawkabani, C. Nicolet, J. J. Simond, A. Schwery, and P. Allenbach, "Investigation of control strategies for variable-speed pump-turbine units by using a simplified model of the converters," *IEEE Trans. Ind. Electron.*, vol. 57, no. 9, pp. 3039–3049, 2010.

[7] SIMSEN, "Simulation software for the analysis of electrical power networks, adjustable speed drives and hydraulic systems." [Online]. Available: http://www.powervisioneng.ch/Simsen_hydro/MainFeatures/SIMSEN_leaflet_14_detailed.pdf.

[8] P. Kundur, N. J. Balu, and M. G. Lauby, "Power system stability and control." p. 1176, 1994.

[9] R. G. de Almeida and J. A. P. Lopes, "Participation of doubly fed induction wind generators in system frequency regulation," *Power Syst. IEEE Trans.*, vol. 22, no. 3, pp. 944–950, 2007.

High flexible Hydropower Generation concepts for future grids

Johann Hell

ANDRITZ HYDRO GmbH, Vienna, Austria
E-mail: Johann.Hell@andritz.com

Abstract. The ongoing changes in electric power generation are resulting in new requirements for the classical generating units. In consequence a paradigm change in operation of power systems is necessary and a new approach in finding solutions is needed. The presented paper is dealing with the new requirements on current and future energy systems with the focus on hydro power generation. A power generation landscape for some European regions is shown and generation and operational flexibility is explained. Based on the requirements from the Transmission System Operator in UK, the transient performance of a Pumped Storage installation is discussed.

1. Flexibility in power system operation

All the impacts of high penetration of RES in the production mix were already discussed in the last years with the conclusion that all remaining classical generating units in grid must be more flexible in the future. The current power generation and consumption is nowadays characterized by the following features:

- **Steeper Ramping :** The rate of increase and decrease of dispatchable capacity of generation
- **Lower Turn downs:** To operate units at part or minimal load
- **Shorter peaks:** The peaks of generated power from RES are shorter in duration

As a consequence the whole power system including all different sources is faced to new challenges. The overall Power System has to provide in the future more Flexibility in Generation, Transmission, Demand side management and System operation.

1.1. The Impact of inflexibility to power system operation

Missing flexibility is resulting in operational limitations and can also reduce the system stability. In case of a severe system fault, the system can become unstable and this could result - under worst case condition - in a system black out.

Indications of reduced or missing electrical system flexibility are:

- **Problems to balance demand and supply:** As a result the grid frequency deviates from the nominal value and exceed the defined limits under worst case conditions
- **Curtailment of RES:** Due to restrictions in transmission capacity the available generated power from RES cannot be transferred to the consumers
- **Violations of area balancing:** These are deviation from schedule in power area balance
- **Negative market prices:** This can be a consequence of any kind of inflexibility. E.g. power plants cannot reduce the output or load cannot absorb the generated power
- **High Price volatility:** Any kind of inflexibility can result changes the energy price

IOP Conf. Series: Journal of Physics: Conf. Series **813** (2017) 012007 doi:10.1088/1742-6596/813/1/012007

1.2. Graphical representation of Flexibility
The power generation mix in different regions of Europe is quite different. In the following diagrams 5 different power sources are considered. This representation was first introduced by Yasuda in [4] and is shown in Figure 1 below. Significant is the power generation in Denmark, with a high contribution of wind power and interconnection. Nordic countries having a high contribution of Hydro Power in the generation mix. The power generation in Portugal is characterized by hydro and thermal generation.

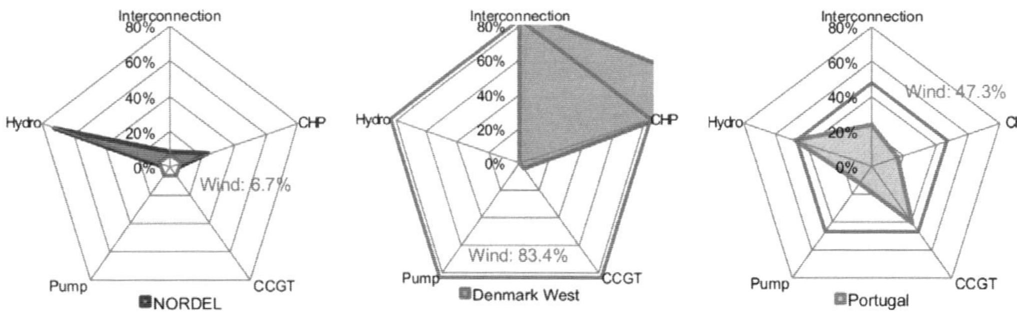

Figure 1: Diagrams showing the power generation mix in Europe in 2013 [4]

1.3. Definition of operational Flexibility in Power Systems
The term "Flexibility" is frequently used in power system discussions but not clear defined. This can result in misunderstanding and differences in expectations in communication about system stability and future grid topologies. Below a physical/mathematical approach is shown [1].

For measurement of operational flexibility of the different power generating units in the grid, the following methodology is established.

- **Power Capacity [MW]**: which is the dispatchable range of generation power output of a unit
- **Power ramp rate [MW/min]:** the rate for increase and decrease of the unit generation power
- **Storage Energy [MWh]:** The energy must be stored needed for power area balancing

1.4. Modelling of operational Flexibility
Operational flexibility can be defined as the interaction of the flexibility needed by the system and the generation available flexibility. For successful operation of an electric power system, the needed Flexibility must be fulfilled by the Flexibility offered by the generating units.

1.4.1. Available generation Flexibility
In Figure 2 below, the Flexibility of different power generation technologies is shown.

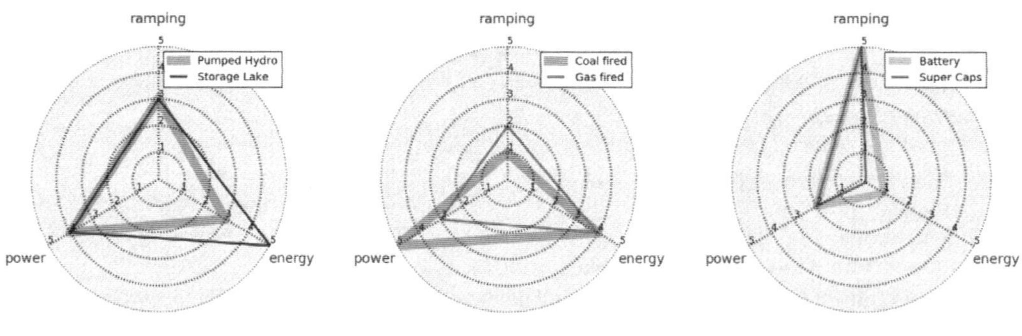

Figure 2: Flexibility chart for different power generation technologies

1.4.2. Needed operational Flexibility
For the needed operational Flexibility no general rules exists. Basically the influencing factors are: contribution of RES in the power generation mix, the stored energy in the grid (inertia) and the degree of interconnection to other systems.

2. Requirements from the Grid
The European electrical grid comprised different Transmission System Operators (TSO) between Portugal and Poland and between Greece and Norway. The installed generation capacity in the entso-e grid is about 994 GW (*2015 - Data provided by ENTSO-E*) with an increasing penetration of renewable power generation, like wind and solar. The ongoing phase out of nuclear power generation in Germany is resulting in new challenges for the TSO's because of reduction of system inertia as well as reduction of system damping.

3. The role of Hydropower generation in the power system
There is no need to explain the importance of Hydro Power generation in the total electrical energy production in the world. In the following sections, the technical features in terms of flexibility of hydro power will be discussed.

3.1. Existing installations
Most of the units currently in commercial operation are running with a constant rotational speed. New installations in Portugal and Switzerland with variable speed concepts are currently in the commissioning phase. Depending on the type of installed turbine, the concepts are offering the following features:

- Units equipped with **KAPLAN** turbines are not so flexible in transient operation and currently are not used for Primary Control.
- **PELTON** units are very flexible and can operate in a wide range. The ramping time must be coordinated with the permissible pressure rise and drop in the often long penstock.
- **FRANCIS** Turbines are used for high power application up to 1000MW, with a limitation in the operating range
- **Reversible Pump Turbines** are more specific designed, because of operation in two directions of rotational speed with different signs of output
- **TERNARY** sets are combination of a Pelton or Francis turbine and a Pump with only one direction of rotational speed. The hydraulic machines can be mechanically separated by a clutch. This solution is showing the highest transient performance.

3.2. New Technologies – Power Electronics / Speed variation
By means of Power Electronics the generator can be separated from the grid and operate with a speed different from the synchronous speed. The specific size of the converter varies on supplier and is in the range of 3-4 m^3/MW. The losses are in the range of 1.5% - 2% for a 100MW converter.

These drawbacks are compensated by additional features of the entire system which makes the unit now much more flexible in operation.

3.2.1. Controlled Power flow
So-called Full Size Frequency Converter (FSFC) is connecting two systems with different voltages and frequency by using power electronics. Reactive power can be controlled independently on both sides. By control of the speed of the generating unit, the hydraulic turbine or pump can operate on the best efficiency point depending on applied load and hydraulic head.

The main advantage of such a concept is the fast control of the power flow between the grid and the electrical machine. For that reason disturbances in the grid (in voltage and frequency) are not influencing the generator.

3.2.2. Synthetic Inertia

Due to decoupling of grid and generating unit, the system is insensitive for frequency excursions in the grid. This behavior can be observed in solar power applications and is not welcome by the TSO.

On the other hand by means of control now units equipped with FSFC can provide additional power depending on frequency gradient. The energy stored in the rotating mass will be used for the fast power response. This feature is called "Synthetic Inertia", when the power is higher than the natural reaction of a synchronous machine with the same size.

4. Ongoing developments in Electrical Power Systems

4.1. Current situation in the Great Britain grid

The electrical grid in GB is characterized by a high penetration of RES and low system inertia. At the moment the frequency control in the grid is provided by primary control with a response time of 9 seconds (compared to a response time of 30 seconds in Continental Europe).

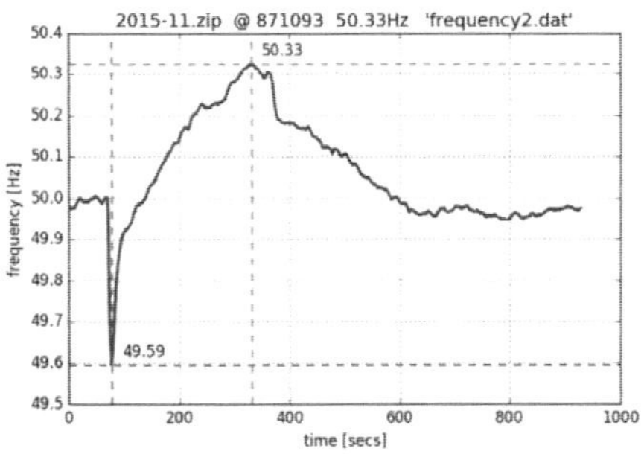

Figure 3: Grid-Frequency in GB record from November 2015

Frequency measurements with a sampling rate of 1 second for the years 2014 and 2015 are provided by national**grid** [6]. In the diagram left an example of a period of 1000 secs, from November 2015 is shown. The drop of the frequency is a result of loss of generation. The following overswing is caused by activation of primary and secondary control.

4.2. Enhanced Frequency Response (EFR)

For fast dynamic grid stabilization national**grid** is asking for new service called "Enhanced Frequency Response" EFR. In Figure 4 three power output curves depending on frequency deviation are defined. The actual value of control power should be within the RED and BLUE boundary curve.

With this new fast responding service the grid can be stabilized much better and as a consequence the amount of primary and secondary control reserve can be reduced.

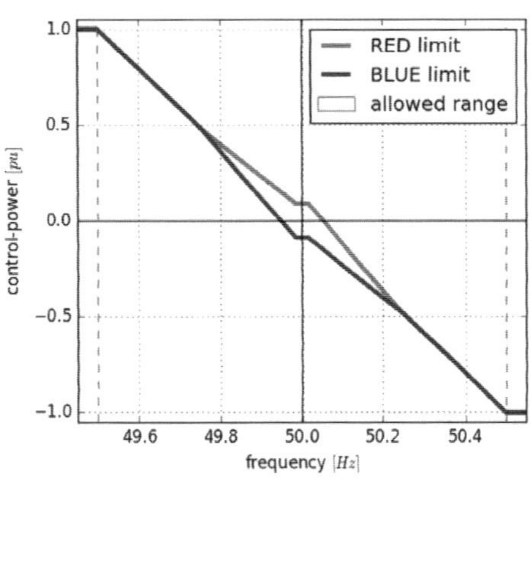

Figure 4: national**grid** definition for Enhanced Frequency Response EFR [6]

The service must be provided symmetrical for a duration of 15 minutes.

The requested response time should be less than 1 second.

.

IOP Conf. Series: Journal of Physics: Conf. Series **813** (2017) 012007 doi:10.1088/1742-6596/813/1/012007

4.3. FIXED speed system response

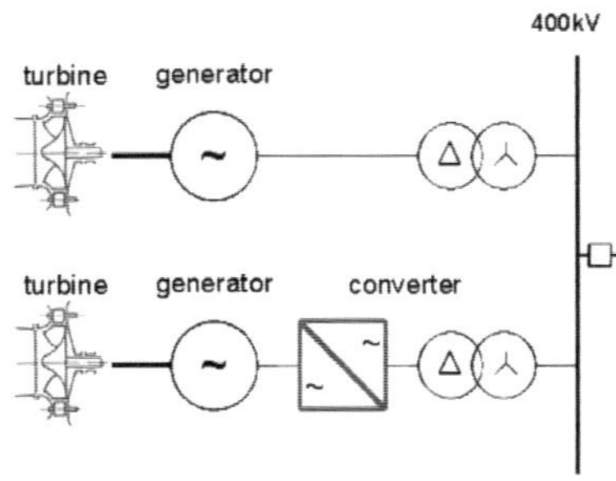

Figure 5: investigated concepts – FIXED speed (upper part) – VARIABLE speed with FSFC (lower part)

Due the requested short response time for the EFR service and the wide operating range classical Pumped Hydro Storage concepts with Synchronous machines and FIXED speed are not in the position to fulfill the requirements. Especially in Pump operation control of the power of the unit is not possible. The simulation results are shown in Figure 6 below.

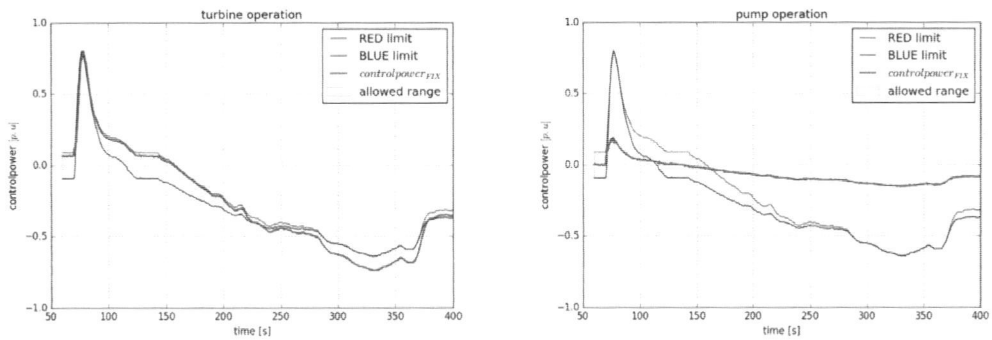

Figure 6: Transient simulation for EFR request of a FIXED speed installation with variable grid frequency (as shown in Figure 3) LEFT: Turbine operation; RIGHT: Pump operation

4.4. VARIABLE speed system response

One of the options to fulfil the requirements is to use a system, which allows a very fast and accurate power control. This can be done by using a variable speed power generation system, equipped with a Full Size Frequency Converter (FSFC). The fast power response now can be provided by using the flywheel effect of the generator (Synthetic Inertia) and for static response the ability of the hydraulic pump-turbine of power variation by changing the rotational speed. The results of the simulation of a variable speed system are presented below in Figure 7. The requirements can be fulfilled with any restrictions. The simulation results showed below based on an example for 300MW Pumped Storage installation.

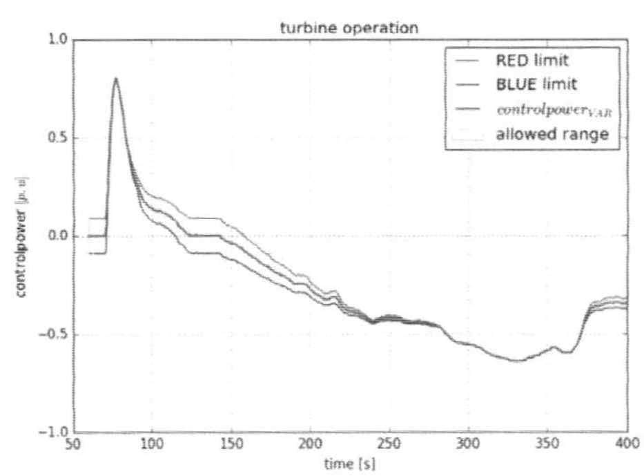

Figure 7: Transient simulation of a VARIABLE speed installation and variable grid frequency

As already mentioned above, with a VARIABLE speed system the requested performance can be achieved easily. The control power is inside the defined range for the considered case.

5. Summary & Conclusion

In the presented paper the current and future situation in electric power grids in terms of Flexibility was discussed. In section 4.1, an example based on the requirements for "Enhanced Frequency Response" published by the TSO of UK – national**grid** was given.

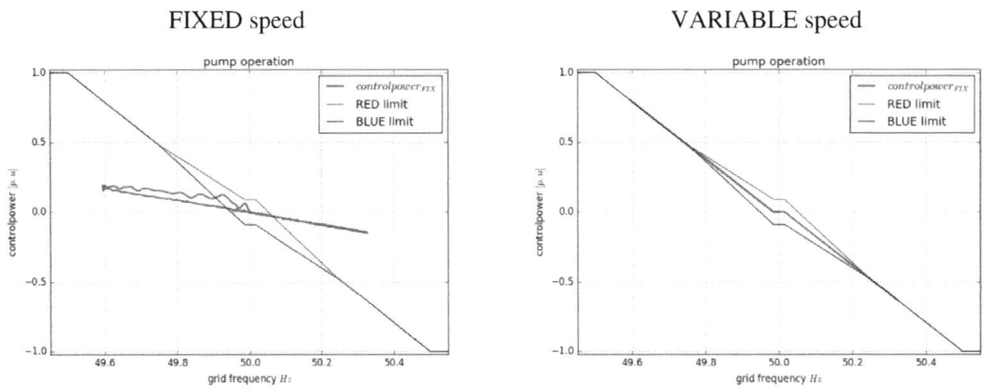

Figure 8: Comparison of the EFR requirements for a FIXED speed unit (left), and a VARIABLE speed unit (right) in pump operation

The results of the simulation for a FIXED and VARIABLE speed Pumped Storage unit can be summarized for the different features as follows:

- To provide a fast responding service with hydro power generating units, variable speed solutions must be used. By using the stored energy of the rotating mass of the rotor, power can be supplied or consumed from the grid within parts of seconds (see Figure 8)
- In combination with the flywheel effect, the variable speed concept also allows to operate the unit in an extended range of output
- Especially in pump operation mode, the power can be controlled by speed variation
- Due to decoupling of grid and generating unit, the system allows a wide operating range for grid frequency and voltage. So the new requirements for grid connection from the Transmission System Operator can be fulfilled easily
- In most of the cases the drawback of a needed converter will be compensated by the additional new features described above

6. References

[1] Andreas Ulbig, Operational Flexibility in Electric Power Systems, DISS. ETH NO. 21882, ETH Zurich 2014
[2] Y. Makarov, C. Loutan, J. Ma, and P. de Mello, Operational Impacts of Wind Generation on California Power Systems, IEEE Transactions on Power Systems, vol. 24, no. 2, pp. 1039 - 1050,May 2009
[3] Electric Power Research Institute (EPRI), Electric Power System Flexibility - CHALLENGES AND OPPORTUNITIES, subject 3002007374, 2016
[4] Yoh Yasuda, Flexibility Chart - Evaluation on diversity of flexibility in various areas, 12th Wind Integration Workshop, Kansai University, Osaka, Japan 2014

[5] P. Kundur, „Power System Stability and Control," McGraw Hill, New York, 1993
[6] Specification Enhanced Frequency Response, national**grid**, UK, 2016 http://www2.nationalgrid.com/Enhanced-Frequency-Response.aspxNationalGrid
[7] J. Hell, Grid Frequency Response – Contribution of Hydro-Power for Grid Stabilization, HYDRO conference Cernobbio, 2014
[8] G.A. Munoz-Hernandez, Modelling and Controlling Hydropower Plants, Springer 2013
[9] HYPERBOLE project EU FP7-ENERGY-2013-1 Grant 608532
web: https://hyperbole.epfl.ch/SitePages/Home.aspx

7. Acknowledgments

The research leading to the results published in this paper is part of the HYPERBOLE research project, granted by the European Commission (ERC/FP7-ENERGY2013-1-Grant 608532).

Performances achieved to the Grid by a Full Power Converter Used in a Variable Speed Pumped Storage Plant

Claude, Jean-Michel

GE Power Conversion, Parc Techn'Hom, 90018 Belfort France

jean-michel.claude@ge.com

Abstract. The growth of renewable energies likes wind and solar requires pumped-storage plants to increase their performances to stabilize grid frequency and voltage. The introduction of a full-power converter constitutes the ultimate step forward to meet the requirement in a safe, reliable and sustainable manner. This article quickly introduces the converter topology and technology before describing the performances it aims to deliver to the grid. Finally, converter bypass is discussed.

1. Introduction

Pumped-storage schemes play a growing role in grid stability to help balance the increasing amount of unpredictable power brought by renewable energies like wind and solar.

However, in Europe — and in many other places in the world — the development of pumped storage is refrained by several counter factors:

- ✓ Less available locations to accommodate new schemes.
- ✓ Lower hydraulic properties affecting the return on investment.
- ✓ Higher environmental pressure from a population that is against the new implementations.
- ✓ Decentralization with smaller power generation plants requires decentralized balancing schemes.

Today, most utility companies have some pumped-storage schemes in their portfolios that were constructed in the 1960s and 1970s, housing units of unitary power in the range of 60-200 megawatts (MW).

After large variable-speed, pumped-storage power plant were launched, based on doubly fed induction machines, the next step (coming up) introduces fully fed power converters in new builds or as upgrades to the above mentioned existing schemes to increase their performances.

This article provides an update on converter technology and describes what performances could be achieved by taking the whole electrical part of the plant, from pump/turbine shaft to the connection to the grid, into account.

2. Case Study – 160-MW Pumped Storage

This case study is based on an existing plant having two units of the following optimized main features:

- Rated operating point in pump mode : 160 MVA; 13.8 kV; 55Hz; 550 rpm
- Rated operating point in turbine mode : 155MVA; 13.2kV; 52.6Hz; 526 rpm

- Generator/motor power factor during operation : 1
- Generator/motor number of poles : 12
- Generator/motor subtransient reactance : 22 percent (%) unsaturated
- Generator/motor + pump/turbine inertia : 200 T.m²
- Grid voltage and short-circuit power : 220kV – 10GVA

The single line diagram for one unit is given in figure 1 below:

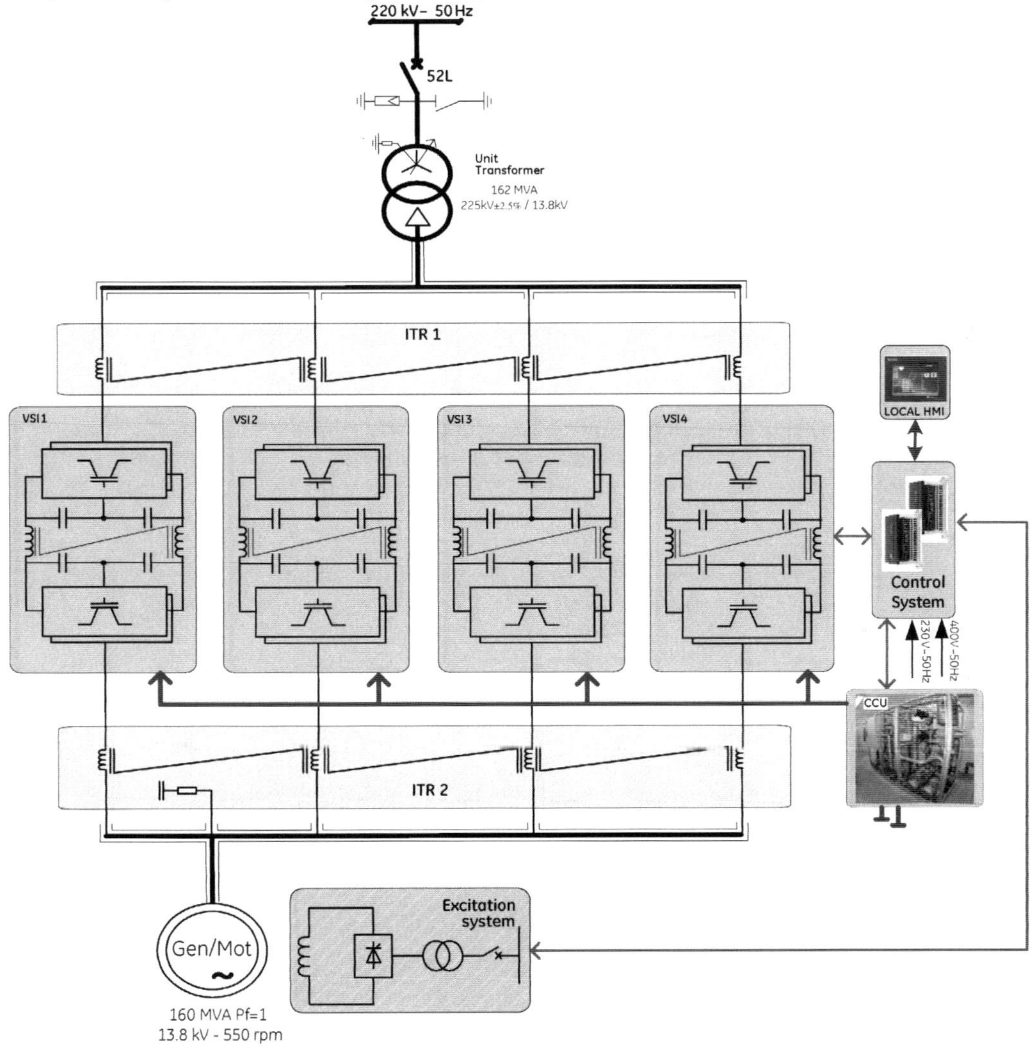

Figure 1

The full converter is modular and made of four identical threads connected in parallel through interlacing reactors. It is deionized water cooled with a water-to-water heat exchanger.

The generator/motor static excitation system is closely interfaced with the converter controller.

There is no generator circuit breaker. The whole unit is protected by the line breaker (52L) and a converter feature allows the unit transformer to be softly magnetized, avoiding inrush current on line side every time it is energized.

The speed rotation direction as per turbine or pump mode is naturally achieved by the converter so there is no need for phase reversal switches.

IOP Conf. Series: Journal of Physics: Conf. Series **813** (2017) 012008 doi:10.1088/1742-6596/813/1/012008

3. Converter Technology

The converter is made of a five level pulse width modulation (PWM) drive utilizing high power injection-enhanced gate transistor (IEGT). The picture 2 below shows a power stack subassembly in which six components are tightened together:

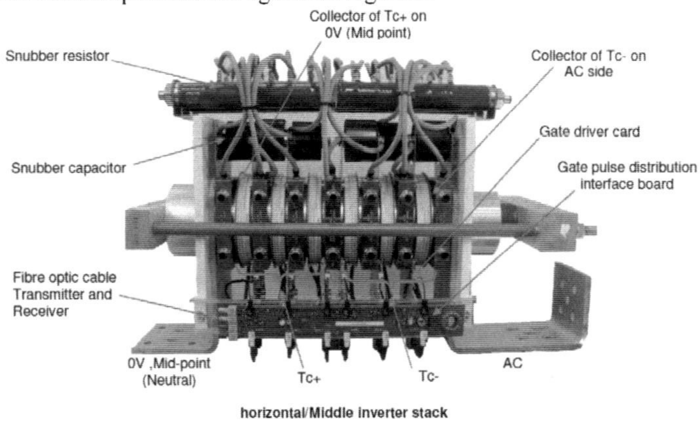

Figure 2

The converter is four quadrants fully reversible with two identical inverters: one on machine side and another one on grid side — named Active Front End (AFE).

The five-level PWM carriers are shifted from one module to the other through the interlacing reactors, providing 33 levels of PWM phase-to-phase reaction at converter output as shown below:

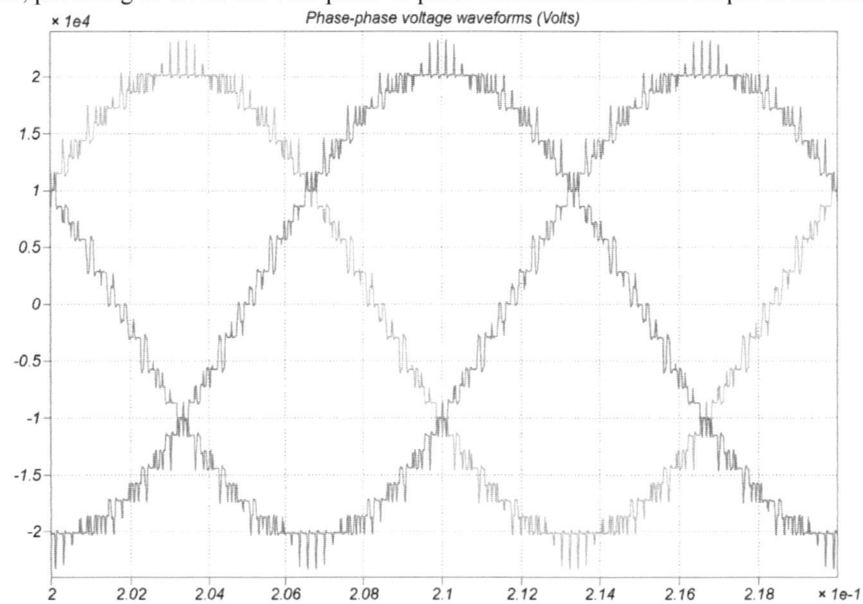

Figure 3

Then, a complementary small output filter suppresses the remaining voltage spikes and notches of short amplitude and high frequency. The resulting output voltage seen by the synchronous machine becomes pure sine wave, having less than 0.5 percent harmonic contain.

4. Performances

4.1. General statement:

Variable speed drive (VSD) brings about key benefits to such a plant:

- The unit speed is adjusted so that the pump/turbine operates at its best efficiency, whatever the water head.
- The water head operating range is extended, allowing the plant to capture more power.
- VSD ensures soft start and stop, smoothing hammer effect on draft tubes when spherical valves are open/closed — avoiding inrush current even on transformers and, thus, increasing the operation flexibility of the plant.
- Reactive power exchanged with the grid is fully controlled, and the plant can satisfy the most stringent requirements from the grid, such as fault condition, frequency and power fine tracking.
- The synchronous machine is optimized: Its rating is reduced by 15 percent as it is continuously operating at power factor 1. Its rotor is built with a lower pole number and it no longer needs to provide large inertia for pump/turbine stability.

4.2. Power factor control on the grid side:

The converter takes care of the power factor control as it is able to control reactive power flow and direction independently from the active power flow and direction. Its capability is expressed through the P/Q diagram below (left side):

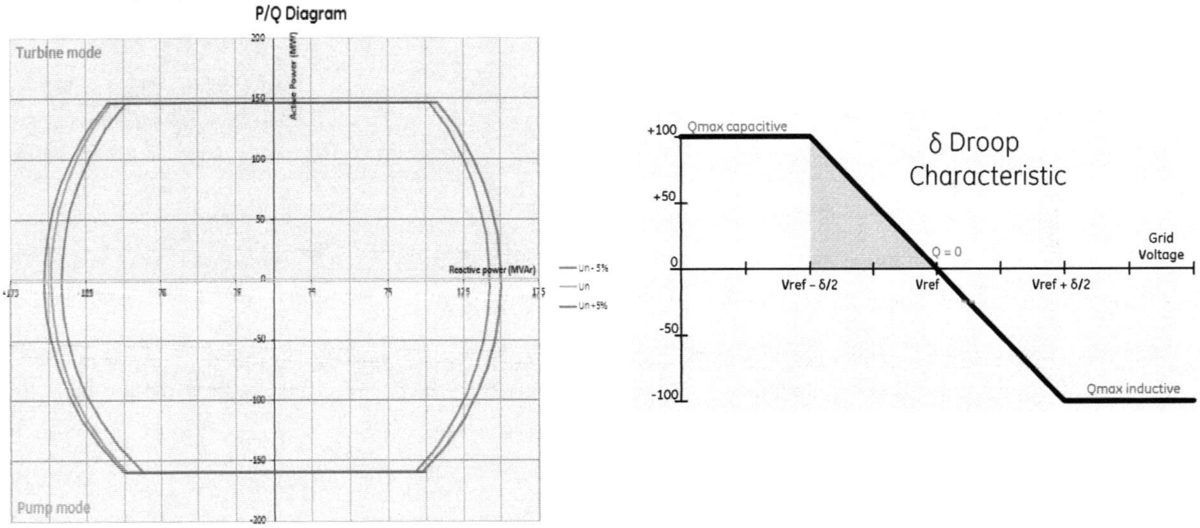

Figure 4

The converter reactive power capacity is dependent of the grid voltage as shown by the Green curve for rated voltage, the Blue one (+5 percent) and the Red one (-5 percent).

An automatic voltage regulator (AVR) function is implemented inside the converter controller. Reactive power set point is derived from voltage droop (or slope) characteristic. The droop (δ) is defined as the reduction in voltage at the point of connection as the converter output varies from fully inductive to fully capacitive. As a consequence, when the grid voltage is low, the VSD supplies capacitive reactive power. When the grid voltage is on high, the VSD consumes inductive reactive power. This function is illustrated on figure 4 above (on the right).

4.3. Power control in pump mode

Power control is the main function devoted to the converter since the pump/turbine could only operate in pump mode with its distributor wide open. Therefore, any power consumption control could only results from speed variation achieved independently of the water net head variation.

4.4. Starting in pump mode – mode transfer

With the converter having the same rating as the pump/turbine, dewatering the runner to reduce its resistive torque is no longer needed. Starting could be achieved within a dozen of seconds instead of a couple of minutes, according to the design.

Mode transfer time from pump to turbine, or vice versa, becomes very fast and is illustrated through the curves of figure 5 below:

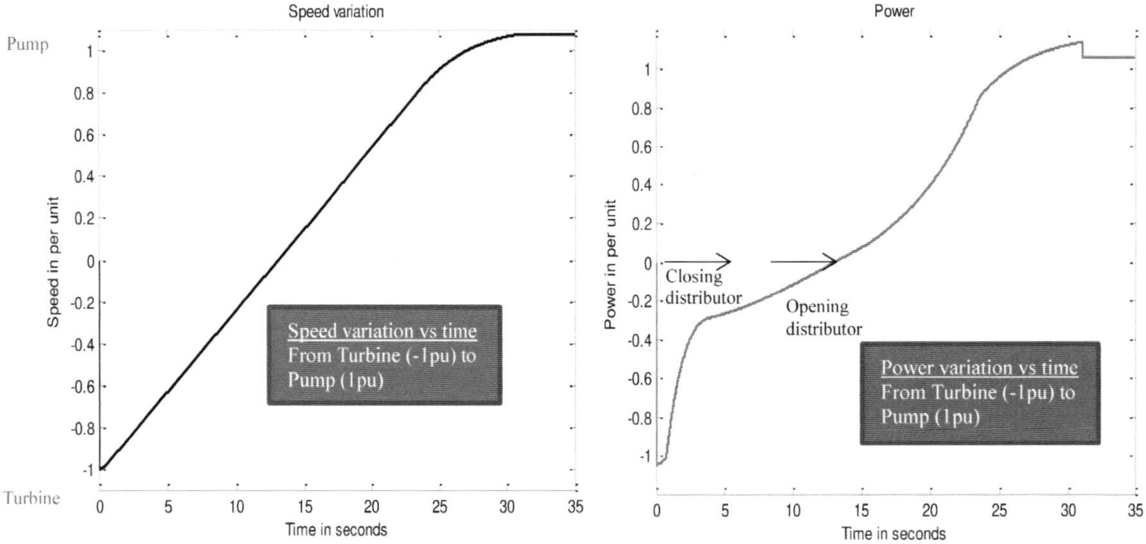

Figure 5

Notwithstanding what would be the hydraulic constraints, the electrical circuit mainly made of the converter is able to change the mode of operation full power in the pump to full power in the turbine within 32 seconds.

IOP Conf. Series: Journal of Physics: Conf. Series **813** (2017) 012008 doi:10.1088/1742-6596/813/1/012008

4.5. Fault ride through
The plant can withstand transient fault condition in the power grid without disconnecting. When a short circuit occurs on the grid, for a limited time — and significant voltage dip of up to 100 percent — during 150 to 500 milliseconds (ms) as shown in figure 6 below (left), the converter is able to interrupt active power transfer with the pump/turbine and then resume operation once the voltage is back.

On the grid side, the converter could simply shut down during transient and then restart automatically. When necessary, it also could provide its rated current into the short circuit to contribute to clearing the fault at first instant and then helping the grid to recover.

The figure 6 below (right) shows the converter phase currents on grid side before, during and after a short circuit, having 100 percent voltage dip. After 60 ms transient, the converter provides back its rated current even if the voltage did not recover.

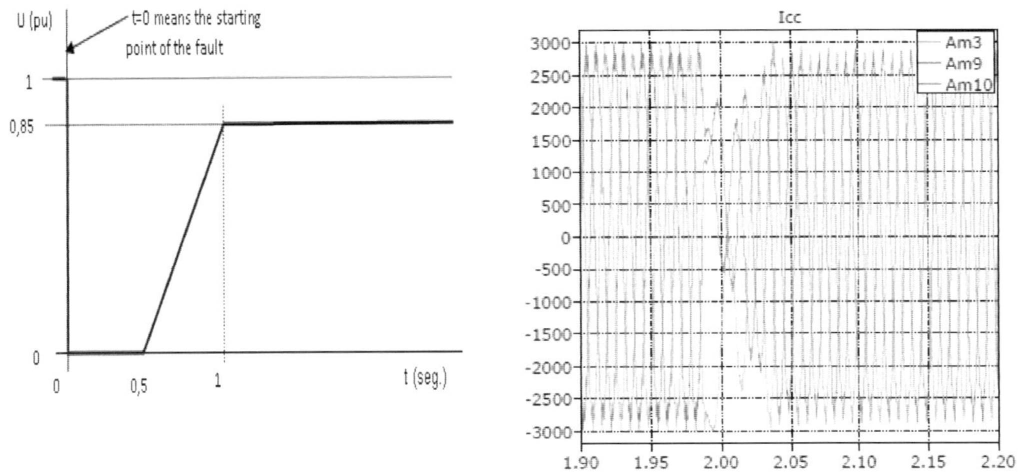

Figure 6

4.6. Pump/turbine stability – virtual inertia
The converter is decoupling the pump/turbine shaft speed from the grid frequency. The fluctuation or oscillation of one is no longer influencing the other, causing the generator/motor inertia to be enlarged for stability purposes.

The converter could claim providing virtual inertia of infinite value.

5. Conclusion:
The full power converter provides dynamic and stable speed and power variation so that the power plant performs fast grid frequency control in both pump and turbine mode.

Thanks to the converter, the operation is fully flexible; the stored volume of water in the reservoirs is better exploited. The plant can achieve a large number of start/stops per day without additional fatigue and stress for its components.

The converter is add-on equipment compared to a conventional pumped-storage plant, but its presence allows for suppressing of the generator circuit breaker, the phase reversal switches and a huge portion of the isolated phase busbars. The generator/motor is optimized and the protections and control systems are simplified.

Coordinated optimization of weekly reserve, day-ahead and balancing energy trade in hydropower

Marte Fodstad[1], **Eirik Schou Grytli**[2] **and Magnus Korpås**[2]

[1] SINTEF Energy Research, Post box 4761 Sluppen, N-7465 Trondheim, Norway
[2] Norwegian University of Science and Technology, Dept. of Electric Power Engineering, N-7491 Trondheim, Norway

E-mail: Marte.Fodstad@sintef.no

Abstract. We present a model for optimal trade in a weekly power reserve market under day-ahead and balancing market price uncertainty. The model takes the perspective of a price-taking hydropower producer and a case study for the Norwegian market design and a Norwegian multi-reservoir water course for a winter week is presented. We demonstrate how a bid curve for the reserve market can be established through a sensitivity analysis on reserve prices, and observe that the optimal trade in the succeeding day-ahead and balancing market is highly sensitive to the reserve obligation.

1. Introduction

Ever since deregulation of the electricity market, the day-ahead market has been the main market in short-term bidding and scheduling. While day-ahead still being by far the largest market, an increasing focus has the last years been given to other short-term markets, such as reserve and balancing markets. This is usually motivated by increasing variability in the power system due to the introduction of more intermittent production.

There is a large research literature on short-term bidding and scheduling in hydropower production, see [1] for a survey. The day-ahead[1] bidding and scheduling problems are tightly linked since the production cost is dominated by the opportunity cost of the water. Also trading of reserve capacity and balancing energy are dependent on the day-ahead bidding and scheduling, since delivery of these services utilizes the same production system as day-ahead delivery. Several papers look into the coordinated optimization of multiple sequential markets. For instance [2] reviews literature on day-ahead, intraday and balancing market trading, but points out that the literature is incomplete and few results show substantial gain in coordinated bidding as opposed to bidding in the markets separately. Also [3] find a limited value of coordinated bidding in the day-ahead and the balancing market, but points out that this might increase as the renewable intermittent production increases causing the value of balancing energy to increase. While [2] and [3] discuss energy markets with different timing, [4, 5] combine bidding in the secondary reserve market trading reserve capacity, the day-ahead market and the intraday market. In a case study they observe less aggressive day-ahead trade and an increased expected profit for a hydro-thermal producer conducting coordinated bidding in the three markets relative to day-ahead trading only.

[1] In literature 'day-ahead' is usually not stated explicitly when referring to this model class.

We present a short-term stochastic programming model for coordinated trading in three markets, the reserve market (RKOM), the day-ahead market (DA) and the balancing market (BM). The model follows the Norwegian market design and a case study for a Norwegian hydropower producer is presented. In the next section the model with data for the case study is presented. In Section 3 the results are presented and discussed before some concluding remarks are given in the last section.

2. Model and data description

We use a stochastic mixed integer program [6] representing a one week horizon with hourly resolution in this work. It is an extension of the one-day model described in [7] with further details given in [8]. In the following we give a short description of the model, emphasizing the model extensions.

2.1. Model description

The model represents three markets, the balancing market (BM), the day-ahead market (DA) and the reserve market RKOM. BM is the tertiary energy reserve market for both up and down regulation, which is traded on an hourly basis and cleared real-time. Remuneration in BM is according to the activated energy. DA is an energy market with hourly resolution and it is cleared the day-ahead. RKOM is a reserve market that is established by the Norwegian TSO to make sure sufficient up-regulation capacity is made available in BM. It is cleared the week before operation for two time blocks; night (00-05 Monday-Sunday) and day (05-24 Monday-Sunday), where only the day-block is included in the model due to limited data availability. A RKOM contract commits the bidder to reserve the offered capacity and bid it in the BM market. [9] Since no pump-capacity or fixed delivery obligations are modelled, the producer can sell, but not buy in the DA market. The modelled producer is a price-taker in all markets.

The markets are represented in a multi-stage scenario tree with 15 stages, each stage having a 24 hour resolution. An illustration of the first day is given in Figure 1, and for each of the remaining six days stages similar to the two last in Figure 1 is added. In the stages where DA is cleared the scenario tree contains DA prices, in the stages with BM activation there are BM prices, and in the first stage a deterministic RKOM price is given. Activation of upward balancing energy is assumed whenever the BM price exceeds the DA price, and visa versa for downward regulation. To reduce the computational burden all BM bids are in one stage, rather than 24 stages that would better reflect the hourly bidding.

The bidding in DA and BM is modelled as in [3] with bidding variables, $x^m_{b,t,s}$, cleared obligations $y^m_{t,s}$ and predefined bid price points $b \in 1 \ldots B$. Here $t \in \mathcal{T}^{\mathrm{RKOM}}$ is the hour from the set of hours in the RKOM day block, $s \in \mathcal{S}$ is the scenario and $m \in \{\mathrm{DA, BM\text{-}, BM\text{+}}\}$ is the market, either DA, BM down-regulation or BM up-regulation. In RKOM the price, P^{RKOM} is assumed known, and the model optimizes the reserve capacity, y^{RKOM}. The capacity reserved for RKOM cannot be sold in DA, which is enforced in Eq. (1). $\overline{Q}$ is the maximum production capacity. Further, the reserved capacity should be offered in the BM market, which implies that the BM bid for the highest bid price at least should be equal to the reservation, as in Eq. (2). The RKOM income, $P^{\mathrm{RKOM}}y^{\mathrm{RKOM}}$, is added to the objective function.

$$y^{\mathrm{DA}}_{t,s} \leq \overline{Q} - y^{\mathrm{RKOM}} \qquad , t \in \mathcal{T}^{\mathrm{RKOM}}, s \in \mathcal{S} \tag{1}$$

$$x^{\mathrm{BM+}}_{B,t,s} \geq y^{\mathrm{RKOM}} \qquad , t \in \mathcal{T}^{\mathrm{RKOM}}, s \in \mathcal{S} \tag{2}$$

The production system used in this study is a Norwegian cascade with 11 reservoirs and 14 generators, with a overall storage capacity of 1332 Mm^3 and a production capacity of 990 MW. The production functions are piecewise linear and concave, and the water head is assumed

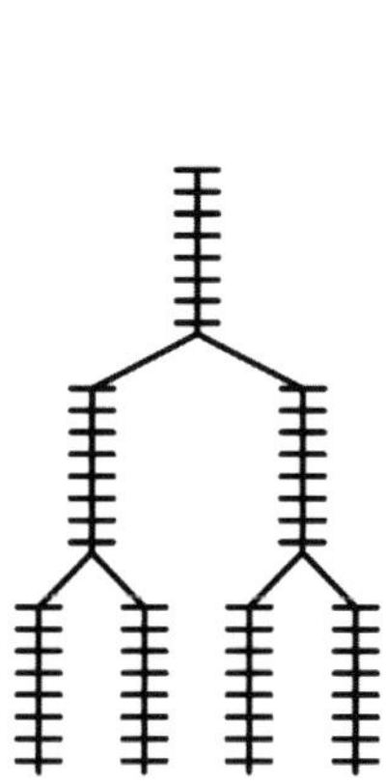

Figure 1. Illustration of the stage structure for the first day. The illustration shows a small scenario tree with two scenarios in each branching. For simplicity only eight hours are illustrated per day.

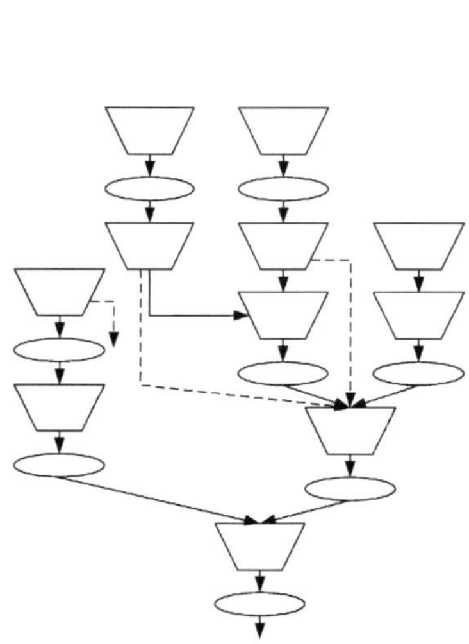

Figure 2. The production system. Trapezoids are reservoirs, ellipsoids are plants, arrows are flows while dashed arrows are spillage or bypass when deviating from the regular water way.

constant. The start/stop of generators is represented with binary variables with related start-up costs and minimum production levels. Between the reservoirs there are water ways for discharge, spillage and bypass with pre-set time delay and minimum and maximum limitations. The marginal water values, which represents the opportunity cost of water, is represented by a set of multi-dimensional linear cuts calculated by a medium-term scheduling model [10]. The case study represent a winter week, with initial level of 75% for all reservoirs. Price scenarios are generated from historical prices for price area NO2 in Norway during the winter of 2014, with an average DA price at 33.7 €/MWh, see [11] for a description of the scenario tree generation procedure. The model is tested for 12 RKOM prices in the range 0.6-13.7 €/MW, all prices within the range of observed prices for the last three years.

3. Results

We focus the case study on how variation in the RKOM price affects the position in the different markets and the overall operation of the production system. The results provide an example of realistic trades and operations, but are dependent on the given DA and BM prices in the scenario tree, and are not necessarily valid for an other scenario tree. Figure 3 shows how the individual components of the objective function change with increasing RKOM price. The overall objective value and RKOM income are increasing, which should be expected, while the income from DA is decreasing. This decrease is due to the required reservation of capacity, but the whole loss in DA is not offset by the increased RKOM income. Parts of the loss is compensated by an increasing BM income since the reserved capacity is offered for up-regulation in BM. Since up-regulation is only activated approximately half the time, the water consumption is reduced when DA trade is replaced with BM trade. This can be seen by the increase in 'Change in EFI' (expected future income), which represents the changed expected value of the stored water from the beginning to the end of the model horizon. The negative values indicate that the production during the week is larger than the inflow, which is typical in the winter season in Norway.

The optimal capacity reservation in RKOM for increasing RKOM prices is presented in Figure 4. This corresponds to the optimal bid curve for this production system for the given week of study and the given RKOM prices. It shows how the model stays out of the RKOM

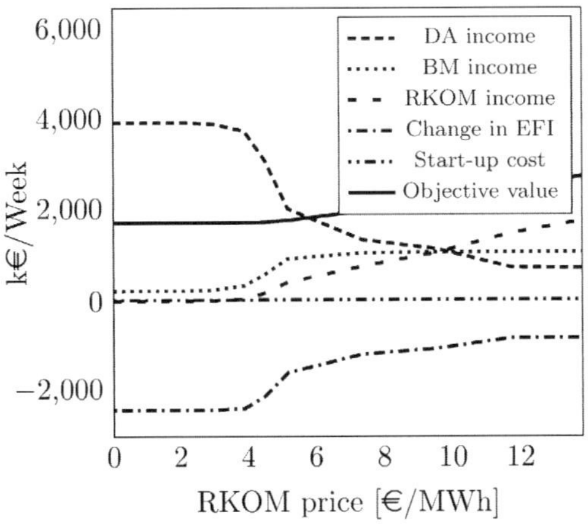

Figure 3. Objective function value for different RKOM prices

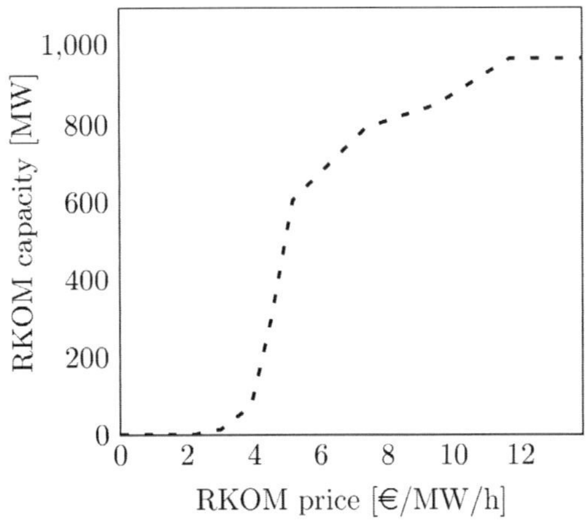

Figure 4. RKOM bid curve given by optimized capacity for different RKOM prices

market at very low prices, and monotonically increases until almost all capacity is offered to the RKOM market at high prices. In real life the higher levels are unlikely to be reached when clearing the RKOM market, since the capacity is more than twice the largest total amount observed in the market the last two years. Still this does not invalidating the bidding curve which for a price taker represents the expected marginal cost of delivering the service.

The expected marginal cost is dominated by the opportunity cost as the only direct cost, the start-up cost, is very small relative to values achieved in the markets, as seen in Figure 3. In the multi-market situation described in this model, the opportunity cost is defined by the combination of trades in DA and BM and the water value representing the storing of water for later weeks. The market rules and production system, as modelled in Eq. (1)-(2), makes the sources for opportunity cost dependent on each other, which causes the optimal position in each market to change as the RKOM position changes. This is illustrated in Figure 5 where expected DA, BM up- and BM down-regulation dispatch volumes as a function of RKOM price is given. As previously observed we see how the DA volume and the total dispatch is reduced as the RKOM price increases. Further, it can be observed that the amount of BM down-regulation is reduced, since down-regulation cannot exceed the DA position. On the other hand, the BM up-regulation is relatively stable over the whole RKOM price range, which is surprising taking into account that the aim of the RKOM market is to secure sufficient BM up-regulation capacity. Zooming in on the expected BM up-regulation dispatch in Figure 6 it shows that the BM up-regulation is decreasing in parts of the RKOM region. A possible explanation for this non-obvious result is that a small DA commitment, due to reserving the capacity in RKOM, requires stopping some generators, and the cost of restarting the generator to provide up-regulation reduces the BM up-regulation bid and thereby the activation of up-regulation for the hydropower producer. This study shows the situation from a single price-taking producer's perspective, while in a market perspective where the need for up-regulation is inelastic and if multiple producers observe the same incentive, this phenomenon could cause the BM price to be higher at a high RKOM price than at a low RKOM price all other equal.

4. Concluding remarks

We have presented a model for optimizing a hydropower producer's trade in three markets; the weekly reserve market (RKOM), the day-ahead market and the balancing energy market. The

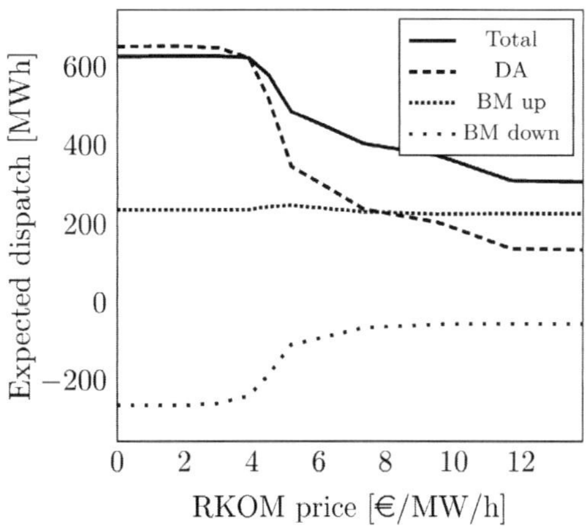

Figure 5. Expected DA and BM dispatch volume for different RKOM prices

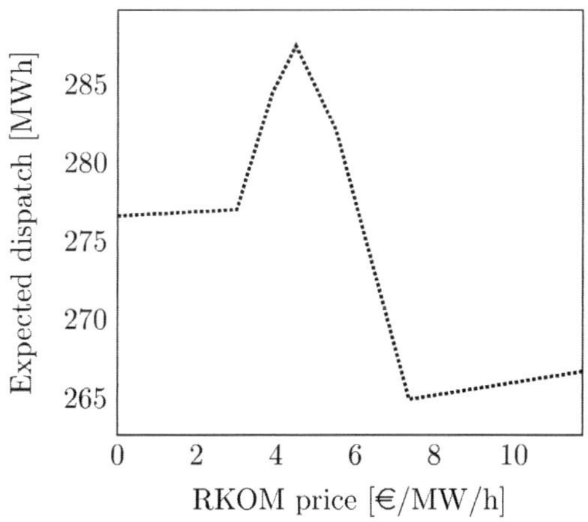

Figure 6. Expected up-regulation dispatch for different RKOM prices

market modeling follows the Norwegian market design. It is illustrated how a sensitivity study on the RKOM price can be used to produce a bid curve for the RKOM market. A case study for a Norwegian producer in a winter week is presented. The results show that both operation and trade in the day-ahead and balancing down-regulation market significantly change as the price and quantity in RKOM change, which supports the usefulness of model-based decision support for RKOM bidding. Further, the study show that for the given day-ahead and balancing market prices the balancing market up-regulation dispatch only moderately changes as the RKOM price and quantity increase, and that the changes are negative in parts of the RKOM price range. This is a surprising observation that should be studied further, since the objective of the RKOM market is to secure sufficient amount of upward balancing energy.

Acknowledgments
The authors thank The Norwegian Research Council for funding through grant no. 502000928.

References

[1] Steeger G, Barroso L A and Rebennack S 2014 *IEEE Transactions on Power Systems* **29** 1758–1766
[2] Klæboe G, Eriksrud A L and Fleten S E 2015 *Energy Systems* **6** 43–61
[3] Boomsma T K, Juul N and Fleten S E 2014 *European Journal of Operational Research* **238** 797–809
[4] Ugedo A, Lobato E, Franco A, Rouco L, Fernández-Caro J and Chofre J 2006 *IEE Proceedings Generation, Transmission & Distribution* **153** 431–442
[5] Ugedo A and Lobato E 2010 Validation of a strategic bidding model within the spanish sequential electricity market *Probabilistic Methods Applied to Power Systems (PMAPS), 2010 IEEE 11th International Conference on* pp 396–401
[6] Birge J R and Louveaux F 1997 *Introduction to Stochastic Programming* (Springer)
[7] Rasmussen C, Hansen J B, Korpås M and Fodstad M 2016 *Energy Procedia* **87** 141–148
[8] Fodstad M, Helseth A and Henden A L 2015 Hydropower scheduling in day-ahead and balancing markets *Proceedings of 12th European Energy Markets, Lisbon*
[9] URL `http://www.statnett.no/en/Market-and-operations/`
[10] Gjelsvik A, Belsnes M M and Haugstad A 1999 An algorithm for stochastic medium-term hydrothermal scheduling under spot price uncertainty *Proceedings 13th Power System Computation Conference, Trondheim, 28. June - 2. July, 1999*
[11] Grytli E S 2016 *Optimal Spot Market Bidding Considering Reserve Power and Reserve Power Options* Master's thesis Norwegian University of Science and Technology

On the Profitability of Variable Speed Pump-Storage-Power in Frequency Restoration Reserve

Jorge Filipe[1,2], Ricardo Bessa[1], Carlos Moreira[1,2], Bernardo Silva[1]
INESC TEC[1] and Faculty of Engineering[2], University of Porto
Porto, Portugal
E-mail: {jorge.m.filipe, carlos.moreira, ricardo.j.bessa, bernardo.a.silva}@inesctec.pt

Abstract. The increase penetration of renewable energy sources (RES) into the European power system has introduced a significant amount of variability and uncertainty in the generation profiles raising the needs for ancillary services as well as other tools like demand response, improved generation forecasting techniques and changes to the market design. While RES is able to replace energy produced by the traditional centralized generation, it cannot displace its capacity in terms of ancillary services provided. Therefore, centralized generation capacity must be retained to perform this function leading to over-capacity issues and underutilisation of the assets. Large-scale reversible hydro power plants represent the majority of the storage solution installed in the power system. This technology comes with high investments costs, hence the constant search for methods to increase and diversify the sources of revenue. Traditional fixed speed pump storage units typically operate in the day-ahead market to perform price arbitrage and, in some specific cases, provide downward replacement reserve (RR). Variable speed pump storage can not only participate in RR but also contribute to FRR, given their ability to control its operating point in pumping mode. This work does an extended analysis of a complete bidding strategy for Pumped Storage Power, enhancing the economic advantages of variable speed pump units in comparison with fixed ones.

1. Introduction

Reversible hydro power plants with reservoir (Pumped Storage Power - PSP) are the most mature storage technology with high integration levels in countries such as Portugal, Spain and Switzerland. Typically, the fixed speed PSP participate in the day-ahead (DA) electricity market for price arbitrage (i.e., pump in low price periods and generate in high price periods) and provision of downward replacement reserve (RR) in specific cases. Due to technical limitations, the current fixed speed PSP technology does not participate in the frequency restoration reserve (FRR). In contrast, variable speed PSP units can not only participate in RR but also in FRR, given their flexibility to quickly change the generating/consumption operating point.

Several works have begin to study the added value of variable speed PSP in the electrical grid in terms of network stability, but only few explore the market perspective. In [1] it is analysed the technical benefits achieved by replacing the existing fixed speed pump units by variable speed ones in an Indian PSP unit, leading to significant benefits since adds frequency regulation in a location where the nominal frequency is constantly below 50Hz. [2] and [3]

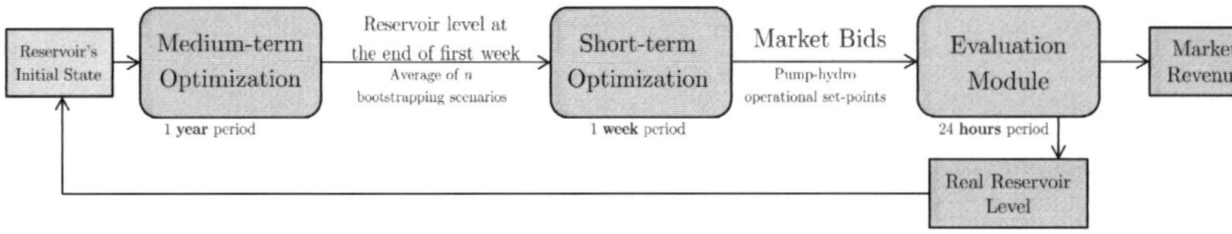

Figure 1: Architecture of the Optimization Framework

explore the use of a ternary unit operated in hydraulic short circuit to provide active power control in pumping mode and concludes that it improves the Virtual Power Plant output power and thus the power network frequency stability in case of significant solar power variations. In [4] it is presented an overview of the trends and challenges in PSP, which clearly states that ancillary services markets, particularly those related to balancing supply and demand, emerge as a valuable source of revenue.

To the authors knowledge, the recent work of [5] was the first to formulate an optimization problem for variable-speed PSP participation in the FRR market, which is then used to estimate the maximum theoretical revenue of the system. The work assumes perfect knowledge of market prices and water inflows and that the amount of energy to be used from the total reserve band available is known at the time of the optimization. In [6] a complete bidding strategy is presented where no prior knowledge of the market operation is assumed, all inputs are acquired through forecasting techniques, allowing its implementation in a real world scenario. The economic analysis was made between a PSP that participates only in the DA market and one that also bids in the FRR market.

In this paper it is presented an extended analysis of the complete bidding strategy previously presented in [6] for operating a PSP unit in the day-ahead market, by performing price arbitrage, but also bidding in the ancillary services market. Compared to [6] the present paper has the following original contributions: a) the possibility to operate either with one or two groups of machines, b) representation of electrical and hydraulic losses in the I/O curves used and c) a fairer economic analysis, by comparing two PSP that bid in both DA and FRR markets, one with a fixed speed pump and the other variable.

2. Bidding Strategies Framework

In Figure 1 is depicted the overall bidding strategy framework for the participation of a PSP unit in the DA and FRR markets, which is divided into three modules: (i) a Medium-term Optimization where is performed an optimal allocation of the water that naturally flows into the reservoir; (ii) a Short-term Optimization that defines the PSP operational set points aiming to maximize its revenue, achieved by combining inputs originating from forecasting techniques (water inflows, DA and FRR market prices) with the technical characteristics of the pump-hydro unit; (iii) an Evaluation Module that, using realized data from the electricity market and water inflows, is able to emulate the PSP real operation and them calculate the actual revenue of the system. As it is clear from the interpretation of Figure 1, the strategy implementation is cyclical. After calculating the updated level in the reservoir, this value is then used as the initial state and the strategy restarts.

2.1. Medium-term Optimization Model

The operation of a PSP unit relies in the correct management of its only resource, i.e. water. The medium-term optimization model, inspired in the work of [7], uses the concept of water value by taking into account the seasonality of the natural water inflows combining it with weekly day-ahead market prices forecasts. Thus finding the optimal allocation of the water resource throughout the year, considering weekly periods (i.e., 52 weeks). The detailed description of this model can be found at [6].

2.2. *Short-term Optimization*
The short-term optimization is responsible to optimally define the operating set-points of the PSP, in other words, the moments where it will operate as pump or turbine and as well as the portion reserved to FRR provision. The formulation of this optimization model presumes that it is not possible to determine *a priori* the amount of electrical energy that will be activated used from the available reserve band. Thus, the revenue for the FRR is represented in the objective function only for the reserve capacity price. Following the same logic, the amount of reserve band made available to perform FRR does not affect the rate of discharge/pumping water flow or the amount of water available in the reservoir. To avoid situations where the energy used in the frequency restoration would result in the violation of a reservoir level it is assumed that a portion of the reservoir is allocated solely for the purpose of frequency restoration.

In order to define the set-points the algorithm relies on a set of forecasted data, such as DA and FRR market prices (π_{DA}, π_{FRR}) and natural water inflows, as well as the technical characteristics of the unit. The objective function (1) aims to maximize the revenue of the system in both markets, DA and FRR, for a period of one week ($t = 168$ hours).

$$\max \sum_{t=1}^{t=168} \left[(g^{d}_{DA,t} - g^{p}_{DA,t}) \cdot \pi_{DA,t} + (us^{d}_{t} + us^{p}_{t} + ds^{d}_{t} + ds^{p}_{t}) \cdot \pi_{SEC,t} \right] \tag{1}$$

Several constraints related with the technical boundaries of the machine as well as equality balance equations for the reservoir are imposed and exhaustively detailed in [6]. Furthermore it is also enforced the Portuguese transmission system operator (TSO) rule, which defines that the amount for upward secondary reserve must account for 2/3 of the total band, leaving 1/3 for downward regulation.

2.3. *Evaluation Module*
To assess the performance of the bidding strategy an evaluation module emulates the real operation, by applying the markets bids determined by the optimization problem to real market conditions. Observed prices from the electricity market are used to evaluate the economic performance and observed natural water inflows used to update reservoir level. Despite the weekly period of the short-term optimization model, the MIBEL only accepts bids for the day-ahead, so in this evaluation module only the first 24 hours bids are considered. Since FRR remuneration is made in two fractions, reserve band and energy activated, it is necessary to estimate the latest based on the ratio of the overall energy used and the total amount of FRR reserve contracted, detailed in (2) for the upward activation but the same method is applied to downward. The amount of energy used is also relevant to update the level of the reservoir, since, the use of the PSP unit to perform downward regulation results in more water stored in the reservoir.

$$\% \uparrow_t = \frac{\text{upward FRR energy used } (t)}{\text{upward FRR band contracted } (t)} \ (\%) \tag{2}$$

3. Case Study
To accurately represent the conversion between water flow and power, we applied a methodology adapted from [8] and [5] to the s-shape characteristic curves (normalized for the point of max efficiency and fixed head - Figure 2a) for all the gate openings, provided by the manufacture of the machine. These curves were affected by two quadratic curves in order to represent both the hydraulic and electrical losses of the system. By restricting the operation to the machine's

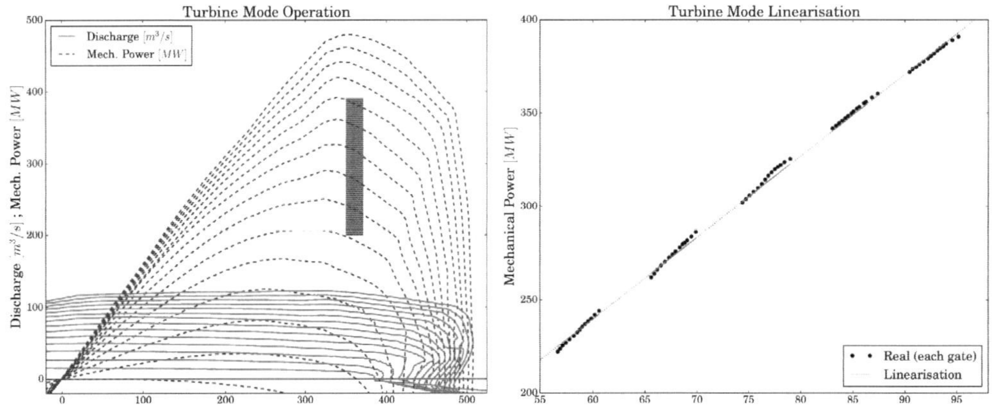

(a) (b)

Figure 2: Turbine mode: (a) Normalized s-shape curve (for max efficiency) (b) Linearization

technical limits (highlighted by the shaded rectangle) the outcome is a linearisation of said curves (Figure 2b) to be able to use them as input in a mixed integer linear optimization problem.

The bidding strategy was applied to a real PSP unit located in the north of Portugal, with two groups of machines, each with the ability to operate in turbine mode between 200MW and 390MW, and in pump mode between 300MW and 390MW. The reservoir has 15.8hm^3(92.1-76.3hm^3) of water available to for the PSP operation. Real data for the MIBEL market prices[1,2] and natural water inflows were also used. The period between Jan 2014 and Jan 2015 was considered.

Figures 3a and 3b represent two typical optimization days, the former considering a PSP with fixed speed pump unit while the last regards an identical PSP but with variable speed pumping capabilities. The dotted lines correspond to the day-ahead market bids and the shaded area to the FRR reserve band, both obtained by the short-term optimization. Bold lines represent the actual power generated or consumed by the PSP unit, being the difference between the DA bid and the actual power the activated energy for FRR provision. Analysing both figures, it is possible to observe that in some situations it is more profitable to operate with only one group of machines (hour 7 in F.3a and hour 18 in F.3b) instead of two. This is a consequence of the reservoir level constraint, either imposed by the medium-term optimization or by its physical limits. In some periods, due to the nature of the market prices, the bidding strategy favours the DA bid and opts by not providing FRR (hour 5 in F.3b) or reducing its offer (hour 4 in F.3b).

An economic analysis was made by comparing the revenues of variable speed PSP with a fixed one, as well as comparing the operation with just one group of machines versus two groups, in the Iberian Market environment. Results are described in Table 1 and Figure 4a.

As expected, with variable speed PSP the revenue from the day-ahead market suffers a small decrease, but it is largely compensated by the revenue from the FRR market. The overall revenue of a PSP with variable speed achieves an increase of 12% in comparison with a fixed one. Additionally we can observe that doubling the amount of machines does not guarantee a proportional gain in terms of revenue, predominantly due to the limited size of the reservoir.

Since the amount of energy used from the reserve band was not considered in the short-term optimization model, it is possible that during the operating day the reservoir level could be

1 www.mercado.ren.pt

2 www.esios.ree.es

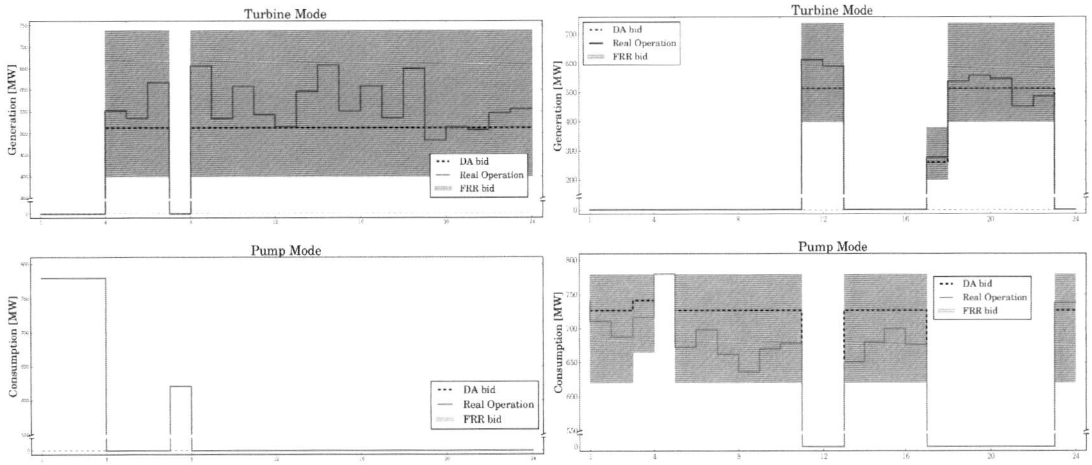

(a) (b)

Figure 3: DA and FRR bids considering a PSP with fixed speed (a) and variable speed (b) pumping capabilities

Table 1: System Revenue from DA and FRR, considering pumping with variable and fixed speed

	Single Group Mach.		2 Groups Mach.		
	Fixed Speed	Variable Speed	Fixed Speed	Variable Speed	
Day-ahead	33.10	31.34	40.28	36.27	(EUR 10^6)
Frequency R. R.	37.45	44.01	63.68	80.03	(EUR 10^6)
Total	70.55	75.35	103.96	116.3	(EUR 10^6)

above or below its technical limits. Furthermore, forecasting errors on the natural water inflows could also result in discrepancies between expected and real amount of water stored. Figure 4b shows a histogram of the difference between the actual level after the operation and the amount of water expected after the optimization process. As depicted, in this case study, rarely the real reservoir level is greater than the expected, while the opposite is frequent. However, during all the simulation, only 1.9% of the hours the real reservoir level was above its technical limit, registering an absolute minimum level of 74.83hm^3. This deviation indicates that if we increase the minimum reservoir level by 1.5hm^3, it is expected that the technical constraints would not be violated.

4. Conclusion

In [6] was demonstrated the validity of the bidding strategy as well as the economic benefit of adding the FRR to the PSP market portfolio. In this paper the same strategy was applied with two added features: representation of losses, both hydraulic and electrical, and the possibility to operate with one or two groups of machines.

It was shown that a PSP with a variable pump unit could increase its revenue by 12% in comparison with a fixed unit. Furthermore, the amount of groups of machines does not led to a proportional gain in revenue, since it is constraint by the stasis of the storage capacity. It was also concluded that the consequences arising by discrepancy between expected and real reservoir level could be mitigated by increasing the reservoir's minimum level constraint by 1.5hm^3.

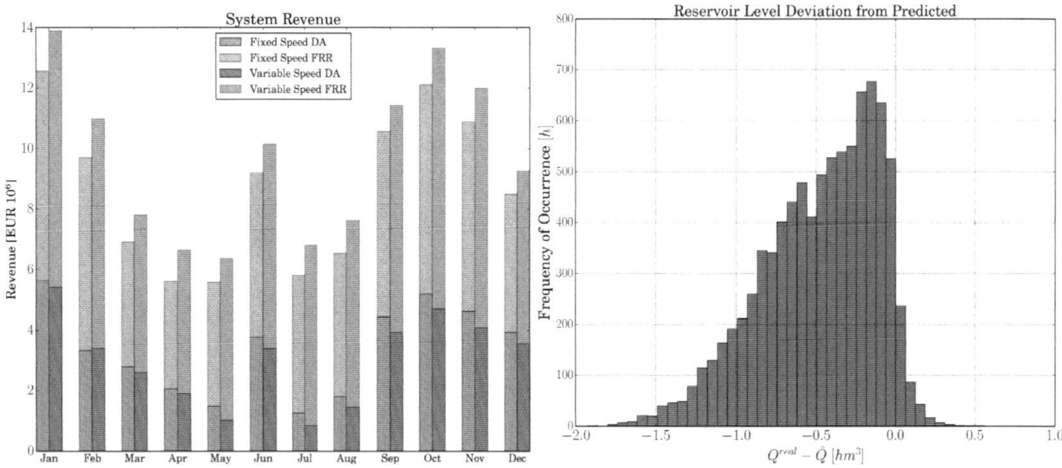

(a) (b)

Figure 4: (a) System Revenue from DA and FRR, considering variable and fixed speed and (b) Reservoir level deviation from the predicted

Acknowledgments

The authors graciously thank the partners from the FP7 European Project HYPERBOLE for the valuable comments about the pumped storage power modelling and EDP Produção for the provision of technical data and natural water inflow time series. A special mention to Christophe Nicolet from Power Vision Engineering for the contribution in modelling the I/O curves and Jiri Koutnik from Voith for providing the s-shape curve.

References

[1] N. Sivakumar, D. Das, and N. Padhy, "Variable speed operation of reversible pump-turbines at kadamparai pumped storage plant – a case study," *Energy Conversion and Management*, vol. 78, pp. 96—104, Feb. 2014.

[2] A. Beguin, C. Nicolet, B. Kawkabani, and F. Avellan, "Virtual power plant with pumped storage power plant for renewable energy integration," in *Proc. of the XXI International Conference on Electrical Machines (ICEM 2014)*, Berlin, Germany, Sept. 2014.

[3] C. Nicolet, A. Beguin, B. Kawkabani, Y. Pannatier, A. Schwery, and F. Avellan, "Variable speed and ternary units to mitigate wind and solar intermittent production," in *Proc. of the 2014 HydroVision International*, Nashville, TN, Jul. 2014.

[4] J. I. Pérez-Díaz, M. Chazarra, J. García-González, G. Cavazzini, and A. Stoppato, "Trends and challenges in the operation of pumped-storage hydropower plants," *Renewable and Sustainable Energy Reviews*, vol. 44, pp. 767—784, Apr. 2015.

[5] M. Chazarra, J. I. Perez-Diaz, and J. Garcia-Gonzalez, "Optimal operation of variable speed pumped storage hydropower plants participating in secondary regulation reserve markets," in *Proc. of the 11th International Conference on the European Energy Market (EEM14)*, Krakow, Poland, May 2014.

[6] J. Filipe, C. Moreira, R. Bessa, and B. Silva, "Optimization of the variable speed pump storage participation in frequency restoration reserve market," *International Conference on the European Energy Market, EEM*, vol. 2016-July, 2016.

[7] J. C. Sousa, V. T. Mendes, and J. T. Saraiva, "Estimation of the remuneration of hydro plants in a market environment using an iterative under-relaxation approach," in *Proc. of the IEEE PowerTech Conference 09*, Bucharest, Romania, Jun. 2009.

[8] J. García-González, E. Parrilla, A. Mateo, and R. Moraga, "Building optimal generation bids of a hydro chain in the day-ahead electricity market under price uncertainty," in *Proc, of the 9th International Conference on Probabilistic Methods Applied to Power Systems (PMAPS 2006)*, Stockholm, Sweden, Jun. 2006.

IOP Conf. Series: Journal of Physics: Conf. Series **813** (2017) 012011 doi:10.1088/1742-6596/813/1/012011

Off-stream Pumped Storage Hydropower plant to increase renewable energy penetration in Santiago Island, Cape Verde

Inês Barreira[1], **Carlos Gueifão**[2] **and J. Ferreira de Jesus**[1]

[1] Área Científica de Energia, Instituto Superior Técnico, Av. Rovisco Pais 1, 1049-001 Lisboa, Portugal

[2] Gesto Energy, Av. Cáceres Monteiro 10 1º Sul, 1495-131 Algés, Portugal

E-mail: ines.barreira@tecnico.ulisboa.pt

E-mail: carlos.gueifao@gestoenergy.com

E-mail: jose.jesus@tecnico.ulisboa.pt

Abstract. In order to reduce the high dependence on imported fuels and to meet the ongoing growth of electricity demand, Cape Verde government set the goal to increase renewable energy penetration in Santiago Island until 2020. To help maximize renewable energy penetration, an off-stream Pumped Storage Hydropower (PSH) plant will be installed in Santiago, in one of the following locations: Chã Gonçalves, Mato Sancho and Ribeira dos Picos. This paper summarizes the studies carried out to find the optimal location and connection point of the PSH plant in Santiago's electricity network. This goal was achieved by assessing the impact of the PSH plant, in each location, on power system stability. The simulation tool PSS/E of Siemens was used to study the steady-state and dynamic behavior of the future (2020) Santiago MV grid. Different scenarios of demand and renewable resources were created. Each hydro unit of the PSH plant was modeled as an adjustable speed reversible turbine employing a DFIM. The results show that Santiago's grid with the PSH plant in Chã Gonçalves is the one that has the best performance.

1. Introduction

The high dependence on fossil fuels is a matter of concern to any country. In 2009, more than 95% of electricity produced in Cape Verde was fossil fuel-based, and the high costs of this resource represent a significant expenditure in Electra (company responsible for the electricity sector) budget. Moreover, electricity consumption is expected to double in Cape Verde until 2020, which will lead to a higher production based on fossil fuels if there is no investment in other types of generating electricity [1].

In order to make the service less costly, more reliable and to meet the growing trend in energy consumption, Cape Verde government launched an ambitious action program that aims to make 50% of Cape Verde's electricity consumption, by 2020, renewable-based. One of the main axis of the program relies on promoting the investment in renewable energy by independent power producers and public-private partnerships [2]. In [1] Gesto Energy identifies renewable energy projects. Some of them were selected due to their competitiveness when compared with the existing fossil fuel-based generating units. The selected projects are assumed to become fully operational by 2020.

Santiago is the Cape Verde Island where the investment on renewable generation will be bigger. To maximize renewable energy penetration (wind, solar and waste), one of the selected projects is a 20 MW rated off-stream Pumped Storage Hydropower (PSH) plant. A technical, economical and environmental feasibility study carried out by Gesto Energy pointed out three potential geographical locations for the PSH plant: Chã Gonçalves, Mato Sancho and Ribeira dos Picos.

The subject of this work came from the need to overcome the integration challenges of the PSH plant in 2020 Santiago's electricity network. The main goal is to find the best location and connection point of the PSH plant, assessing the impact of this energy storage system, in each location, on power system stability. The main contribution of this work is to help the integration of renewable energy in Santiago Island.

This paper is composed of 5 Sections, including the present one. Section 2 introduces the case study of this work. A brief description of the current (2015) Santiago MV distribution and transmission grid, existing generating units and peak consumption is performed. Then, the main aspects of the future (2020) electricity network are introduced, such as the expected load growth rate (2015 to 2020), the new renewable generating units specified in [1] and the main project specifications of the PSH plant. Finally, the suggested connection points of the PSH plant, in each location, in the 2020 grid single line diagram are presented. Section 3 begins with a summary of the methodology used to accomplish the main goal of this study. Then, the different scenarios of demand and renewable resources created to study the grid are presented, followed by a description of the main aspects concerning steady-state and dynamic modeling of the grid in PSS/E 33 of Siemens. In Section 4 the results of the simulations performed are presented and analyzed. Section 5 draws the conclusions of this work.

2. Case Study

The current power system of Santiago Island is comprised by three thermal power plants, one wind power plant, and one solar power plant. Palmarejo thermal power plant has 11 fuel oil generating units with a total installed capacity of 76.34 MW; Gamboa thermal power plant, 3 diesel generating units with a total installed capacity of 7.45 MW and Assomada thermal power plant, 4 diesel generating units with a total installed capacity of 3.92 MW. The wind power plant has an installed capacity of 9.35 MW and consists of 11 wind turbines Vestas V-52 (DFIG technology) of 850 kW each. Finally, the solar power plant has a maximum power of 5 MWp.

The MV electricity network consists of a 20 kV distribution system and a 60 kV transmission system. All generating units are directly connected to the distribution system. Since the production is mostly gathered in south, the transmission system transmits power from the south to the center and north of the island, and it is comprised of three substations (SS): Palmarejo 20/60 kV SS; New S.Filipe 20/60 kV SS and Calheta 60/20 kV SS. The transmission system single line diagram is shown in Figure 1.

In 2015, the peak load was 29.33 MW. Based on a demand growth assessment developed by Gesto Energy [1], it is expected for load to grow 27% by 2020. Furthermore, according to Electra, two other loads are to be considered: a 7 MW desalination plant and 2 MW corresponding to the installed capacity of a casino. In this study, it was assumed that these two loads consume their installed capacity. Thus, for 2020 the expected peak load is 46.15 MW.

Concerning production, 2020 grid scenario has five more wind power plants (2 with an installed capacity of 3.40 MW, 2 of 6.80 MW and 1 of 18.70 MW) comprising Vestas V-52 850 kW and a 5 MW waste power plant. Summing up, 2020 Santiago's power system case study has a total installed capacity of 145.9 MW, from which 87.70 MW are thermal units, 48.5 MW of wind turbines, 5 MW of solar power and the remaining 5 MW of waste power.

IOP Conf. Series: Journal of Physics: Conf. Series **813** (2017) 012011 doi:10.1088/1742-6596/813/1/012011

The PSH plant in project for 2020 Santiago Island is of off-stream type i.e. both reservoirs are independent of a natural stream flow so stored potential energy relies entirely on water that has been previously pumped. It will have an installed capacity of 20 MW distributed over two 10 MW hydro units, each one with a reversible Francis turbine sharing the same penstock. For the generator/motor technology, Gesto Energy proposed conventional single-speed synchronous machines complemented with flywheels to compensate the slow response time of conventional hydro power plants with regard to the fast power fluctuations associated with intermittent renewables, such as wind and solar. In this study this combination was replaced by DFIMs.

Taking into account the identified potential geographical locations for the PSH plant, the connection points to the MV grid are the ones shown in Figure 1. For the PSH plant in Chã Gonçalves or Mato Sancho, two possible connection points were considered. For PSH Chã Gonçalves, both suggested connection points are in 60 kV side of Palmarejo SS and New S.Filipe SS, respectively. Both connections are made with an overhead line 16 km long. For PSH Mato Sancho, one of the connection points is in 60 kV side of Calheta SS, with a 21.56 km overhead line; the remaining one is a 8 km double cable connection to the 20kV distribution system. For the PSH plant in Ribeira dos Picos, the suggested connection point is in the existing 60 kV overhead line that connects Palmarejo SS to Calheta SS. The projected overhead line is 1.5 km long, and the connection to the transmission system is 30 km distant from Palmarejo SS and 8 km distant from Calheta SS.

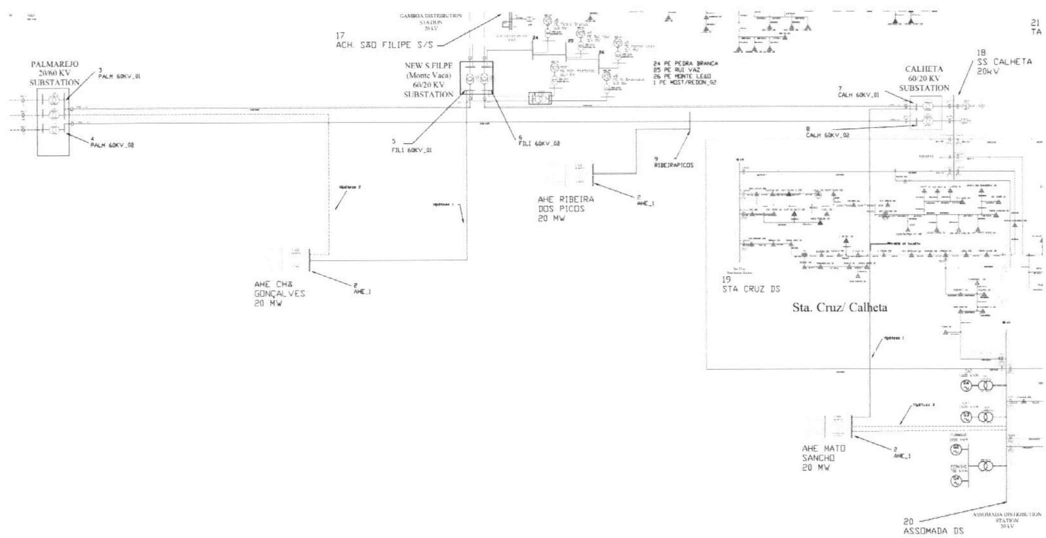

Figure 1. Suggested connection points of the PSH plant in Santiago's MV grid single line diagram.

3. Simulation Scenarios and Models

The grid case study was dimensioned in PSS/E 33 of Siemens to represent all branches, generating units and consumption points of the current island electricity grid together with the new 2020 projects. Load and renewable resources scenarios were created so that grid performance could be studied in different load and for different renewable power levels.

For power flow purposes, all grid components were sized according to Electra's data set. To study the dynamic performance of the grid, it was associated a dynamic model for the different generating technologies so their specific response to a grid disturbance could be accounted for.

The models were properly sized to best represent the existing generating technologies in Santiago Island. The new 2020 generating technologies were sized based on PSS/E grid test cases and other relevant studies. More information about the models' parameters can be found in [3].

3.1. Load and Renewable power Scenarios

A daily load and available renewable power forecast for 2020 was made based on a load curve with the generation dispatch of a day of January 2015. The main assumptions taken were that consumption behavior would not change from 2015 to 2020 and the wind and solar conditions of that day would occur again. According to Electra, there are no substantial differences between summer and winter peak load in Santiago Island. In January, Santiago Island has high wind speeds and yearly average solar irradiances. The load and available renewable power scenarios selected to study the grid are summarized in Table 1. The 4 hours of the day intend to represent different load levels (off peak, mid peak and on peak) and different renewable resources scenarios (wind and solar). The waste power plant generates its installed capacity in every scenario.

Table 1. Load and available renewable production scenarios.

Scenario	Load [MW]	Available renewable power		
		Wind [MW]	Solar [MW]	Waste [MW]
06:00h	26.592	38.864	0	5
09:00h	34.926	41.973	0.839	5
13:00h	38.542	43.527	2.320	5
20:00h	46.146	37.309	0	5

3.2. Simulation Models

From a power flow perspective, all generating units are modeled as a conventional source with a specified active power output, a reactive power range, and a bus voltage to regulate (PV bus). For thermal units, one of them was specified as the swing bus; the others, their real power output (or de-commitment) is stated when the economic dispatch is performed. For the renewable generating units, their real power output is defined based on the scenario being studied.

For the short-term dynamic analysis, the thermal units are modeled by the generator GENSAL (salient pole synchronous machine), the DC excitation system IEEET1 and the diesel machine with associated speed governor DEGOV1. The waste power plant is also modeled by GENSAL and IEEET1, but no governor was associated, so this plant does not supply primary frequency regulation. To model the dynamic performance of the DFIG wind turbines the WT3 generic wind turbine model is used. For the solar power plant, the model used is the one available in PSS/E model library. Both wind and solar power plants have under/over voltage and frequency relays. For the wind power plants, under voltage relays' parameters were set to simulate fault ride-through capability.

Concerning the PSH plant, when both hydro units are dispatched, they run in same direction: both are in generating mode or in pumping mode, according to system conditions. Also, mode transitions were not a subject in dynamic simulation studies. Thus, the generating and pumping modes are studied separately. In a steady-state approach, the two hydro units are modeled each as a conventional source connected to the same PV bus. Their real power output is set when the dispatch is performed (positive sign power if in generating mode, negative sign power if in pumping mode). For the dynamic analysis were used the new generic models launched in January 2016 by Siemens PTI for adjustable speed PSH plants employing DFIMs. For both generating and pumping models, there are two types of models: model that simulates the dynamic performance of a single hydro unit and penstock; and model that simulates the dynamic performance of a common penstock shared by up to four hydro units.

For pumping mode, the plant is modeled by PSHPM1 (PSH plant in pumping mode with a single hydro unit). This model was chosen due to misbehave of the model PSHPM4 that accounts for up to four units. So, for pumping mode scenarios, the PSH plant was modeled as a single hydro unit equivalent to the lumped unit, both in steady-state and dynamic simulations. This incident brings some constraints, e.g. the inability to simulate load shedding disconnecting only one hydro unit. The speed adjustment considered for this operation mode ranges from +6% to -4% of the synchronous speed. The wicket gates are controlled to an optimum position in accordance with the current power command. This optimum position is given by a characteristic that provides the optimal gate position as a function of the desired MW output. The optimal power/gate relationship used in this model is the one provided as an example by Siemens PTI.

In generating mode the hydro units are modeled each by PSHGN4. No modeling of a bifurcation into two individual penstocks (each feeding one unit) was made so the response of the two units should be equal to that of the lumped unit. Prime mover governor controls the wicket gates so the desired mechanical power can be achieved. The speed is adjusted for the optimum speed to generate the current power command. This optimum speed is given by a characteristic that provides the optimal speed as a function of the desired MW output. The optimal power/speed relationship used in this model is the one provided as an example by Siemens PTI. This characteristic admits speed adjustments from +15% to -10% of the synchronous speed.

These models developed by Siemens PTI are still in their infancy and further development is under way. Design of these models and details of their functionality can be found in [4]. Note that these characteristics introduced above are very much specific of the turbine-pump design and can be obtained by the manufacturers. No such data was available, justifying the use of the examples provided by Siemens PTI.

For both generating and pumping models, it was necessary to state the water column time constant T_w as it accounts for the delay imposed by the hydraulic circuit to the PSH plant response time. Table 2 summarizes the T_w values relative to the operation of the PSH plant at rated power. This parameter is the only one that differs in the modeling of the PSH plant for each geographical location since the penstock length and diameter are specific to the project location. For pumping mode, T_w has negative value due to the sign convention adopted for this operation mode.

Table 2. Water column time constant of the PSH plant.

PSH plant site	T_w **[seconds]**	
	Pumping mode	Generating mode
Chã Gonçalves	-1.80	2.50
Mato Sancho	-1.69	2.31
Ribeira dos Picos	-1.73	2.35

Frequency relays were modeled to disconnect the PSH plant instantly when wide frequency deviations occur. It was also modeled for PSHPM1 an under frequency relay to simulate load shedding when frequency is below 48 Hz for more than 0.2 seconds.

4. Results and Analysis

Both 06:00h, 09:00h and 13:00h scenarios have more renewable power available than the required to satisfy consumption. However, this is not true concerning reactive power, as the reactive power injected by the renewable sources revealed to be insufficient to meet the required by the load.

Hence, thermal units were dispatched to help in reactive power generation. In order to maximize renewable energy penetration, their real power output was specified to the minimum required to operate the unit. Concerning the PSH plant, both hydro units were dispatched to maximize its contribution to reactive power generation (thus minimizing commitment of thermal units). Since there is surplus renewable power, the hydro units are in pumping mode, consuming -7.5 MW each. This dispatch gives flexibility to the operation of thermal units since it allows the PSH plant to increase consumption when it is necessary to decrease generation. Wind power was curtailed to eliminate the surplus renewable power and to accommodate the thermal units. In 20:00h scenario, the PSH plant was defined to be in generating mode, injecting each hydro unit 5 MW. Thermal units were also dispatched due to lack of reactive power generation. Only two types of thermal units were dispatched, both from Palmarejo thermal power plant: 11.350 MW rated Wärtsilä units and 5.582 MW rated Caterpillar units. Table 3 summarizes the thermal units commitment.

Table 3. Thermal units commitment.

Load & Renewable Scenario	Grid Scenario		
	Chã Gonçalves	**Mato Sancho**	**Ribeira dos Picos**
06:00h	WART 5	WART 5	WART 5
09:00h	WART 5	WART 5, 6	WART 5
13:00h	WART 5, CAT 1	WART 5, 6	WART 5, CAT 1
20:00h	WART 5, 6	WART 5, 6, 7	WART 5, 6

4.1. Steady-state simulations

The connection of the PSH plant in 2020 Santiago's grid revealed overloaded branches only with the PSH plant in Mato Sancho connected to the 20 kV distribution system, which is why this connection hypothesis for location Mato Sancho was excluded for further studies. With all other PSH plant connection points, no overloaded branches were registered.

Regarding bus voltage profile, in grid with PSH plant in Chã Gonçalves a 3.6 Mvar capacitor bank had to be connected to the 20 kV side of Calheta SS so it could be possible to keep bus voltages in peak hour (20:00h) above 0.95 p.u., in Santa Cruz and Santa Catarina (regions electrically far from the main island's production stations). With PSH plant in Mato Sancho or Ribeira dos Picos, this voltage limit violation no longer occurs, due to the fact that these two PSH plant sites are near to Santa Cruz and Santa Catarina, increasing bus voltages with the injection of reactive power near the regions. Although steady-state studies didn't point out the need to install a capacitor bank in grid scenarios with PSH plant in Mato Sancho or Ribeira dos Picos, its use has proven to be beneficial for the grid voltage profile and especially for the reduction of reactive power losses, which are significantly high in Santiago Island.

The grid with the PSH plant in Mato Sancho revealed a lower renewable energy penetration when compared to grid scenarios with PSH plant in Chã Gonçalves or Ribeira dos Picos. The connection point of PSH Mato Sancho to the grid, such as the voltage drop on its connection line limit the reactive power injected by the PSH plant, which leads to the need to have more or bigger rated thermal units operating.

4.2. Dynamic simulations

For dynamic studies two types of grid disturbances were applied: simulation of a 3-phase fault on the 20 kV side of WART 5 transformer - the fault is cleared tripping the transformer which leads to the disconnection of the thermal unit; and simulation of the loss of one of the three 60 kV transmission lines.

IOP Conf. Series: Journal of Physics: Conf. Series **813** (2017) 012011 doi:10.1088/1742-6596/813/1/012011

The simulation of the 3-phase fault led to the tripping of the PSH plant, during the short circuit, for the plant sites Mato Sancho (generating mode, 20:00h) and Ribeira dos Picos (both operating modes, all scenarios[1]). The tripping of the PSH plant was performed by the under frequency relay, as a result of the frequency decay at the connection bus of the PSH plant to the grid.

In pumping mode, the loss of the PSH plant can be managed by the power system, mainly due to the wind turbines pitch control, which decreases the wind power injected into the grid. Figure 2 shows the response of the frequency at Palmarejo thermal power plant 20 kV bus for the grid case with PSH plant in Ribeira dos Picos, 13:00h. As can be seen, the event causes an exceeding of mechanical power, thus causing a positive frequency excursion. Figure 3 shows the response of mechanical power and electrical power of one of the 49 wind turbines operating, where the action of the pitch control is illustrated. The system reaches an acceptable steady-state operation point.

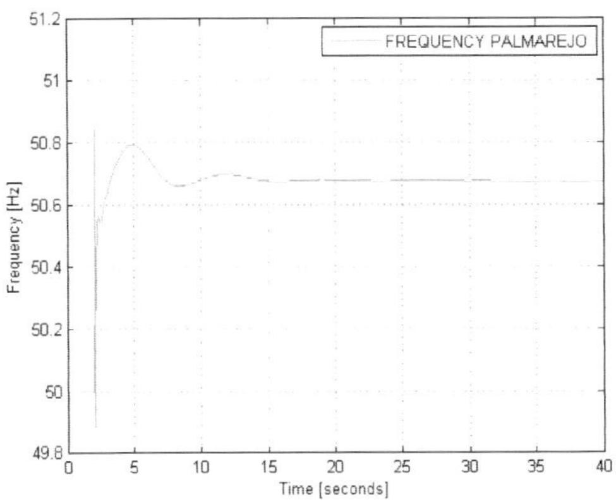

Figure 2. Frequency at Palmarejo PS bus, 13:00h RP.

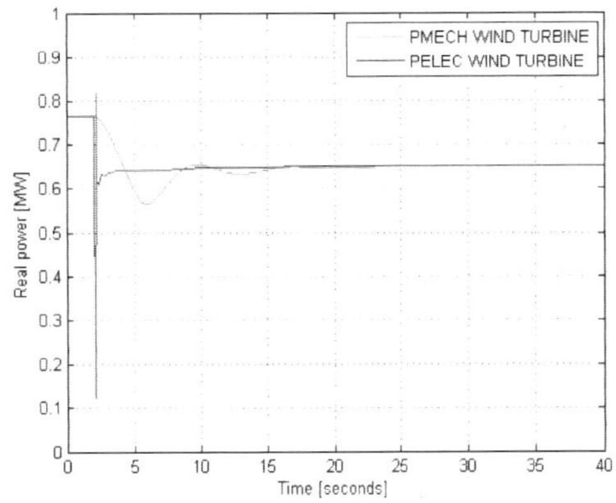

Figure 3. Wind turbine mechanical power and electrical power, 13:00h RP.

In generating mode, the loss of the PSH plant led to critical grid scenarios. Because of the major generation loss, frequency gets below the minimum acceptable value 48.5 Hz in less than 10 seconds. The loss of the PSH plant also causes a major reactive power deficiency, leading to a marked decrease in system voltages. Figure 4 and Figure 5 illustrate the fundamental simulations results. In grid case with the PSH plant in Ribeira dos Picos (Figure 4), the remaining thermal unit WART 6 gets overloaded and system voltages get below the minimum acceptable value 0.90 p.u.. Under voltage relays start disconnecting wind turbines at 12.1 seconds of simulation. At t= 22 seconds the system collapses. For the grid case with the PSH plant in Mato Sancho (Figure 5), the system reaches a steady-state operation point – albeit unacceptable – mainly because the wind turbines can keep their terminal voltage above 0.90 p.u., thus preventing wind turbines' under voltage relays from tripping.

With the PSH plant in Chã Gonçalves, the simulated fault no longer causes the loss of the PSH plant, for both connection points suggested for this plant site. As a result, the system reaches an acceptable steady-state operation point, both in pumping mode and generating mode scenarios.

[1] For 09:00h and 13:00h scenarios, the simulated fault also causes the tripping of the solar power plant by the performance of the under voltage relays.

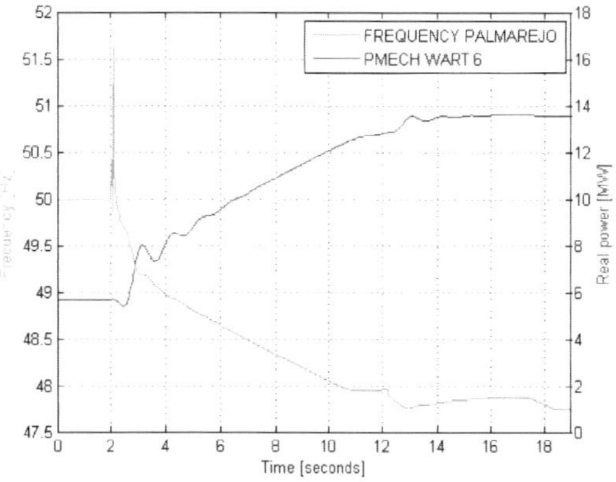

Figure 4. Frequency at Palmarejo PS bus and WART 6 mechanical power, 20:00h RP.

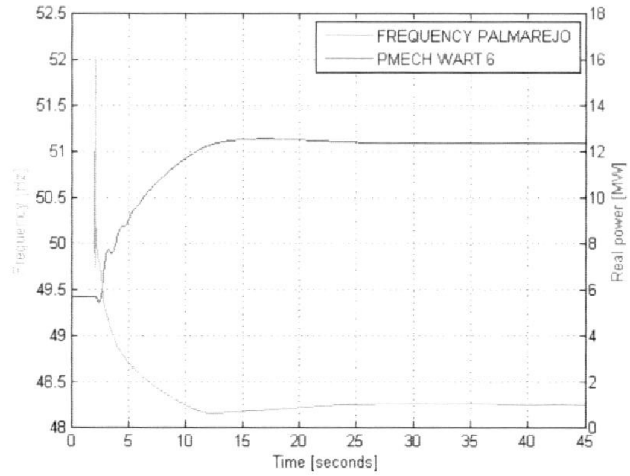

Figure 5. Frequency at Palmarejo PS bus and WART 6 mechanical power, 20:00h MS.

Figure 6 and Figure 7 illustrate the main system response results when the PSH plant Chã Gonçalves is in pumping mode, 13:00h. As can be seen, the PSH plant supplies primary frequency regulation (Figure 6), as the mechanical power required by the pumps decreases (less power drawn from the system) to control the under-frequency. An important contribution to arrest the frequency decay is also given by the stored kinetic energy in the rotating wind turbines that transiently supply additional power generation (Figure 7). However, frequency reaches 48 Hz as the wind turbines inertial response fades. To arrest the frequency decay the tripping of the PSH plant is performed.

Figure 8 and Figure 9 illustrate the main system response results when the PSH plant Chã Gonçalves is in generating mode. The frequency decay is arrested by the response of the governors of the remaining generators capable of supplying primary frequency regulation, which act to increase the mechanical power produced by their prime movers. Figure 9 only illustrates the mechanical power response of one hydro unit. The other hydro unit has identical response.

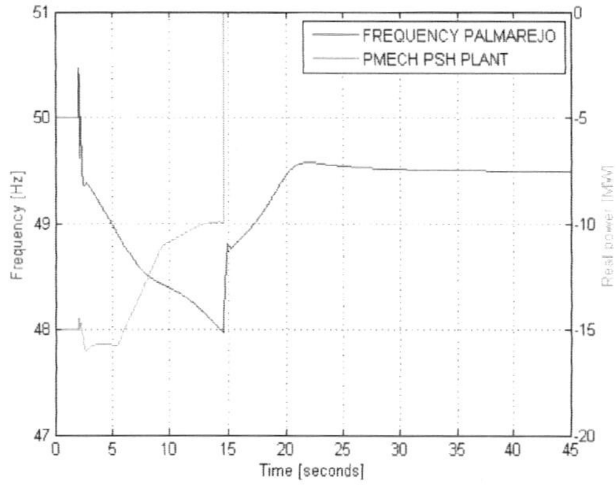

Figure 6. Frequency at Palmarejo PS bus and PSH plant mechanical power, 13:00h CG.

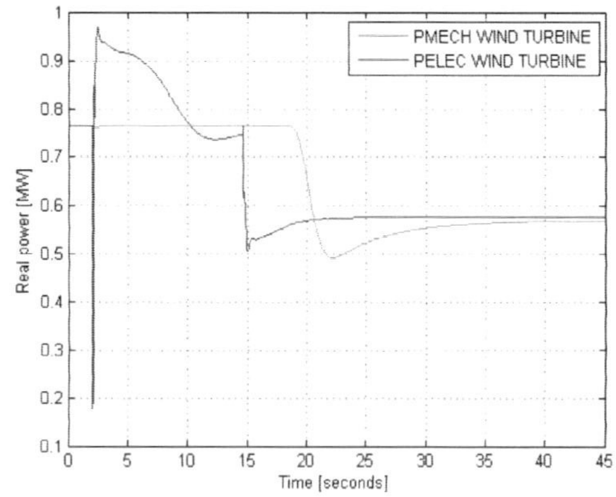

Figure 7. Wind turbine mechanical power and electrical power, 13:00h CG.

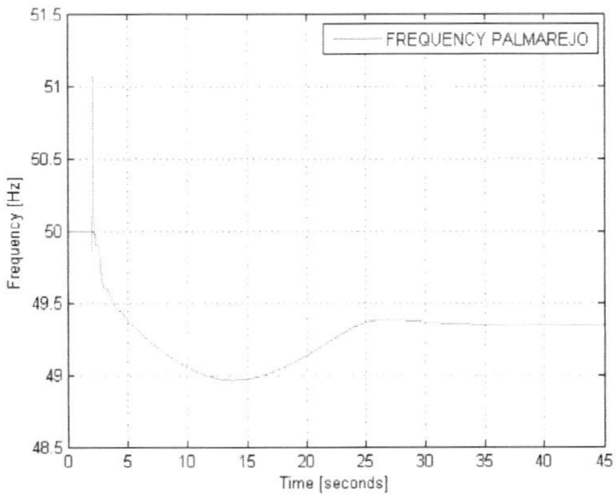

Figure 8. Frequency at Palmarejo PS bus, 20:00h CG.

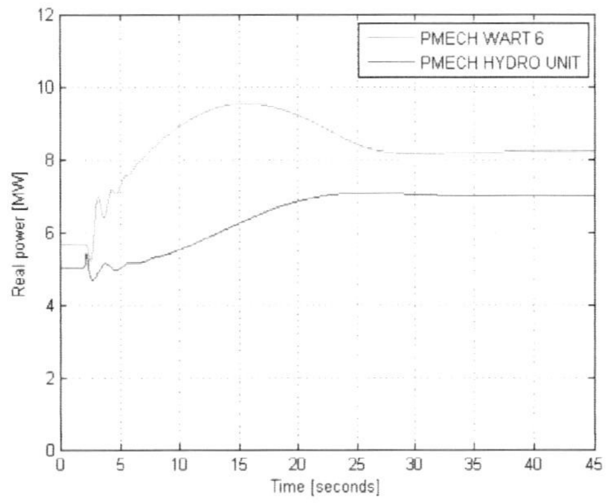

Figure 9. Hydro unit mechanical power and WART 6 mechanical power, 20:00h CG.

Overall, the results obtained indicate that the point where the PSH plant is connected to the grid can lead to the loss of this resource when in the system a 3-phase fault on the 20 kV side of WART 5 transformer occurs.

In the simulation of an outage in the transmission lines, some violations in grid operation were detected with the PSH plant in Mato Sancho and Ribeira dos Picos, both in pumping mode scenarios. The outage of the New S.Filipe – Calheta 60 kV overhead line causes a significant overload in Calheta SS transformers (around 140% and 190% in the worst case scenario) with the PSH plant in Mato Sancho, and some minor line overloads with PSH plant in Ribeira dos Picos. The outage of the Palmarejo – Calheta 60 kV overhead line causes a significant overload in Calheta SS transformers (around 130% and 180% in the worst case scenario) with PSH plant in Ribeira dos Picos. In the grid with the PSH plant in Chã Gonçalves, an outage in one of the three 60 kV transmission lines does not cause any overload or voltage limits violation, for both connection points suggested for this plant site. This result indicates that the power system with the PSH plant in Mato Sancho or Ribeira dos Picos is not n-1 secure.

5. Conclusions

In conclusion, the results of the steady-state and dynamic studies show that 2020 Santiago's grid with the PSH plant in Chã Gonçalves is the one that has the best performance. Even though in steady-state simulations the PSH plant in Ribeira dos Picos also confirmed to be a good option, it revealed clear disadvantages in dynamic simulations, such as the involuntary plant tripping and the failure to assure the n-1 security criteria in a grid disturbance likely to happen. The PSH plant in Mato Sancho revealed similar problems.

For the PSH plant in Chã Gonçalves there were studied two connection points to the grid, both with an overhead line 16 km long that did not reveal any significant differences in the results obtained. The grid with the PSH plant in Chã Gonçalves has an average of 82% of renewable energy penetration. An average of 3% of injected active power and 9% of injected reactive power are losses. It is necessary to install a capacitor bank of 3.6 Mvar to keep Santa Catarina and Santa Cruz bus voltages above 0.95 p.u. in peak hour.

References

[1] Energy G 2012 *Plano Energético Renovável de Cabo Verde (The Renewable Energy Master Plan of Cape Verde)* p 133

[2] Carvalho J 2013 *Cape Verde National Rep. for the Proc. 3rd Int. Conf. on Sustainable Development in Small Island States in Development (Apia)* (Praia) p 42

[3] Barreira I P 2017 *Aproveitamento Hidroeléctrico de Bombagem Pura na ilha de Santiago, Cabo Verde para aumento de penetração de Energias Renováveis* Msc thesis Instituto Superior Técnico

[4] Koritarov V, Guzowski L, Feltes J, Kazachkov Y, Gong B, Trouille B and Donalek P 2013 *Modeling Adjustable Speed Pumped Storage Hydro Units Employing Doubly-Fed Induction Machines* p 50

IOP Conf. Series: Journal of Physics: Conf. Series **813** (2017) 012012 doi:10.1088/1742-6596/813/1/012012

Contribution of variable-speed pump hydro storage for power system dynamic performance

B. Silva, C. Moreira

INESC TEC – INESC Technology and Science, Porto, Portugal

E-mail: {bsilva, carlos.moreira}@inesctec.pt

Abstract. This paper presents the study of variable-speed Pump Storage Powerplant (PSP) in the Portuguese power system. It evaluates the progressive integration in three major locations and compares the power system performance following a severe fault event with consequent disconnection of non-Fault Ride-through (FRT) compliant Wind Farms (WF). To achieve such objective, a frequency responsive model was developed in PSS/E and was further used to substitute existing fixed-speed PSP. The results allow identifying a clear enhancement on the power system performance by the presence of frequency responsive variable-speed PSP, especially for the scenario presented, with high level of renewables integration.

1. Introduction

On the last decades, European countries have been adopting renewable generation to cope the European Union ambitious targets on green-house gases reduction and on the share of renewable energy in the electricity consumption. From a chronological perspective, hydro power has been in-line with these objectives for many decades. However, the need of resource diversification and the unavailability of significant conditions for exploring hydro resources in many countries have been pushing for the integration of different technologies such as solar photovoltaic and wind energy.
In that particular field, Portugal is an interesting test-case regarding wind power integration having 4937 MW (2015 data) [1] of installed power and record-level renewable energy in the electricity consumption having supplied the entire load during 107 hours by renewable energy [2] In addition, the Portuguese power system recently accommodates the biggest variable speed pump/generators, 400 MVA units at the Venda Nova III power plant [3]. The work presented in this paper is focused in the evaluation of the Portuguese power system transient stability with the penetration of frequency-responsive variable-speed PHS. In order to achieve such fulfillment, a reduced-order model has been developed and implemented in PSS/E over the Portuguese power system in the valley hours' scenario. The analyzed cases comprise the actual scenario (comprising only fixed-speed PSP) and progressively takes into account the interconnection of new variable-speed PSP by considering firstly the aforementioned Venda Nova III and then the substitution of fixed-speed PSP for frequency-responsive variable-speed PSP. The analysis was performed for major disturbances such a fault event and the disconnection of a large conventional thermal generation unit. The results show the benefits overcoming from the extra flexibility provided by variable-speed PSP on participating in the primary frequency regulation.

2. Modelling of variable-speed PHS for transient stability studies in large power systems

The modelling of hydro powerplant is a topic of extreme importance for the performance of transient stability studies. The accuracy on describing the parameters and defining the hydraulic time constants are crucial for obtaining representative results. The work carried out in the FP7 Hyperbole Project, consisted on the developing and evaluating the performance of reduced-order models for representing variable-speed PSP based on Doubly-fed Induction Generator (DFIG) and on Full Scale Frequency Converter (FSFC) technologies. The control flexibility provided by power converters in both technologies fully decouples machine speed and grid frequency, thus allowing fast responses with respect to active power control. In the context of Hyperbole project a deep analysis was performed between the full-order model in SIMSEN and a reduced order model in Matlab/Simulink. The major outcome from this task consisted on the identification that the reduced-order model represents the major PSP dynamics, being sufficient for transient stability studies.

2.1. The variable-speed PSP model

For the sake of the work presented in this paper, the original idea consisted on transposing the reduced-order model to PSS/E via an add-on called Graphical Model Builder (GMB) that basically should help on the transposition of the aforementioned model to PSS/E. Unfortunately, the GMB proved to be ineffective on providing support on this specific target. On alternative, the variable-speed PSP has been represented by a frequency responsive load model, an approach commonly used to represent electronically interfaced sources (the FSFC case) on transient stability studies where the objective is to study the impacts that this new generation sources will bring to the power system dynamic response.

Have stated that, the variable-speed PSP was represented by the load model illustrated in Figure 1.

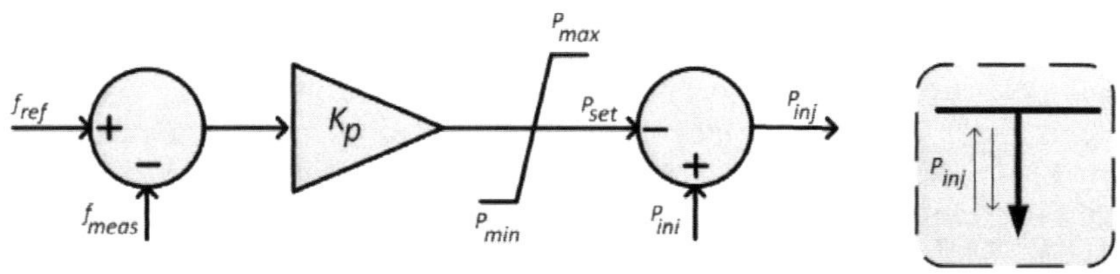

Figure 1: Power/Frequency proportional controller

Basically, the load model is composed by a proportional controller, responsible for setting the active power at the load. The Active Power/Frequency proportional controller is responsible for defining the load active power variation proportionally to the frequency deviation measured at the machine terminals, adapted by the K_p droop gain. In order to better characterize the typical variable speed PSP behaviour, a ramp rate was also included limiting fast responses, based on the outcomes from the modelling work presented on the Hyperbole Project deliverable 6.2 [4].

2.2. The Power System model

The power system base case model was kindly provided by REN (the Portuguese Transmission System Operator – TSO). Considering that the major objective of the work here presented consisted on evaluating the performance of the power system dynamics in the presence of variable-speed PSP, namely in pump mode, the scenario adopted for the elaboration of the simulations was the valley hours' scenarios which contained some hydro power plants operated in pump mode (despite being fixed-speed PSP). Due to confidentiality aspects, REN could not provide the Spanish grid-side. Thus, dynamic equivalents were developed and placed in the interconnections for enabling the provision of inertia and frequency response following disturbances. After these modifications, the final case was used as the base case that will be presented hereafter.

In order to accomplish the test-cases, the variable-speed frequency responsive PSP have been integrated in PSS/E by coding the model presented in Section 2.1. in Fortran and compiling it for PSS/E use. That way, this model can be associated to any busbar of any PSS/E simulation. For the sake of the study here presented, the model was used on different locations of the Portuguese power system as presented in Section 2.3.

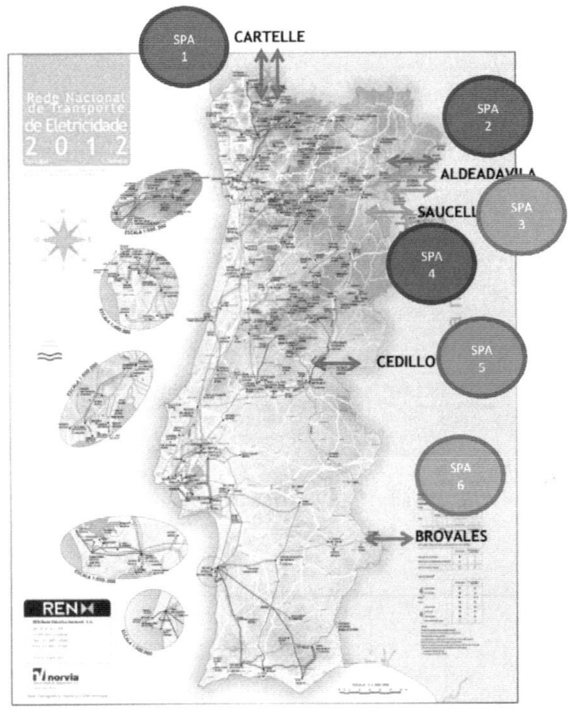

Figure 2: Portuguese power system with Spanish dynamic equivallents – adapted from [5]

2.3. Test Case definition

The test cases analysed in this paper consists on the incremental integration of the variable-speed PSP units in the Portuguese power system, by substituting fixed-speed pumping units with the previously presented variable-speed PSP frequency responsive model. The rationale consisted on creating a base-case scenario, CASE A for allowing further comparisons. Then, the approach consisted on evaluating the system performance with the first Portuguese variable-speed PSP, Venda Nova III (CASE B). Finally, CASE C and CASE D were created to allow assessing what would be the impacts of retrofitting existing fixed-speed PSP units for variable speed ones, namely Aguieira PSP (CASE C) and finally Alqueva PSP (CASE D). The cases and associated details are presented in Table 1.

Table 1: Cases description according with the PSP integration

CASE #	Variable-speed PSP	Pump mode Active Power
CASE A	None	BASE CASE
CASE B	Venda Nova III	1 x 400 MW
CASE C	Venda Nova III Aguieira	1 x 400 MW 3 x 91 MW
CASE D	Venda Nova III Aguieira Alqueva	1 x 400 MW 3 x 91 MW 4 x 110 MW

IOP Conf. Series: Journal of Physics: Conf. Series **813** (2017) 012012 doi:10.1088/1742-6596/813/1/012012

In order to analyse the effects of the primary frequency support provided by the variable-speed PSP, a fault event located at Lavos substation 400 kV busbar with a duration of 500ms was considered. The fault event in this specific location should lead to the tripping of wind farms without Fault-ride through capability, which is a severe situation for the power system operation in terms of generation loss and thus, frequency disturbance. Simultaneously, during the simulation an algorithm will monitor the Portuguese WF generation in order to determine the amount of WF disconnected following the fault event.

3. Simulation Results

As previously presented, the Portuguese power system performance was analysed for several cases, under the same disturbance, differing in the frequency-responsive variable-speed PSP integration level as described in Table 1. The disturbance considered for the sake of analysis consisted on a three-phase fault at Lavos 400kV busbar. The major results are illustrated in Figure 3 to Figure 7. The results allowed observing that during the fault event, severe voltage sag occurred and, in the aftermath, the voltage was able to recover to pre-fault level as illustrated in Figure 3.

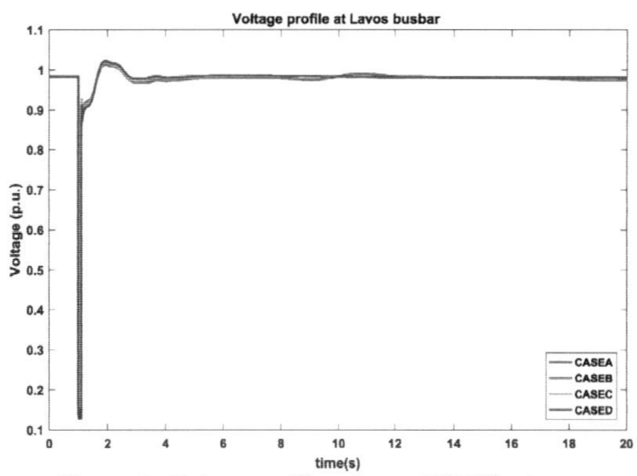

Figure 3 : Voltage profile at Lavos 400kV busbar

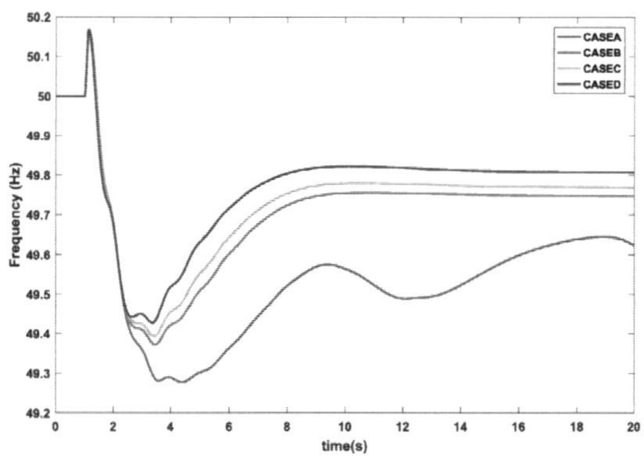

Figure 4 : Portuguese Power System frequency

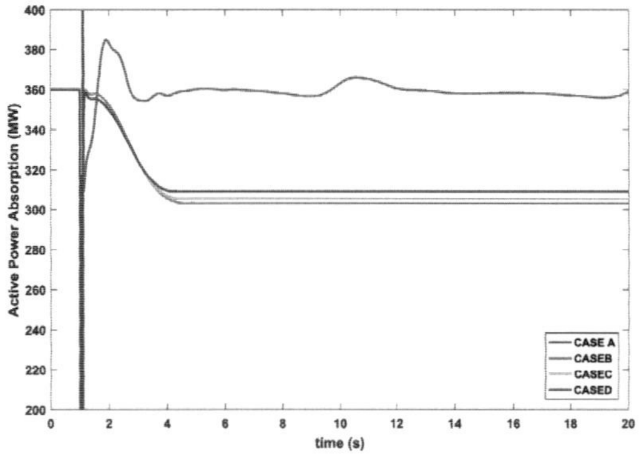

Figure 5 : Venda Nova III PSP Active Power consumption

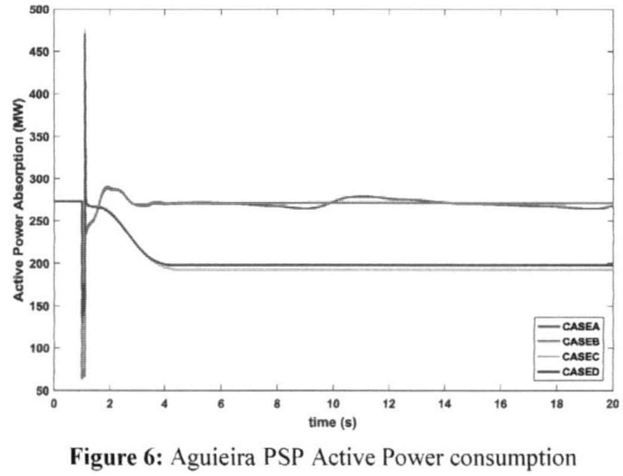

Figure 6: Aguieira PSP Active Power consumption

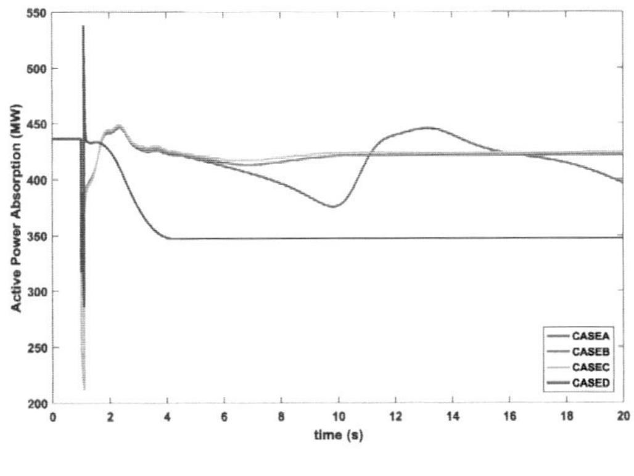

Figure 7 : Alqueva PSP Active Power consumption

In terms of frequency, the fault event and the consequent disconnection of wind power led to under-frequency deviation with large excursion as illustrated in Figure 4. For the case where no variable-speed PSP contribution was taken in consideration (CASE A), large oscillations can be observed. The integration of variable-speed PSP mitigates the oscillatory behaviour helping the power system stabilization. Furthermore, the progressive increase on the variable-speed PSP integration level reduces the frequency deviation, as depicted in Figure 4.

Regarding the active power response, the results for each power plant are illustrated in Figure 5, Figure 6 and Figure 7. For the CASE A, there is no variable-speed PSP on the grid thus, the generation units that have been monitored do not respond to frequency deviation. However, an oscillatory behaviour is observed on every power plant monitored. This behaviour is related with voltage fluctuation and with frequency disturbance, which affects the active power consumption. The inclusion of variable-speed PSP mitigate these effects thus, the active power response is smoother. It is also possible to verify that the generation units are able to provide primary frequency support by reducing the absorbed power while operating as pump. The observed active power reductions in the simulation results have a permanent characteristic. This fact is related with the inexistence of secondary control implemented in PSS/E Portuguese grid model. As a matter of fact, this is also observable in the frequency response (see Figure 4) which do not recovers back to the 50Hz but stabilizes in a value below. The aforementioned results are common for all variable-speed PSP as can be observed in Figure 5, Figure 6, Figure 7, for each analysed case.

4. Conclusions

The work presented in this paper consisted on evaluating the performance of the Portuguese Power system facing the progressive integration of frequency responsive variable-speed PSP in key locations, substituting fixed-speed PSP. A frequency responsive load model was developed in PSS/E, and parameterized for allowing the representation variable-speed PSP major grid-side dynamics. A severe disturbance consisting on a fault event that led to the disconnection of a significant amount of non-FRT compliant WF. This fault event led to load-generation imbalance, affecting the frequency, which decayed. The progressive integration of frequency-responsive variable-speed PSP showed to be beneficial regarding the provision of frequency support. The results show that the higher integration level, the lower frequency decay. In addition, the presence of variable-speed PSP helped on smoothing the power system frequency on what regards to oscillations. The effective active power contribution was achieved by the proportional active power consumption reduction at the variable-speed PSP, depending on the cases. It is also important to stress that no problems at the power system level were observed from the integration level performed. For the future, it would be important to analyze the

IOP Conf. Series: Journal of Physics: Conf. Series **813** (2017) 012012 doi:10.1088/1742-6596/813/1/012012

complete Iberian power system, namely in scenarios with high share of renewable and variable-speed PSP operating in pump mode in order to assess the power system stability namely in what regards the reduced inertia levels'.

References

[1] APREN – Associação Portuguesa de Energias Renováveis – Yearbook 2016.
[2] The Guardian, UK – *"Portugal runs for four days straight on renewable energy alone"*
[3] M. Matos Fernandes, "Connection of the First Pump-Storage Variable Speed Machines in Portugal: TSO Requirements"- *11th International Workshop on Large-Scale Itegration of Wind Power as well as on Transmission Networks for Offshore Wind Power Plants,* 13-15 November 2012 – Lisbon, Portugal
[4] HYPERBOLE Project– "Deliverable D6.2", *7th Framework Programme, European Comission*
[5] REN – Redes Energéticas Nacionais – Centro de informação url: http://www.centrodeinformacao.ren.pt/

IOP Conf. Series: Journal of Physics: Conf. Series **813** (2017) 012013 doi:10.1088/1742-6596/813/1/012013

Economic Viability of Pumped-Storage Power Plants Equipped with Ternary Units and Considering Hydraulic Short-Circuit Operation

Manuel Chazarra[a], Juan I. Pérez-Díaz[a] and Javier García-González[b]

[a] Department of Hydraulic, Energy and Environment Engineering, Escuela de Ingenieros de Caminos, Canales y Puertos, Technical University of Madrid, 28040, Spain
[b] Institute for Research in Technology, ICAI School of Engineering, Comillas Pontifical University, 28015, Madrid

Corresponding author's e-mail address: manuel.chazarra@upm.es

Abstract. This paper analyses the economic viability of pumped-storage hydropower plants equipped with ternary units and considering hydraulic short-circuit operation. The analysed plant is assumed to participate in the day-ahead energy market and in the secondary regulation service of the Spanish power system. A deterministic day-ahead energy and reserve scheduling model is used to estimate the maximum theoretical income of the plant assuming perfect information of the next day prices and the residual demand curves of the secondary regulation reserve market. Results show that the pay-back periods with and without the hydraulic short-circuit operation are significantly lower than their expected lifetime and that the pay-back periods can be reduced with the inclusion of the hydraulic short-circuit operation.

1. Introduction

Pumped storage hydropower plants (PSHPs) are considered worldwide as a mature technology to store large quantities of energy and to improve the flexibility of the power systems [1]. Therefore, they can play an important role in the context of a high penetration of intermittent renewable energies such as wind and solar power. However, PSHPs are characterised by high capital costs and even higher if they are equipped with ternary units and the hydraulic short-circuit operation is enabled.

Several papers have been found in the technical literature analysing the economic viability of conventional PSHPs participating in the day-ahead energy market. [2] compares the profits and the investment costs in six electricity markets such as Great Britain or Nordpool, among others, using a heuristic algorithm based on the well-known price-arbitrage to operate the PSHP. Results obtained in [2] show that the recovery of the investment costs is not possible almost in all the analysed markets. Results in [3] show that PSHPs are profitable in almost all the seven real-time markets analysed in the United States, such as ERCOT (Texas) or NYISO (New York). However, results can be considered slightly optimistic as they are obtained for 2008, when there was a significant difference between peak and off-peak energy prices and also reaching in some months the highest average prices of the last decade. [4] analyses the economic viability of conventional PSHPs participating also in the regulation market. Results in [4] show that the considered PSHP cannot recover the investment costs if it only participates in the day-ahead energy market and can recover the investment costs only for a limited range of the storage size if it also participates in the regulation market. According to the best of our knowledge, there is no published paper in which the economic viability of PSHPs equipped with ternary units and considering the hydraulic short-circuit operation (HSC-PSHPs) is analysed [5].

The main goal and contribution of this paper is to evaluate the economic viability of closed-loop and daily-cycle PSHPs equipped with ternary units and considering hydraulic short-circuit operation. The considered ternary unit is composed by a Pelton turbine, a fixed-speed pump and an electric machine. Closed-loop means that no natural water inflows are received in the upper reservoir and that no water outflows are released from the lower reservoir of the hydro system.

The economic viability is evaluated using the so-called pay-back period, i.e. the number of years that the investment costs are expected to be recovered. The pay-back period is estimated from the maximum theoretical income (MTI) and the investment costs of the PSHP. The MTI is obtained from the results of a deterministic day-ahead energy and secondary reserve scheduling model, based on mixed integer quadratic programing. The model is sequentially run day by day for a time period of one year (2014). The investment costs of the PSHP are estimated from available data in the technical literature of existing, to be commissioned and projected PSHPs. The pay-back period is estimated with and without considering hydraulic short-circuit operation.

The paper is organised as follows: the MTI is presented in Section 2 whereas the estimated investment costs are described in Section 3. The discussion of the economic viability is shown in Section 4. Finally, conclusions are drawn in Section 5.

2. Maximum theoretical income

The scheduling model used to obtain the MTI of the PSHP is based on [6]. The PSHP is assumed to participate in the day-ahead energy market and in the secondary regulation service (SRS) of the Spanish electricity system. The SRS comprises the participation in the day-ahead secondary regulation reserve market and the deployment of the reserves in real-time, i.e. the secondary regulation energy, managed by the Transmission System Operator. Note that both the secondary regulation reserves and energy are remunerated in the Spanish system. The tertiary regulation service is not included as it is outside the scope of the model that is used to obtain the maximum theoretical income of the PSHPs[1].

In [6], the PSHP is modelled as a price-taker in both, the day-ahead energy and reserve markets. The assumption that the PSHP is a price-taker in the reserve market can significantly overestimate the MTI obtained in the SRS. Due to this, in this paper, the model from [6] is adapted to consider the PSHP as a price-maker only in the secondary regulation reserve market. The formulation used to model the PSHP as a price-maker is based on the one presented in [7], using a linear approximation of the residual demand curves of the secondary regulation reserve market. Note that the PSHP is modelled as a price-taker in the day-ahead energy market because the maximum power of the plant (594.8 MW) represents 1.3% of the maximum demand of the Spanish system.

We use the term MTI to refer to the income that the PSHP would obtained assuming perfect information of the next day hourly energy prices, the residual demand curves of the reserve market, the amount of the committed reserves effectively used in real-time and the regulation energy prices.

The technical data of the PSHP can be seen in Table 1. The data are the same as those used in [6] for the case of PSHP with ternary units. Note that the plant is composed by two hydro units whereas the data in Table 1 is for the entire plant. Efficiencies in generating mode at maximum and minimum water discharge are 90% and 80%, respectively. Efficiency in pumping mode is 90%. Start-up costs in generating and pumping modes are obtained following the guidelines of [8]. As the maximum variation of the water level in the upper reservoir can be negligible in comparison with the available gross head, the head dependency in the mathematical formulation is neglected.

---

[1] The required modifications to consider the tertiary service are the following: 1) a new term in the objective function with the income/cost due to the upward/downward tertiary energy, 2) the available upward tertiary energy is the difference between the maximum power minus the committed power in the day-ahead market and the upward secondary reserve and 3) the available downward tertiary energy is the committed power in the day-ahead market minus the downward secondary reserve.

Table 1. Technical data of the PSHP. g refers to power, q refers to flow and cSU refers to start-up cost. Superscript d refers to generating mode whereas p refers to pumping mode. Flows are expressed in m^3/s, power in MW and start-up cost in €.

$\bar{g}^d$	$\bar{q}^d$	$\underline{g}^d$	$\underline{q}^d$	$\bar{g}^p$	$\bar{q}^p$	cSU^d	cSU^p
594.8	231.4	52.9	23.2	779.6	231.4	3301.2	2643

The MTIs of the PSHP with (HSC) and without (Conv) considering the operation in hydraulic short-circuit mode are presented in Table 2. It shows the net income in the day-ahead energy market (DM Income), i.e. income for selling energy in generating mode minus cost for purchasing energy in pumping mode, the income in the day-ahead secondary regulation reserve market (SM Income), the income for the real-time use of the upward reserves (ER2UP Income), the cost for the real-time use of the downward reserves (ER2DW Cost) and the start-up costs in generating mode (SupT Cost) and in pumping mode (SupP Cost). Note that in the Spanish power system, the upward/downward secondary regulation reserve which is effectively used in real-time is paid/bought at the price of the tertiary regulation reserve.

Table 2. Maximum theoretical income of the PSHP and in each market and service considered. All data are expressed in €.

PSHP	DM Income	SM Income	ER2up Income	ER2dw Cost	SupT Cost	SupP Cost	Total
Conv	10 636 427	25 839 584	8 662 665	-7 552 568	-3 060 212	-1 694 163	32 831 733
HSC	1 178 941	37 698 356	15 118 981	-8 149 052	-2 069 852	-1 647 910	42 129 462

The inclusion of the operation in HSC mode increases the total MTI around 28%, especially thanks to the participation in the SRS. Regarding each market and service, it increases the SM Income around 46%, decreases the DM Income around 88%, increases the net income for the secondary regulation energy more than five times and reduces the start-up costs around 22%.

The methodology used to obtain the MTI assumes that the upper reservoir starts and finishes each day with the same amount of stored water (half of the storage capacity). The authors are currently working to include a look-ahead period and relax the end of day target volume in the scheduling model which will hopefully enlarge the MTI.

3. Investment costs

Investment costs of PSHPs are strongly site-dependent. In the technical literature, to the best of our knowledge, there are no papers where the investment costs of PSHPs with ternary units are analysed. However, several papers can be found regarding investment costs of PSHPs equipped with binary units. [2] proposes a range between 470-2170 €/kW from projects in countries such as Spain, Portugal or Switzerland, among others. [9] proposes a range between 775-1280 €/kW from projects in Germany and Luxemburg. Or [10], which proposes a range between 2000-4300 $/kW. Note that the latter is expressed in $ and has been changed to € using the exchange rate of 1.12 $/€ at 29/07/2016. Therefore, the considered cost ranges between 1786-3839 €/kW.

There is an absolute lack of information about investment costs of HSC-PSHPs. However, hydro industry experts estimate an increase of 30%–40% of the investment costs of a PSHP equipped with binary units [11]. In this paper, an extra cost of 35% is considered for the HSC-PSHP. In addition to this, we assume that the extra investment cost related to permit a PSHP equipped with ternary units to operate in HSC mode is 3%. This extra investment cost is due to both 1) the reinforcement of the pipes in the short-circuit link and 2) the more complex design of the said pipe section in order to, respectively, resist pressure oscillations of a higher amplitude and to reduce the hydraulic losses.

Several ranges of the investment costs have been considered according to the information available in the literature and the above-mentioned hypothesis (Table 3).

Table 3. Investment cost ranges considered in the paper according to the information available in the technical literature. All data are expressed in millions of €.

PSHP	[2]		[9]		[10]	
	Min	Max	Min	Max	Min	Max
Conv	366	1690.2	603.6	996.9	1390.9	2990.4
HSC	377.4	1742.5	622.3	1027.8	1433.9	3082.9

4. Economic viability

The economic viability is analysed using the so-called pay-back period, i.e. the number of years that the investment costs are expected to be recovered according to the expected income of the PSHP. The calculation of the pay-back period assumes that the interest rate and the expected increase of the MTI due to an increase of the demand will be 0% in the whole pay-back period. The estimated pay-back periods corresponding to the proposed investment cost ranges are shown in Table 4.

Table 4. Pay-back period with the considered investment cost ranges. All data are expressed in years.

PSHP	[2]		[9]		[10]	
	Min	Max	Min	Max	Min	Max
Conv	11.2	51.5	18.4	30.4	42.4	91.1
HSC	9	41.4	14.8	24.4	34	73.2

According to the investment costs proposed in [2], the mean pay-back period of the Conv PSHP and of the HSC-PSHP would be 31.3 and 25.2 years, respectively. Assuming a lifetime of the plant between 40 years (see for example [2]) and 60 years (see for example [1]), or even 100 years [12], the economic viability is not discarded.

Even better results are obtained when the investment costs proposed in [9] are considered: the mean pay-back period of the Conv PSHP and of the HSC-PSHP would be 24.4 and 19.6 years, respectively.

By contrast, with the investment costs proposed in [10] the PSHP is not economically viable, as the mean pay-back period of the Conv PSHP and of the HSC-PSHP would be 66.7 and 53.6 years, respectively.

A promising result of the paper is that the inclusion of the HSC operation substantially reduces the pay-back periods in all cases as the increase in the MTI is significantly larger than the increase in the investment costs.

Despite the foregoing, further research must be carried out because, in this paper, perfect information is assumed in all uncertain data and therefore the income is expected to be reduced due to the effects of imperfect information. To give an idea, the value of perfect information of the price in the day-ahead energy market can be around 25% of the MTI according to the results presented in [7].

5. Conclusions

The economic viability of a closed-loop and daily-cycle PSHP equipped with ternary units with and without considering the operation in hydraulic short-circuit mode has been preliminarily evaluated in this paper. The PSHP is assumed to participate in the day-ahead energy market as a price-taker and in the day-ahead secondary regulation reserve market as a price-maker. In addition, the net income from the real-time use of the committed reserves has been also taken into account. The pay-back period of the PSHP has been estimated considering different ranges of investment costs proposed in the technical literature. Results show that the pay-back periods of the PSHP are in most cases lower than the lifetime of the PSHP and, therefore, the economic viability is not discarded. Besides, the inclusion of the hydraulic short-circuit operation reduces the pay-back periods in all cases. Nonetheless, these results have been obtained assuming perfect information of all the uncertain data in the day-ahead energy and reserve scheduling. Further work is necessary to estimate the economic viability of the PSHP taking into account uncertainty. In addition, the evaluation of the economic viability of PSHPs equipped with variable speed binary units is also deemed as promising future work. Finally, it should be taken into account that the consideration of other markets, such as the intraday markets and the tertiary regulation service, would contribute to increase the revenues of PSHPs in the short-term whereas the dissemination of other storage technologies and the penetration of the electric vehicle might cause in the long-term a reduction in the price spread and, therefore, the revenues.

Acknowledgments
This work was supported by the Spanish Ministry of Economy and Competitiveness under the project "Optimal operation and control of pumped-storage hydropower plants" of The National Scientific Research, Development and Technological Innovation Plan 2008-2011 (Ref. ENE2012-32207).

References
[1] Evans A, Strezov V, Evans T. Assessment of utility energy storage options for increased renewable energy penetration. Renew Sustain Energy Rev 2012;16:4141–7.
[2] Connolly D, Lund H, Finn P, Mathiesen BV, Leahy M. Practical operation strategies for pumped hydroelectric energy storage (PHES) utilising electricity price arbitrage. Energy Policy 2011;39:4189–96. doi:10.1016/j.enpol.2011.04.032.
[3] Bradbury K, Pratson L, Patiño-Echeverri D. Economic viability of energy storage systems based on price arbitrage potential in real-time U.S. electricity markets. Appl Energy 2014;114:512–9.
[4] Ekman CK, Jensen SH. Prospects for large scale electricity storage in Denmark. Energy Convers Manag 2010;51:1140–7. doi:10.1016/j.enconman.2009.12.023.
[5] Pérez-Díaz JI, Chazarra M, García-González J, Cavazzini G, Stoppato A. Trends and challenges in the operation of pumped-storage hydropower plants. Renew Sustain Energy Rev 2015;44:767 81.
[6] Chazarra M, Pérez-Díaz JI, García-González J. Optimal Energy and Reserve Scheduling of Pumped-Storage Power Plants Considering Hydraulic Short-Circuit Operation. IEEE Trans Power Syst 2017;32:344–53. doi:10.1109/TPWRS.2016.2545740.
[7] Chazarra M, Pérez-Díaz JI, García-González J. Value of Perfect Information of Spot Prices in the Joint Energy and Reserve Hourly Scheduling of Pumped Storage Plants. Eur. Energy Mark., 2016.
[8] Nilsson O, Sjelvgren D. Hydro Unit Start-up Costs and Their Impact on the Short Term Scheduling Strategies of Swedish Power Producers. IEEE Trans Power Syst 1997;12:38–44.
[9] Steffen B. Prospects for pumped-hydro storage in Germany. Energy Policy 2012;45:420–9. doi:10.1016/j.enpol.2012.02.052.
[10] Barbour E, Wilson I, Radcliffe J, Ding Y, Li Y. A review of pumped hydro energy storage development in significant international electricity markets. Renew Sustain Energy Rev 2016;61:421–32. doi:10.1016/j.rser.2016.04.019.
[11] Botterud A, Levin T, Koratirov V. Pumped Storage Hydropower: Benefits for Grid Reliability and Integration of Variable Renewable Energy. 2014.
[12] Ardizzon G, Cavazzini G, Pavesi G. A new generation of small hydro and pumped-hydro power plants: Advances and future challenges. Renew Sustain Energy Rev 2014;31:746–61. doi:10.1016/j.rser.2013.12.043.

IOP Conf. Series: Journal of Physics: Conf. Series **813** (2017) 012014 doi:10.1088/1742-6596/813/1/012014

Dynamic stresses in a Francis model turbine at deep part load

Wilhelm Weber, Florian von Locquenghien, Philipp Conrad and Jiri Koutnik

Voith Hydro Holding GmbH & Co. KG, Alexanderstraße 11, 89522 Heidenheim, Germany

E-mail: wilhelm.weber@voith.com

Abstract. A comparison between numerically obtained dynamic stresses in a Francis model turbine at deep part load with experimental ones is presented. Due to the change in the electrical power mix to more content of new renewable energy sources, Francis turbines are forced to operate at deep part load in order to compensate stochastic nature of wind and solar power and to ensure grid stability. For the extension of the operating range towards deep part load improved understanding of the harsh flow conditions and their impact on material fatigue of hydraulic components is required in order to ensure long life time of the power unit.

In this paper pressure loads on a model turbine runner from unsteady two-phase computational fluid dynamics simulation at deep part load are used for calculation of mechanical stresses by finite element analysis. Therewith, stress distribution over time is determined. Since only few runner rotations are simulated due to enormous numerical cost, more effort has to be spent to evaluation procedure in order to obtain objective results.

By comparing the numerical results with measured strains accuracy of the whole simulation procedure is verified.

1. Introduction

More and more new renewable energy sources like wind and solar power contribute to the European energy production. Since a certain amount of stochastic is attached to these energy sources highly variable operation of common power plants is required to compensate their fluctuations and to preserve the electrical grid stability. For this task hydraulic turbines and pump-turbines are especially suited. Therefore, they are no longer used only for electrical power generation at best efficiency condition but also for stand in if electrical energy from wind and solar drops spontaneously. As a consequence, the operating range of hydraulic turbines is extended towards deep part load. However, there are drawbacks to overcome. Since the turbine is operated further away from the design point in off-design conditions the fluid flow becomes more complex resulting in vortices with local pressure oscillations and cavitation phenomena, which can be harmful to the runner structure and lead to material fatigue. Therefore, it is essential to predict the loads on the turbine runner and the stresses for a fatigue assessment during design phase. Therewith, long operation time without material failure and time effort for repairs is ensured.

Recently, a model Francis turbine was equipped with strain gauges and measurements were performed for various operating points at the EPFL Lausanne [1]. Among other topics the influence of deep part load (DPL) operation on runner stress was investigated.

According to this measurement the loading pressures at one DPL operating point (Q_{ed} = 0.05445, n_{ed} = 0.2881, σ = 0.11) on the turbine structure was determined by computational fluid dynamics (CFD) simulation [2]. Herein, the model includes the whole water passage from spiral case inlet to draft tube outlet. Since cavitation was expected at the operating point under investigation two-phase modeling was performed.

The pressures distributions of 2000 time steps at a sampling rate of 5334 Hz are mapped onto the structural model for further processing. Finite element analysis (FEA) is carried out for calculation of strains and stresses. Therewith, time history of the strains and stresses especially at the locations of strain gauges in the experiment are determined. Since only numerical data for five runner rotations is available, comparison in the time domain is difficult. Therefore, numerical results will be transferred to frequency domain in order to obtain comparable information.

In the following, numerical setup for stress analysis, evaluation procedure as well as comparison between numerical results and measured strains will be shown.

2. Numerical stress analysis

A finite element model of the entire turbine runner (model scale) was generated, see figure 1. It was meshed with 1.57 million quadratic tetrahedron elements resulting in 2.70 million nodes. At the flange connection it is supported in axial and circumferential direction. On crown and band areas as well as runner blades pressures from CFD simulation are mapped. In the runner side chambers, which were not included in the fluid flow simulation, pressure is prescribed according to general distribution between pressure at vaneless space and draft tube considering the seal location. Modal analysis of the runner in air and water shows that the natural frequencies are

Figure 1. FEA model.

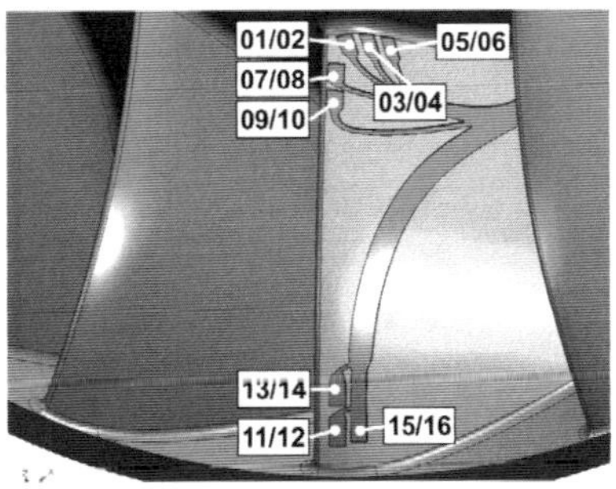

Figure 2. Locations of strain gauges.

much higher than the frequency range of interest. Therefore, a series of static computations is carried out to determine a pseudo-time series of the stresses.

3. Numerical and experimental results

From the results of the finite element analysis directional strain at locations of strain gauges (see Fig. 2) in their orientation direction is extracted for all blades. The strains are scaled to stresses by Young's modulus in order to get more descriptive results. In the same way the strains of the test rig measurement are scaled.

Since the measurement data are slightly affected by the measurement system, frequency domain is chosen instead of time domain. In order to compare the data of 5 simulated runner rotations with the data of 800 measured runner rotations equivalent frequency spectrums have to be determined. These are finally used for the comparison of calculated and measured strains.

3.1. Numerical results

In figures 3 and 5 an excerpt of numerically obtained time series of scaled strains at pressure side and suction side at crown are shown. Here, the dynamics of rotor stator interaction (RSI) is clearly observable. Beside this, time signals show some irregular behavior, which differs also form blade to blade.

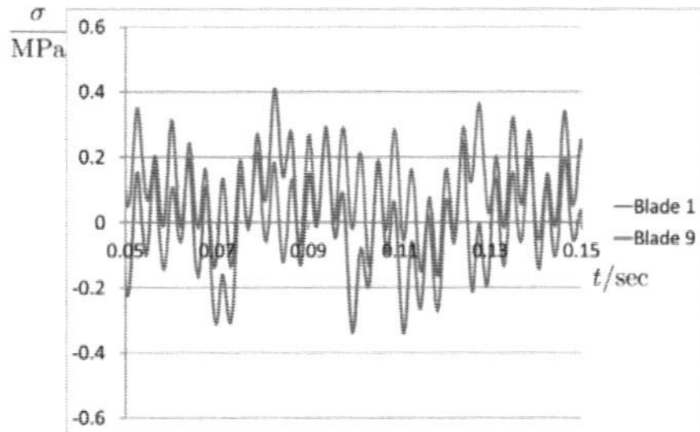

Figure 3. Time history at pressure side close to crown

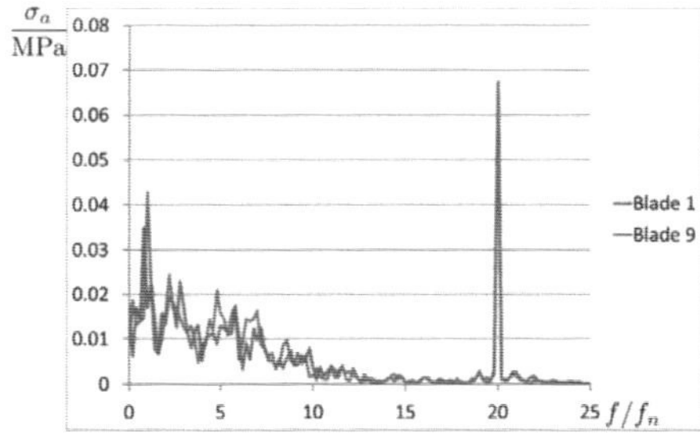

Figure 4. Frequency spectrum of figure 3

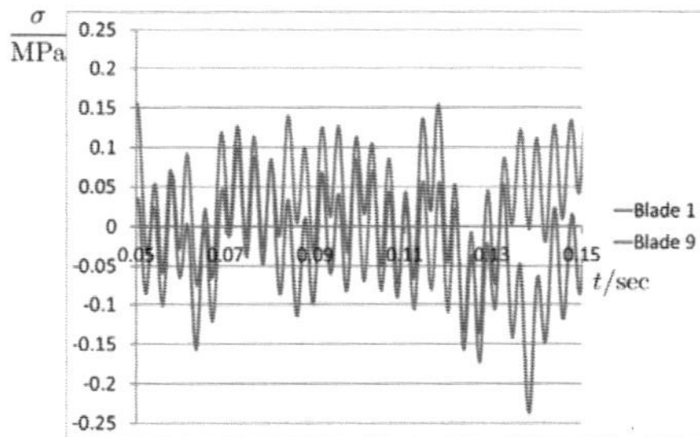

Figure 5. Time history at suction side close to crown

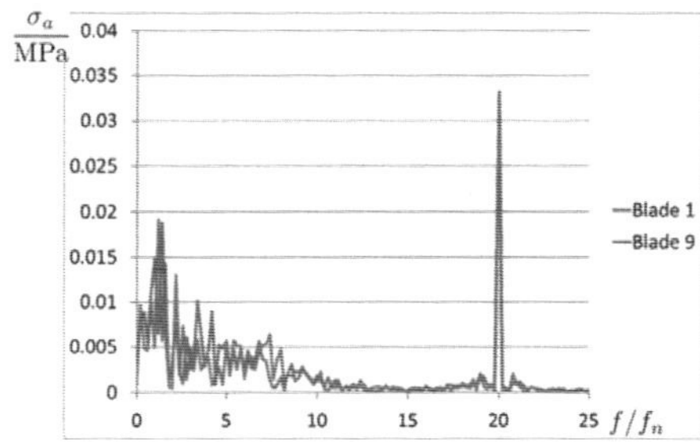

Figure 6. Frequency spectrum of figure 5

In figures 4 and 6 the corresponding frequency spectrums are shown. Here, a good agreement for the stress amplitude at gate passing frequency is observed for different blades. For the range of 0 to $10 \times f_n$ the general behavior seems similar, but locally there are large differences in the stress amplitude at one frequency. This characteristic arises from the fact that only a time series of 5 runner rotations is used for the Fourier transform, which is quite short. Therefore, an analysis to get objective values is performed in the next section.

3.2. Equivalent frequency spectrum

For the determination of frequency spectra that are comparable between numerical and measured data, one experimental data set of a strain gauge at the crown is analyzed in detail.

In figure 7 the frequency spectrums of 1875 samples that correspond to 5 rotations, of 30000 samples (80 rotations) and of the whole time signal are illustrated. Due to the different lengths of the time signals the frequency resolution of the spectrums is also different. The stress amplitudes that are related to a discrete excitation (e.g. RSI at $20f_n$) are comparable for all spectrums. For the whole frequency range, the amplitudes depend on the resolution of the spectrum, because stress amplitude of the coarsest resolution is distributed on different frequencies for the finer resolutions due to the stochastic nature of the stresses. Therefore,

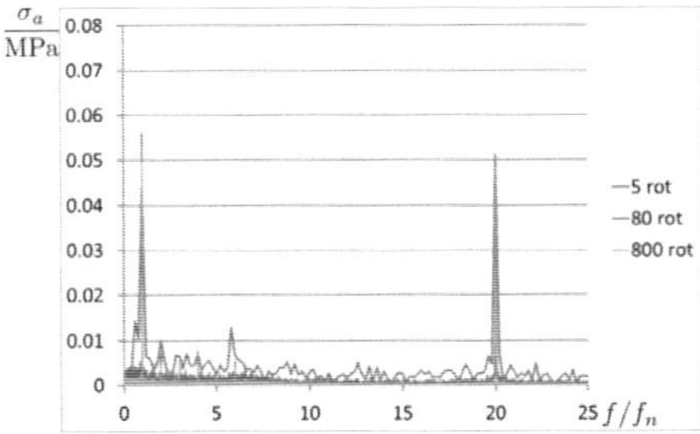

Figure 7. Frequency spectrum for different length of time history

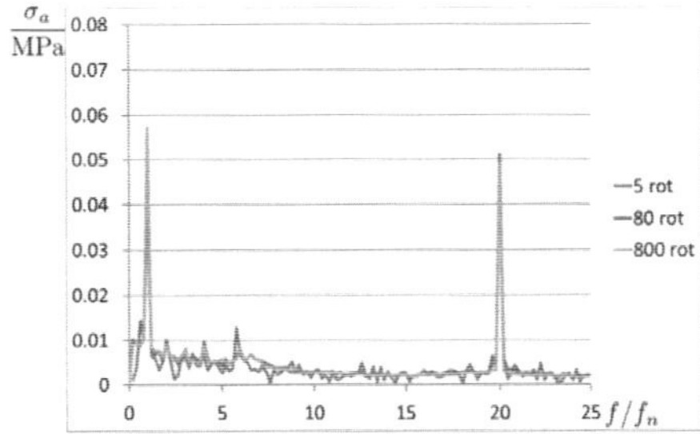

Figure 8. Post-processing of frequency spectrums to same resolution as for 5 runner rotations

all spectrums are brought to the same resolution by computing the energetic sum within each frequency range. Thereby, the spectrums of 80 runner rotations and 800 rotations are almost identical, see figure 8. However, the spectrum with the lowest number of samples still differ. In figure 9 a total of 16 spectrums with 1875 samples at different times of the measurement data are shown. Additionally, upper (Max) and lower (Min) bound as well as the root mean square

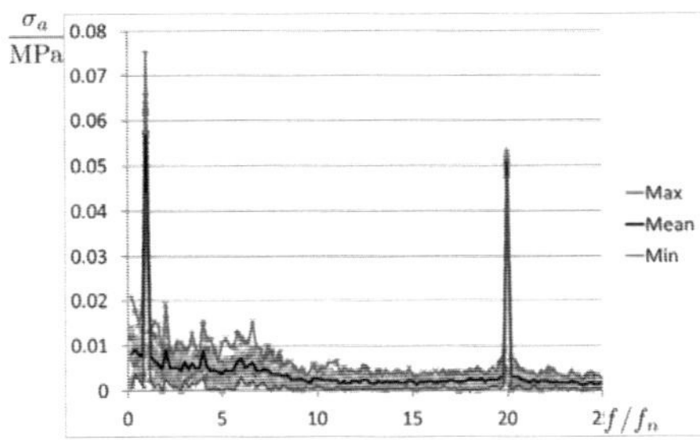

Figure 9. Frequency spectrums for different short time histories

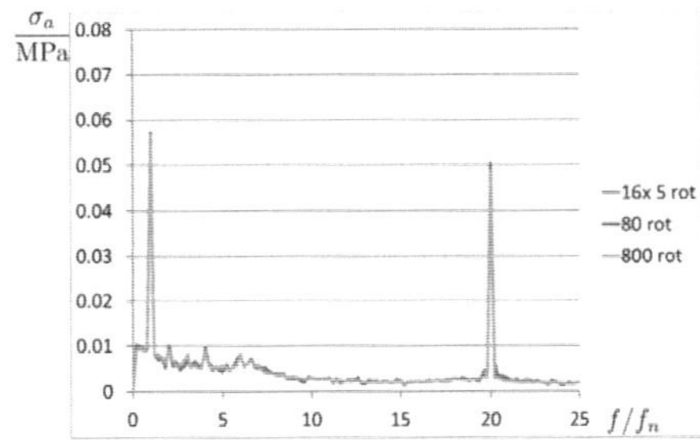

Figure 10. Comparable frequency spectrums for different data lengths

value (Mean) is shown. It can be observed, that the spread of the amplitudes is quite high. Therefore, one single data set of 5 runner rotations is too short to be representative for the entire dynamic. If one compares the root mean square value of 16 data sets with the frequency spectrums of longer time histories as shown in figure 10 a very good agreement between all data is observed. Therefore, the calculated stresses at the strain gauge locations of the 16 runner blades are used for the comparison with the experimental data. Here, an average frequency spectrum is determined. From measurements, the resolution of the frequency spectrum is coarsened to obtain same resolution as from simulation.

3.3. Comparison of numerical with experimental results

In figures 11 and 12 the frequency spectrums of the stresses at the pressure and suction side in the vicinity of the crown are shown. In detail, strain gauges 05 and 06 are analyzed, cf. figure 2. In figure 13 the comparison for the strain gauge 12 at the suction side close to the band is shown. Unfortunately, all strain gauges at the pressure side were broken before this test. Overall, a good agreement between experimental and numerical results can be observed in the frequency range of interest which is up to 10 times f_n. For higher frequencies low strain ampltiudes are a result of low pressure oscillations that have already been observed in the CFD

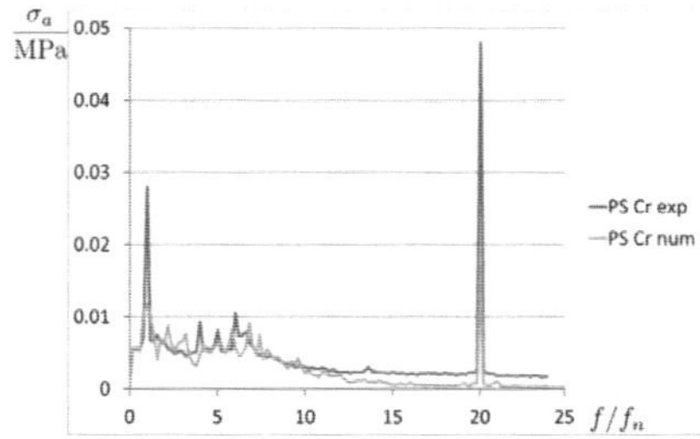

Figure 11. Measured and calculated strains at pressure side near crown

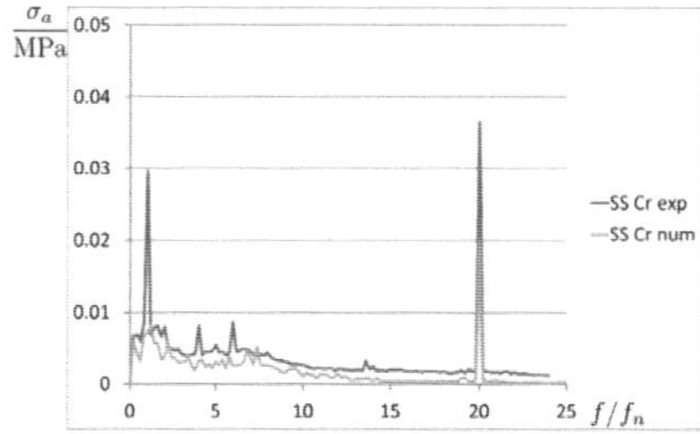

Figure 12. Measured and calculated strains at suction side near crown

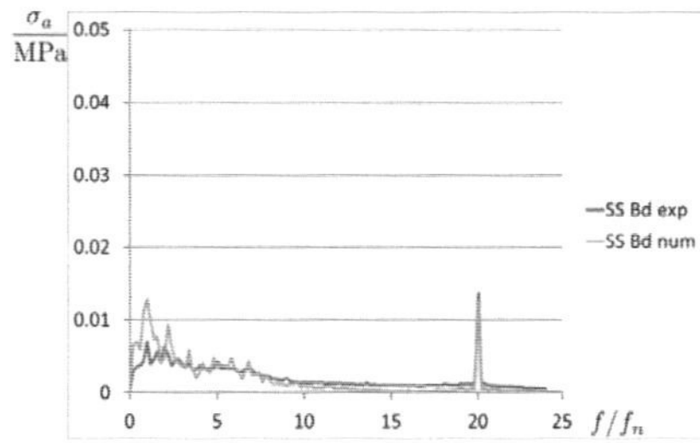

Figure 13. Measured and calculated strains at suction side near band.

simulation, see [2]. The difference in the RSI-stresses is not relevant, since it is out of scope in this examination. This phenomenon is captured well by other simulation approaches [3].

For the frequency f_n there are some interferences from the measurement system that affect the measured amplitudes at this frequency. Also, the first harmonics are slightly influenced, as can also be seen from green line in figure 7.

4. Conclusion

A comparison of numerical obtained stresses in a Francis model turbine at DPL operation with experimental ones has been performed in frequency domain. Average frequency spectrums for all blades were used for the numerical data. The frequency spectrums of the measurement data were coarsened to the resolution of numerical ones in order to obtain objective spectrums.

A good agreement is observed for the numerical results at pressure side crown and suction side band between f_n and $10f_n$. At the suction side near crown the amplitudes are slightly underestimated. The deviations above $10f_n$ are linked to the low numerical pressure oscillations from computational fluid flow simulation. Below f_n longer numerical time series are necessary for more reliable amplitudes.

Acknowledgments

The research leading to the results published in this paper is part of the HYPERBOLE research project, granted by the European Commission (ERC/FP7-ENERGY-2013-1-Grant 608532)

References

[1] Hyperbole - hydropower plants performance and flexible operation towards lean integration of new renewable energies fP7-ENERGY-2013-1
[2] Conrad P, Weber W and Jung A 2017 Deep part load flow analysis in a francis model turbine by means of two-phase unsteady flow simulations (Porto: HYPERBOLE Conference)
[3] Seidel U, Hübner B, Löfflad J and Faigle P 2012 *IOP Conference Series: Earth and Environmental Science* **15** 052010 URL `http://stacks.iop.org/1755-1315/15/i=5/a=052010`

IOP Conf. Series: Journal of Physics: Conf. Series **813** (2017) 012015 doi:10.1088/1742-6596/813/1/012015

Developments in multiple-valve pipeline column separation control

A Bergant[1], U Karadžić[2], A Tijsseling[3]
[1]Litostroj Power d.o.o., 1000 Ljubljana, Slovenia
[2]University of Montenegro, 81000 Podgorica, Montenegro
[3]Eindhoven University of Technology, 5600 MB Eindhoven, The Netherlands

E-mail: anton.bergant@litostrojpower.eu

Abstract. This paper investigates novel multiple-valve pipeline column separation control. Experimental investigations have been performed in a laboratory pipeline apparatus. The apparatus consists of an upstream end high-pressure tank, a horizontal steel pipeline (total length 55.37 m, inner diameter 18 mm), four valve units positioned along the pipeline including the end points, and a downstream end tank. The transient event is induced by downstream-end or/and upstream-end valve closures (single- or multiple-valve action). A multiple-valve induced closure event may significantly attenuate the severity of pressure oscillations in the pipeline system. Discrete gas cavity model (DGCM) predictions and measured results for two typical cases including single- and multiple-valve closures are compared and discussed.

1. Introduction

State-of-the-art hydraulic pipelines undergo a broad range of operating regimes. Unsteady pipe flows may induce severe loads in the piping systems, including large pressure pulsations and pipeline vibrations [1]. Water hammer is the propagation of pressure waves along liquid-filled pipelines (water, oil), and it is induced by a change in flow rate (flow velocity). Vaporous cavitation occurs when the pressure drops to the liquid vapour pressure [2]. There are two distinct types: (1) *column separation*, where a localized (discrete) vapour cavity with a large void fraction separates two liquid columns (or one column and a dead end) and (2) *flashing*, where a region of distributed vaporous cavitation with a small void fraction extends over long sections of the pipeline. A discrete vapour cavity may form at a boundary (valve, pump, water turbine), at a knee or at a high point in the pipeline. Distributed vaporous cavitation occurs when a rarefaction wave progressively drops the pressure in an extended region of the pipe to the liquid vapour pressure. The collapse of vapour cavities may induce large pressures. Engineers try to avoid column separation, although there are some systems in which column separation is accepted. Most of the column separation research has been done for a classical case with single-valve closure in a simple reservoir-pipeline-valve system [1], [2]. However, there are a number of typical industrial pipelines with multiple-valves, at least with two of them (upstream- and downstream-end valves). Multiple closing or opening of valves may induce very large or low pressure waves due to superposition of the waves [3], [4]. The associated phenomena are well understood, but accurate predictions are problematic because of uncertainties in wave speeds and reflection

[1] To whom any correspondence should be addressed.

coefficients; therefore engineers try to avoid multiple valve action in the system. There are reports, for example, on the usage of single or even multiple-check valves to attenuate column separation effects and this has been addressed theoretically by Karney and Simpson [5] and practically applied by Dudlik et al. [6]. To even better understand the phenomena associated with these types of transient events, and to create an experimental database, a large number of tests have been performed in a laboratory pipe system at the University of Montenegro [4]. The small-scale apparatus consists of an upstream end high-pressurized tank, horizontal steel pipeline (total length 55.37 m, inner diameter 18 mm), four valve units positioned along the pipeline including the end points, and a downstream end tank (outflow tank). This paper investigates the effects of vaporous cavitation caused by the single-valve closure (downstream-end valve) and by the multiple-valve closure (nearly simultaneous closure of upstream-end and downstream-end valves). Comparisons between the results of two distinct column separation tests and numerical simulations using a discrete gas cavity model (DGCM) are presented and discussed.

2. Discrete gas cavity model

The *classical* discrete gas cavity model (DGCM) is based on simplified water hammer equations (the convective transport terms are neglected) [1], [2]. The DGCM allows gas cavities (with the vapour pressure set as bottom pressure) to form at all computational sections in the method of characteristics (MOC) grid. A liquid phase with a constant wave speed a is assumed to occupy the reaches connecting the computational sections. A discrete gas cavity is described by two water hammer compatibility equations, the continuity equation for the gas cavity volume, and the ideal gas equation. Detailed description of the DGCM is given in [1]. The DGCM model can be successfully used for the simulation of unsteady pure liquid pipe flow by utilizing a very low gas void fraction and for transient cavitating flow with void fractions up to 1. The MOC based DGCM algorithm in this paper incorporates a convolution-based (with quasi-2D weighting function) unsteady friction model [7].

3. Laboratory pipeline apparatus

A small-scale pipeline apparatus has been designed and constructed at the University of Montenegro for investigating rapid water hammer events including column separation. The apparatus is comprised of a horizontal pipeline that connects the upstream end high-pressurized tank to the outflow tank (steel pipe of total length L = 55.37 m; internal diameter D = 18 mm; pipe wall thickness e = 2.0 mm; maximum allowable pressure in the pipeline $p_{max,\ all}$ = 25 MPa) – see Fig. 1. The symbols are defined as they first appear in the paper. Four valve units are positioned along the pipeline including the end points (valves V*/3* at positions 0/3, 1/3, 2/3 and 3/3). The air pressure in the upstream end tank can be adjusted up to 800 kPa. The pressure in the tank is kept constant during each experimental run by using a high-precision fast-acting air pressure regulator (precision class: 0.2%) in the compressed air supply line. Four dynamic high-frequency pressure transducers are positioned within the valve units along the pipeline including the end points (see Fig. 1). Pressures $p_{0/3}$, $p_{1/3}$, $p_{2/3}$ and $p_{3/3}$ are measured by Dytran 2300V4 high-frequency piezoelectric absolute pressure transducers (pressure range: from 0 to 6.9 MPa; resonant frequency: 500 kHz; acceleration compensated; discharge time constant: 10 seconds (fixed)). The water temperature is continuously monitored by the thermometer installed in the outflow tank. The water hammer wave speed was determined as a = 1340±10 m/s. The test procedure is as follows. The steady-state flow conditions (in advance of a dynamic test) are controlled by a set pressure in the upstream end tank and by a set opening of the downstream end control needle valve (valve V3/3C in Fig. 1). The water level in the upstream end pressurized tank can be adjusted. From initial steady flow conditions (flow, no-flow), a transient event is initiated by downstream-end (V3/3H) or/and upstream-end (V0/3U) valve closures (single- or multiple-valve action).

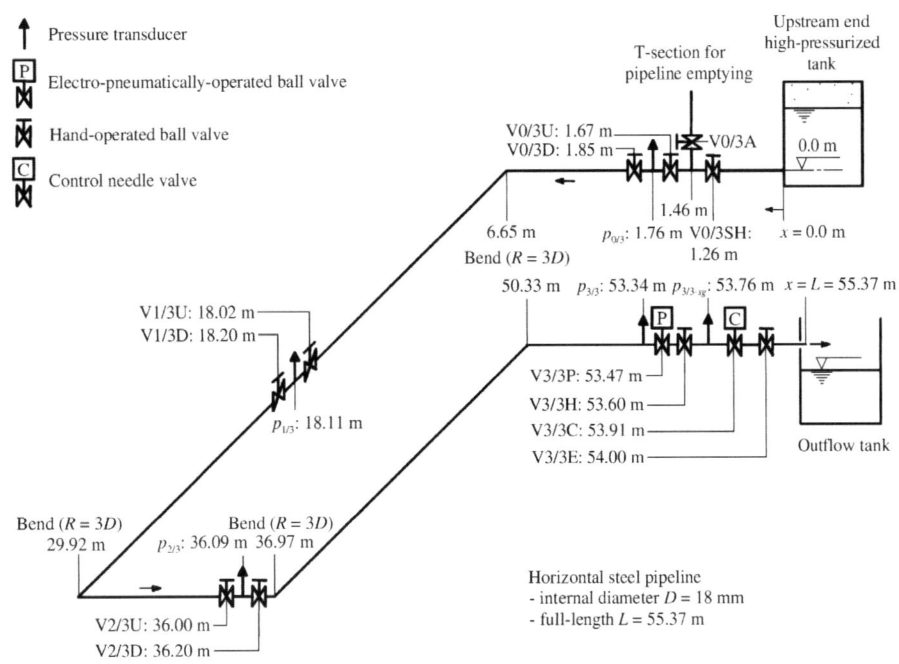

Fig. 1 Layout of small-scale pipeline apparatus

4. Numerical and experimental results

The following two distinct experimental runs in the test apparatus (Fig. 1) are presented and analyzed theoretically in this section: (1) the first case study investigates single-valve (V3/3H) closure only and (2) the second case study investigates the closure of two valves (first V3/3H and then V0/3U with a time delay of $9.5L/a$). Numerical results from the DGCM (see Section 2) are compared with results of laboratory measurements at the upstream end-valve (pressure $p_{0/3}$ in Fig. 1), along the pipeline (pressures $p_{1/3}$ and $p_{2/3}$) and at the downstream end valve (pressure $p_{3/3}$). The computational time step for all runs is $\Delta t = 0.0004$ s. Water and surrounding temperatures were about 25 and 30 °C, respectively.

4.1 Single-valve closure induced column separation test

Numerical and experimental results for the single-valve closure (valve V3/3H in Fig. 1) are shown in Fig. 2. This case is a classical example of passive column separation [8] with a maximum head rise equal to the Joukowsky head rise caused by valve closure (and *not* by cavity collapse). The initial flow velocity was $V_0 = 2.19$ m/s. The effective valve closure time of 20 ms is much less than the wave reflection time $2L/a$ of 0.08 seconds, and it is about 50% of the total closure time. A rapid valve closure generates a column separation event with severe repeated column separations at the valve. The valve closure first induced a Joukowsky head rise at the valve and subsequently, at time $t = 0.09$ s, first severe column separation at the valve. The negative pressure wave travels along the pipeline and makes the head drop to the vapour head at all measured positions along the pipeline. The maximum measured head at the valve of $H_{3/3} = 295$ m after the first cavity collapsed at $t = 0.4$ is less than the Joukowsky head of $H_{3/3} = 330$ m (including friction) just before first column separation. The datum level for all heads in the pipeline and at the tank is at the top of the horizontal steel pipe (elevation 0.0 m in Fig. 1). Pressure histories along the pipeline enable accurate tracing of distributed vaporous cavitation zones and intermediate large cavities. The computational results give pressure histories that are in good agreement with the measured results.

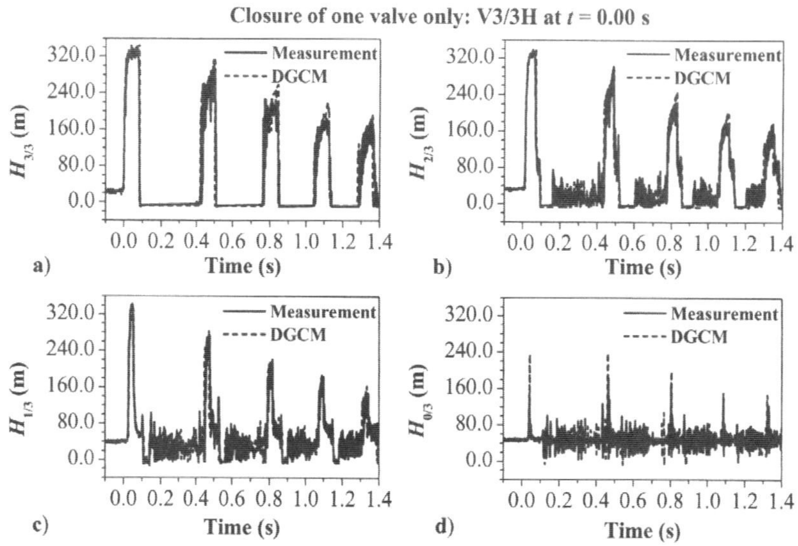

Fig.2 Comparison of measured and calculated heads at the end valves ($H_{3/3}$ and $H_{0/3}$) and along the pipeline ($H_{2/3}$ and $H_{1/3}$) for single-valve closure case; V_0 = 2.19 m/s [8].

4.2 Multiple-valve closure induced column separation test

Figure 3 presents measured and computed head traces in the system ($H_{3/3}$, $H_{2/3}$, $H_{1/3}$ and $H_{0/3}$) for the fast but not simultaneous closure of the two valves V3/3H and V0/3U for the same initial flow velocity V_0 = 2.19 m/s as for the single-valve closure case [4]. Again, the valve closure times were less than the wave reflection time $2L/a$ = 0.08 seconds (V3/3H: 60 ms and V0/3U: 20 ms). The initial fast closure of the downstream end valve (V3/3H) produces the classical Joukowsky head of $H_{3/3}$ = 330 m. The second but delayed fast closure of the downstream end valve (V0/3U) produces the pressure head drop from the reservoir head to the vapour head of -10.1 m. The time delay between the two end valve closures is 0.38 s. At the time of the pressure drop at V0/3U the water moves towards the downstream end (away from the reservoir). The pressure heads along the pipeline (at least at the $H_{3/3}$, $H_{2/3}$, $H_{1/3}$ and $H_{0/3}$ positions) remain practically constant at the vapour head for a longer period. We may conclude that for the case considered the system's dynamic response in terms of long-term pressure pulsations is less violent (one pressure pulse) than the response of a similar classical column separation case with the downstream end valve closure only (large number of pressure pulses; Section 4.1). The number of pressure pulses during a transient event is important for the assessment of the system lifetime. The effect of reopening of one of the valves is the topic of current authors' research. Again, the results obtained using the DGCM give pressure histories that are in favourable agreement with the experimental results.

5. Conclusions

Experimental investigations have been performed in a laboratory pipeline apparatus. Dynamic pressures were investigated at the pipe end points and at one- and two-third points along the pipe. Transient events were induced by downstream-end or/and upstream-end valve closures (single- or multiple-valve action). Comparison between single- and multiple-valve closures showed that multiple-valve induced closure event may significantly attenuate the severity of column separation induced oscillations in the pipeline system. Discrete gas cavity model (DGCM) predictions and measured results for the two typical cases including single- and multiple-valve closures are in good agreement.

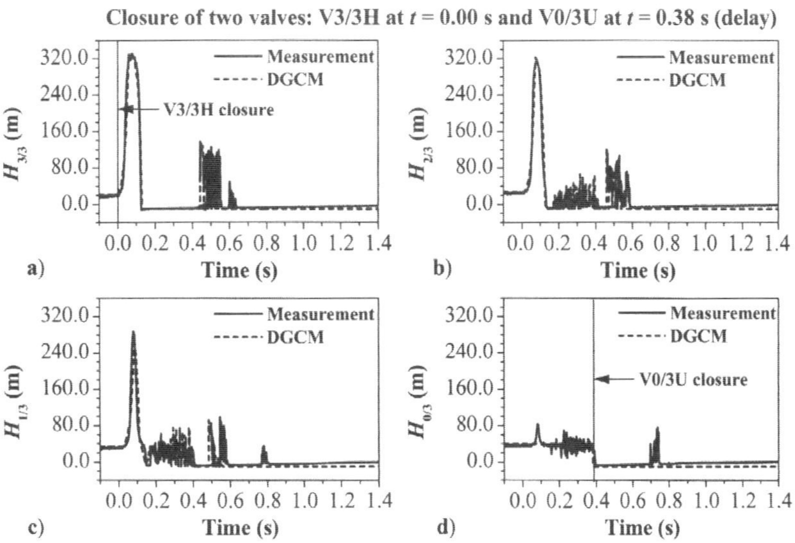

Fig.3 Comparison of measured and calculated heads at the end valves ($H_{3/3}$ and $H_{0/3}$) and along the pipeline ($H_{2/3}$ and $H_{1/3}$) for multiple-valve closure case; V_0 = 2.19 m/s [4].

6. References

[1] Wylie E B and Streeter V L 1993 *Fluid Transients in Systems* (Englewood Cliffs: Prentice-Hall)
[2] Bergant A, Simpson A R and Tijsseling A S 2006 Water hammer with column separation: A historical review *Journal of Fluids and Structures* **22** pp 135-171
[3] Bergant A, van 't Westende J M C, Koppel T, Gale J, Hou Q, Pandula Z and Tijsseling A S 2010 Water hammer and column separation due to accidental simultaneous closure of control valves in a large scale two-phase flow experimental test rig *Proceedings of the ASME Pressure Vessels and Piping Division Conference (Bellevue, Washington, USA)* Paper PVP2010-26131
[4] Bergant A and Karadžić U 2015 Developments in valve-induced water-hammer experimentation in a small-scale pipeline apparatus *Proceedings of the 12th International Conference on Pressure Surges (Dublin)* (Cranfield: BHR Group) pp 639-652
[5] Karney B W and Simpson A R 2007 In-line check valves for water hammer control *Journal of Hydraulic Research* **45** pp 547-554
[6] Dudlik A, Schönfeld S B H, Schlüter S and Prasser M-H 2004 ABS-Armaturen für Rohrleitungen. Hydraulisches Bremssystem vermeidet Druckstöβe und Kavitationschläge. *P&A Kompendium 2004* (München: Publish-Industry Verlag) pp 203-205
[7] Zielke W 1968 Frequency-dependent friction in transient pipe flow *Journal of Basic Engineering* **90** pp 109-115
[8] Bergant A and Karadžić U 2015 Numerical and experimental investigations of transient cavitating pipe flow *Journal of Energy Technology* **8** pp 31-42

ACKNOWLEDGEMENTS

The authors gratefully acknowledge the support of the Slovenian Research Agency (ARRS) and of the Ministry of Science of Montenegro (MSM) conducted through the projects BI-ME/14-15-016 (ARRS, MSM) and L2-5491 (ARRS).

Prediction of Francis Turbine Prototype Part Load Pressure and Output Power Fluctuations with Hydroelectric Model

S Alligné[1], C Nicolet[1], A Béguin[1], C Landry[1], J Gomes[2], F Avellan[2]

[1] Power Vision Engineering Sàrl, Ch. Des champs courbes 1, CH 1024 Ecublens, Switzerland
[2] Ecole Polytechnique Fédérale de Lausanne, Laboratory for Hydraulic Machines, CH 1007 Lausanne, Switzerland

Abstract. The prediction of pressure and output power fluctuations amplitudes on Francis turbine prototype is a challenge for hydro-equipment industry since it is subjected to guarantees to ensure smooth and reliable operation of the hydro units. The European FP7 research project Hyperbole aims to setup a methodology to transpose the pressure fluctuations induced by the cavitation vortex rope from the reduced scale model to the prototype generating units. A Francis turbine unit of 444MW with a specific speed value of $\nu = 0.29$, is considered as case study. A SIMSEN model of the power station including electrical system, controllers, rotating train and hydraulic system with transposed draft tube excitation sources is setup. Based on this model, a frequency analysis of the hydroelectric system is performed for all technologies to analyse potential interactions between hydraulic excitation sources and electrical components. Three technologies have been compared: the classical fixed speed configuration with Synchronous Machine (SM) and the two variable speed technologies which are Doubly Fed Induction Machine (DFIM) and Full Size Frequency Converter (FSFC).

1. Introduction

In the framework of the European FP7 research project Hyperbole, a methodology is setup to predict pressure and output power fluctuations on prototype induced by the cavitation vortex rope based on experimental measurements on the reduced scale model. The developed methodology relies on an advanced modelling of the draft tube cavitation flow which main parameters are the cavitation compliance, the dissipation and the excitation source [1]. Specific measurements on reduced scale model to quantify model parameters including dissipation are required [2]. First, this paper presents the methodology and focuses on the transposition to the prototype of the draft tube model parameters identified on the reduced scale model. Then, a numerical model of the power station including electrical system, controllers, rotating train and hydraulic system with transposed draft tube excitation sources is setup. Based on this model, frequency response of the electro-mechanical, the hydro-mechanical and the hydroelectric systems are compared to analyse the influence of the different modelling approaches to predict both pressure and output power fluctuations induced by the cavitation vortex. Three technologies for electrical part have been compared: the classical fixed speed configuration with Synchronous Machine (SM) and the two variable speed technologies which are Doubly Fed Induction Machine (DFIM) and Full Size Frequency Converter (FSFC).

2. Methodology

The methodology to predict the pressure fluctuations on prototype is illustrated in figure 1. The first step is to identify the hydroacoustic characteristics of the draft tube cavitation vortex rope on a Francis turbine reduced scale model installed on a test rig. To achieve this, the test rig hydraulic circuit is excited by an external periodical discharge source and the system response is compared to the response of a numerical model of the test rig [2]. An identification process comparing experimental and numerical hydraulic responses enables to identify the parameters of an advanced model of the draft tube cavitation flow. Then, these reduced scale model parameters are transposed to the prototype and used in the numerical model of the prototype power plant for the prediction of the resulting pressure and output power fluctuations.

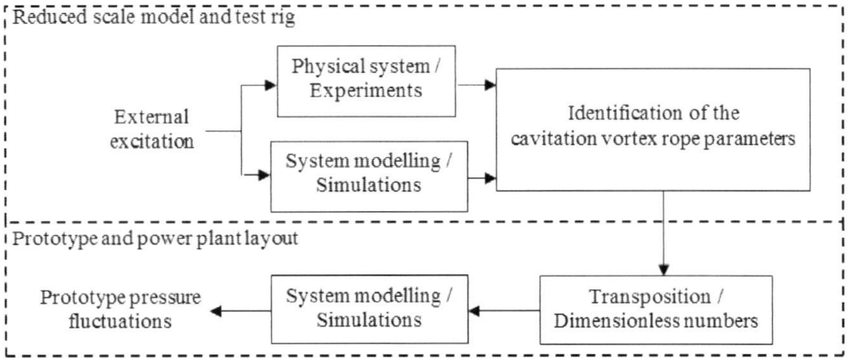

Figure 1. Methodology for prediction of pressure fluctuations on prototype.

The modelling of the draft tube cavitation flow is described by continuity and momentum equations including the convective terms and the divergent geometry [1]. For this investigation, two cavitation vortex rope parameters of this model have been identified experimentally at the reduced scale model:

- the local wave speed a defined implicitly by the cavitation compliance C_c;
- the second viscosity μ'' introducing dissipation induced by the phase change during cavitation volume fluctuations.

The wave speed and the second viscosity identified at the reduced scale model are normalized by the outlet pressure level of the draft tube, leading to two dimensionless numbers respectively Π and M^{*} [2]. These dimensionless numbers can be approximated by a power function of the void fraction β which are not dependent on the operating point of the hydraulic machine in the range of part load conditions. To use these dimensionless numbers for transposition purpose, the void fraction β must be known and is derived from a cavitation curve $\beta = f(\sigma)$ depending on the operating point and the Froude number obtained from experiments. By assuming the Froude similitude the prototype void fraction β^{P} is equal to the reduced scale model void fraction β^{M} allowing deriving the transposed draft tube flow parameters at the prototype scale.

3. Hydraulic layout modelling of the power plant

The power plant of interest features four Francis type turbines rated at power output of 444 MW under the 171m rated net head. Each unit is supplied by individual penstock. A numerical model of the power plant is set up with the SIMSEN software including reservoirs, penstock, the two quadrant characteristic of the turbine, the rotating inertia and the advanced cavitation draft tube model [3]. The draft tube model parameters at reduced scale model have been derived for two part load operating points named PL1 and PL2, defined in table 1a). A distributed model of the draft tube is used considering several control volumes along the draft tube length where continuity and momentum equations are applied [3]. Constant wave speed and second viscosity parameters are considered along the draft tube length. The table 1b) shows the influence of the operating point on the transposed prototype draft tube parameters and on the resulting first eigenmodes of the power plant defined by frequency and damping $s = \alpha + j(2\pi f)$. By changing the operating point from PL1 to PL2, the void fraction is increased. Hence, the first eigenfrequency value f_1 is decreased from 0.30 to 0.18 times the runner frequency n and the eigendamping value α_1 is increased towards positive values.

Table 1. a) Investigated operating points b) Influence of the operating point on the draft tube parameters and on the resulting two first eigenfrequencies.

a)

	PL1	PL2
GVO	15.2	12
nED / nED BEP	1	1
QED / QED BEP	0.8	0.64
Fr	5.6	5.6
σ	0.11	0.11

b)

	PL1	PL2
σ (–)	0.11	0.11
β (–)	0.0124	0.0421
a (m/s)	76.9	45.3
μ'' (Pa.s)	3.06E+05	6.14E+04
$\alpha 1$ (s^{-1})	−0.40	−0.10
f1/n (-)	0.30	0.18
$\alpha 2$ (s^{-1})	−0.86	−1.02
f2/n (-)	0.78	0.69

4. Electrical layout modelling of the power plant

For the electrical layout modelling, the current three motor-generator technologies are considered: the classical fixed speed configuration (SM) with PSS2B and the two variable speed technologies which are Doubly Fed Induction Machine (DFIM) and Full Size Frequency Converter (FSFC). To have detailed information about the different components models and control strategies, the reader can refer to [4]. The connection to the grid is modelled with an infinite power three phase voltage bus behind a short-circuit impedance, to represent the short-circuit power limitation of the connection point. The hydroelectric models are shown in figure 2.

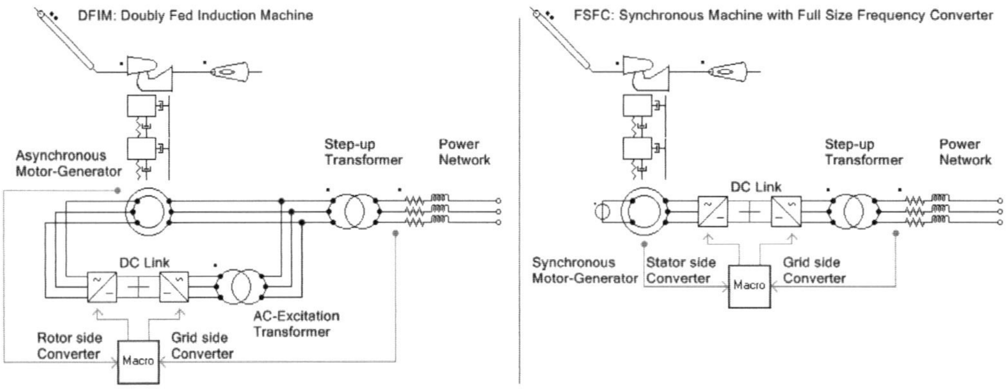

Figure 2. Hydroelectric SIMSEN models of DFIM (Left) and FSFC technologies.

5. Frequency analysis of the hydroelectric powerplant

To characterize the dynamic system response in the frequency domain, transfer functions are computed by performing a time domain simulation with a white noise excitation modelled by a Pseudo Random Binary Sequence (PRBS) considering two different types of excitations:

- a momentum excitation source in the draft tube for the hydro-mechanical and the hydroelectric systems located at one pressure node in the draft tube;
- an external torque on the rotating masses for the electro-mechanical system.

With a period of $dT = 0.1\,s$, the energy spectrum of the PRBS signal is distributed uniformly in the range 0 to 5Hz, covering the excitation range of the helical vortex rope being between 0.2 and 0.4 times the runner frequency n corresponding respectively to 0.43 Hz and 0.86 Hz. The normalized transfer functions are defined as the ratio between the output of interest Y and the excitation source.

$$G(s)_{hydro/hydroel} = \frac{Y(s)/Y_n}{S_h(s)/H_n} \;;\; G(s)_{electro} = \frac{Y(s)/Y_n}{T_{ext}(s)/T_n} \tag{1}$$

The amplitude of the normalized transfer function of the hydro-mechanical system, defined as the ratio between the draft tube pressure cone and the momentum excitation source in the draft tube, is represented in figure 3 for the two investigated operating points. Due to the higher void fraction for PL2, the second viscosity is lower and therefore, damping values of the eigenmodes and system response amplitude are higher. The vortex rope frequency being between 0.2 and 0.4 times the runner frequency n (yellow area in figure 3), a match with the first eigenfrequency is only possible for PL1 where amplitudes are rather small for resonance conditions. For PL2, the amplitude response obtained for the second eigenmode at this location is higher than the one obtained for PL1 despite a higher damping value α for PL2. This effect is due to the difference of cavitation parameters between PL1 and PL2 that affects the spatial distribution of pressure amplitudes along the piping system and also to the difference of relative position of the excitation source in the eigenmode. These transfer functions, for both operating points, are unchanged using hydroelectric models whatever the electrical system technology considered. However, for the prediction of the output power fluctuations, it depends on the level of modelling approach. In figure 3, the amplitude of the normalized transfer functions defined as the ratio between the output power and the external excitation source are plotted for the fixed speed technology and comparison is performed between electro-mechanical, hydro-mechanical and hydroelectric models considering synchronous machine with PSS2B and without turbine governor.

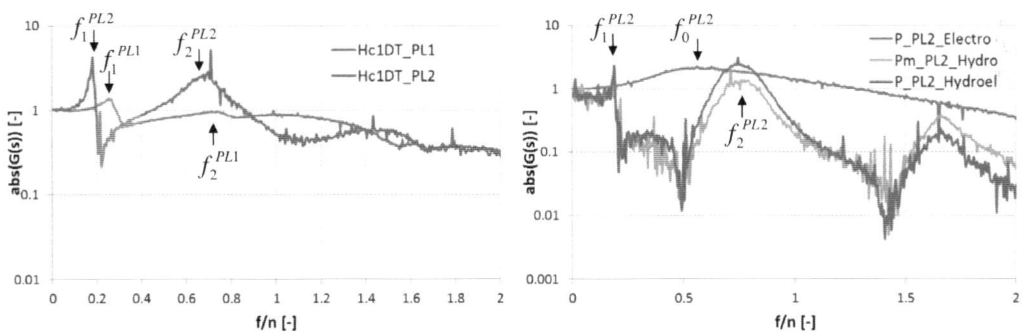

Figure 3. (Left) Hydro-mechanical transfer function of draft tube pressure cone – (Right) Comparison of active power transfer functions between electro-mechanical system and hydroelectric system.

The "local eigenmode" of the synchronous machine representing the rotor oscillations against the power grid is found at 1.2Hz, i.e. $f_0/n = 0.56$ times the runner frequency n. This mode is clearly observed with the transfer function of the electro-mechanical system. With the hydroelectric model, the transfer function of the active power is influenced by the hydraulic system, since hydraulic eigenfrequencies can be observed. Hence, with an electro-mechanical model, the modelling of the vortex rope excitation source by just an external source torque is not representative of the hydraulic system dynamics. Indeed, hydraulic eigenfrequencies, which may interact with the vortex rope precession frequency, are not visible in the frequency spectrum. It has been mentioned above that the hydroelectric model predicts the same pressure fluctuations in the hydraulic system as the hydro-mechanical model. However, for prediction of output power fluctuations, the modelling of the electrical part is necessary and the hydro-mechanical model is not sufficient anymore. Indeed, the hydroelectrical transfer function is the result of the multiplication between the hydro-mechanical and the electro-mechanical transfer functions. Hence, the transfer function of the electro-mechanical model, featuring the synchronous machine local eigenmode, amplifies or reduces the prediction of the mechanical power fluctuations of the hydro-mechanical model. For the PL2 operating point, this local eigenmode amplifies the second hydraulic eigenfrequency and results in prediction of higher amplitude than the electro-mechanical model. The active power transfer functions of the electro-mechanical models with variable speed technologies are compared to the fixed speed case for PL1 in figure 4. Since the control strategy of the variable speed technologies imposes a constant active power at the point of connection to the grid, the transfer functions feature low gain compared to the fixed speed technology. Hence, this low gain cancels the mechanical power fluctuations of the hydro-mechanical model as shown in figure 4: amplification of a hydraulic eigenmode is not possible.

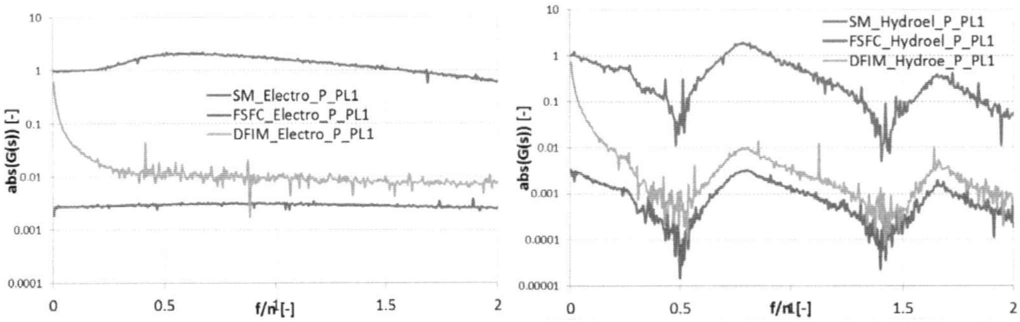

Figure 4. (Left) Comparison of electro-mechanical (Left) and hydroelectric (Right) transfer functions between technologies.

6. Conclusions

Prediction of pressure and output power fluctuations induced by cavitation vortex rope at part load conditions has been compared by means of frequency analysis between the classical fixed speed configuration and the variable speed technology.

For the fixed speed technology, it has been shown that in the low frequency range, hydro-mechanical models are sufficient for prediction of pressure fluctuations in the hydraulic system. This could be different for weaker or isolated power networks. However, for prediction of output power, the hydroelectric model is necessary. Compared to the electro-mechanical model, the detailed hydraulic modelling enables to take into account potential hydraulic resonances and anti-resonances resulting from the interaction of the cavitation vortex rope precession frequency with the hydraulic system that influences potential power fluctuations transmitted to the power network. On the other hand, compared to the hydro-mechanical model, the dynamics of the electrical machine can amplify or reduce the mechanical power fluctuations.

Due to the control of the variable speed technology, rotational speed fluctuations are experienced to ensure constant active power at the grid connecting point which could influence the dynamics behaviour of the cavitation vortex rope. However, the amplitudes of rotational speed fluctuations are very low inducing a weak feedback on the hydraulic response. Hence, the use of a hydroelectric model is not necessary for prediction of pressure fluctuations and they are similar to the hydro-mechanical model with constant speed. This could be different for weaker or isolated power networks. Measurements on prototype are foreseen to validate these results.

7. References

[1] Alligné S., Nicolet C., Tsujimoto Y., Avellan F, Cavitation surge modelling in Francis turbine draft tube *J. of Hydraulic Research.* 2014, **52** (3).

[2] Landry C., Favrel A., Müller A., Nicolet C., Avellan F., Experimental identification of the local wave speed and the second viscosity in cavitating draft tube flow *J. of Hydraulic Research* 2014.

[3] Alligné S., Landry C., Favrel A., Nicolet C., Avellan F., Francis turbine draft tube modelling for prediction of pressure fluctuations on prototype. In Proc. of 9th International Symposium on Cavitation, 2015.

[4] Béguin A., Nicolet C., Assessment of power step performances of variable speed pump-turbine unit by means of hydro-electrical system simulation. Hyperbole Conference, Porto, 2017.

Acknowledgments

The research leading to the results published in this paper is part of the HYPERBOLE research project, granted by the European Commission (ERC/FP7-ENERGY-2013-1-Grant 608532). The authors would like to thank BC Hydro for making available the reduced scale model and the prototype for validation, in particular Dany Burggraeve and Jacob Iosfin. Finally, parameters of the electrical model have been provided by Mr. Johann Hell from ANDRITZ Hydro who authors would like to thank.

IOP Conf. Series: Journal of Physics: Conf. Series **813** (2017) 012017 doi:10.1088/1742-6596/813/1/012017

Analysis of pressure pulsations during the fast transition of a pump turbine from pumping to generating mode

O Braun[1] **and N Ruchonnet**[2]
[1]Andritz AG, Pumps, Graz, Austria
[2]Andritz Hydro AG, turbine physics, Vevey, Switzerland

Abstract. Variable speed motor-generators based on full size converter offer new options for operation of reversible pump-turbines. To explore the challenges related with this new technology a specific test rig configuration is developed to perform such transition with a PT reduced scale model. Transition from pump to turbine and vice versa are performed with various guide vanes opening and transition time. In order to validate the predictive capabilities of state-of-the-art CFD methods to predict the flow field under fast transition through off design operating conditions, pressure pulsations are measured as well in the rotating as in the stationary frame. Since some phenomena are likely to occur only for short periods of time, wavelet transformation methods are used to compare the pressure pulsations occurring during the transitions for different times of transition ranging from 20s to 4s.

1. Introduction

Variable speed motor-generators based on full converter solutions offer new options for operation of reversible pump-turbines (PT) [1], including fast transition from pump to turbine mode, where the pump is operating in pump-brake mode for a while. To explore the challenges related with this new technology a specific test rig configuration is developed to perform such transition with a PT reduced scale model. Transition from pump to turbine and vice versa are performed with various guide vanes opening and transition times ranging from 4s to 20s. Dynamic pressure sensors have been installed to capture transient flow phenomena and various mechanical components have been instrumented to evaluate their loadings. Amongst the available measurements, pressure fluctuation measurements represent the most direct means of validation for Computational Fluid Dynamics (CFD) simulations of such transitions. Furthermore, the hydraulic phenomena occurring during the fast transition through the pump brake quadrant of the 4-Quadrant characteristic are the main source of dynamic loads on the machine components and direct pressure measurements constitute the first step in understanding the characteristics of the loads before its consequences on the structure are further investigated.

2. Experimental setup

The overall setup of the test is obtained by modifications on an universal test rig at ANDRITZ Hydro Hydraulic Laboratory in Linz, which was initially designed for testing hydraulic machinery in steady conditions in all four quadrants. This setup is allowing for a direct transition from pumping to generating mode and vice versa while maintaining a head which is comparable to what would be expected on a prototype implementation as described in [2] The setup and its validation by 1-D simulations is described in more detail in [3].

Challenges in implementation encountered are the time-accurate measurement of flow rate, where finally a correction based on numerous pressure measurements available had to be considered for further data analysis, as well as the implementation of the variable rotation rate in the digital control system which is initially aimed at constant condition operation. Finally the implemented control parameters induced strong torque fluctuations in the shaft line, that are clearly visible in the overall representation of a typical transition manoeuvre in Figure 1

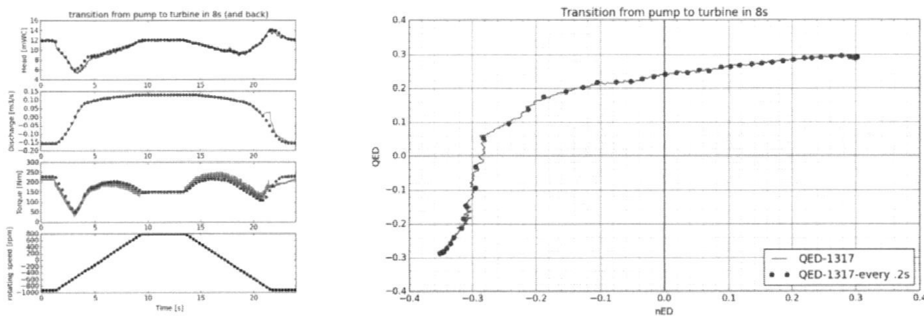

Figure 1. Transition from pumping to generating mode as measured and simulated. Left: Time history of Head, Discharge Torque and rotation rate for a transition forth and back, right: Transition from pumping to generating mode in the dimensionless 4-quadrant representation

The instrumentation considered in this paper consists of:

- Five miniature piezoresistive pressure sensors located on the runner as shown in Figure 2, sensors 16, 19 failed early , sensor 20 also before completing measurements
- Four piezoresistive pressure sensors located between the guide vanes as shown in Figure 3 (left)
- Four quartz crystal pressure sensors in the draft tube cone as shown in Figure 3 (right)

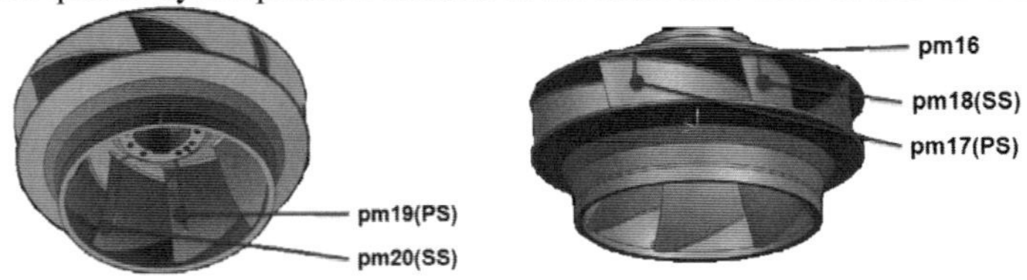

Figure 2. Position of the dynamic pressure sensors on the pump turbine runner

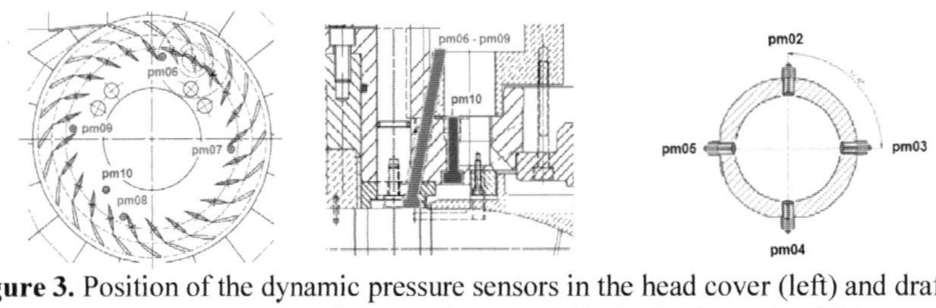

Figure 3. Position of the dynamic pressure sensors in the head cover (left) and draft tube (right), spiral case flange to the upper left, draft tube to the right

The data of all channels in rotating and stationary part are recorded simultaneously and stored with a raw sampling frequency of 5kHz respectively 1kHz for the rotating frame.

3. Results

In the following, the measurements obtained from transitions at constant guide vane opening of 15° and under 12 m Head are presented. A slow transition (20s), which is relatively close to quasi steady, the transitions which have been shown to be corresponding to what could be feasible at prototype scale with respect to the waterways (8s) and a very fast transition (4s) are compared.

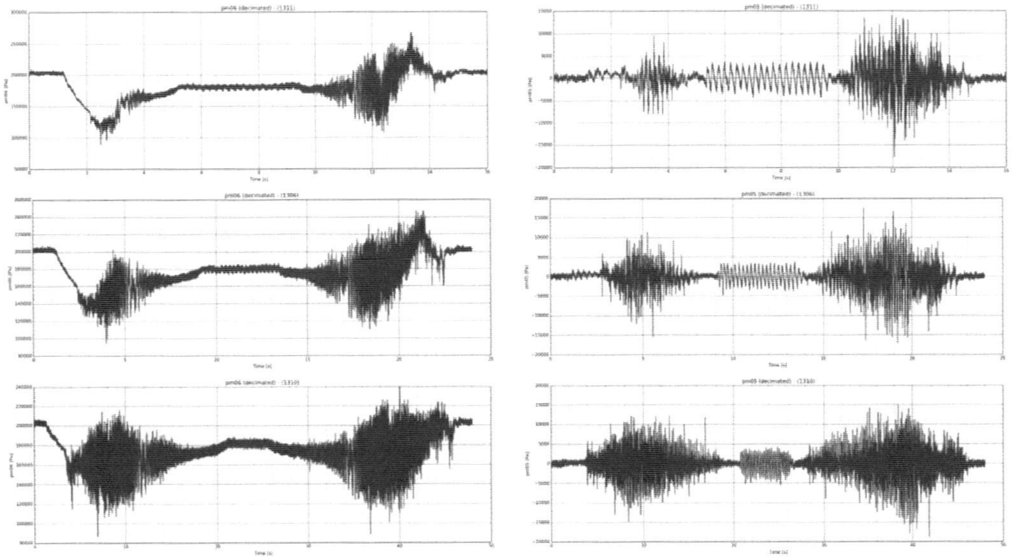

Figure 4. Pressure in stationary components for different transitions from pump to turbine and vice versa. Top: 4s, Middle: 8s, Bottom: 20s, left:Between guide vanes pm06, right: Draft tube, pm 05

Figure 4 displays the time history of pressure in stationary parts and shows that, although qualitatively similar there is a clear difference in pressure fluctuation occurring during a fast transition versus a slower one, and not only by difference in the number of cycles the structure is undergoing. In the fastest transition, it seems that the nonuniformities in the flow do not build up as intensely than in the slower and quasi-steady case.

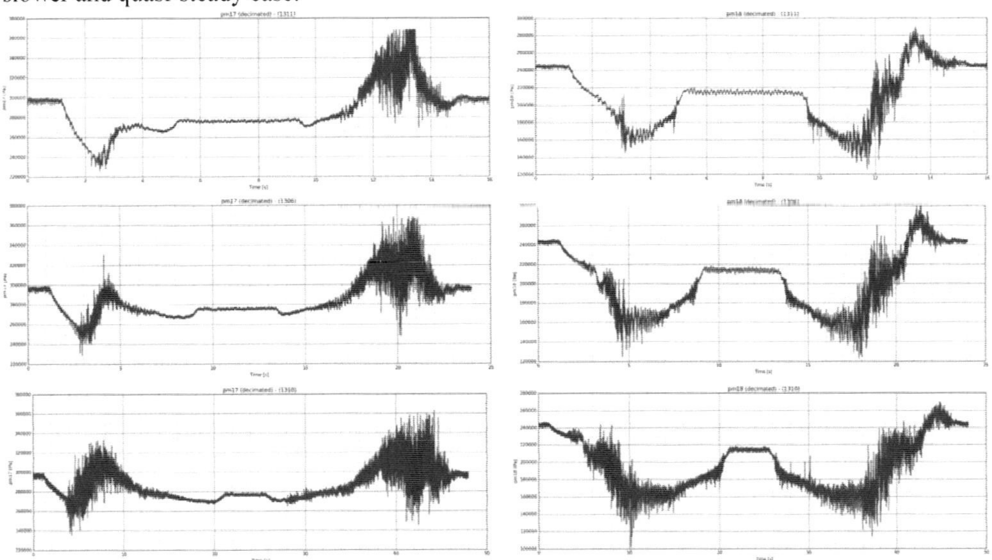

Figure 5. Pressure on runner blades for different transitions from pump to turbine and vice versa. Top: 4s, Middle: 8s, Bottom: 20s, left:pm17 pressure side , right: pm 18 suction side

Figure 5 displays that pressure fluctuations on the runner leading edge display a similar behaviour showing that remaining in the off design conditions by an extremely slow transition induces more cycles and even of higher amplitude than a faster transition.

4. Wavelet Analysis of pressure fluctuations

For time signals with quickly varying frequency content, the proven tools for frequency analysis such as simple Fourier transforms and even short time Fourier transforms lack flexibility in tuning of frequency versus time resolution. Wavelet transforms give more flexibility, but are still lacking a common best practice in the use of various mother wavelet functions and interpretation. The following figures have been obtained using a Morlet 5 continuous wavelet transform with the aforementioned time series data. Figure 6 is showing that the pressure recorded between the guide vanes (pm06) is dominated by the interaction with the runner when the unit operates in pump brake conditions, the linear slopes of the runner revolution rate as imposed is outstanding all other phenomena. The impression from timeseries that the interaction is more intense in slower transition scenarios than in faster ones is confirmed by the obtained convolution.The draft tube pressure fluctuations represented on the right show only slight differences, as could be interpreted in the time series already.

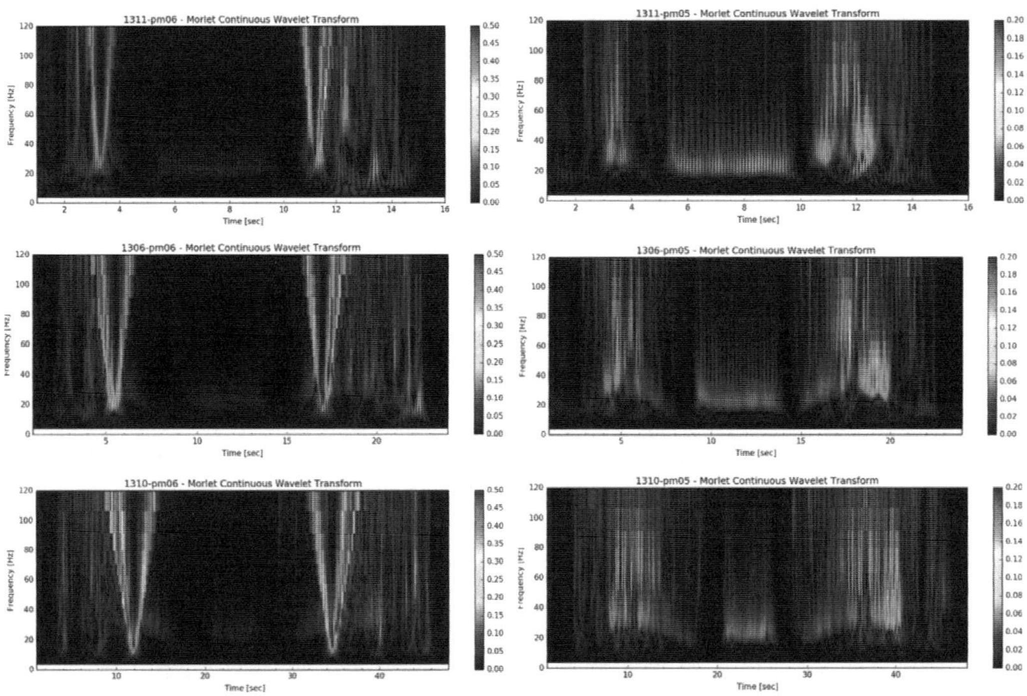

Figure 6. Morlet Continuous wavelet transform of pressure during mode transitions
Top: 4s, Middle: 8s, Bottom: 20s, left:Between guide vanes pm06, right: Draft tube, pm 05

Figure 7 displays the pressure fluctuations on the runner periphery, consistently with the results obtained between the guide vanes, excitations with frequencies evolving with the runner rotation rate are seen. In the acceleration towards pump mode (last part of the sequence) high contribution from higher frequencies are seen. The used Morlet 5 wavelet (and wavelet methods in general) resembles a strongly windowed sine of only 5 periods so its frequency resolution is rather poor. Using a variable type of wavelet for different frequencies could improve this situation, and reaching a good resolution of low frequency phenomena presenting only very few periods and higher frequency excitations often showing a more consistent periodicity over a significant duration.

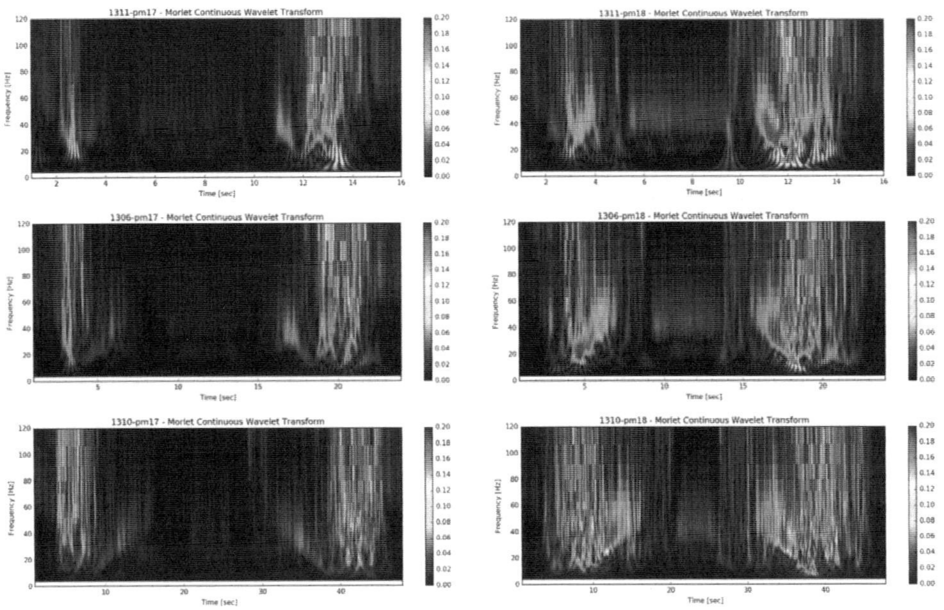

Figure 7. Morlet Continuous wavelet transform of pressure during mode transitions. Top: 4s, Middle: 8s, Bottom: 20s, left:pm17 pressure side , right: pm 18 suction side

5. Conclusion and outlook

The measurements of pressure fluctuations during fast transition and their simple display as timeseries as well as their wavelet transformed representation show that the intensity of the pressure fluctuation in the hydraulic unit does not increase with the speed of the manoeuver and that there is even a trend to lower pressure fluctuations the faster the manoeuver is executed. The presented measurements were made at constant guide vane opening for sake of simplicity of the first of its kind measurement in reduced model scale. Including a variation of a guide vane control sequence into such kind of transition adds a consequent number of additional degrees of freedom but from the difference in pressure fluctuations that are seen between fast and slow transitions, it seems conceivable that optimum sequences can be found that are allowing for fast transition while reducing the dynamic load on the components and providing fast response time and also inertia response to the grid. Numerical simulations of the transition manoeuvres represent a challenge in terms of computational resources and analysis, first implementations, results and interpretations were presented shortly after the measurements [4]. Given the non-stationary character of the results, further detailed analysis will be needed to fully assess the capabilities and shortcomings of numerical methods to predict flow under quickly varying off-design conditions.

6. References

[1] Hell J., Egretzberger M., Lechner A., Schürhuber R., Vaillant Y. Full size converter solutions for pumped storage plants – a promising new technology. *Hydro2012, Bilbao*. 2012.

[2] Nicolet, C., Braun, O., Ruchonnet, N, Béguin, A., Hell, J., Avellan, F., "Full Size Frequency Full Size Frequency Converter for Fast Francis "Full Size Frequency Full Size Frequency Converter for Fast Francis Pump-Turbine Operating Mode Transition", HYDROVISION International, Minnesota, 2016

[3] Ruchonnet, N., Braun, O., "Reduced scale model test of pump-turbine transition". 6th IAHR International Meeting of the Workgroup on Cavitation and Dynamic Problems in Hydraulic Machinery and Systems, Ljubljana. 2015.

[4] Stens, C., Riedelbauch, S. "Investigation of a fast transition from pump mode to generating mode in a model scale reversible pump turbine, 28th AIRH Symposium on Hydraulic Machinery and Systems, Grenoble. 2016

Acknowledgments
The research leading to the results published in this paper is part of the HYPERBOLE research project, granted by the European Commission (ERC/FP7-ENERGY2013-1-Grant 608532). The author would like to thank the colleagues in Linz for performing the test rig measurement.

IOP Conf. Series: Journal of Physics: Conf. Series **813** (2017) 012018 doi:10.1088/1742-6596/813/1/012018

Experimental Hydro-Mechanical Characterization of Full Load Pressure Surge in Francis Turbines

A Müller[1,2], A Favrel[1], C Landry[1,3], K Yamamoto[1] and F Avellan[1]

[1] EPFL Laboratory for Hydraulic Machines, 1007 Lausanne, Switzerland
[2] Nippon Koei Co., Ltd., Hydro Plant Planning Office, 962-8508 Sukagawa, Japan
[3] Power Vision Engineering LLC, 1024 Ecublens, Switzerland

E-mail: andres.mueller@alumni.epfl.ch

Abstract. Full load pressure surge limits the operating range of hydro-electric generating units by causing significant power output swings and by compromising the safety of the plant. It appears during the off-design operation of hydraulic machines, which is increasingly required to regulate the broad integration of volatile renewable energy sources into the existing power network. The underlying causes and governing physical mechanisms of this instability were investigated in the frame of a large European research project and this paper documents the main findings from two experimental campaigns on a reduced scale model of a Francis turbine. The multi-phase flow in the draft tube is characterized by Particle Image Velocimetry, Laser Doppler Velocimetry and high-speed visualizations, along with synchronized measurements of the relevant hydro-mechanical quantities. The final result is a comprehensive overview of how the unsteady draft tube flow and the mechanical torque on the runner shaft behave during one mean period of the pressure oscillation, thus defining the unstable fluid-structure interaction responsible for the power swings. A discussion of the root cause is initiated, based on the state of the art. Finally, the latest results will enable a validation of recent RANS flow simulations used for determining the key parameters of hydro-acoustic stability models.

1. Introduction

When operating a hydraulic machine outside its best efficiency point (BEP), the flow leaving the runner carries a significant residual swirl. The resulting inhomogeneous pressure distribution leads to the formation of cavitation. At full load ($Q > Q_{\mathrm{BEP}}$), the latter takes the shape of an axisymmetric cavitation vortex rope attached to the runner hub. As the cavity grows with a decreasing pressure level in the draft tube, the hydro-acoustic Eigenfrequencies may sink to a critical level and the system enters self-oscillation. Violent pressure fluctuations are then observed throughout the system, which translate into dangerous power swings on the generator side. Their peak-to-peak amplitude typically reaches several percentage points with respect to the nominal output. This is naturally unacceptable in large units, that produce several hundred Megawatts and are possibly connected to long electric transmission lines.

The phenomenon was reported as early as the 1940s [1]. The first experimental descriptions from reduced scale model testing are from the late 1980s [2, 3]. Analytical [4, 5, 6, 7], 1-D hydro-acoustic [8, 9] as well as numerical [10, 11] analyses of the swirling flow or/and the associated unstable behavior followed over the years (this list being by no means exhaustive). How the physics are represented in these models is crucial, if one day the behavior of the machine is to

IOP Conf. Series: Journal of Physics: Conf. Series **813** (2017) 012018 doi:10.1088/1742-6596/813/1/012018

be accurately predicted at any given scale. Therefore, a thorough investigation of the sustaining mechanisms and the underlying causes of the self-oscillation is necessary. This was performed in one of the 6 scientific work packages of the HYPERBOLE research project, funded by the European Commission in its 7th Framework Program. The general objective of the project is to move towards a smooth integration of a large portion of renewable energies in the existing power network by taking advantage of the regulatory capacities of hydropower and pump-storage plants. The project coming to an end, this article provides a condensed overview of the research (see Section 2), the main results (Section 3 and Section 4), the problems to be solved and the corresponding perspectives (Section 5) in the field of full load surge.

2. Methodology

2.1. Experimental setup and measurements

Firstly, the draft tube flow was extensively characterized for stable and unstable operating points. The term *unstable* refers to the occurrence of full load pressure surge. The notion of stability is further discussed in Section 4. The instrumented model installed on the EPFL test rig PF3 for the Laser Doppler Velocimetry (LDV) and the Particle Image Velocimetry (PIV) measurements is shown in Figure 1. More details are found in [12, 13] for the PIV and [14, 15] for the LDV configuration, respectively. For the draft tube flow visualizations in the top row of Figure 2, the setup was similar to the one in Figure 1(b), with a different camera and an additional uniform LED backlight source on the other side of the Plexiglas cone in order to enhance the contrast between the liquid and the gaseous phase. For the visualization of the inter-blade flow from below, shown in the bottom row of Figure 2, the high-speed camera was tilted and synchronized with a stroboscopic light. A different water-box was used to minimize the optical distortion, for which the flat surface faced the inter-blade channels. The corresponding setup is not shown here but described in [16].

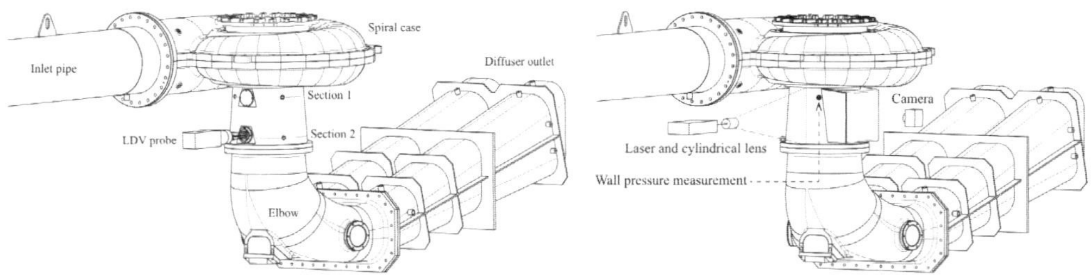

(a) LDV setup (b) PIV setup

Figure 1: Draft tube flow survey on a 1:16 reduced scale model of a Francis turbine.

The operating conditions are defined by the non-dimensional speed factor $n_{\mathrm{ED}} = (n{\cdot}D)/\sqrt{E} = 0.288$ and the discharge factor $Q_{\mathrm{ED}} = Q/(D^2\sqrt{E}) = 0.26$, where Q is the discharge, n the rotational frequency, D the runner outlet diameter and E the specific hydraulic energy. The non-dimensional cavitation number, expressing the pressure in the draft tube as a function of the turbine's level setting, corresponds to the one on the real size machine ($\sigma_{\mathrm{plant}} = 0.11$) for the unstable operation. It can be adjusted with a vacuum pump over the free surface of the downstream tank. For the stable configuration, the pressure in said tank was set to atmospheric. The measurements were synchronized by sharing a common trigger.

2.2. Data processing

All quantities were averaged with respect to the wall pressure signal in the draft tube cone. The procedure is described in detail in [17] and consists of isolating the different cycles of a signal according to the locations of the characteristic pressure peaks. The cycles are then superposed and the data averaged in 10^2 sub-windows over one period. This offers a comprehensive impression of how the system behaves during one typical period.

The LDV measurements taken during the self-oscillation were averaged in the same manner. Thanks to the synchronized start, each velocity measurement has a unique time stamp within the data acquisition. Based on its relative position with respect to the period of the pressure cycle it was recorded in, the measurement can be attributed to one of the defined sub-windows and then the mean value of all the other measurements that fall into that same sub-window is calculated. The result is a pressure-phase averaged evolution of the tangential and the axial velocity components Cu and Cm at a given radial position. The latter can then be integrated over all the radial measurement positions on each of the two streamwise locations shown in Figure 1(a) to yield the axial fluxes of angular and axial momentums, the ratio of which shall be defined as the swirl number for further considerations (for details please referred to [15]).

3. Results

As a starting point, Figure 2 shows the characteristic breathing motion of the cavitation vortex rope as well as the formation and collapse of cavitation on the runner blades at several equally spaced instants over one pressure oscillation period. It is noted that the cavitation on the runner blades develops seriously between the instants (c) - (g), corresponding to the collapsing phase of the vortex rope. The blade cavitation also starts shedding into the draft tube flow in the form of small bubbles, which darken the images in the top row of Figure 2. Once the blades have been cleared at instant (h), the vortex rope reshapes again from above.

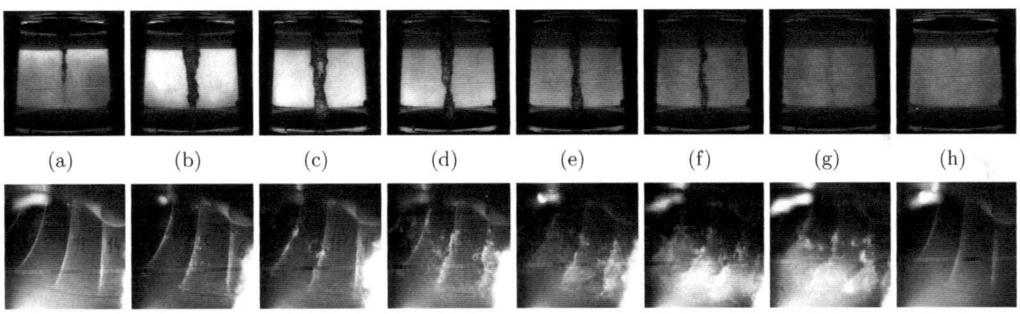

Figure 2: Dynamics of the cavitation vortex rope in the draft tube cone (top) and the blade cavitation (bottom) at 8 equally spaced instants during 1 cycle of the instability.

The instants (a) - (h) in Figure 2 may be put into relation with the hydro-mechanical data based on their relative position within the period of the pressure cycle they were recorded in. These are represented by vertical dashed lines in Figure 3(a) and Figure 3(b), which contain the pressure-phase averaged evolutions of the swirl number S_n in the draft tube at the 2 streamwise locations shown in Figure 1(a) as well as the mechanical torque factor $T_{\mathrm{ED}} = T_{\mathrm{m}}/\left(\rho D^3 E\right)$, together with the wall pressure factor $c_{\mathrm{p}} = (p - \bar{p})\,/\,(\rho E)$. Figure 3 offers a compact view on how the hydraulic and mechanical parts of the machine evolve during a mean period $\bar{T}$ of the surge. This clears the important task of establishing the fluid-structure interaction mechanisms, but does not answer the question of the root cause for the instability.

For the purpose of finding out what ultimately causes full load pressure surge, the transition is studied numerically with a RANS simulation [18]. Figure 4 shows a stable draft tube flow at an atmospheric pressure in the downstream tank, which was characterized by PIV in order to validate the numerical model. It is noted that a slender, axisymmetric cavitation vortex rope

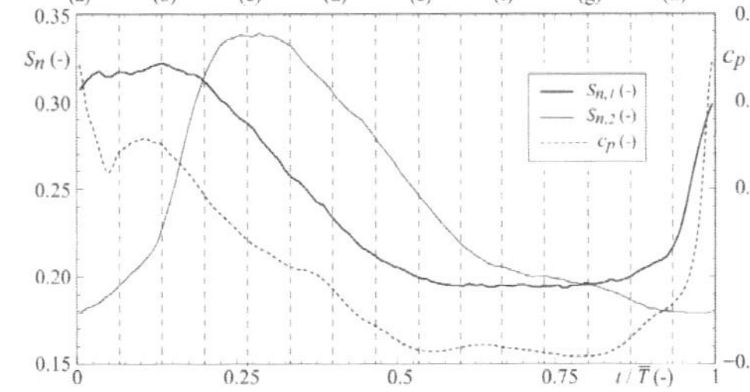

(a) Swirl (S_n) and wall pressure (c_p)

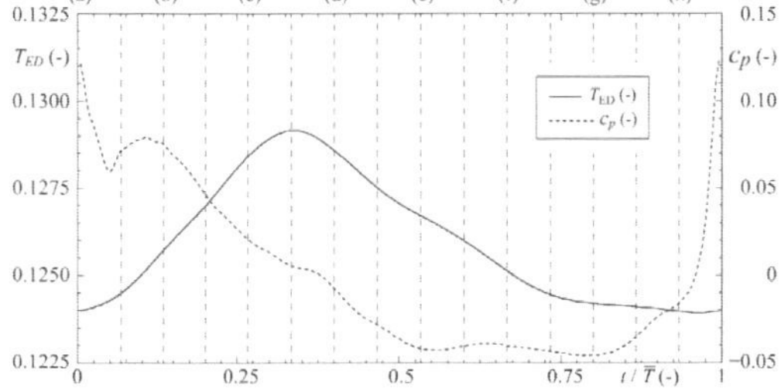

(b) Torque (T_ED) and wall pressure (c_p)

Figure 3: Pressure-phase averaged hydro-mechanical system behavior during full load surge.

is already present at this point. When the pressure in the draft tube is decreased to reach σ_plant, the cavity will grow in its diameter, start fluctuating and eventually enter self-oscillation as shown in Figure 2. The color chart in Figure 4 to the right displays the magnitudes of the mean velocity field resulting from 600 measurements in a vertical (meridional) plane of the draft tube. Even if the light sheet generated by the Laser partly crosses the vortex rope and vectors are calculated on the other side, only the region between the left cone wall and the vortex rope edge is deemed trustworthy. Measurements were also taken in two horizontal cross-sections of the draft tube, at the same streamwise locations as the LDV measurements. The processing of the PIV data is however an ongoing task at the time of writing of this paper and will be the object of future publications.

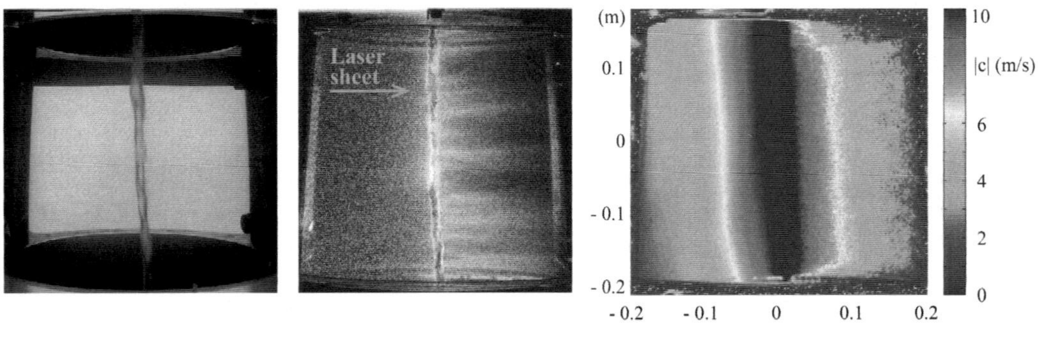

Figure 4: Draft tube flow visualization (left), single image from the PIV camera (middle) and amplitudes of the mean velocity field (right) at σ = atm. for the validation of RANS models.

4. Discussion

As previously indicated, the breathing motion of the vortex rope is governed by the swirl variation, which is in turn caused by the modification of the relative flow angle at the runner exit due to the formation of significant volumes of cavitation on the blades. The blade cavitation also changes the hydrodynamic properties of the runner and thus the transfer of momentum to the shaft, represented by the mechanical torque. The torque peak between instants (c) and (d) appears shortly after the vortex rope reaches its maximum volume. T_ED then decreases while the cavitation on the runner blades becomes dominant and starts growing again with the vortex rope. In other words, the mechanical torque appears to be well synchronized with the vortex rope volume, considering a slight delay due to the mechanical inertia of the system.

The cause for the instability has been debated by several authors in the past and a destabilizing role of the pressure recovery is brought up regularly [19, 20, 9]. Indeed, since the pressure in the downstream tank is constant, any discharge increase from the runner would have to be compensated by a drop in the draft tube pressure and thus by a cavity growth. And in the vicinity of one of the hydro-acoustic Eigenfrequencies, the slightest perturbation might cause this positive feedback to trigger the kind of self-excited hydro-mechanical oscillation shown in Figure 2 and Figure 3. The further experimental and numerical examination of the transition shall hopefully bring more clarity to this discussion.

5. Conclusion and perspectives

The unstable fluid-structure interaction mechanisms causing dangerous power swings in Francis turbines at full load were identified experimentally in the course of the HYPERBOLE research project. Based on the obtained data, RANS models are being validated for the closer numerical study of the pressure surge onset, its root cause as well as for the identification of the key parameters of 1-D hydro-acoustic flow models used for stability analysis.

Acknowledgments

The HYPERBOLE research project is funded by the European Commission [ERC/FP7-ENERGY-2013-1-Grant 608532]. The authors would also like to thank BC Hydro for making available the reduced scale model, in particular Danny Burggraeve and Jacob Iosfin. Finally, the commitment of the laboratories technical staff is kindly acknowledged, especially Georges Crittin, Maxime Raton, Victor Rivas, Alain Renaud, Philippe Faucherre and Vincent Berruex.

References

[1] Rheingans W J 1940 *Transactions of the ASME* **62** 171–184
[2] Jacob T, Prenat J and Maria D 1988 *Houille Blanche* **3-4** 293–300
[3] Jacob T 1996 Francis turbine surge: discuss. and data base *Proc. of the 18th IAHR Symp., Valencia, Spain*
[4] Dörfler P 1985 Francis turbine surge prediction and prevention *Waterpower 85* (ASCE) pp 952–961
[5] Susan-Resiga R, Muntean S, Avellan F and Anton I 2011 *Applied Mathematical Modelling* **35** 4759–4773
[6] Rudolf P and Štefan D 2012 Decomposition of the swirling flow field downstream of Francis turbine runner *IOP Conference Series: Earth and Environmental Science* vol 15 p 062008
[7] Kuibin P, Pylev I and Zakharov A 2012 Two-phase models development for description of vortex-induced pulsation in Francis turbine *IOP Conference Series: Earth and Environmental Science* vol 15 p 022001
[8] Koutnik J and Pulpitel L 1996 Modeling of the Francis turbine full-load surge *Proc. of Modeling, Testing and Monitoring for Hydro Power Plants, Lausanne, Switzerland*
[9] Alligné S, Nicolet C, Tsujimoto Y and Avellan F 2014 *Journal of Hydraulic Research* **52** 399–411
[10] Alligné S, Nicolet C, Ruchonnet N, Hasmatuchi V, Maruzewski P and Avellan F 2009 Numerical simulation of non-linear self oscillations of a full load vortex rope *Proc. of the 3rd Meet. of the IAHR Workgroup on Cavitation and Dynamic Problems in Hydraulic Mach. and Syst., Brno, Czech Republic* vol 2 pp 325–338
[11] Chirkov D, Avdyushenko A, Panov L, Bannikov D, Cherny S, Skorospelov V and Pylev I 2012 CFD simulation of pressure and discharge surge in Francis turbine at off-design conditions *IOP Conference Series: Earth and Environmental Science* vol 15 p 032038
[12] Müller A, Dreyer M, Andreini N and Avellan F 2013 *Experiments in Fluids* **54** 1–11
[13] Favrel A, Müller A, Landry C, Yamamoto K and Avellan F 2015 *Experiments in Fluids* **56** 215
[14] Favrel A, Müller A, Landry C, Yamamoto K and Avellan F 2016 *Experiments in Fluids* **57**
[15] Müller A, Favrel A, Landry C and Avellan F 2017 *Journal of Fluids and Structures* **69** 56–71
[16] Müller A, Favrel A, Landry C, Yamamoto K and Avellan F 2014 *IOP Conference Series: Earth and Environmental Science* **22** 032034
[17] Müller A, Yamamoto K, Alligné S, Yonezawa K, Tsujimoto Y and Avellan F 2016 *ASME J. Fluids Eng.* **138**
[18] Alligné S, Decaix J, Müller A, Nicolet C, Avellan F and Münch C 2016 RANS computations for identification of 1-D cavitation model parameters: application to full load cavitation vortex rope *Proc. of the 28th IAHR Symposium on Hydraulic Machinery and Systems, Grenoble, France*
[19] Chen C, Nicolet C, Yonezawa K, Farhat M, Avellan F and Tsujimoto Y 2008 *ASME J. Fluids Eng.* **130**
[20] Dörfler P, Keller M and Braun O 2010 Francis full-load surge mechanism identified by unsteady 2-phase CFD *IOP Conference Series: Earth and Environmental Science* vol 12 p 012026

IOP Conf. Series: Journal of Physics: Conf. Series **813** (2017) 012019 doi:10.1088/1742-6596/813/1/012019

Numerical simulation of a cavitating draft tube vortex rope in a Francis turbine at part load conditions for different σ-levels

J Wack and S Riedelbauch

Institute of Fluid Mechanics and Hydraulic Machinery, University of Stuttgart, Pfaffenwaldring 10, 70555 Stuttgart, Germany

E-mail: jonas.wack@ihs.uni-stuttgart.de

Abstract. In the present study, numerical investigations of a Francis turbine at model scale are performed. A part load operating point is selected for the analysis with a focus on the cavitating draft tube vortex rope. For three σ-levels, different aspects are investigated. First, the extension of the vapor volume is analyzed and compared to high-speed visualization recordings from the measurements. Furthermore, the pressure fluctuations for different σ-levels are analyzed and compared to experimental data at a measuring point in the draft tube cone. Finally, the impact of the different cavitation volumes on the axial and circumferential velocity distribution is investigated. In addition, the SST and SAS turbulence models are compared for two pressure levels.

1. Introduction

The safe integration of renewable energies like photovoltaics is a key issue of nowadays energy market. Hydro power plants play an important role for this topic as they can be easily regulated. Nevertheless, this comes across with the fact that turbines are more and more operated at off-design conditions which are characterized by strongly swirling flows and the occurrence of cavitation.

In a Francis turbine, different cavitation phenomena can occur depending on the operating point [1]. At part load conditions, a helically shaped vortex rope forms in the draft tube cone. For low σ-levels, the pressure in the vortex core falls below the vapor pressure which results in cavitation. In experiments the appearance of cavitation allows to visualize the shape of the vortex rope (e.g. [2, 3]), which can be used for comparison with simulation results. Typically the frequency of the vortex rope rotation is between 0.2 and 0.4 times the runner frequency [4]. Favrel et al. [5] observed that the precession frequency is a function of the discharge factor while the σ-level has no significant impact. The occurring pressure oscillations depend on the discharge, the σ-level and the Froude number [5, 6]. Furthermore, the presence of a cavitation volume significantly reduces the natural frequency of the system [7]. When the natural frequency and the precession frequency coincide it comes to resonance conditions with high pressure fluctuations.

2. Modelling and numerical setup

The simulation of cavitating flows is more complex compared to single phase flows. In the scope of this study, the homogenous model is applied for the two-phase treatment which assumes that the liquid and vapor phase share a common pressure and velocity field [8]. As cavitation model, the Zwart model is used and the four model parameters are set according to Zwart et al. [9]. Furthermore, for URANS simulations the widely used SST turbulence model is applied to close the system of governing equations. Additionally, two simulations are performed with the SAS-SST model which can operate in scale resolving simulation mode [10].

The simulations are performed with Ansys CFX on a Francis turbine at model scale. The simulation domain is ranging from the spiral case inlet to the draft tube outlet. For all investigations a structured mesh with approximately 25 million cells is used. Most of the cells are located in the draft tube as the main flow phenomenon is located there.

For the treatment of the interfaces between stationary and rotating domains, a transient rotor stator approach is applied. As inlet boundary condition, a constant total pressure is set and at the outlet an opening boundary condition is prescribed. The total pressure at the inlet and static pressure at the outlet are set that both the required discharge and pressure level in the turbine are met. For the spatial discretization, a high resolution scheme is used for all simulations performed with the SST turbulence model and a bounded central differencing scheme is applied for the simulation with the SAS model. As temporal discretization, a second order backward Euler scheme is used for all setups.

3. Numerical results

In the scope of this study, an operating point at part load conditions is investigated for three different σ-numbers (0.06, 0.09 and 0.11). The discharge of the examined operating point is 81 % compared to best efficiency point conditions.

3.1. Cavitating vortex rope

The regions where cavitation occurs can be visualized with an isosurface of the void fraction. In figure 1 the simulation results for different cavitation numbers and turbulence models are displayed and can be compared to high-speed visualization recordings from the measurements. Additionally, in the draft tube cone two lines are shown that are used for the evaluation of the velocity profiles.

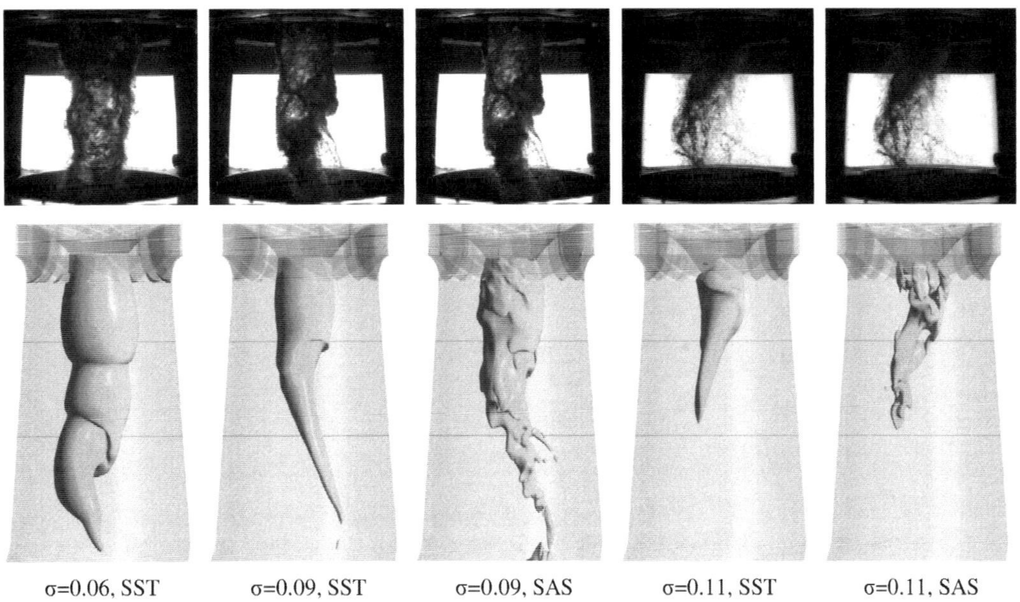

Figure 1. Cavitating vortex rope visualized with an isosurface of the void fraction (value 0.25) top: measurement – bottom: simulation results

As expected, the cavitation volume increases with reduction of the σ-number. For the lowest pressure level, a large vapor volume forms in the draft tube cone which is almost axisymmetric in the upper part for both simulation and measurement. The typical helical shape for part load is only observable in the lower region of the cone where the diameter of the vortex rope is already significantly smaller. In addition, cavitation occurs on the suction side of the runner blades close to the trailing edge due to the very low pressure level.

For σ=0.09, a comparison of the results obtained with the different turbulence models shows that the SAS model resolves smaller turbulence structures that are noticeable in the shape of the cavitation volume. Especially in the lower region of the draft tube cone where the vapor phase decays, differences are clearly observable to the SST results. Nevertheless, in terms of vapor volume size only minor differences are present between both turbulence models. This is in contrast to observations made for a deep part load operating point where the SAS model predicted a significantly bigger cavitation volume of the inter-blade vortices in the runner compared to the SST model [11]. This can be explained by the fact that the occurring vortex rope at part load conditions has a much bigger diameter compared to the inter-blade vortices. Consequently, a different resolution of the pressure minimum in the vortex core caused by the turbulence model plays a less significant role at part load condition.

For the highest pressure level similar observations like for σ=0.09 can be made. In the measurement it can be noticed that the cavitating vortex rope decays into small structures and vapor bubbles. This can be better reproduced by the SAS model. Nevertheless, the chosen two-phase modeling approach is not suitable to resolve single vapor bubbles.

3.2. *Pressure fluctuations*

The pressure fluctuations for σ=0.09 and σ=0.11 are evaluated at a measuring point in the draft tube cone. The results are displayed in figure 2. Both turbulence models can reproduce changes in amplitude that are present especially for σ=0.09. While for the lower σ-level the amplitude of the measurement is well met except for the last cycle, for σ=0.11 the simulation results slightly underestimate the occurring pressure variations. In terms of frequency, both models are in good agreement with the measurements over a wide range. However, the SAS model tends to overestimate the frequency for cycles with lower amplitude.

A comparison of the cases with different σ-numbers shows that the pressure level has a significant impact on the amplitude of the pressure fluctuations that has also been observed by Favrel et al. [6]. Due to the fact that the investigated σ-levels are below the resonance conditions, the pressure fluctuations decrease with the σ-number. Between σ=0.09 and σ=0.11 the amplitude changes by a factor of approximately two. This is on the one hand caused by getting away from resonance conditions. On the other hand the increased cavitation volume leads to a damping of the pressure fluctuations. The results suggest that in the presence of a cavitating vortex rope, two-phase simulations are necessary to correctly predict the occurring pressure amplitudes.

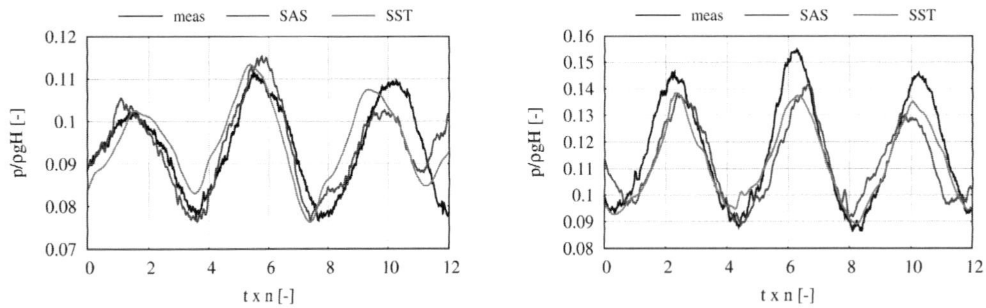

Figure 2. Pressure at measuring point in the draft tube cone for σ=0.09 (left) and σ=0.11 (right)

3.3. Axial and circumferential velocity distribution in the draft tube cone

The influence of the σ-level on the axial and circumferential velocity distribution in the draft tube cone is investigated along two lines that are displayed in figure 1. Line 1 and 2 are located 0.39 and 1.02 times the outer runner outlet diameter below the runner outlet. Measurement data is available for two σ-numbers close to the investigated pressure levels. The results are presented in figure 3.

At line 1, the results of the averaged axial velocity c_{ax} for σ=0.11 show a good agreement with the measurements over a wide range. Deviations can be found close to the draft tube wall which might be a result of neglecting the flow through the runner side gap on the crown side and in the stagnant region in the center. There, the simulation is forming a plateau while for the measurement at σ=0.12 the backflow increases to the center of the draft tube. This can be explained by the occurrence of a region with permanent cavitation in the middle of the measuring line for the slightly lower pressure level in the simulation. At σ=0.10 the measurement shows a similar course compared to the simulation. Close to the center no data is available due to measuring problems caused by the presence of vapor.

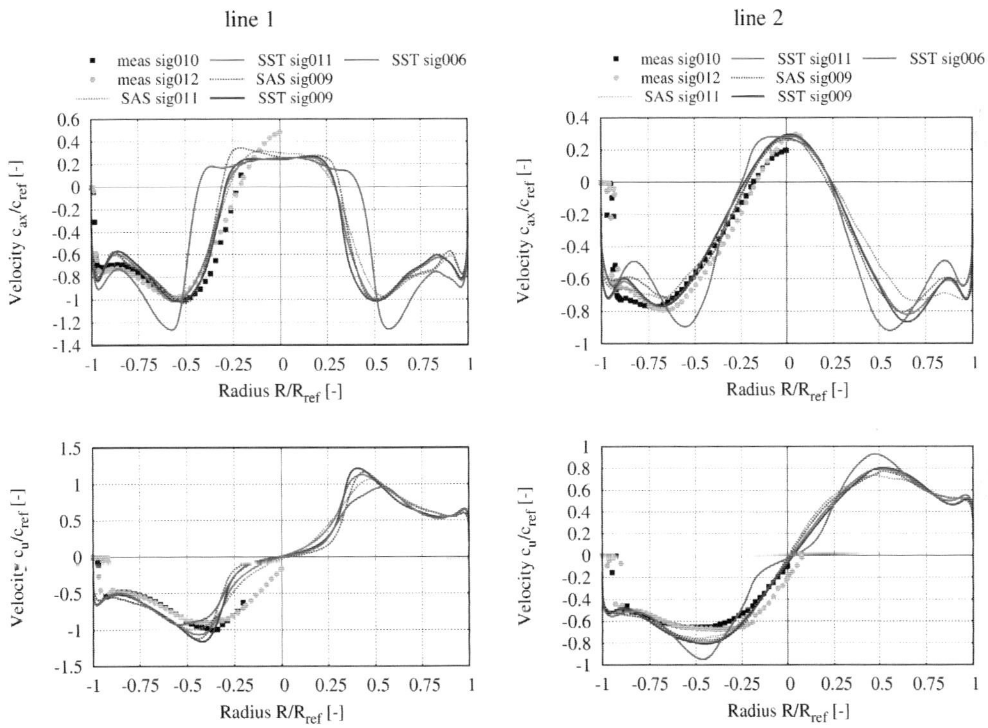

Figure 3. Axial (top) and circumferential (bottom) velocity distribution along two lines in the draft tube cone averaged over ten vortex rope revolutions

With a reduction of the pressure level, the extension of the stagnant region increases as the cavitation volume grows. For σ=0.06, the region of backflow is huge and occupies almost the inner half of the draft tube cone. This results in a significantly higher axial velocity in the outer part of the draft tube cone which can be explained by the conservation of mass. A comparison of the different turbulence models shows that the biggest differences between the SAS and SST model are observed in the outer region close to the draft tube wall.

Further downstream (at line 2), for σ=0.09 and σ=0.11 a good agreement of the axial velocity with the measurements can be stated. Due to the helical shape and smaller diameter of the vortex rope in that area, there is no region with permanent cavitation which results in the different course of the velocity distribution in the stagnant region. The biggest deviation to the measurement again can be found close to the draft tube wall. Contrary to the cases at higher pressure levels, the axial velocity distribution for σ=0.06 looks different. The reason is that a region with permanent cavitation persists which has an impact on the velocity field.

In the circumferential velocity distribution, the impact of cavitation can also be determined. All simulation results at line 1 show a maximum circumferential velocity at a dimensionless radius of around 0.5. Going to the center of the draft tube, there is first a steep velocity gradient which significantly reduces around a dimensionless radius of approximately 0.25. Contrary to this, the measurement with σ=0.12 has a higher velocity gradient in the center and no change in slope. At line 2 the results for σ=0.09 and σ=0.11 show the same characteristics like the measurements while the low pressure level still shows the change in slope in the center due to the presence of permanent cavitation.

4. Conclusion and outlook

Cavitation simulations were carried out on a Francis turbine at model scale for a part load operating point at different σ-levels. As expected, the vapor volume increases with decreasing pressure level. The axial and circumferential velocity distribution is affected by the σ-level. With increasing cavitation volume the stagnant region is getting bigger. Furthermore, the occurrence of regions with permanent cavitation significantly influences the shape of the velocity profiles. Moreover, the σ-level affects the amplitude of the pressure fluctuations which necessitates the usage of two-phase simulations. Below resonance conditions the amplitude decreases with reduction of the σ-number. The results of the SST and SAS model show a similar size of the cavitation volume. Nevertheless, the SAS model can resolve smaller vortex structures which results in differences in the shape of the vortex rope especially in the region of decaying vapor volume.

For further improvement of the simulation results the use of more complex two-phase models should be investigated in the future.

Acknowledgments

The research leading to the results published in this paper is part of the HYPERBOLE research project, granted by the European Commission (ERC/FP7-ENERGY-2013-1-Grant 608532). Special thanks to EPFL, in particular Keita Yamamoto, Andres Müller, Arthur Favrel, Christian Landry and François Avellan, for providing all experimental data enabling comparisons presented here. Furthermore, the authors would like to thank the HLRS Stuttgart within the Bundesprojekt framework.

References

[1] Avellan F 2004 *6th Int. Conf. on Hydraulic Machinery and Hydrodynamics* Timisoara Romania
[2] Favrel A, Müller A, Landry C, Yamamoto K and Avellan F 2015 *Exp. in Fluids* **56**(215)
[3] Houde S, Iliescu M S, Fraser R, Lemay S, Ciocan G D and Deschênes C 2011 *Proc. ASME-JSME-KSME 2011 Joint Fluids Engineering Conf.* pp 171-182
[4] Nicolet C, Zobeiri A, Maruzewski P and Avellan F 2010 *IOP Conf. Series: Earth and Env. Sci.* **12** 012053
[5] Favrel A, Müller A, Landry C, Yamamoto K and Avellan F 2016 *Proc. 28th IAHR Symp. on Hydraulic Machinery and Systems* Grenoble France
[6] Favrel A, Landry C, Müller A, Nicolet C, Yamamoto K and Avellan F 2014 *IOP Conf. Series: Earth and Env. Sci.* **22** 032035
[7] Landry C, Favrel A, Müller A, Yamamoto K and Avellan F 2014 *IOP Conf. Series: Earth and Env. Sci.* **22** 032037
[8] Bouziad Y A 2006 *Ph.D. Thesis EPFL Lausanne*
[9] Zwart P J, Gerber A G, Belamri T 2004 *Int. Conf. on Multiphase Flow* Yokohama Japan
[10] Menter F R, Schütze J and Gritskevich M 2012 *Progress in Hybrid RANS-LES Modelling* ed S Fu, W Haase, S H Peng and D Schwamborn (Berlin Heidelberg: Springer) pp 15–28
[11] Wack J, Riedelbauch S, Yamamoto K and Avellan F 2016 *Proc. 9th Int. Conf. on Multiphase Flow* Firenze Italy

IOP Conf. Series: Journal of Physics: Conf. Series **813** (2017) 012020 doi:10.1088/1742-6596/813/1/012020

Unsteady hydraulic simulation of the cavitating part load vortex rope in Francis turbines

J Brammer[1], C Segoufin[1], F Duparchy[1], P Y Lowys[1], A Favrel[2], F Avellan[2]

[1]GE Renewable Energy, Hydro Solutions, 82 Avenue Léon Blum – BP75, 38041 Grenoble Cedex 9, France
[2]EPFL Laboratory for Hydraulic Machines, Avenue de Cour 33 bis, 1007 Lausanne, Switzerland

E-mail: james.brammer@ge.com

Abstract. For Francis turbines at part load operation a helical vortex rope is formed due to the swirling nature of the flow exiting the runner. This vortex creates pressure fluctuations which can lead to power swings, and the unsteady loading can lead to fatigue damage of the runner. In the case that the vortex rope cavitates there is the additional risk that hydro-acoustic resonance can occur. It is therefore important to be able to accurately simulate this phenomenon to address these issues. In this paper an unsteady, multi-phase CFD model was used to simulate two part-load operating points, for two different cavitation conditions. The simulation results were validated with test-rig data, and showed very good agreement. These results also served as an input for FEA calculations and fatigue analysis, which are presented in a separate study.

1. Introduction

The stochastic nature of production of new renewable energy technologies such as wind and photovoltaic has created the need for hydro power plants to be operated in a more flexible way to stabilise the grid. For Francis turbines, this increases number of hours at off-design conditions. At part load operation – typically between 50 % and 85 % of the optimal power of the turbine, the flow exiting the runner and, therefore, entering the draft tube has an increasing tangential component. This swirling flow creates instability in the flow field, and thus a helical vortex rope is formed in the draft tube cone. This vortex rope induces pressure fluctuations that can not only cause power swings but also increase fatigue loading on turbine components due to dynamic stresses. This vortex rope exists regardless of the presence of cavitation – however as already identified in the model tests conducted as a part of the Hyperbole project, the level of cavitation can have a dramatic effect on the pressure fluctuation levels as the natural frequency of the hydraulic circuit changes [1]. In a worst-case scenario, the excitation frequency of the vortex rope can match with the natural frequency of the test platform, and a resonance may occur. The same phenomena may occur on the power plant for possibly a different cavitation level and where the resonance can reach in rare cases an elevated level.

To study the part load vortex rope two operating points were identified. These have been named PL1 (corresponding to 80 % optimum discharge) and PL2 (corresponding to 64 % optimum discharge) respectively. Both were at rated head, and two different cavitation (sigma) numbers were chosen to study the influence of cavitation, as well as the accuracy of the CFD.

2. Single-phase calculations

2.1. Numerical setup

The part load vortex rope is a low frequency phenomenon – the rotation frequency is typically around $0.25 \cdot f_0$, where f_0 is the runner rotation frequency. Therefore, to have at least 10 revolutions of the vortex rope 40 runner revolutions are required. This necessitates a very large number of iterations during the CFD simulations. To achieve these numerical simulations in a practical timeframe a simplified computation domain was used. This domain neglected the spiral case and distributor. Instead a single guide vane was modelled upstream of the runner, using a prescribed flow angle at the domain inlet, and it was assumed that the flow was identical in each guide vane channel. The total number of mesh nodes used was 7.3 M (4.3 M in draft tube, 2.8 M in the runner and 0.2 M in the guide vane domain). To connect the guide vane domain with the runner, a stage interface therefore had to be used. This interface provides a circumferentially averaged flow field to the runner inlet, and therefore no effect of Rotor-Stator-Interaction (RSI) was modelled. This simplification was deemed acceptable for this analysis as the vortex rope is the main source of excitation at part load. This simplification not only dramatically reduced the total mesh size, but by not modelling the RSI a larger time step could be used; this was necessary to simulate a long enough time to ensure convergence. The runner and the draft tube domains were coupled using a transient-rotor-stator interface. The mass flow rate was specified at the inlet of the guide vane domain, along with a constant flow angle. At the outlet of the draft tube a constant static pressure was applied. All wall conditions were set using a no-slip condition and a smooth roughness model. The SAS-SST turbulence model was selected. The time step used was equivalent to half a blade passage of rotation. This is a rather large value; it was however deemed sufficient due to the domain simplification previously described.

2.2. Results

Initially a high cavitation number (sigma = 0.38) was investigated, where no cavitation was observed on profiles during the model test, and therefore single phase CFD could be used. The results showed discrepancies in prediction of both the torque and head values. This is mainly due to the spiral case and distributor not being considered, but also could be improved with mesh refinement. However, as the purpose of the study was to investigate the part load vortex rope, the velocity profiles in the draft tube cone were also compared with LDV measurements [2] in order to confirm that from a hydraulic perspective the correct operating point was being modelled.

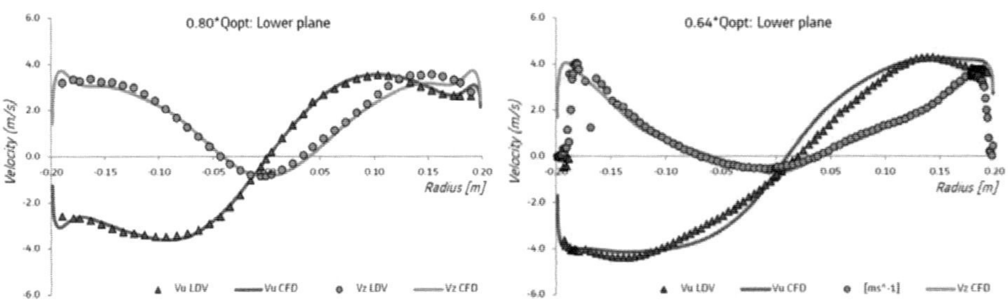

Figure 1: Time-averaged velocity profiles in the draft tube cone
(red series = circumferential velocity, green series = axial velocity, solid line = CFD, Points = LDV measurements)

This comparison is shown in Figure 1 for both part load operating points. An excellent agreement is observed between the CFD and experimental results, which confirms that the correct operating point was being modelled with a reasonable level of accuracy. It is also interesting to note that both the experimental and CFD results show a slight asymmetry in the flow field. The pressure fluctuations in the runner and draft tube were then analysed. Virtual pressure sensors were placed in the CFD model, in the same positions as on the scale model used for experimental measurements on the test rig. This included multiple sensors in the draft tube cone, elbow and liner, as well as on the runner blades.

Figure 2 shows the results for a sensor placed in the draft tube cone for both operating points. Comparing the CFD predictions with the experimental measurements a very good correlation is observed, the form of the time series matches well with the measurements, suggesting that the dynamics of the vortex rope is well predicted. The experimental signals show a higher frequency content, which was not expected in the simulation due to the relatively large time step used, and to the fact that the stage interface was used at the runner inlet.

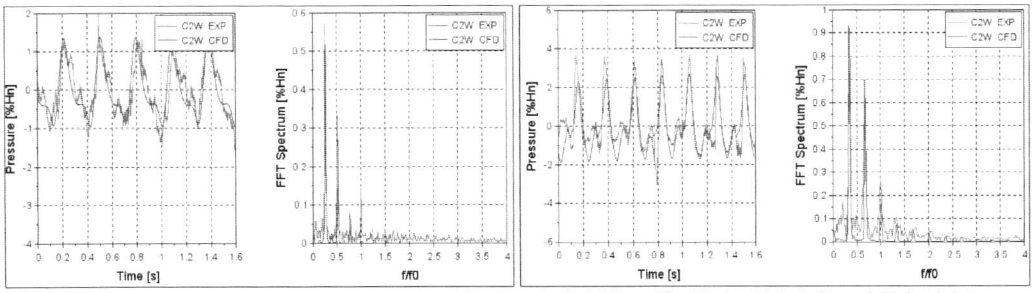

Figure 2: Pressure versus time and associated FFT analysis ($\Delta H_{RMS}/Hn$ %) for PL1 (left) and PL2 (right)

The pressure fluctuations measured by four rotating sensors mounted on the instrumented runner are shown in Figure 3. Two sensors were located on the pressure side of one blade close to the trailing edge and two others located on the suction side of one blade close to the trailing edge. Looking at the results the CFD and experimental frequencies match, and the low frequency fluctuations are the signature of the part load vortex rope. The CFD does however appear to under predict the magnitude of the pressure fluctuations on the runner blades.

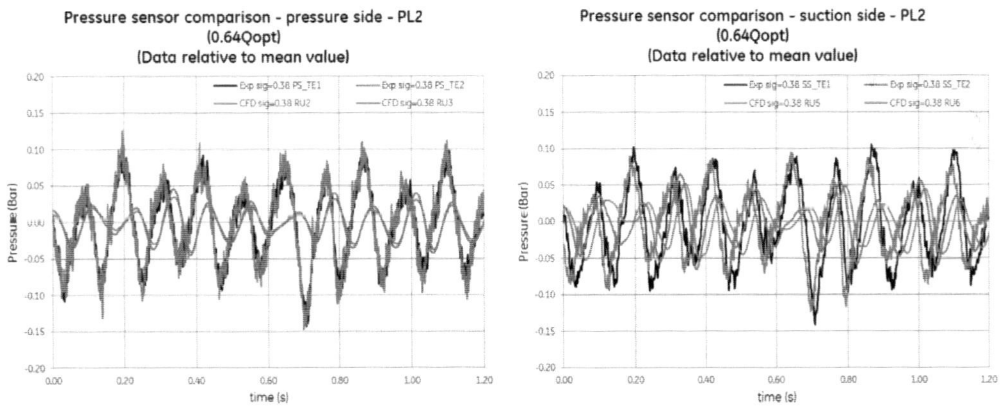

Figure 3: Pressure sensors mounted on the instrumented runner for PL2
(left: pressure side of blade, right: suction side of blade)

This is made more evident on the pressure side sensors (left graph) where the RSI signal values are higher. Therefore, it is likely that this under prediction of the pressure fluctuations is due to the stage interface, and the large time step used. Further analysis of the on-board sensor measurements is detailed in [3].

3. Multi-phase calculations

Complimentary studies of this runner have shown that the cavitation number has a significant influence on the pressure fluctuations throughout the hydraulic system [1], [4] and [5]. At sigma = 0.16 there is a dramatic increase in the level of pressure fluctuations despite the cavitation volume being smaller than at the plant sigma condition (plant sigma = 0.12). This is due to the natural frequencies of the hydro-acoustic system and the turbine test-rig being equal at this point, and therefore resonance occurs. As this resonance is an interaction of the hydraulic circuit with the test rig circuit, it is not possible to model this phenomenon using only a CFD calculation of the single turbine, even with multiple phases. A representation of the test rig would be required in the simulation, and this could potentially be achieved by coupling a CFD calculation with a 1D hydro-acoustic model. Therefore, for the following CFD results this resonance point was avoided, and a cavitation number sigma = 0.12 was chosen, with this point corresponding to the plant value.

3.1. Numerical setup

The numerical setup was the same as the single-phase calculations, except that a multi-phase model was implemented and that the static pressure at the outlet was adjusted iteratively until the correct cavitation number was calculated. For this study the homogenous cavitation model was used [6], such that the two phases shared a common flow field. This was deemed appropriate given that the high flows in the turbine would dominate the advective processes. The two phases were defined as water and water vapour respectively. The phase-change process was modelled using the Rayleigh-Plesset model.

3.2. Results

Figure 4 shows a comparison of the visualized cavitating vortex rope and the CFD prediction. In the CFD calculation the vortex is visualized by two iso-volumes of the vapour volume fraction, with values of 10 % and 90 % used to capture the full extent of the cavitation volume. Comparing the two visualizations, the vortex rope has been well predicted in terms of the cavity shape. It was observed, by both CFD and the experiments, that as the vortex rope contacts the elbow wall, it collapses and travels upwards towards the runner.

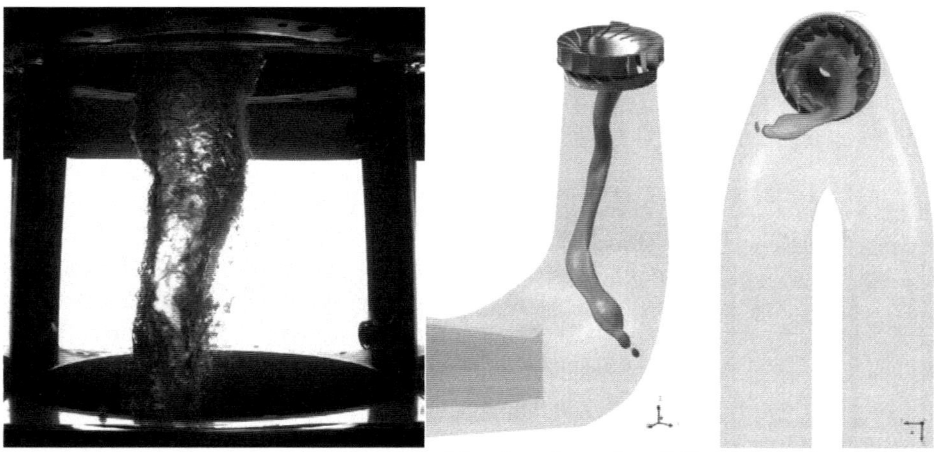

Figure 4: Comparing the observed vortex rope from laboratory observations (left) and CFD calculations (right).

The time-averaged values of the pressure fluctuations in the draft tube cone and on the runner are shown in Figure 5 for both cavitation conditions (i.e. sigma = 0.12 and 0.35). Both calculations show good agreement with the experimental results, with the FFT results showing good coherence. For sigma = 0.12 the cone sensors (left image) are particularly accurate, whereas for the embedded sensors the levels are slightly under predicted, as was the case for the first simulation in cavitation free conditions.

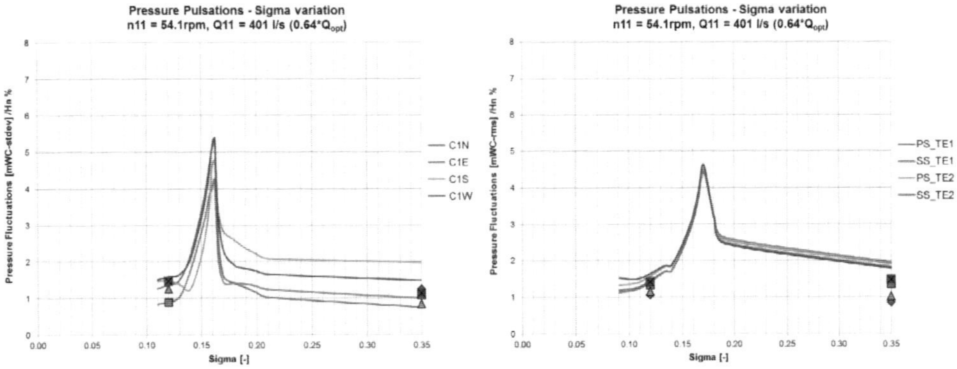

Figure 5: Cavitation number (sigma) influence on the pressure fluctuations (ΔH_{RMS}/Hn %) in the cone (left) and rotating sensors mounted on the instrumented runner (right) for PL2– Points are CFD results, line series are experimental data

4. Conclusion

Single and multi-phase CFD calculations were launched to investigate the part load cavitation vortex rope. The results were validated with standard integral values, LDV flow field surveys, and wall pressure sensors both on the stationary parts of the turbine model as well as the runner, all of which were measured as part of the Hyperbole project.

For the case of hydro-acoustic resonance, the current methodology is not able to capture this phenomenon as the entire hydraulic circuit would need to be modelled [7], therefore two sigma values were chosen outside of this resonance – The first one at sigma = 0.38 and the second one at sigma = 0.12, which corresponds to the plant value.

A simplified computational domain was used to reduce the CPU time. The CFD predictions showed an excellent correlation with the experimental test rig measurements and was therefore capable to predict the complex phenomena of the part load vortex rope outside of the resonance condition.

The influence of the cavitation number had little effect on the pressure fluctuations, and therefore a single-phase simulation would have been sufficient to predict the dynamic loads. However, other operating conditions may need to use multi-phase models if the phenomena are dependent on cavitation. This type of calculation allows us to not only gain an understanding of the hydraulic behaviour at off-design conditions, but can also serve as an input to mechanical calculations so that the fatigue and runner lifetime can be evaluated [8].

Acknowledgments

The research leading to the results published in this paper is part of the HYPERBOLE research project, granted by the European Commission (ERC/FP7- ENERGY-2013-1-Grant 608532).

References

[1] Favrel, A., Müller, A., Landry, C., Yamamoto, K. and Avellan, F., Experiments in Fluids, vol. 56(12), 2015. doi:10.1007/s00348-015-2085-5.

[2] Favrel, A., Müller, A., Landry, C., Yamamoto, K. and Avellan, F., Experiments in Fluids, vol. 57(11), 2016. doi: 10.1007/s00348-016-2257-

[3] Duparchy, F., Favrel, A., Lowys, P.Y., Landry, C., Müller, A., Yamamoto, K. and Avellan, F., 2016 Journal of Physics: Conference Series 656 012061 (CAV2015, Lausanne, Switzerland)

[4] Müller A., Dreyer M., Andreini N. and Avellan F., Experiments in Fluids, vol. 54, num. 4, p. 1-11, 2013.

[5] Yamamoto, K., Müller, A., Favrel, A et al. 2014, In IOP Conference Series: Earth and Environmental Science (Vol. 22, No. 2, p. 022011). IOP Publishing.

[6] ANSYS CFX, Release 16.0, ANSYS CFX-Solver Theory Guide, Chapter 5: Multiphase Flow Theory.

[7] Landry, C., Favrel, A., Müller, A., Nicolet, C., Yamamoto, K. and Avellan, F., 27th IAHR Symposium on Hydraulic Machines and Systems in *IOP Conf. Ser.: Earth Environ. Sci.* 22 032037, 2014.

[8] Brammer, J., Duparchy, F., Lowys P.Y., Thibaud, M, Wheeler, K. and de Colombel, T., "*Increasing the operating range of Francis turbine by considering dynamic phenomena at partial load*", Hydro 2016, Montreux, Switzerland.

Instance Analysis for the Error of Three-pivot Pressure Transducer Static Balancing Method for Hydraulic Turbine Runner

Hanli Weng[1], **Youping Li**[2]

[1] College of Electrical Engineering & New Energy, China Three Gorges University, Yichang, China
[2] Technology and Research Center, China Yangtze Power Co., Ltd, Yichang, China

Abstract. The working principle, process device and test procedure of runner static balancing test method by weighting with three-pivot pressure transducers are introduced in this paper. Based on an actual instance of a V hydraulic turbine runner, the error and sensitivity of the three-pivot pressure transducer static balancing method are analysed. Suggestions about improving the accuracy and the application of the method are also proposed.

1. Introduction

Static balancing test should be accomplished to turbine runners during shop tests after a new runner has been assembled or an old runner has been repaired by repair welding according to relative rules[1,2]. Presently, four methods are mainly applied to the static balancing test of hydraulic turbine runner, including the methods of balancing with steel ball on plate, balancing on hydro static bearing[3], balancing with measuring rod and balancing with three-pivot pressure transducers[4]. References [3] and [5] detailedly analysed the error and sensitivity of the methods of balancing with steel ball on plate and balancing on hydro static bearing respectively. Additionally, reference [3] analysed the method of balancing with measuring rod and raised doubt about its actual reliability. However, researches about the error and sensitivity analyses for the method of balancing with three-pivot pressure transducers for hydraulic turbine runner are still rare. In this circumstances, combining an actual instance of a V hydraulic turbine runner, the error and sensitivity analyses of the three-pivot pressure transducer static balancing method are conducted in this paper.

2. Working principle of runner static balancing method by weighting with three-pivot pressure transducers

2.1. Working principle

The three-pivot pressure transducer static balancing method is based on the principle of plane moment balance. As seen in Figure 1, suppose that the forces at the three vertexes of an equilateral triangle are F_A, F_B and F_C respectively, which are normal to the paper plane and inward. On the other hand, there exists an unknown force F in the plane(referring to the weight deviation), which is normal to the paper plane and outward. In this case, for the plane to remain static without rotation, the moments of the four forces about x and y axes are both 0. Assuming F is in the first quadrant, equation (1) comes into existence.

$$\begin{cases} F_X + F_C\cos 30° - F_B\cos 30° = 0 \\ F_Y - F_A + (F_B + F_C)\sin 30° = 0 \end{cases} \tag{1}$$

By virtue of equation (1), F and α(the weight deviation angle) can be obtained.

$$\begin{cases} F=\left(F_X^2+F_Y^2\right)^{1/2} \\ \alpha=\arctan(F_X / F_Y) \end{cases} \tag{2}$$

Then the unbalanced moment M can be calculated by

$$M = F \times R \tag{3}$$

Where R is the distribution circle radius of three pressure transducers.

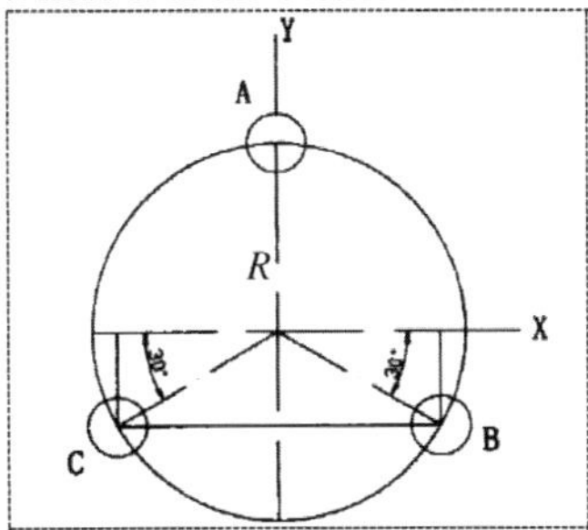

Figure 1. Principle diagram of the three-pivot pressure transducer static balancing method.

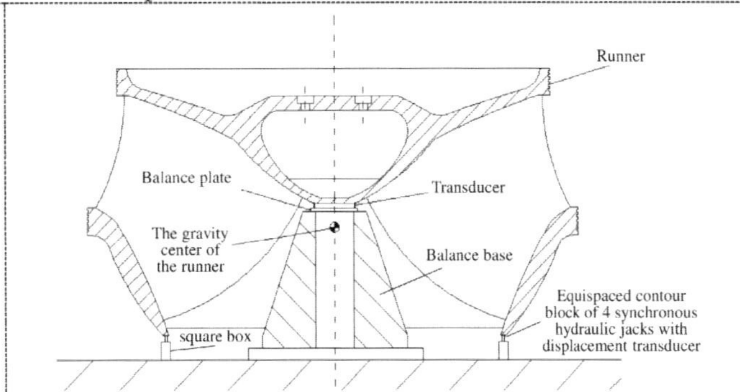

Figure 2. Runner static balancing system diagram based on three-pivot pressure transducer static balancing method.

2.2. The process device and test procedure

Take the Francis turbine runner as an example, the common test process device of the three-pivot pressure transducer static balancing method is shown in Figure 2.

The steps of the test procedure are as follows:

① Three pressure transducers are disposed on the balance plate according to Figure 1.

② The runner is placed according to Figure 2. The runner weight is born by the four jacks firstly. As the runner stabilized, the jacks lower slowly and the runner weight is born by the three transducers finally.

③ According to Figure 1, F_A, F_B and F_C can be read from the three transducers. Furthermore, the unbalanced moment M and its position can be calculated from equations (1)-(3).

④ The weight balancing for the runner is conducted by virtue of the calculation results in step ③ (including rough counter weight and fine counter weight).

3. Parameters of runner and process device

3.1. The main parameters of V runner

The runner weight: m=178982kg;

The rotational speed of the turbine: n=125rpm;

The static balancing has to meet the requirements defined by ISO 1940/1 Grade G6.3 and the corresponding allowable residual unbalanced moment $E_r \leq 86.14$kg·m.

3.2. The main parameters of the process device

The balance plate of the process device is located by applying the spigot at the water outlet end of runner crown. The sizes of the spigot and the balance plate, as well as their fit tolerance, are required to be 1380H7/g6. The concentricity of the spigot(δ_1) is 0.05mm. The balance plate weight(m_b) is 2000kg. The levelness(δ_3) is required to be 0.05mm/m and the installation accuracy should meet the requirement of the eccentricity(δ_2) being less than 0.05mm.

The height between the theoretical gravity center of the runner and the mounting plane of pressure transducers(h) is 791mm.

The measurement range, non-linearity error and sensitivity of each pressure transducer are 100T, 0.02%FS and 2.85mV/V respectively. The distribution circle radius of three pressure transducers(R) is 690mm and the resolution of the corresponding weight display instrument is 0.01uV/V.

4. Error analysis

It can be seen from the working principle and process of the three-pivot pressure transducer static balancing method analysed in section 2, errors are mainly caused by the following factors:

4.1. The accuracy error caused by the pressure transducer E_1

As the non-linearity error of each pressure transducer is 0.02%FS, the accumulative error of the three transducers can be regarded as the repeated measurement error[6], which equals to $\sqrt{\dfrac{0.02\%^2+0.02\%^2+0.02\%^2}{3-1}}=0.0245\%$ FS. Consequently, the weight measurement error Δm=100T×0.0245%=24.5kg. In addition, the distribution circle radius of three pressure transducers R is 690mm. Hence, the unbalanced moment error can be calculated as: $E_1=\Delta m\times R\times 10^{-3}=16.90kg{\bullet}m$.

4.2. The error caused by the deviation between the gravity center of the runner and the distribution circle center of three pressure transducers E_2

As mentioned in section 3.2, the concentricity of the spigot at the water outlet end of runner crown(δ_1) is 0.05mm and the eccentricity of the balance plate(δ_2) is less than 0.05mm according to the installation accuracy requirement. In this case, the maximum deviation between the theoretical gravity center of the runner and the distribution circle center of three pressure transducers μ_1=0.075mm, which is calculated according to $\mu_1=\delta_2+\delta_1/2$, as seen in Figure 3.

As a result, the unbalanced moment error can be calculated as: $E_2=m\times\mu_1\times 10^{-3}=13.42kg{\bullet}m$.

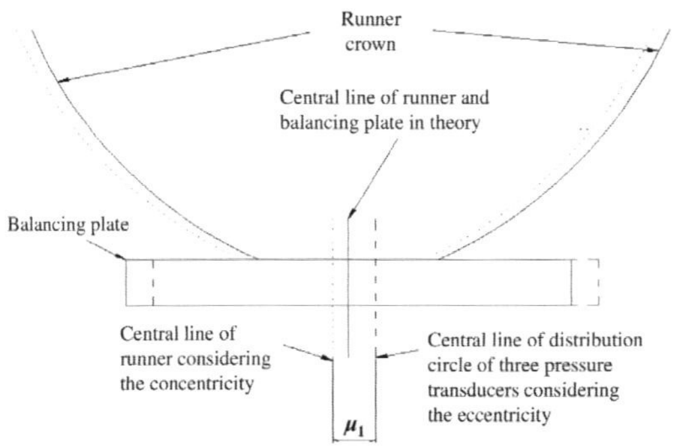

Figure 3. Calculation schematic diagram for the error caused by the deviation between the gravity center of the runner and the distribution circle center of three pressure transducers.

4.3. The error caused by the levelness of the runner E_3

By virtue of Figure 4, $\mu_2=h\times\tan\alpha$ according to the plane geometry knowledge and $\tan\alpha=\delta_3$(the levelness of the runner, which is 0.05mm/m). In this case, the unbalanced moment error caused by the levelness of the runner can be calculated as: $E_3=m\times\mu_2\times 10^{-3}=m\times h\times\delta_3\times 10^{-3}=7.08kg{\bullet}m$.

4.4. The error caused by the levelness of the balancing plate E_4

As illustrated in Figure 5, $\mu_3=R-R\cos\beta=R(1-\cos\beta)$ and $\tan\beta=\delta_3$(the levelness of the balancing plate, which is 0.05mm/m). As $1-\cos\beta=2\sin^2(\beta/2)$, when β is quite small, $\sin(\beta/2)\approx\tan(\beta/2)\approx\tan\beta/2=\delta_3/2$ and hence $\mu_3=R\times\delta_3^2/2$. In this case, the unbalanced moment error caused by the levelness of the balancing plate can be calculated as: $E_4=m_b\times\mu_3\times 10^{-3}=m_b\times R\times\dfrac{\delta_3^2}{2}\times 10^{-3}=1.725\times 10^{-6}kg{\bullet}m$.

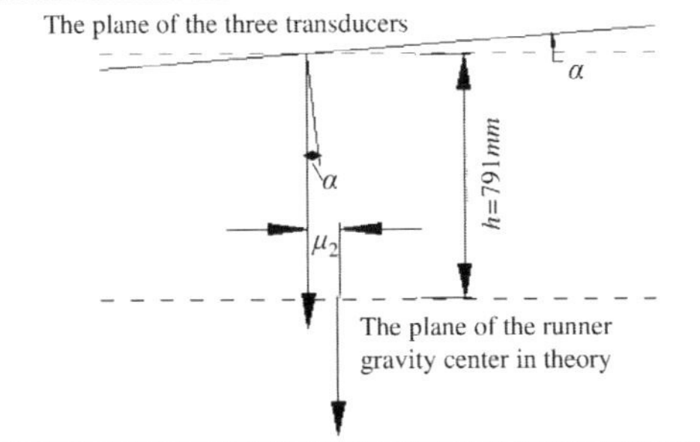

Figure 4. Calculation schematic diagram for the error caused by the levelness of the runner.

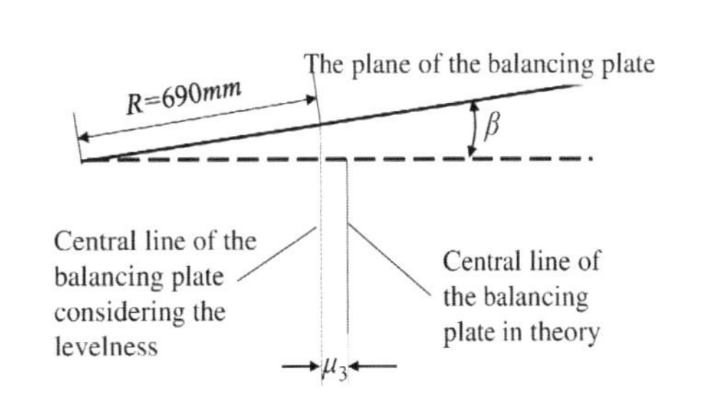

Figure 5. Calculation schematic diagram for the error caused by the levelness of the balancing plate.

In conclusion, the total error of the balancing system can be calculated by summing the above four errors: $E_S = E_1 + E_2 + E_3 + E_4 = 37.4kg{\bullet}m$. And the ratio of E_S to the allowable residual unbalanced moment E_r is: $E_S / E_r = 37.4 \div 86.14 \times 100\% = 43.42\%$.

5. Sensitivity analysis

For the steel ball on plate based method, the sensitivity of the static balancing test system is represented by the vertical distance between the gravity center of the runner and the center of the balancing ball. It is calculated from the deflection of the runner band when a weight block with standard mass is placed to the runner band[5].

For the three-pivot pressure transducer static balancing method, as the contact between the transducers and the runner is rigid, a weight block placed at any position of the runner will not result in the descend of the runner band. Therefore, the sensitivity calculation method mentioned in reference [5] is not applicable to the three-pivot pressure transducer static balancing method.

The runner manufacturer pointed out that, the minimum unbalanced moment of the runner G_{min} can be calculated by the resolution of the pressure transducers and the digital weight display, and the sensitivity of the method can be represented by G_{min}. Presetting an unbalanced moment for the runner after the final test, G_{min} can be verified by comparing the theoretical value and the actual measured value of this unbalanced moment.

The minimum weight W_{min} that can be accurately measured and displayed by the digital weight display is 1kg. The runner manufacturer pointed out that the display is unstable in unit digits in actual application due to the environment factors, but is stable in tens digits. Therefore, W_{min} equals to 10kg. Assuming the unbalanced mass is located at the point with the longest distance from the center line of the runner, G_{min} can be calculated as: $G_{\min} = W_{\min} \times R = 10 \times 690 \times 10^{-3} = 6.90kg{\bullet}m$.

A 4kg weight block was placed at the point with the distance of 1.7m from the center line on the runner assembling joint after the final balancing test. The theoretical value of the unbalance moment is 6.80kg·m(4kg×1.7m) which is close to G_{min} and the measured value is 7.59kg·m. The error ratio between the theoretical value and the actual measured value of the unbalance moment equals to $(7.59\text{-}6.8) \div 6.8 \times 100\% = 11.62\%$, which is less than the ratio of the system error to the allowable residual unbalanced moment(43.42%). Consequently, the three-pivot pressure transducer static balancing method can basically recognize the G_{min} of 6.90kg·m.

6. Conclusion and discussion

6.1. Judgment of test results

Considering that the system error of the three-pivot pressure transducer static balancing method is considerably large, the final residual unbalanced moment after fine balancing E_f is supposed to be less than the difference between Er and E_S (satisfying inequality (4)) instead of just less than Er during site acceptance tests, Where Er is the allowable residual unbalanced moment according to the contract and E_S is the system error based on the theoretical analyses in section 4.

$$E_f < E_r - E_S \tag{4}$$

The test result is E_f=39.8$kg \cdot m$, which is less than $E_r - E_S$=86.14 − 37.4 = 48.74$kg \cdot m$. Consequently, inequality (4) is satisfied and the static balancing test of the V runner is considered to pass the acceptance.

6.2. Further improvement of measurement accuracy

On the basis of above analyses, the test system error of the three-pivot pressure transducer static balancing method mainly results from the machining and installation errors of the process device, as well as the transducer error. The accuracy of the transducer applied in the test system has already reached 0.02%FS and is difficult to be further improved.

However, in the respect of machining accuracy, the related requirement for the levelness according to Chinese standards[1,5] is 0.02mm/m, while the requirement of the manufacturer for the levelness is 0.05mm/m. The concentricity deviation between the process device and the runner is required to be 0.07mm according to Chinese standards[1,5], while 0.075mm according to the requirement of the manufacturer. Obviously, the requirements of the manufacturer are below the Chinese national standards. Therefore, the machining and installation accuracies of the manufacturer have some room for improvement and the measurement accuracy of the test system is hopefully to be improved.

6.3. Discussion

The principle of the three-pivot pressure transducer static balancing method discussed in this paper is similar to that of the method of balancing with measuring rod mentioned in reference [3]. By virtue of reference [3], the system error of the measuring rod based method may reach 30-40% of the allowable residual unbalanced moment, which is basically in accordance with the result of the three-pivot pressure transducer static balancing method analysed in this paper. Moreover, similar to the doubt about the actual reliability of the measuring rod based method raised in reference [3], the actual reliability of the three-pivot pressure transducer static balancing method is supposed to be investigated. Although, the three-pivot pressure transducer static balancing method has been repeatedly adopted by several hydraulic turbine manufacturers[4], the relevant international standard[7] has not covered this method in the static balancing parts yet. This method should be researched and further improved by the hydraulic turbine industry and manufacturers. In this case, the relevant standards can be developed.

References

[1] General Administration of Quality Supervision 2004 *Inspection and Quarantine, GB8564-2003: Specification Installation of Hydraulic Turbine Generator Units* (Beijing: China Standards Press)

[2] Yu L J 1995 *Installation and Maintenance of Hydraulic Turbine Generator Units* (Beijing: China Water & Power Press)

[3] Chen L Y 2011 Comparison and Analysis of Vertical Static Balancing Methods for Large Hydro-turbine Runner *China. J. L. Elec. Mach. and Hydr. Turb.* **2** 34

[4] Xu H and Wang W P 2009 Static Balancing Method by Weighting with Three-Pivot Pressure Transducers and its application *China J. Dongfang Elec. Mach.* **3** 13

[5] Ministry of Industry and Information Technology of the People's Republic of China 2013 *JB/T6752-2013: Runner static balancing test procedures of medium and small turbines* (Beijing: China Machine Press)

[6] Measurement Adjustment Academic Committee of Wuhan University 2013 *Error theory and foundation of surveying adjustment* (Wuhan: Wuhan University Press)

[7] International Organization for Standardization 2003 *ISO 1904-1: Mechanical vibration-Balancing Quality requirements for rotors in a constant (rigid) state-Part 1: Specification and verification of balance tolerances* (Switzerland)

Acknowledgments

Supported by the China Three Gorges University Talent Scientific Research Foundation, Grant No. KJ2014B043.

IOP Conf. Series: Journal of Physics: Conf. Series **813** (2017) 012022 doi:10.1088/1742-6596/813/1/012022

Experimental investigation of the mass flow gain factor in a draft tube with cavitation vortex rope

C Landry[1], A Favrel[2], A Müller[2], K Yamamoto[2], S Alligné[1] and F Avellan[2]

[1]Power Vision Engineering Sàrl, Chemin des Champs Courbes 1, 1024 Ecublens, Switzerland

[2]EPFL Laboratory for Hydraulic Machines, Avenue de Cour 33bis, 1007 Lausanne, Switzerland

E-mail: christian.landry@powervision-eng.ch

Abstract. At off-design operating operations, cavitating flow is often observed in hydraulic machines. The presence of a cavitation vortex rope may induce draft tube surge and electrical power swings at part load and full load operations. The stability analysis of these operating conditions requires a numerical pipe model taking into account the complexity of the two-phase flow. Among the hydroacoustic parameters describing the cavitating draft tube flow in the numerical model, the mass flow gain factor, representing the mass excitation source expressed as the rate of change of the cavitation volume as a function of the discharge, remains difficult to model. This paper presents a quasi-static method to estimate the mass flow gain factor in the draft tube for a given cavitation vortex rope volume in the case of a reduced scale physical model of a $\nu = 0.27$ Francis turbine. The methodology is based on an experimental identification of the natural frequency of the test rig hydraulic system for different Thoma numbers. With the identification of the natural frequency, it is possible to model the wave speed, the cavitation compliance and the volume of the cavitation vortex rope. By applying this new methodology for different discharge values, it becomes possible to identify the mass flow gain factor and improve the accuracy of the system stability analysis.

1. Introduction

The integration of the renewable energy resources has a significant impact on the stability of the electrical grid. The fast changes in power generation by renewable energy resources require a flexible operating range of the hydro units. At off-design operating conditions of hydraulic machines, the onset of a cavitation volume attached to the runner hub is often observed. At part load condition, a draft tube surge and electrical power swings [7], [5] occurs when the vortex rope frequency matches the eigenfrequencies of the hydraulic system. At full load operation, self-oscillations are experienced as an axial pulsation of the cavitation volume corresponding to the one of the eigenfrequency of the system and may cause negative damping [9], [14]. The behavior of full load surge in model test is reported by Prenat and Jacob in 1986 [16]. Koutnik et al. [10] successfully simulate full load surge using a simplified 1D mathematical model of the full-load vortex rope and using the cavitation compliance and the mass flow gain factor (MFGF) parameters [4]. More recently, by deriving the compliance and the MFGF parameters from CFD simulations of the complete machine, Flemming et al. [6] predicted instability of a prototype installation based on eigenvalues and eigenmodes analysis.

IOP Conf. Series: Journal of Physics: Conf. Series **813** (2017) 012022 doi:10.1088/1742-6596/813/1/012022

In this paper, a quasi-static method estimates the MFGF using experimental data and a one-dimensional draft tube model developed by Alligné et al. [1]. On the one hand, the methodology is based on an experimental identification of the natural frequency of a ν = 0.27 Francis turbine for different Thoma numbers and speed factor n_{ED}. On the other hand, the numerical model of the draft tube is derived from momentum and continuity equations including the convective terms and the divergent geometry of the draft tube to reproduce the dynamic behavior of the two-phase flow. This model requires knowledge of the wave speed, the bulk viscosity induced by the cavitation vortex rope and the MFGF. Landry [12] [13] developed a methodology and dimensionless power laws to model the wave speed and the bulk viscosity for a given Francis turbine at part load condition.

2. One dimensional modeling of the cavitation vortex rope dynamics

A one dimensional numerical model to simulate the behavior of the cavitation vortex rope is used, taking into account the dissipation induced by the cavitation vortex rope, the divergent geometry of the draft tube and the convective term of the momentum equation. The test rig hydroacoustic model for the numerical simulations is made with the SIMSEN software developed by EPFL. The modeling of the hydraulic components is based on an equivalent electrical scheme representation, see Figure **1**.

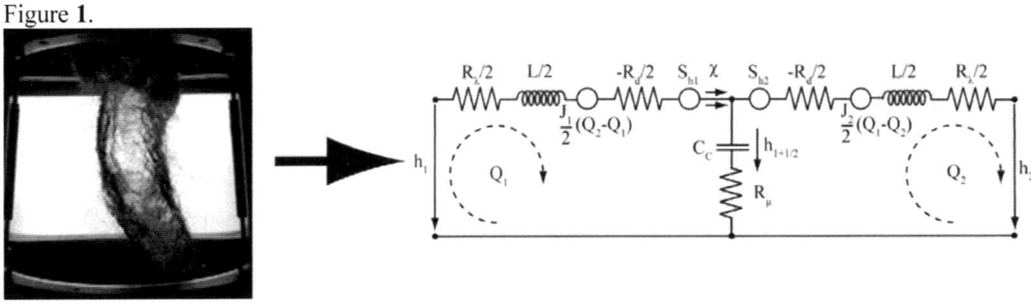

Figure 1. Visualization of the cavitation vortex rope in the draft tube cone and the corresponding equivalent circuit.

The momentum and the continuity equations are solved using the Finite Difference Method with 1st order centered scheme discretization in space and a scheme of Lax for the discharge state variable. This discretization yields a system of ordinary differential equations that can be represented as a T-shaped equivalent circuit. For the hydroacoustic model of the cavitation vortex rope, uniform pressure and velocity distributions in a cross section A are assumed. The parameters of the equivalent circuit are widely described by Alligné et al. [1]:

- The mathematical development of the cavitation volume fluctuations yields a dynamic parameter called cavitation compliance C_c. This dynamic parameter represents the variation of the cavitation volume V_c with respect to a variation of pressure and implicitly defines the local wave speed a in the draft tube, influencing the traveling time of pressure waves.

$$C_c = -\frac{\partial V_c}{\partial h_{i+\frac{1}{2}}} = \frac{gAdx}{a^2} \tag{1}$$

- The MFGF represents the excitation mass source of the hydraulic system and is defined as the cavitation volume variation as a function of a discharge variation Q_1 at the runner outlet.

$$\chi = -\frac{\partial V_c}{\partial Q_1} \tag{2}$$

The flow mechanism determining the MFGF is not clear and is therefore difficult to model.

3. Methodology

The methodology is based on the direct link between the natural frequency of the hydraulic system and the local wave speed in the draft tube [8], [12]. The first step is to identify the natural frequency of the hydraulic system for a given operating point. To experimentally identify the natural frequency, an excitation system was designed to inject or extract a periodical hydroacoustic power at a given frequency. This excitation system is an upgraded version of the rotating valve used by Blommaert [3] to actively control the hydraulic fluctuations of a Francis turbine at part load operating conditions. The excitation system and the spectral analysis were widely described in [12]. With pressure sensors located on the hydraulic system, the natural frequency can be identified with a spectral analysis of the forced harmonic response. Then, the wave speed value can be adjusted in the draft tube modeling to obtain a numerical natural frequency similar to the experimental natural frequency. Numerically, the natural frequency is defined by conducting an eigenvalue study of the non-linear hydraulic system. The eigenvalues are based on the linearization of the set of equations around the operating point [2]. Damping and oscillation frequency of the eigenmodes are respectively given by the real part and the imaginary part of the eigenvalues.

After the identification of the wave speed, the second step is to compute the value of the cavitation vortex rope V_c, defined as the volume limited by the vapor pressure iso-surface. With equation (1), the cavitation compliance can be rewritten as a function of the Thoma number [11]:

$$C_C = -\frac{\partial V_C}{\partial h_{i+\frac{1}{2}}} \approx -\frac{\partial V_C}{\partial h_{\bar{i}}} = -g\frac{\partial V_C}{\partial NPSE} = -\frac{\partial V_C}{\partial NPSH} = -\frac{\partial V_C}{\partial\left(\sigma \cdot H_{Turb}\right)} \Rightarrow -\frac{\partial V_C}{\partial\left(\sigma \cdot H_{Turb}\right)} = \frac{gAdx}{a^2} \qquad (3)$$

Moreover, for a given operating point, *i.e.* Q_{ED} and n_{ED} are constant, the cavitation volume can be computed by integrating the equation (3). With this information, it becomes possible to compute the cavitation volume and the void fraction β, defined as the fraction of a reference volume V_{ref} that is occupied by the gas phase volume V_c. The reference volume is defined by the numerical elements where the cavitation compliance is adjusted and corresponds to the cone and the first half of the elbow volumes of the draft tube modeling. The void fraction values are finally validated by analyzing the visualization of the cavitation vortex rope with a high-speed camera.

The last step to determine the value of the MFGF needs the identification of cavitation volume for different discharge values. Thus, for a constant guide vane opening and discharge factor, the discharge value is modified by increasing the rotational speed of the hydraulic turbine. By identifying the cavitation volume for 3 different speed factors, the MFGF can be computed with equation (2) by using the Finite Difference Method.

4. Results and discussion

For a fixed set of values of the discharge factor Q_{ED}, the Froude and the Thoma numbers, see Table 1, the hydraulic system is experimentally excited in a frequency range around the natural frequency, in order to determine the eigenmode of the test rig. For each excitation frequency, the forced harmonic response of the hydraulic system is measured with the 27 pressure sensors located throughout the test rig. Figure 2 (Left) shows the relation between the Thoma number and the first eigenfrequency of the hydraulic system obtained by applying the same methodology for all Thoma numbers, Froude numbers and Q_{ED} values for operating conditions *PL1* and *PL2*. Globally, the natural frequency increases with the Thoma number when the volume of the cavitation vortex rope decreases. In addition to that, the Froude number influence remains small and its decrease causes a slight reduction of the eigenfrequency.

Table 1. Selected Francis turbine operating conditions

		$n_{ED}/n_{ED,BEP}$	$Q_{ED}/Q_{ED,BEP}$	*Fr*	σ
FL 1	Full Load	[0.91, 0.95, 1]	1.3	5.79	[0.12 – 0.22]
PL 1	Part Load	[0.95, 1, 1.04]	0.80	[6.56, 7.66, 8.75, 9.85]	[0.06 - 0.15]
PL 2	Part Load	[0.95, 1, 1.04]	0.64	[7.66, 8.75, 9.85]	[0.10 - 0.20]
PL 3	Part Load	[0.95, 1, 1.04]	0.53	7.66	[0.08 - 0.14]
PL 4	Part Load	[0.95, 1, 1.04]	0.725	6.56	[0.06 - 0.13]

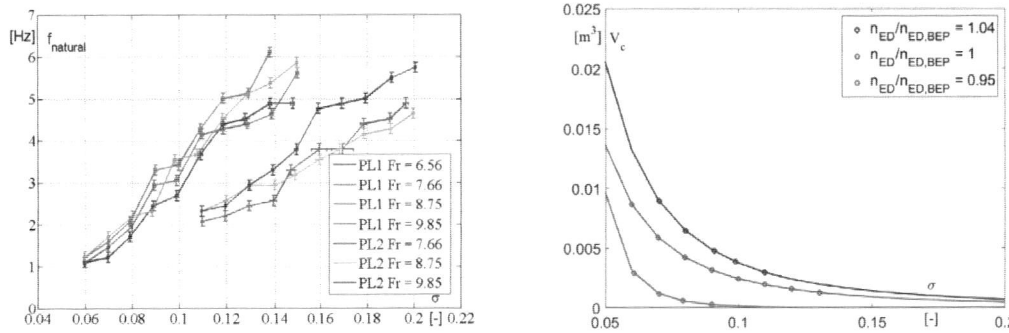

Figure 2. Influence of the Thoma number on the natural frequency for the operating point PL1 and PL2 (Left) and evolution of the cavitation volume for the operating point PL1 (Right).

By analyzing the evolution of the cavitation compliance as a function of the Thomas number, it was observed that the values of the cavitation compliance can be interpolated with a power regression [11]. By integrating this power regression, the cavitation volume and therefore the void fraction can be directly computed. The void fraction obtained is compared with the void fraction estimated with a high-speed visualization of the cavitation vortex rope in the Plexiglas cone. The experimental method based on the image processing of a high-speed flow visualization was already used by Müller et al. [15] to determine the cavitation volume oscillation on a microturbine featuring a self-oscillating vortex rope in its conical draft tube. The experimental estimation of the vortex rope volume is based on the detection of the edges of the cavity. The accuracy of this analysis relies on proper lighting conditions and, therefore, accuracy depends on sharp contrast between the liquid and the gaseous phase of the flow. The numerical value of the void fraction is in good agreement with the experimental values deduced from the high-speed camera, see Figure 3 (Left).

After validating the cavitation volume with experimental visualization, it is now possible to compute the MFGF with equation (2). For full load conditions, the MFGF is negative, see Figure 3 (Right). Indeed, by increasing the discharge, the tangential velocity in the draft tube cone increases. Therefore, the pressure in the vortex core decreases and the cavitation volume increase. On the contrary, at part load conditions, the MFGF is positive. The tangential velocity in the draft tube cone decreases when the discharge increases. Thus, the pressure in the vortex core increases and the cavitation volume decreases. At deep part load, the cavitation volume increases with the discharge and the MFGF behaves similarly to full load conditions. Finally, these properties of the MFGF are amplified when the Thoma number decreases.

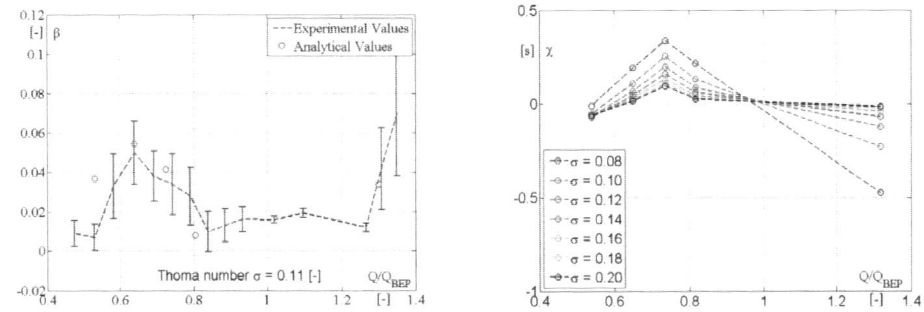

Figure 3. Comparison between the numerical and the experimental values of the void fraction for different discharge factor (Left) and evolution of the MFGF as function of the discharge value for different Thoma numbers (Right).

5. Conclusion

This paper presents a quasi-static method to estimate the MFGF in the draft tube for a given cavitation vortex rope volume in the case of a reduced scale model of a Francis turbine. The methodology is based on an experimental identification of the natural frequency of the hydraulic system for different Thoma numbers and speed factors n_{ED}. With the identification of the natural frequency, it is possible to model the cavitation compliance and the volume of the cavitation vortex rope. This volume was confirmed by analyzing the visualization of the cavitation vortex rope with a high-speed camera. By applying this methodology for different speed factors, it becomes possible to identify the MFGF. Finally, the next step is to define a dimensionless number in order to generalize the MFGF and predict its value for all operating points. This will ultimately pave the way to a more precise stability analysis of hydropower plants.

Acknowledgements

The research leading to the results published in this paper is part of the HYPERBOLE research project, granted by the European Commission (ERC/FP7- ENERGY-2013-1-Grant 608532).

References

[1] Alligné S, Nicolet C, Tsujimoto Y and Avellan F 2014 *Journal of Hydraulic Research* ***52*** *3*
[2] Alligné S, Nicolet C, Allenbach P, Kawkabani B, Simon J.J, Avellan F 2009 *International Journal of Fluid Machinery and Systems*, **2** 4.
[3] Blommaert G 2000 *PhD thesis* EPFL *Lausanne, Switzerland.*
[4] Brennen C and Acosta A J 1973 *Journal Spacecraft*, **10** 3.
[5] Favrel A, Müller A, Landry C, Yamamoto K, and Avellan F 2016 *Experiments in Fluids* **57** 168
[6] Flemming F, Foust J, Koutnik J, Fisher R K, 2008 *24th IAHR Symposium, Foz Do Iguassu.*
[7] Fritsch A and Maria D 1988 *La Houille Blanche 3 4.*
[8] Jacob T 1993 *PhD thesis* EPFL *Lausanne, Switzerland.*
[9] Jenkins A 2013 *Physics Reports*, **525** 2.
[10] Koutnik J and Pulpitel L 1996, *Modeling, Testing and Monitoring for Hydro Power Plants, Lausanne, Switzerland.*
[11] Koutnik J, Nicolet C, Schohl G and Avellan F 2006 *23th IAHR Symposium, Yokohama, Japan.*
[12] Landry C 2015 *PhD thesis* EPFL *Lausanne, Switzerland.*
[13] Landry C, Favrel A, Müller A, Nicolet C, Avellan F 2016 *Journal of Hydraulic Research* **54** 2
[14] Müller A, Dreyer M, Andreini N and Avellan F 2013 *Experiments in Fluids* **54** 4.
[15] Müller A, Yamamoto K, Alligné S, Yonezawa K, Tsujimoto Y and Avellan F 2015 *Journal of Fluids Engineering* **138** 2
[16] Prenat J E, Jacob T 1986 *13th IAHR Symposium, Montreal, Canada.*

Dynamics of the precessing vortex rope and its interaction with the system at Francis turbines part load operating conditions

A Favrel[1], A Müller[1], C Landry[1], J Gomes[1], K Yamamoto[1] and F Avellan[1]

[1]EPFL Laboratory for Hydraulic Machines, Avenue de Cour 33bis, 1007 Lausanne, Switzerland
E-mail: arthur.favrel@epfl.ch

Abstract. At part load conditions, Francis turbines experience the formation of a cavitation vortex rope at the runner outlet whose precession acts as a pressure excitation source for the hydraulic circuit. This can lead to hydro-acoustic resonances characterized by high pressure pulsations, as well as torque and output power fluctuations. This study highlights the influence of the discharge factor on both the vortex parameters and the pressure excitation source by performing Particle Image Velocimetry (PIV) and pressure measurements. Moreover, it is shown that the occurrence of hydro-acoustic resonances in cavitation conditions mainly depend on the swirl degree of the flow independently of the speed factor. Empirical laws linking both natural and precession frequencies with the operating parameters of the machine are, then, derived, enabling the prediction of resonance conditions on the complete part load operating range of the turbine.

1. Introduction

Francis turbines operating at off-design conditions experience the development of unfavorable flow patterns inducing cavitation and pressure pulsations, such as inter-blade cavitation vortices for very low values of discharge [1] or a self-oscillating cavitation vortex in the draft tube at full load conditions [2]. At part load conditions, the precessing cavitation vortex rope acts as a pressure excitation source for the hydraulic circuit, leading to the propagation of pressure fluctuations at the precession frequency in the whole system [3]. The presence of cavitation in the vortex core decreases the local wave speed and therefore the system eigenfrequencies [4] leading to the risk of resonance [5].

In the framework of the HYPERBOLE research project, the dynamics behavior of the part load vortex rope has been extensively investigated by model tests. This article reports the main results and observations, which led finally to the establishment of empirical laws enabling the prediction of resonance conditions at the reduced model scale. First, the influence of the discharge factor on the vortex dynamics in cavitation-free conditions is investigated by means of both pressure and Particle Image Velocimetry (PIV) measurements. In a second part, the interaction between the excitation source and the system eigenfrequencies is investigated in cavitation conditions by identifying the resonance conditions for different values of the speed factor, as well as the influence of the discharge factor on the precession and natural frequencies. Dimensionless laws linking both natural and precession frequencies with the operating parameters are, then, established, enabling the prediction of resonance occurrences on the complete part load operating range of the reduced scale model.

2. Test-case and experimental set-up

The test-case is a reduced scale physical model of a Francis turbine featuring 16 blades and 20 guide vanes installed on the close-loop test rig of EPFL Laboratory for Hydraulic Machines. The upper part of the hydraulic circuit is given in Figure 1, including the upstream pipe, the spiral casing, the draft tube and the downstream reservoir. Pressure fluctuations are measured on the test-rig by flush-mounted piezo-resistive pressure sensors. Their location is given in Figure 1. Four pressure sensors are installed in one horizontal cross-section of the draft tube cone and regularly spaced by an angle of 90°, enabling the decomposition of the pressure fluctuations into convective and synchronous components. In the upstream pipe, two pressure sensors are installed, enabling the direct measurement of the synchronous pressure fluctuation and therefore the estimation of the system hydro-acoustic response.

PIV measurements are also performed in two different horizontal cross-sections of the draft tube cone by using the same set-up than presented in [6]. The optical access for the camera is provided by a Plexiglass window at the bottom of the draft tube elbow. The acquisition of the PIV images is synchronized with pressure measurements, enabling a phase averaging of the flow velocity fields to recover their periodical evolution over one precession period [6, 7].

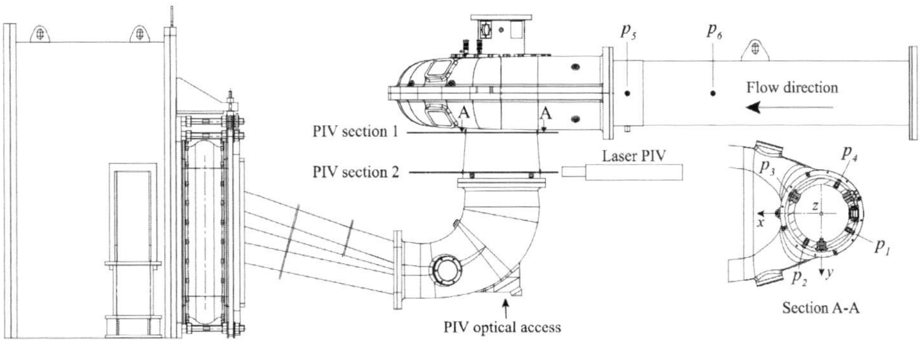

Figure 1. Reduced-scale physical model of a Francis turbine installed on EPFL test rig, together with the location of the pressure transducers and the PIV measurement sections.

3. Vortex dynamics in cavitation-free conditions

3.1. Flow regimes in cavitation-free conditions

Pressure measurements are first performed for a wide range of discharge factor values in cavitation-free conditions. The speed factor is kept constant at the rated value $n_{ED} / n_{ED}^{*} = 1$. The influence of the discharge on the precession frequency and the coherence between pressure signals measured in the cone is reported in Figure 2. Moreover, the corresponding auto-spectrum amplitude is given in Figure 3 for a pressure sensor located in the upstream pipe. This value can be directly linked to the intensity of the pressure excitation source as the measurements are taken in cavitation-free conditions and it is assumed there is no amplification of the synchronous pressure fluctuation as the first system eigenfrequency is sufficiently high.

Three flow regimes are highlighted depending on the value of the discharge factor [6]. The vortex rope starts developing within the first regime and is highly coherent and well-developed within the second regime, for which the precession frequency increases linearly as the discharge is decreased. At the same time, the excitation source intensity increases and reaches its maximum for a value of discharge equal to 65% of the rated value. Beyond 62% of the rated value (regime 3), the vortex rope loses its coherence and periodicity, which is illustrated in Figure 2-b by a drop in the value of C_{xy}. This results in a sudden decrease of the excitation source intensity, as shown in Figure 3.

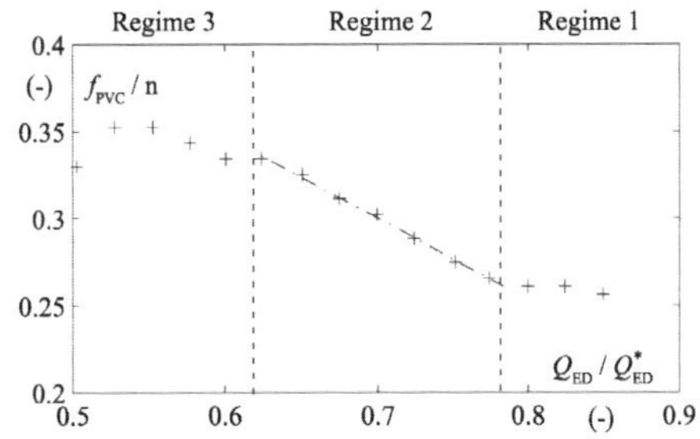

(a) Precession frequency

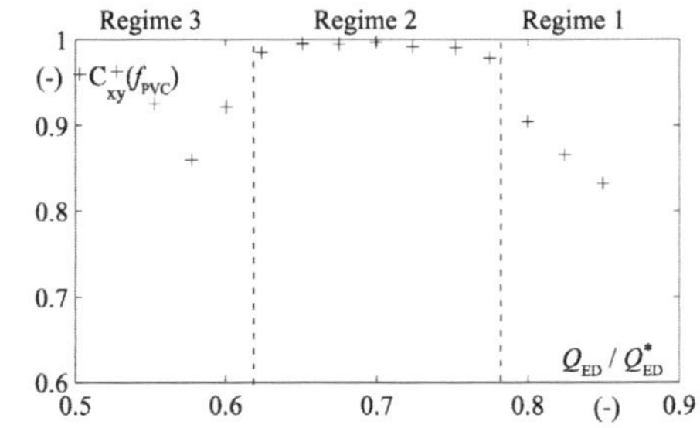

(b) Coherence

Figure 2. (a) Precession frequency made dimensionless by the runner frequency and (b) coherence at the precession frequency between pressure signals measured in the cone (sensors p_2 and p_3) as a function of the discharge factor – n_{ED} / n_{ED}^*= 1.

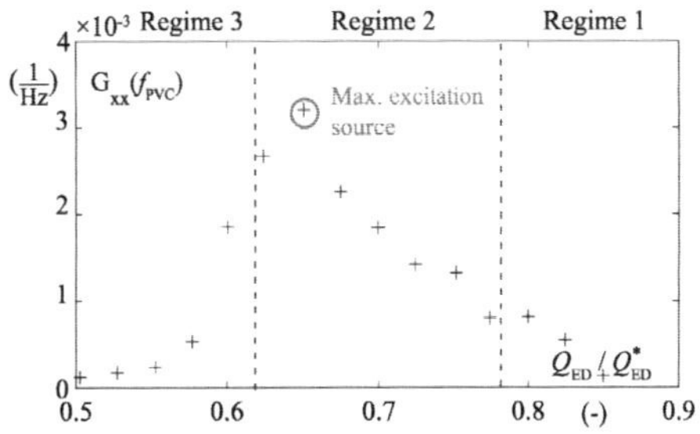

Figure 3. Influence of the discharge factor on the auto-spectrum amplitude at the precession frequency for pressure signals measured in the upstream pipe in cavitation-free conditions – n_{ED} / n_{ED}^*= 1.

3.2. Influence of the flow discharge on the vortex parameters

The tangential flow velocity fields are measured in two horizontal cross-sections of the cone by means of PIV performed in cavitation-free conditions for different values of the discharge factor to link the vortex parameters with the excitation source intensity. For each operating point, the vortex trajectory and circulation are determined by identifying the vortex centre and the vortex core limits with the algorithms proposed by Graftieaux et al. [8]. The influence of the discharge factor on both the vortex trajectory length and the vortex circulation is reported in Figure 4. When the discharge factor is decreased within the flow regimes 1 and 2, the vortex circulation increases while the vortex trajectory widens. They reach their maximum within the flow regime 2 for low discharge factor values when the excitation source intensity is also at its maximum. Beyond the transition from the regimes 2 to 3, the vortex trajectory slightly retracts and the determination of the vortex core limits and the computation of the vortex circulation are not possible anymore [6]. Based on these results, a physical mechanism responsible for the establishment of the pressure excitation source was proposed and discussed in [6]. The flow regime possibly leading to the highest pressure pulsations in case of resonance with the system would correspond to discharge factor values between 62 and 65% of the rated value.

4. Interaction with the system in cavitation conditions

4.1. Resonance conditions in cavitation conditions

In cavitation conditions, similar flow regimes are observed, as shown in Figure 5. The natural frequency is determined by cross-spectral analysis between two pressure signals measured in the cone [9]. Its value strongly decreases in the flow regime 1 when the discharge factor is decreased and is almost constant for low values of discharge factor in the flow regime 2. Resonance conditions occur for a discharge factor value equal to about 78% of the rated value for this n_{ED}-value. By varying the value of the speed factor, similar flow regimes and resonance conditions are observed with, however, a slight shift. This shift corresponds to that of the swirl-free conditions, see Figure 6-b.

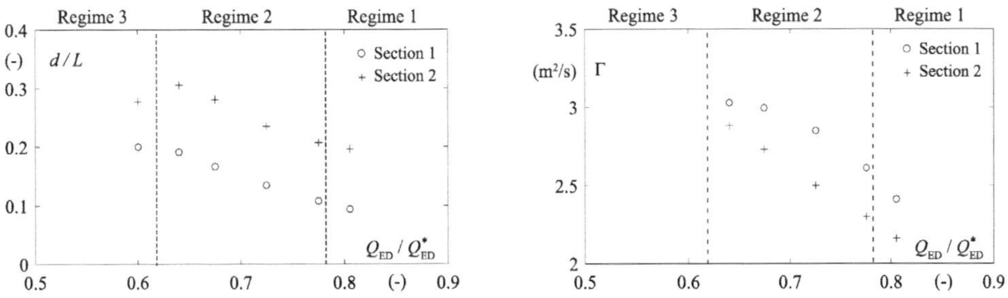

(a) Vortex trajectory length *(b) Vortex circulation*

Figure 4. Influence of the discharge factor on the vortex trajectory and circulation in cavitation-free conditions – $n_{ED} / n_{ED}^* = 1$.

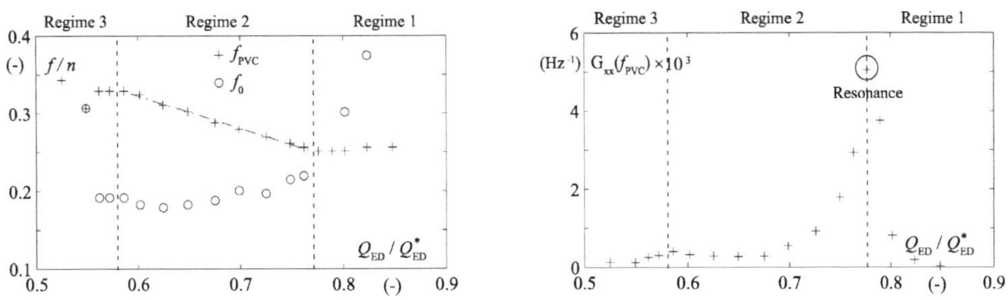

(a) Frequencies *(b) Auto-spectrum amplitude*

Figure 5. Influence of the discharge factor in cavitation conditions on the natural and precession frequencies and on the amplitude of the hydro-acoustic response of the system – $n_{ED} / n_{ED}^* = 1$.

4.2. Resonance prediction at the model scale

The results described above show that the dynamics of the vortex and both precession and natural frequencies seem to be mainly driven by the degree of swirl of the flow. The latter can be characterized by the swirl number *S*, defined as the ratio between the axial flux of angular momentum and the axial flux of axial momentum. By assuming a constant axial velocity $Cm = Q / \pi R^2$ and a solid-body rotation and by using the velocity triangles at the runner outlet, the swirl number can be expressed as a function of the speed and discharge factors:

$$S = n_{ED}\frac{\pi^2}{8}\left(\frac{1}{Q_{ED}} - \frac{1}{Q^0_{ED}}\right) \tag{1}$$

The Strouhal numbers $St_{PVC} = f_{PVC} D^3 / Q$ and $St_0 = f_0 D^3 / Q$ are plotted as a function of the swirl number in Figure 6-a by including all the investigated operating points. The data collapse in one linear function for the precession frequency and in one power function for the natural frequency. The swirl number for which resonances are expected to occur can be identified and is equal to $S = 0.83$. Based on this value, a relation between the speed and discharge factors in resonance conditions is derived:

$$Q^{res}_{ED} = 1 / \left(\frac{0.673}{n_{ED}} + \frac{1}{0.6674 n_{ED} + 0.0516}\right) \tag{2}$$

In this expression, the value of the discharge factor Q^0_{ED} in swirl-free conditions is approximated by a linear fitting, see Figure 6-b. The occurrence of resonances can be finally predicted on the complete part load operating range by using Equation 2. The resonance conditions identified experimentally and the computed ones show an excellent agreement, see Figure 6-b. Further investigations will focus on the transposition of the natural frequencies to the prototype scale by using 1D SIMSEN models. Resonance conditions will be finally predicted at the prototype scale and the results will be validated by on-site measurements performed on the prototype generating unit.

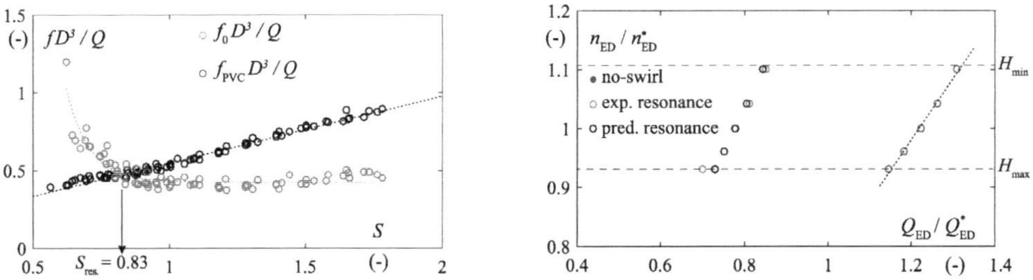

(a) Strouhal numbers versus Swirl number *(b) Computed and observed resonances*

Figure 6. Strouhal numbers versus swirl number and comparison between computed and observed resonance conditions

5. Conclusion

Based on PIV survey coupled with pressure measurements, different flow regimes are highlighted, which depend on the value of the discharge. The link between the vortex dynamics and the pressure excitation source intensity is highlighted. The excitation intensity reaches its maximum when the flow circulation and the vortex trajectory are also at their maximum. Beyond a given value of discharge, the vortex dynamics lose its coherence and periodicity, resulting in a drop in the excitation intensity.

In cavitation conditions, similar flow regimes are observed and resonance conditions are identified by spectral analysis between pressure signals for different values of speed factor. It is shown that both precession and natural frequencies mainly depend on the degree of swirl of the flow exiting the runner. Empirical laws linking both precession and natural frequencies with the operating parameters are established, enabling the prediction of resonance conditions on the complete hillchart at the model scale.

These results demonstrate that the experimental determination of both precession and natural frequencies for a limited number of operating points enables finally the prediction of resonance conditions on the complete part load operating range. The results will be transposed to the prototype scale by using 1D SIMSEN model and the methodology will be validated by on-site measurements.

Acknowledgements

The research leading to the results published in this paper is part of the HYPERBOLE research project, granted by the European Commission (ERC/FP7- ENERGY-2013-1-Grant 608532).

References

[1] Yamamoto K, Müller A, Favrel A, Landry, C and Avellan F 2015 *Proceedings of the 6th IAHR Workgroup, Ljubljana, Slovenia*
[2] Müller A, Dreyer M, Andreini N and Avellan F 2013 *Experiments in Fluids* **54**
[3] Nishi M, Matsunaga S, Kubota T and Senoo Y 1982 *Proceedings of the 10th IAHR Symposium, Tokyo, Japan*
[4] Landry C, Favrel A, Müller A, Nicolet C and Avellan F 2016 *Journal of Hydraulic Research* **54** 185-196
[5] Favrel A, Landry C, Müller A, Yamamoto K and Avellan F 2014 *IOP Conference Series: Earth and Environmental Science*
[6] Favrel A, Müller A, Landry C, Yamamoto K and Avellan F 2015 *Experiments in Fluids* **56** 1-15
[7] Favrel A 2016 *Dynamics of the cavitation precessing vortex rope for Francis turbines at part load operating conditions* EPFL Ph.D. Thesis, Lausanne
[8] Graftieaux L, Michard M and Nathalie G 2001 *Measurement Science and Technology* **12** 1422-1429
[9] Favrel A, Müller A, Landry C, Yamamoto K and Avellan F 2016 *Experiments in Fluids* **57** 168

Influence of guide vane opening on the flow phenomena in a pump turbine during a fast transition from pump mode to generating mode

C Stens and S Riedelbauch

University of Stuttgart, Institute of Fluid Mechanics and Hydraulic Machinery, Pfaffenwaldring 10, 70550 Stuttgart, Germany

christine.stens@ihs.uni-stuttgart.de

Abstract. Due to a more fluctuating energy production caused by renewable energies such as wind and solar power, the number of changes between operating points in pumped storage power plants has increased over the last years. To further increase available regulating power, it is desirable to speed up these changes of operation conditions in Hydro units. Previous studies showed that CFD is well capable of predicting the flow phenomena in the machine under unsteady conditions for a large guide vane opening angle. The present paper investigates the benefits of nearly closed guide vanes during the transition. Results are compared between the two different angles as well as between simulation and measurement.

1. Introduction

Pumped storage power plants can be used to resolve imbalances in the grid by providing energy or storing excessive power as required. For reversible pump turbines, the change from one operating mode to the other is equivalent to a change in rotational direction. With the development and improvement of doubly-fed induction machines, machines can run at variable speed and additional brake torque can be provided for slowing down the rotation in pump mode. Apart from a regulation of power consumption or output via rotational speed, this enables a faster transition from pump mode to generating mode. However, this manoeuvre requires knowledge about flow phenomena in the machine to avoid damages.

CFD has been used by various authors to analyse unsteady conditions in Francis and pump turbines, such as runaway (e.g. [1-3]) start-up [4], speed no load conditions [5] or load rejection [6]. In previous work, the authors investigated the transient for a relatively large guide vane opening angle of 25° and compared simulated results to measured data [7, 8]. Generally, a good agreement between simulation and measurement was achieved. Severe pressure fluctuations were found especially in pump brake mode. In practice, the guide vanes may be closed to a smaller angle before the transient is started. However, this prolongs the time of the process.

The present paper aims at determining whether a smaller guide vane angle significantly reduces the fluctuations. Furthermore, it serves as a second validation case for CFD methodology for highly dynamic flow conditions.

2. Simulation setup

Simulations are carried out using OpenFOAM® 2.3 and guide vane angles of 5° (y5) and 25° (y25). The meshes contain approximately 20M nodes distributed over four domains which are connected via arbitrary mesh interfaces. Time step is constant and k-omega-SST-model is used for turbulence. Flow rate is taken from experiment and prescribed at the respective inlet, i.e. at the draft tube in pump mode and at the spiral case in generating mode. As the measured signal of flow rate was found to be delayed compared to all other acquired data, it was shifted in time by a defined value. However, further investigations of experimental data indicated that this time shift is smaller than assumed in previous work [7, 8]. This paper uses flow rate based on the latest findings, so that results for y25 differ slightly from earlier results.

Transition time is 8 s for both opening angles and change in rotational speed is linear. This leads to a steep decrease of flow rate in pump mode. In pump brake mode, flow rate rapidly increases in the opposite direction. For y5, flow rate stays nearly constant for the rest of the process, independently of the still increasing rotational speed. For y25, flow rate continues to increase slowly until the end of the transient.

For comparison with experiment, pressure is evaluated at certain points in the machine, where pressure sensors were located during measurement. The positions of the sensors are described in [9].

3. Global results

Figure 1 presents the comparison between simulated and measured results in a four quadrant plot. The smaller guide vane angle is characterized by a nearly constant Q_{ED} over a wide range of n_{ED} values in pump brake and generating mode, while the discharge factor Q_{ED} for the larger opening angle continuously increases. The agreement between simulation and experiment is very good for both cases and improves the results found with the different flow rate in [8], supporting the assumption that the previous flow rate was delayed with respect to rotational speed.

Figure 2 shows the evolution of head over time. At the beginning of the transient, the head difference between measurement and simulation is comparable for both guide vane openings. The largest deviations are found in pump brake mode, where CFD for the smaller angle severely overestimates the head rise after the reversal of flow direction. For y5, the deviation in the four quadrant plot translates to a significant difference in simulated head in pump brake and generating mode of approximately 20% of reference head between 3 s and 5 s. For the larger angle, a similar deviation appears at 5 s, but with lower amplitude.

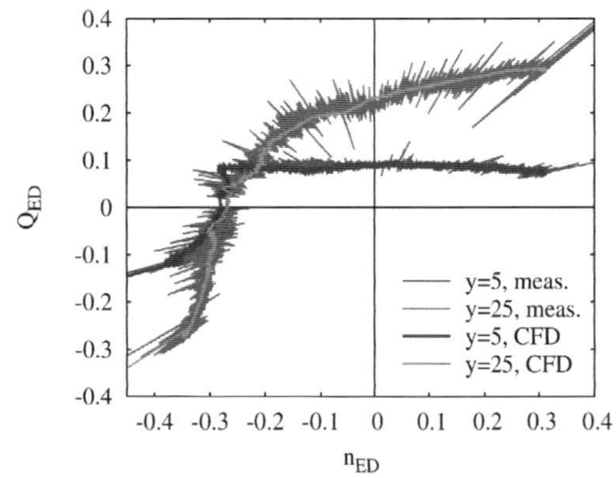

Figure 1. Comparison of simulation and measurement in a four quadrant plot.

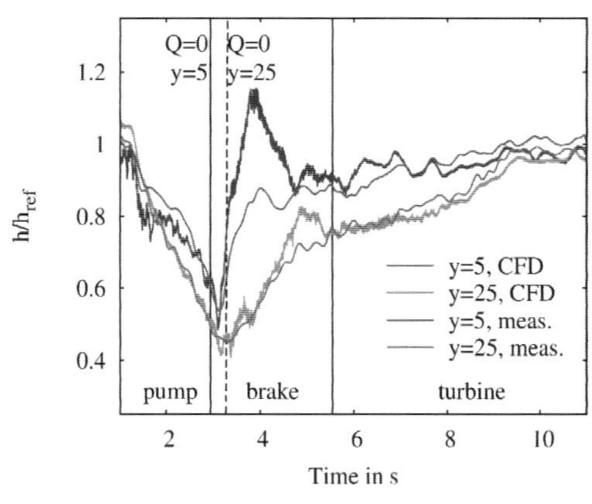

Figure 2. Simulated vs. measured head over time.

4. Results in the guide vane channels

Figure 3 shows the comparison of the pressure signals between simulation and experiment for both opening angles together with a moving average with a sample length of 0.01 s. Over large parts of the transient, simulation accurately captures both mean values and amplitudes of the fluctuations between the guide vanes. The deviation in head for y5 in pump brake mode is reflected in the signal, as well as the differences at the beginning and the end. For the larger angle, amplitudes in measurement are significantly higher than in simulation at the start of generating mode. Compared to flow rate used in previous work [8], agreement is improved until 7.1 s, where the difference between simulation and experiment increases.

Generally, the amplitudes of the pressure fluctuations are visibly larger for the smaller angle than for the larger in pump brake mode. In generating mode, they quickly fade out and reach a level that is below the one for y25. This is correctly captured by simulation. Via FFT, the dominating frequency can clearly be identified to be the blade passing frequency. In pump brake mode, additional frequencies are found for both cases, but they randomly appear and disappear over time.

Higher fluctuations in pressure translate to higher fluctuations in guide vane torque, especially in pump brake mode and at the beginning of generating mode. Additionally, the general level of guide vane torque increases for a smaller opening angle.

In previous work, rotating stall was identified at the end of pump mode for y25. This is not found in the guide vanes for the small guide vane angle, where only a short channel exists between adjacent guide vanes. However, a non-uniform velocity distribution between the channels as well as over the channel height is found in both cases. Near the end of pump mode at 2.6 s and with y5, all channels experience backflow or zero flow near the hub and the shroud, while flow direction in the middle of the channels is outward. This is consistent with the behaviour for the larger angle shortly before flow rate reverses.

At 4 s, time of high deviation from measured head for y5, flow is found to be stable from the stay vanes up to the guide vane channels. In the vaneless space between runner and guide vanes, only small vortices exist near the top and bottom of the twin cascade, while flow in the middle of the channel is stable.

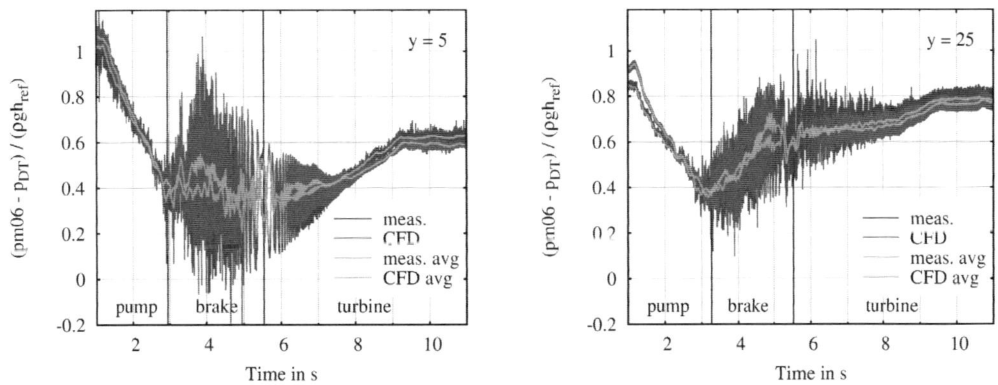

Figure 3. Pressure between the guide vanes for both opening angles.

5. Results in the runner

In the runner, two pressure sensors are positioned near the turbine leading edge of the blades, one on suction side and one on pressure side. The corresponding signals over time are presented in figures 4 and 5. Like the signal from the guide vane sensors, the pressure side signal reflects the findings from the head curve, i.e. the values are overestimated in pump mode, underestimated in generating mode and show the peak in pump brake mode at 4 s for y5.

Results on suction side are in good agreement with measured data. For the larger opening angle, high amplitude fluctuations appear between 3.5 s and 8.5 s which are not reproduced in simulation. For a smaller guide vane opening, fluctuations are reduced, but simulated amplitudes are still slightly lower than measured ones.

Although fluctuations at the pressure sensors are comparable for both openings, a smaller guide vane opening causes large fluctuations in the torque contributions of each runner blade during pump mode, especially at the beginning. Vortices are found in the runner channels close to the suction side at approximately 50% of chord length at t = 1s, the start of the transient, where flow is stable for a larger guide vane opening. Vortices appear and disappear without systematically rotating through the runner channels.

Apart from the single monitor points where a comparison with measurement is possible, simulated pressure on the runner blades is evaluated along a line on pressure side at 50% span and analysed over time. Overall maximum and minimum values for y5 are found at 4 s, the time of maximum deviation in head between simulation and measurement. This leads to a steep pressure gradient from 25% to 45% of chord length for this operating point.

For y25, maxima and minima are less pronounced and the maximum is flatter. The difference in head and pressure on the pressure side for y25 around 4.8 s also coincides with the highest maximum pressure and pressure gradient over time for this configuration. This leads to the conclusion that simulation overestimates the pressure maxima and pressure gradient in the runner under certain flow regimes and maximum deviation from experiment coincides with maximum pressure and maximum pressure gradient on the pressure side of the blades.

For the larger angle, stable flow establishes after the end of the transient. However, simulation overestimates measurement. For y5, simulation predicts fluctuations on pressure side even after the end of the transient, caused by vortices in the runner channels. Although the flow is more complex, a better agreement with measurement is obtained.

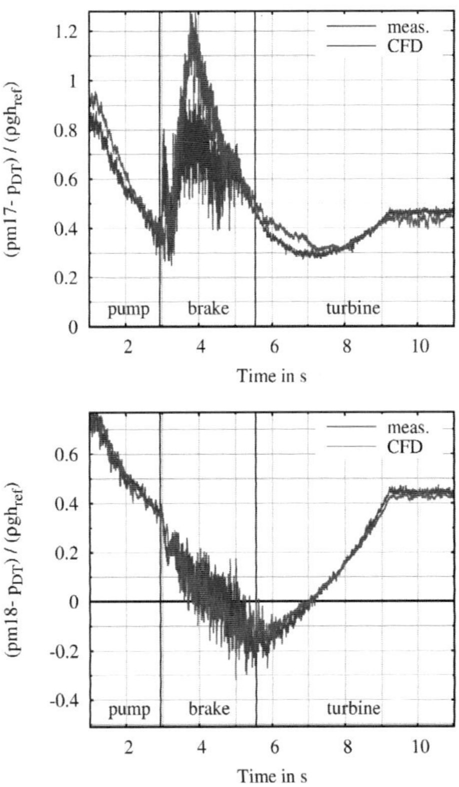

Figure 4. Pressure on the leading edge for y5 on pressure side (top) and suction side (bottom).

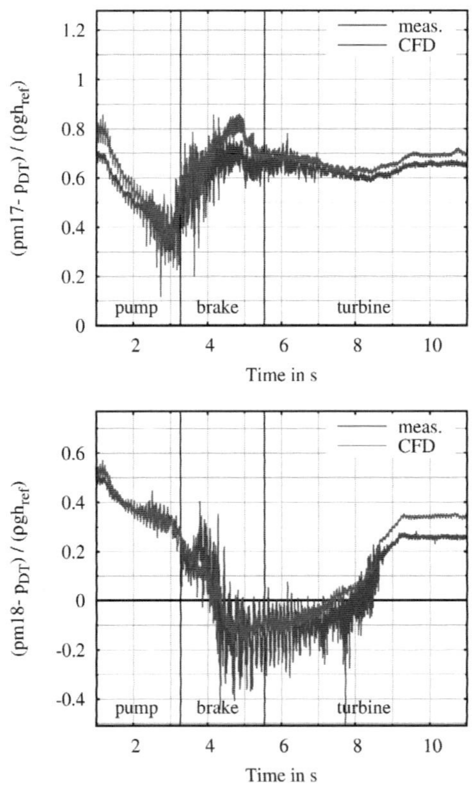

Figure 5. Pressure on the leading edge for y25 on pressure side (top) and suction side (bottom).

6. Results in the draft tube

In pump mode, stable upwards flow establishes in the draft tube under the runner for y25. For y5, flow rate is too low for stable flow, so that flow is detached from the walls with downward flow close to the walls and upward in the middle. The fluid rotates with the runner direction in pump mode and dissipation mode and reverses the direction of rotation at 8 s. At this point of time, the upwards flow shortly collapses and axial flow direction is downward in the complete cross section. As the fluid resumes its rotation, the upwards flow in the middle of the draft tube is re-established.

The detached flow at the beginning of the transition leads to larger fluctuations in the pressure signal. However, the fluctuations seen between 4 s and 8 s for y25 are significantly reduced in both measurement and simulation. With the corrected flow rate, simulation is still not capable of resolving the low frequency fluctuations from the vortex rope seen in measurement for the larger angle; however, some improvement is achieved.

7. Conclusion and outlook

A comparison of a fast transient of a reversible pump turbine from pump mode to generating mode with two different guide vane angles shows how the angle affects pressure fluctuations and flow phenomena in the machine. By taking the additional time to close the guide vanes before the transient, fluctuations in the draft tube are significantly reduced. A small reduction is also visible on the suction side of the runner blades. On the other hand, a smaller guide vane angle increases fluctuations in the guide vanes, especially in pump brake mode. This causes a higher torque on the guide vanes.

Generally, CFD shows a good agreement with measurement for both cases. Differences appear in pump brake mode, where high maximum pressure values and gradients are found on the pressure side of the runner blades. Compared to previous results for y25 obtained with a time shifted flow rate, the current work shows a better agreement. From a CFD point of view, the time shift used in the current work is thus more probable than the older value.

Future work will investigate a coupling of CFD with a 1D method to simulate the complete test rig. This allows deriving flow rate independently of measurement.

References

[1] Cherny S, Chirkov D, Bannikov D, Lapin V, Skorospelov V, Eshkunova I and Avdushenko, A 2010 *IOP Conference Series: Earth and Environmental Science* **12** 012071

[2] Fortin M, Houde S and Deschênes C 2014 *Proc. of 27th Symp. of Hydraulic Machinery and Systems* Montreal Canada

[3] Li J, Yu J and Wu Y 2010 *IOP Conference Series: Earth and Environmental Science* **12** 012001

[4] Nicolle J, Morissette J F and Giroux A M 2012 *IOP Conference Series: Earth and Environmental Science* **15** 062014

[5] Casartelli E, Mangani L, Romanelli G and Staubli T 2014 *Proc. of 27th Symp. of Hydraulic Machinery and Systems* Montreal Canada

[6] Moessinger P, Jung A 2016 *Proc. of 28th Symp. of Hydraulic Machinery and Systems* Grenoble France

[7] Stens C and Riedelbauch S 2015 *Cavitation and Dynamic Problems: 6th IAHR Meeting of the Working Group* ed A Lipej and S Muhic (Ljubljana, Slovenia) p 323

[8] Stens C and Riedelbauch S 2016 *Proc. of 28th Symp. of Hydraulic Machinery and Systems* Grenoble France

[9] Ruchonnet N and Braun O 2015 *Cavitation and Dynamic Problems: 6th IAHR Meeting of the Working Group* ed A Lipej and S Muhic (Ljubljana, Slovenia) p 264

Going Large or Going Small in Plant Design: *Comparison between a P.P. with three small Kaplan turbines and a P.P. with just one Large Kaplan turbine.*

C Castro-Otero
Hydro Manager, BALIÑO, S.A. Spain.

Abstract. Very often small turbine manufacturers are requested to produce sizeable turbines, too large in terms of physical dimensions, power or designing capacity. In these cases clever alternative solutions should be found to meet customers´ needs. For instance: in the old times twin runner Francis turbines were an option instead of one large machine, or if a too large Pelton turbine cannot be manufactured or designed, a good option is to install a medium size Francis and a small Pelton. Likewise, a similar approach needs to be taken should the manufacturer be asked for a *too large Kaplan*. Facing this situation a good option is to install three or more small Kaplan turbines. This particular case was studied in depth and after all the considerations had been made, the following question arouse: Is this a way out for the manufacturer or is it really the best option for the customer? The choice made as a way out for the manufacturer became the *best option* for the customer and a success for both parties. This paper aims to encourage developers and engineering firms to search for more options than the traditional one to find the best option in plant design.

1. Introduction

When tendering power plants or turbines, several questions should be taken into account, for instance: equipment cost, efficiency, energy production, civil works issues, transport of equipment and so on. Generally, in the scope of supply received by manufacturers clients include the type and number of turbines and generators to be delivered. In the present case a "large" Kaplan turbine was requested to work in the Xallas River in Galicia, Spain with the following conditions:

-Hn= 21 m.

-Q= 35 m^3/s.

The author acknowledges that the necessary turbine was not extremely large for the company he represents, however the location of the plant made the installation of such a large machine difficult to carry out, too expensive and challenging. Additionally, new roads and bridges had to be built to transport and erect the turbine and generator. Thus, it was necessary to find an alternative to avoid the aforementioned problems and consequently the flow was divided several times to be diverted to different turbines.

A balance between costs and number of machines was found employing three Kaplan turbines. The entire project using three small turbines was studied *and findings were compared with an alternative one using just a single large turbine.*

IOP Conf. Series: Journal of Physics: Conf. Series 813 (2017) 012025 doi:10.1088/1742-6596/813/1/012

2. Plant and turbine description.

The selected turbines were:

-First option: using three small turbines.

- Vertical Kaplan.
- Steel Spiral Casing partially embedded in concrete.
- Runner diameter D_{22}= 1500 mm.
- Generator stands directly on the turbine. n= 375 rpm.

-Second option: using one large turbine.

- Vertical Kaplan.
- Steel Spiral Casing partially embedded in concrete.
- Runner diameter D_{22}= 2400 mm.
- Generator stands on a concrete floor. n= 230 rpm.

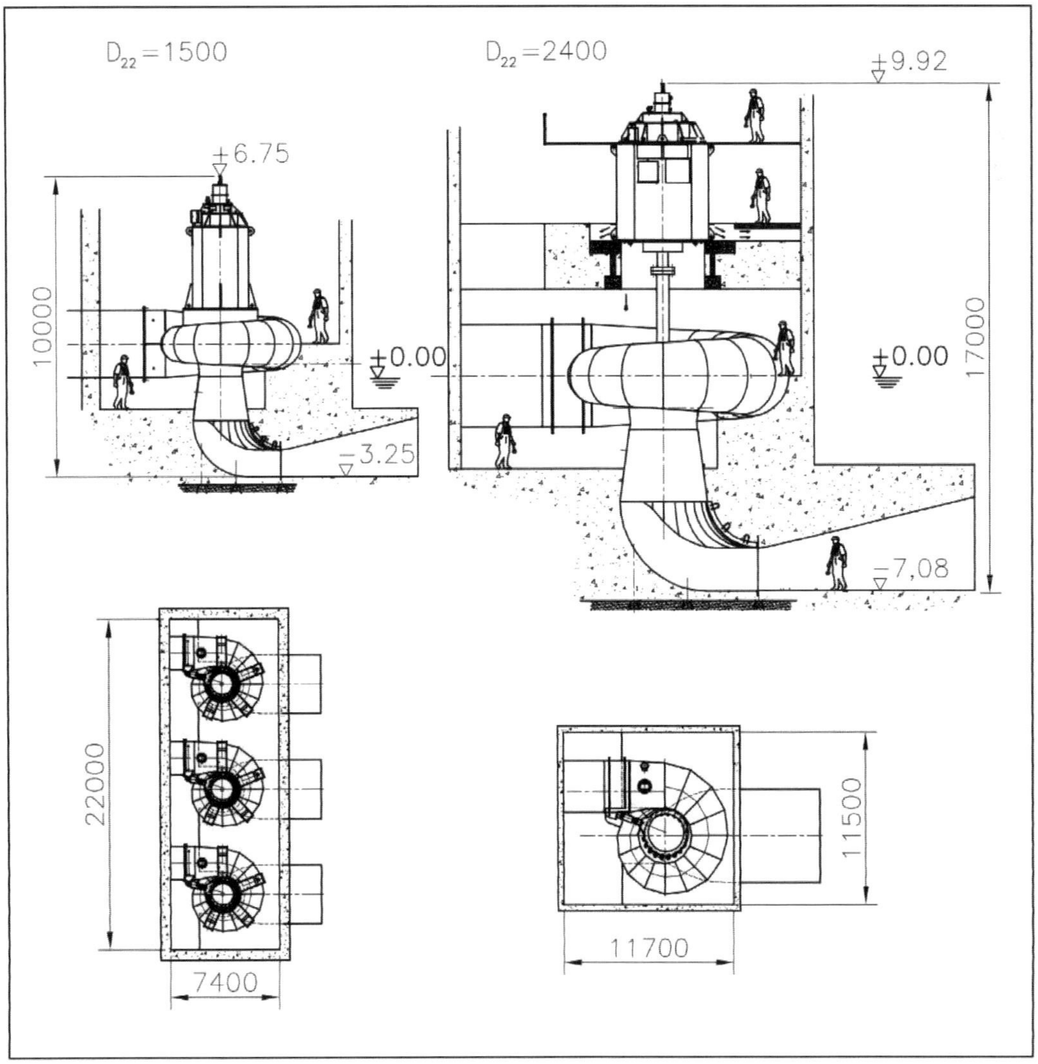

Figure 1. The drawing shows the general disposition of both options.

2.1. Small turbines

As seen in figure 1 the generator for the small turbine, weighting 24 ton, seats directly on the turbine upper cover thus, the distance between generator and the runner is relatively short. Because of the short shaft, the runner can be directly connected to the generator shaft. Additionally, due to the short shaft, using a radial bearing for a runner shaft is no longer required.

The plant is small, 22 m long and 7.4 m wide and the turbine-generator is 10 m high.

For maintenance purposes it is only necessary one crane to lift the generator, which is the heaviest part. The volume of earth removed for the construction of the plant is 529 m^3 and the hole for the draft tube is 3m deep.

2.2. Single large turbine

The large generator cannot stand on the turbine itself due to its weight, 54 ton, and seats on a concrete floor 5 m above the turbine. This means there must be a generator shaft and a runner shaft and also a radial bearing in the turbine to withstand the runner shaft.

The large turbine plant is 11.5 m long and 11.7 m wide and the total height of the turbine and generator is 17 m.

For maintenance purposes two cranes need to be installed; one to lift the generator, which is the heaviest part, and a smaller one over the turbine to manipulate it should it be required.

The volume of earth removed for the construction of the plant is 952 m^3 and the hole for the draft tube is 7m deep.

2.3. Considerations

After studying both turbine options one can conclude that the benefits of utilizing three small turbines outweigh the use of a single larger turbine.

The main advantage of working with the first option is the fact that a small generator does not need a concrete floor to seat the generator on, therefore making the whole plant more cost effective.

On the other hand, the three small turbines plant is 22 m long for the turbines area and requires some more meters for the control room and the loading and unloading area, which is a considerable amount of space, which is not always available.

3. Efficiency

After comparing the dimensions of both options the next point to be checked is their efficiency (see fig.2).

The group of the three smaller turbines has:

-lower peak efficiency but almost flat in a wider flow range.
-higher efficiency at full load.
-higher efficiency at low flow,
-a larger flow range.

The larger turbine has:

-larger peak efficiency,
-slightly lower efficiency at full load.
-lower efficiency at low flow,
-a shorter flow range.

IOP Conf. Series: Journal of Physics: Conf. Series 813 (2017) 012025 doi:10.1088/1742-6596/813/1/01

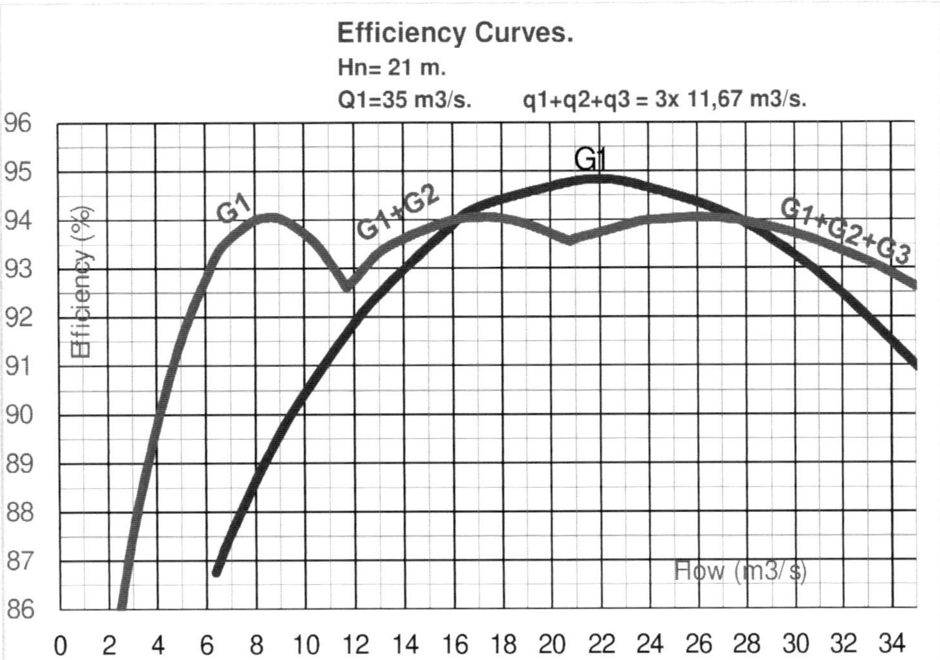

Figure 2. Efficiency curves for a large turbine and a combination of three small ones.

With the three-turbine-option there is more flexibility in the operation of the plant and overall bigger efficiency can be obtained.

In any case, all these differences can be positive or negative depending on the flow duration curve of the river.

The following paragraph will deal with how these differences affect the energy production.

4. Energy Production

The flow duration curve shows the usual shape for small rivers in a rainy region. This means that the flow will fluctuate, having some days of extremely large flow and a few others with very low flow [1]. The energy production including the generator efficiency is as follows:

- -Three small turbines, E= 16.5 GW/year.
- -Single large turbine, E= 14.6 GW/year.

In short, there is an 11% increase in energy production with the first option.

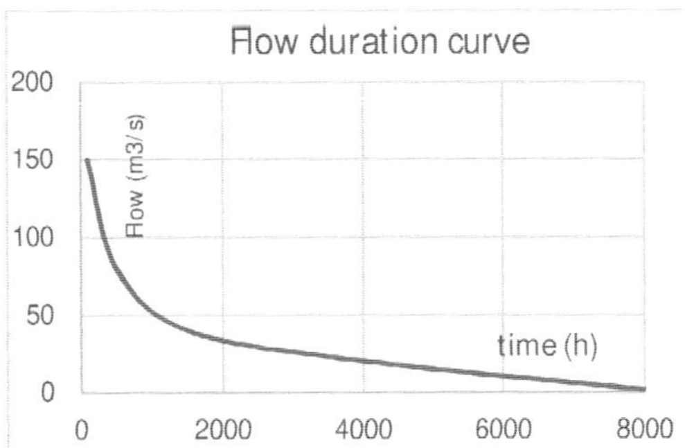

5. Price

Mainly due to the fact that the weight of the three-small-turbine option is lighter than that of the large turbine, the price of the first option is therefore lower.

As the company has the expertise of manufacturing both types of turbines, it has been found that the large turbine option is 35% more expensive. The price for the generator is approximately the same.

6. Maintenance

As it has been previously explained, the plant with three small turbines does not have a separate floor for the generator, making general maintenance tasks easier since all the problems will occur on the same floor. Additionally, the fact of avoiding a runner bearing is an advantage as this is usually a critical part of a power plant.

Using the small-turbine-option has a drawback because there are 3 of them making failures 3 times more likely to happen. Consequently more attention has to be paid especially to oil aggregates and electrical equipment where frequent and small failures tend to arise.
On the other hand, the risk of loss of production related to unit failure or maintenance is reduced by having three machines. In case one stops working the producer still has the other two turbines running.

7. Conclusions

Many variables must be taken into account when designing a power plant. In preliminary plans general decisions are taken and main guidelines are stablished, but things can be changed after a thorough study of the project and doing the calculations.

This is exactly what happened in this case; initially it was apparently clear that the best option was a single large turbine albeit when the developer took the opinion and calculations of the manufacturers into consideration, the outcome surprised both the manufacturer and the client when found that the best choice was different to the original one.

It goes without saying that what it has been proven an advantage in this project might be a problem in another and a single large turbine might be the best option.

References

[1] European Small Hydropower Association. 2004. Guide on How to Develop a Small Hydropower Plant.

IOP Conf. Series: Journal of Physics: Conf. Series **813** (2017) 012026 doi:10.1088/1742-6596/813/1/012026

Valve exploiting the principle of a side channel turbine

Pavel Jandourek[1], František Pochylý[1] and Vladimír Haban[1]

[1]Victor Kaplan Department of Fluid Engineering, Faculty of Mechanical Engineering, Brno University of Technology, Technická 2896/2, 61669 Brno, Czech Republic

Abstract. The presented article deals with a side channel turbine, which can be used as a suitable substitute for a pressure reducing valve. Pressure reducing valves are a source of high hydraulic losses. The aim is to replace them by a side channel turbine. With that in mind, hydraulic losses can be replaced by a production of electrical energy at comparable characteristics of the reducing valve and the side channel turbine. The basis for the design is the loss characteristics of the pressure reducing valve. Thereby create a new kind of turbine valve with speed-controlled flow in dependence of the runner revolution. It is technical innovation and new renewable source of energy, which can be in future used in rehabilitation or projecting of pumped-storage power plants. It also increases the power of the power plant.

1. Introduction

At the pumped-storage power plant Dalešice the water for cooling is taken directly from the penstock, see 'Figure 2'. Pressure of the water is reduced in pressure reduction valve. The water taken for cooling is not used to produce electrical energy. During throttling in the valve, part of the energy is transferred in the form of heat to the cooling water and warms it up. The water pressure is throttled from 8.5 to 4 bars. With this comes the potential of energetic utilization. The flow is 0.145 m^3s^{-1} and the possibility of regulation of the flow to 0.08 m^3s^{-1} is required. Another utilization can be in the water mains, energetics, chemical processes or in the place where the conversion of energy into heat occurs due to throttling.

Figure 1. Pumped-storage power plant Dalešice (Czech Republic) (http://www.cez.cz)

Figure 2. Pressure reducing valve in a cooling system [1]

IOP Conf. Series: Journal of Physics: Conf. Series **813** (2017) 012026 doi:10.1088/1742-6596/813/1/012026

Pressure reducing valves are a source of high hydraulic losses and their replacement is possible thanks to the comparable resistance characteristics with the side channel turbine. Characteristics of the valve flow coefficient K_v on the relative opening of the valve, see 'Figure 4', is analogous to the turbine characteristic of unit flow Q_{11} on unit speed n_{11}, see 'Figure 6'. A medium pressure drop across the valve is needed for energy recuperation.

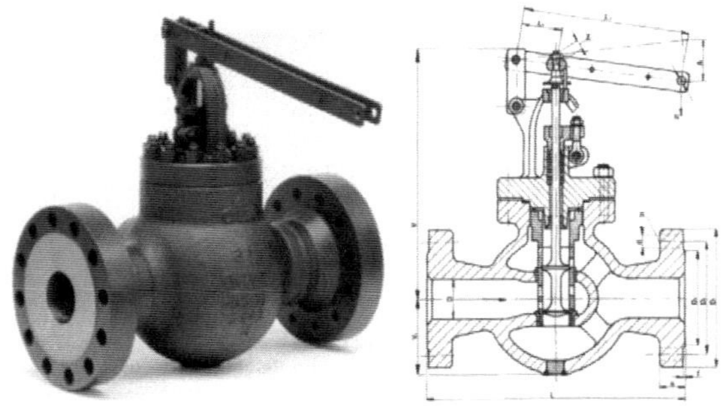

Figure 3. Pressure reducing valve (http://www.ldmvalves.com)

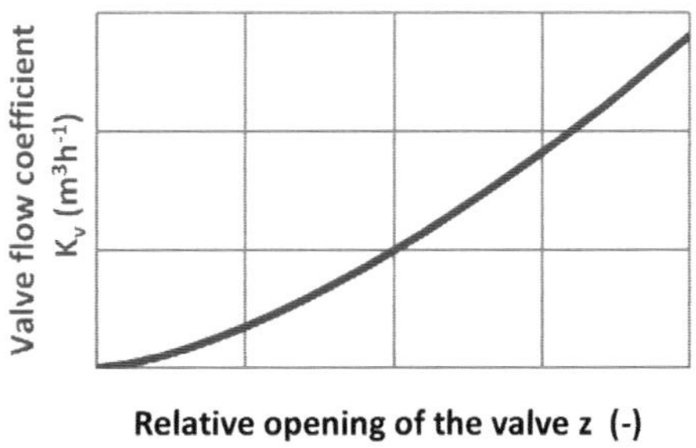

Figure 4. Flow characteristic of a pressure reducing valve

2. Side channel turbines

The side channel turbine falls into the category of reaction turbines and is intended for small discharges and medium heads. The working area is within the specific speed $n_s = 2 \div 30$ min^{-1}. The turbine can only process a part of the head. The main disadvantage is low efficiency, which is usually within the range of 20 to 50%. These machines find their use due to the fact that reaching specific speed $n_s < 35$ min^{-1} with classical turbines (Francis' turbine or multistage pump in turbine mode) at high efficiency is quite problematic. Under these circumstances, a counteracting vortex starts to occur in the channels between the blades of the impeller. This causes the decrease of efficiency which can even get below 30%. In a situation like this, it might be worth using side channel turbines as they have higher efficiency, than the low specific speed turbines.

Side channel turbines are useful for regulation due to their flat curve of efficiency in the area of optimum point and linear dependency of flow on revolutions. Theoretical energy efficiency characteristics η, unit flow Q_{11}, power P_{11} and torque M_{11} are plotted in dependence on unit speed n_{11} in 'Figure 6' [2].

Figure 5. Side channel machine (http://www.delavanagpumps.com)

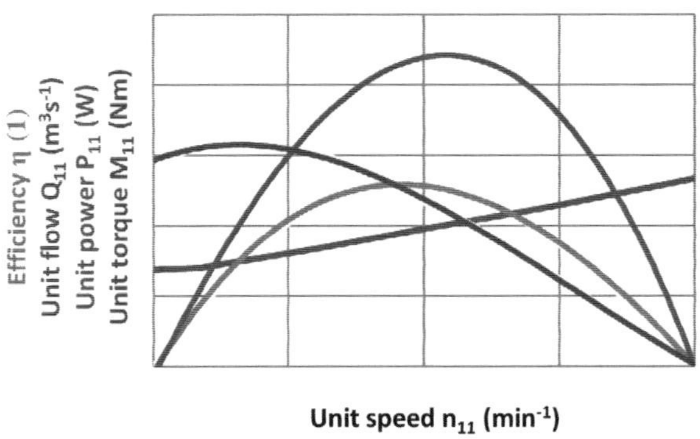

Figure 6. The course of the energy characteristics of side channel turbines

3. Measurement of the side channel turbine

3.1. The pump in turbine mode

It is further examined, whether it is possible to use the side channel pump as a turbine for energy recovery in the piping system. For this experiment, a commercially produced side channel pump was

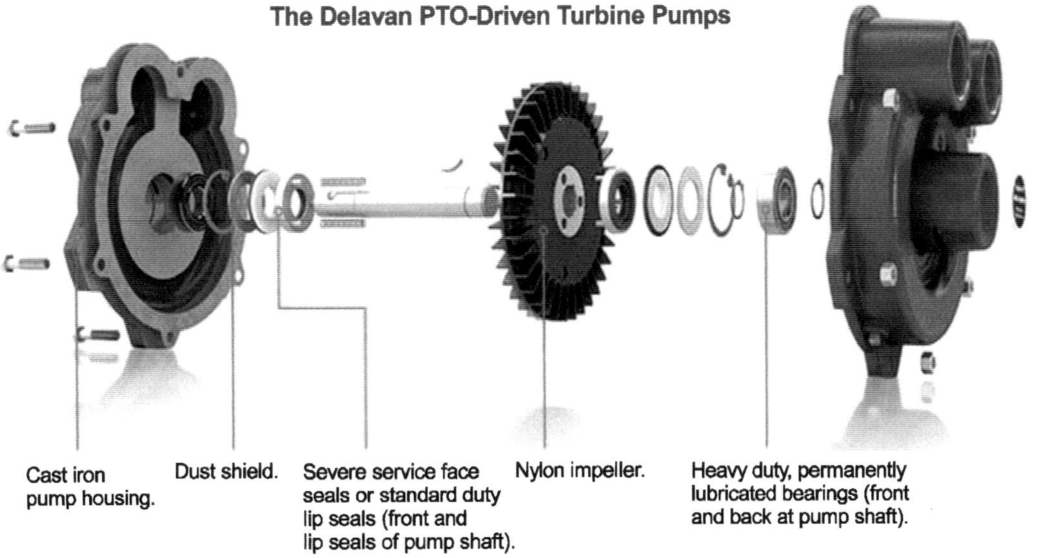

Figure 7. Assembly of side channel machine (http://www.delavanagpumps.com)

purchased from the American company Delavan, labeled Turbo 90, see 'Figure 7'. The pump has a closed type impeller. This means that the water enters and leaks directly into the channel. The impeller is symmetrical and has 36 blades on each side. The opposing blades are rotated by half the angular pitch. The side channel is part of the iron casting cabinet, which according to the manufacturer should endure 7 bar overpressure. The size of the connecting pipe is 3/2 inch. The shaft is sealed with ceramic - carbon mechanical seals. The maximum speed of the wheel is 1200 min^{-1}. The width of the side gap δ is approximately 0.1 mm. The dimensions of the hydraulic parts are shown in 'Figure 8'. The impeller is made from nylon. It is connected with the hub straight-sided splines and can move in the axial direction. It is self-centered during operation.

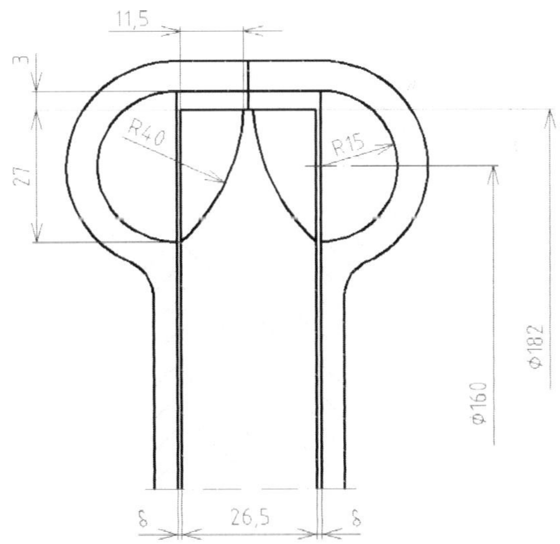

Figure 8. Dimensions of the hydraulic shape

3.2. Measurement of the characteristics of the side channel turbine

The side channel pump is measured in turbine mode. A test rig for the measurement of the energy characteristics of the side channel turbine is located in the laboratory of Victor Kaplan dept. of fluid engineering 'Figure 9 and 10'. Frequency converters are used for change of speed frequency of the turbine runner and feed pump impeller. Values of inlet and outlet pressure, torque, speed and flow rate are recorded by an industrial computer and our program, which is written in Labview.

Figure 9. Test rig

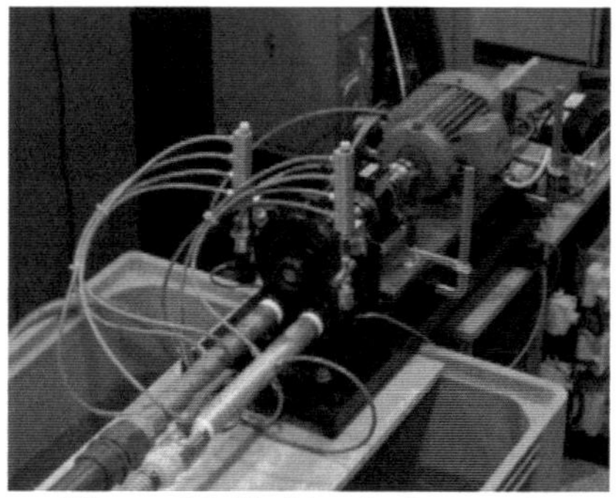

Figure 10. Turbine in the test rig

3.3. Characteristics of the turbine at constant heads with back-pressure

This section investigates the change of the characteristics of the turbine interposed in the piping system as a replacement of a pressure reducing valve. Turbine works only with a part of the pressure energy from the system. The characteristics of the turbine are measured at a constant head 15 meters and a changing speed of the runner. The change in the absolute values of the back-pressure in the suction line during the measurement is shown in the graph 'Figure 11' - $p_b = f(n_{11})$. The back pressure is not constant, but rather depends on the flow rate. Maximal reached value of efficiency under backpressure is 18,9% and 21,5% with the atmospheric pressure on the suction. Flow is maintained.

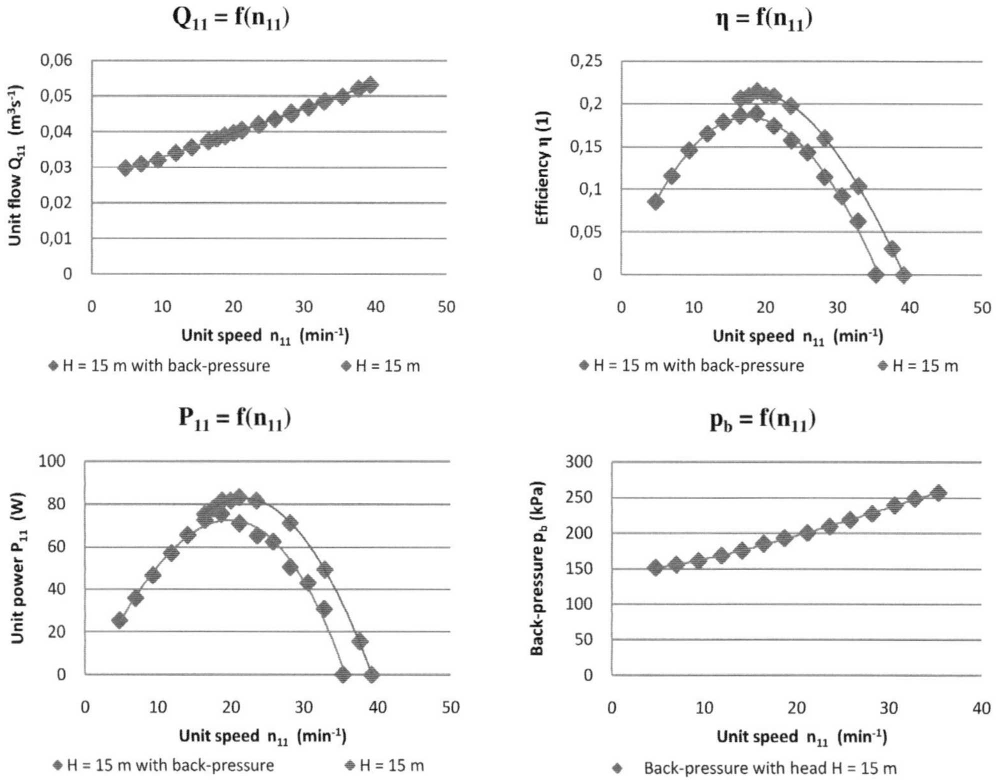

Figure 11. Turbine graphs measured with constant heads and back-pressure in suction

Characteristics of the turbine valve - during operation of the side channel turbine in the capacity of the pressure regulator is expected that only part of the pressure energy in the piping system will be used. Losses-making and flow characteristics of the side channel turbine Delavan Turbo 90 are shown in

'Figure 7'. The characteristic curve of the turbine is comparable as the curve of the pressure reducing valve.

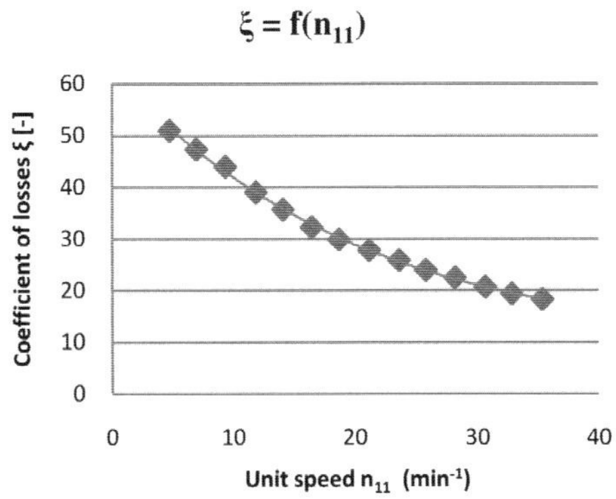

Figure 12. Loss-making characteristic of the turbine

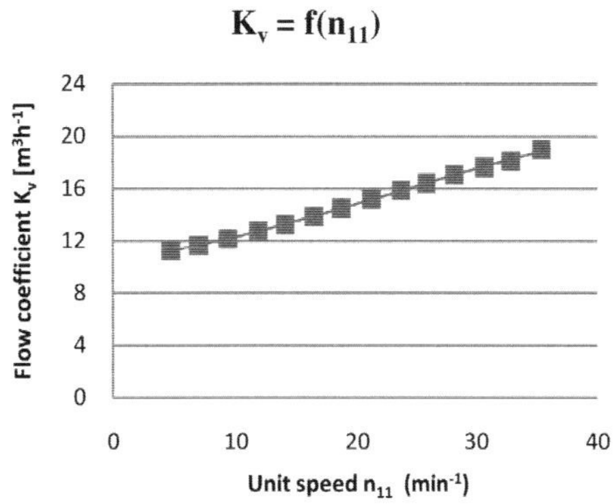

Figure 13. Flow characteristic of the turbine

4. Conclusion

In many technical solutions, it is necessary to decrease the pressure of liquids. Pressure reducing valves are normally used for this pressure reduction. However, it utilizes a considerable part of the pressure energy, which is not used, could be transformed just into electric energy. The pumping technique uses side channel pumps for pumping small quantities of liquids to large transport heights. These pumps are characterized by low-specific speed $n_s = 4 \div 40$ min^{-1} and after a small reconstruction they can be used in a turbine mode. It is possible to replace pressure reducing valves with side channel turbines. Thereby create a new kind of turbine valve with speed-controlled flow dependent on the revolution of the runner.

In order to verify the theory, a commercially produced side channel pump was purchased and used to measure turbine characteristics and verify the function of the partial recovery of energy. The efficiency of recovery is about 19% which opens a space for optimization of the hydraulic design of the turbine valve.

Replacement of the pressure reduction valve by side channel turbine is possible. Turbine valve works on swirl principle. Function of new turbine valve is verified on purchased pump measured in turbine mode. But is necessary to improve hydraulic design and increase hydraulic efficiency. According to the research is expected that custom design can reach efficiency of recovery energy about 50%. It is interesting challenge for future R&D investigation.

References

[1] HALUZA, M.; POCHYLÝ, F.; HABÁN, V.; HUDEC, M.; KUBÁLEK, J. *Study of hydro-potential utilization of cooling water on the hydro-electric power plant Dalešice.* Brno: Brno University of Technology, Faculty of mechanical engineering, 2007. p. 1-16.

[2] KLAS, R.; POCHYLÝ, F.; RUDOLF, P. *Analysis of novel low specific speed pump designs.* In Earth and Environmental Science. IOP Conference Series: Earth and Environmental Science. United Kingdom of Great Britain and Northern Ireland: IOP Science, 2014. p. 1-11. ISSN: 1755-1315.

Acknowledgments

This work is an output of research and scientific activities of NETME Centre, regional R&D centre built with the financial support from the Operational Programme Research and Development for Innovations within the project NETME Centre (New Technologies for Mechanical Engineering), Reg. No. CZ.1.05/2.1.00/01.0002 and, in the follow-up sustainability stage, supported through NETME CENTRE PLUS (LO1202) by financial means from the Ministry of Education, Youth and Sports under the „National Sustainability Programme I".

IOP Conf. Series: Journal of Physics: Conf. Series **813** (2017) 012027 doi:10.1088/1742-6596/813/1/012027

Deep Part Load Flow Analysis in a Francis Model turbine by means of two-phase unsteady flow simulations

Philipp Conrad, Wilhelm Weber and Alexander Jung

Voith Hydro Holding GmbH & Co KG, Alexanderstraße 11, 89522 Heidenheim, Germany

E-mail: philipp.conrad@voith.com

Abstract.
Hydropower plants are indispensable to stabilize the grid by reacting quickly to changes of the energy demand. However, an extension of the operating range towards high and deep part load conditions without fatigue of the hydraulic components is desirable to increase their flexibility. In this paper a model sized Francis turbine at low discharge operating conditions ($Q/Q_{BEP} = 0.27$) is analyzed by means of computational fluid dynamics (CFD). Unsteady two-phase simulations for two Thoma-number conditions are conducted. Stochastic pressure oscillations, observed on the test rig at low discharge, require sophisticated numerical models together with small time steps, large grid sizes and long simulation times to cope with these fluctuations. In this paper the BSL-EARSM model (Explicit Algebraic Reynolds Stress) was applied as a compromise between scale resolving and two-equation turbulence models with respect to computational effort and accuracy. Simulation results are compared to pressure measurements showing reasonable agreement in resolving the frequency spectra and amplitude. Inner blade vortices were predicted successfully in shape and size. Surface streamlines in blade-to-blade view are presented, giving insights to the formation of the inner blade vortices. The acquired time dependent pressure fields can be used for quasi-static structural analysis (FEA) for fatigue calculations in the future.

1. Introduction

The European energy mix undergoes massive changes as more and more new renewable energy e.g. generated by wind and solar penetrate into the energy market. Their unlimited precedence injection and their volatility in power output lead to an imbalance between energy production and consumption. Hydropower plants play an important role in stabilizing the grid by responding quickly to sudden changes in the energy production and demand. An extension of the operating range towards off-design condition including deep-part load is desirable to add flexibility. However, conditions outside the originally indented operating range may lead to harmful loading due to pressure oscillations and cavitation phenomena. Key for long lifetimes and save operations of hydraulic components is the successful prediction of the loads on the prototype. Steady state flow simulations using CFD and structural analyses based on Finite Element Methods (FEM) are applied successfully during the design phase of the hydraulic components, but become inaccurate towards off-design conditions. Single phase-flow simulations for deep part load conditions were published by Magnoli[1]. Wack et al.[2] focused on the cavitation phenomena studying mesh size and outlet boundary treatment. Yamamoto et al.[3]

investigated experimentally the inner blade vortices by means of high speed visualizations at different speed factors.

2. Numerical setup

The computational domain reaches from the spiral case inlet to the draft tube outlet and was discretized by 14.54m hexahedral cells. Runner crown bores and side gaps were neglected to simplify the geometry. Simulations where carried out with the commercial CFD solver ANSYS CFX. Advection and temporal terms were discretized by high-resolution methods and second-order backward Euler, respectively. For two-phase modeling, vapor and liquid water are treated as continuously with constant properties. Because small vapor bubbles are convected with the main flow due to their low inertia the homogeneous multiphase model was employed. The Rayleigh-Plesset cavitation model was implemented to account for phase change. Stochastic flow fields at deep part load conditions demand for advanced turbulence treatment. In this study the so called explicit algebraic Reynolds stress model (BSL EARSM) is used. EARSM models have a comparable computational effectiveness as standard two-equation models and are numerically more robust then Reynold stress models (RSM)[5]. Derived from RSMs they account for many flow phenomena without solving transport equations for all six unknown stresses[4]. The BSL EARSM is an extension of the standard two-equation models and includes the nonlinear relation between the Reynolds stresses and the mean strain-rate and vorticity tensors. Additionally, curvature correction was employed because of the known insensitiveness of eddy-viscosity models to streamline curvature and system rotation[6]. The measured mass flow was prescribed at the spiral case inlet. At the draft tube outflow an opening boundary condition with option "Entrainment" was set and a relative pressure corresponding to the measured Thoma-number was specified.The case was initialized by decreasing the Thoma-number and the time step size linearly ($\Delta t = 0.9°$).

3. Numerical results

The investigated operating points are presented in table 1. Since the mass flow and the wicket gate opening is prescribed the turbine head is a simulation result. Comparison to the measured head yields large discrepancies (an underestimation of 9.87% and 10.12% for condition 1 and 2 respectively). However, the deviations in torque are less then 1.2% for both Thoma-numbers.

Table 1: Operating conditions

	Cond 1	*Cond 2*
Q_{ed}	0.05445	0.05442
n_{ed}	0.2881	0.2883
σ	0.11	0.07
α	5	5

Figure 1: From left to right: Surface streamlines at 10%, 50% & 90% span (hub to shroud) in blade-to-blade view.

3.1. Flow visualization

The inner blade vortices, main source for the dynamic loads acting on the runner blade, originate from the flow misalignment at the runner leading edge for low discharges and small wicket gate openings. In figure 1 surface streamlines of the transient averaged velocity at three constant span surfaces are presented in blade-to-blade view. The inner blade vortices can be clearly identified at approximately one third of the blade chord length and attached to the suction side. From the runner hub to half span height the vortex is axial parallel which becomes obvious when comparing the streamlines at 10% and 50% span. Near the hub the vortex blocks the main flow almost entirely. Near the shroud the vortex turns radial and follows the shroud contour.

3.2. Cavitation phenomena

The channel vortices rotate with high velocities forming low pressure zones in the vortex core. Depending on the Thoma-number the corresponding pressure may drop below the vapor pressure forcing the liquid to evaporate. For the investigated conditions large differences in the vapor volume are noticeable. Figure 2 shows the runner from below for one instance in time. Cavitation is visualized by a red iso-surface of the vapor volume fraction at 10%. For the higher Thoma-number, small cavitation regions are predicted only in one half of the runner channels, wherein for condition 2 almost every channel shows cavitation and the vapor volumes are remarkably larger. The differences in vapor volume inside the channels indicate fluctuations in the flow field and an interaction with the rotating low pressure zones inside the draft tube (cf. figure 5).

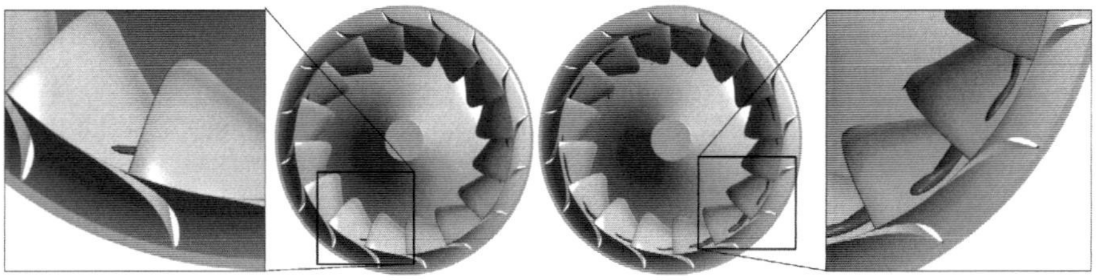

Figure 2: Runner from below with iso-surfaces at 10% vapor fraction colored in red (Condition 1 left, Condition 2 right).

3.3. Pressure oscillations

To compare the simulation results with measurement data, 5.12 simulated runner revolutions, resolved by 2048 time steps, were used. The available measurement data covers 1600 runner revolutions resulting in clear peaks for all deterministic frequencies. For time spans comparable to the simulation time, large differences among them were identified. Therefore an arbitrary time span was selected and used for all comparisons. Figure 3 shows the time signal and the corresponding frequency spectra for pressure sensor SS_TE2, positioned on the blade suction side (cf. meridional view in figure 3). The FFT reveals a broad band of low frequencies between 0-10 times the runner frequency which is resolved successfully by the simulation. While the amplitudes for $\sigma = 0.11$ are slightly underestimated, the simulation overestimates these for the lower Thoma-number. Rotor-stator interaction becomes visible as a peak at $20 \cdot f_{runner}$ and gets damped by the increasing vapor volume, in contrast to the amplitudes of the low frequency which are amplified. Both effects are captured by the simulation. By plotting the pressure amplitude decomposed into distinct frequencies on the runner suction side the origin for the above discussed low frequency fluctuation can clearly be assigned to the occurrence of the inner blade vortex whose trace is clearly visible. Figure 4 shows the results for ten frequencies between 0.2 and 2.0 times f_{runner} at $\sigma = 0.11$. The contour plots show that the highest amplitudes arise upstream of the measuring position SS_TE2, which means that slight positional deviations in the measuring point placement and the prediction of the vortex shape would result in large deviations for the predicted amplitudes.

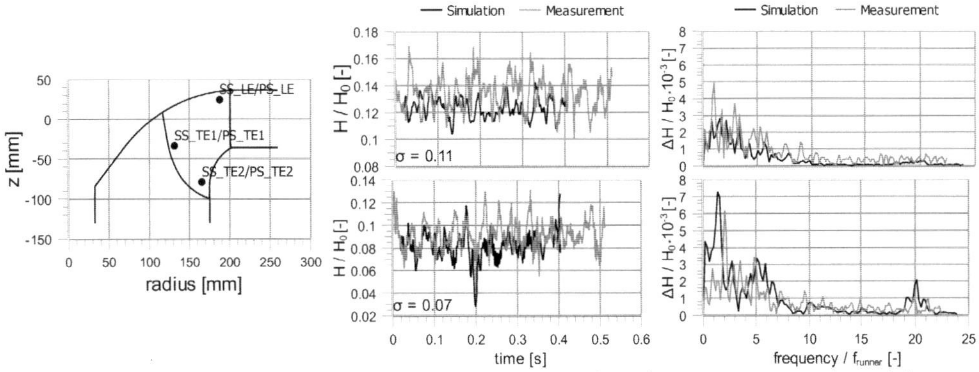

Figure 3: From left to right: Measuring positions in meridional view, time signals and frequency spectra for both conditions (SS_TE2).

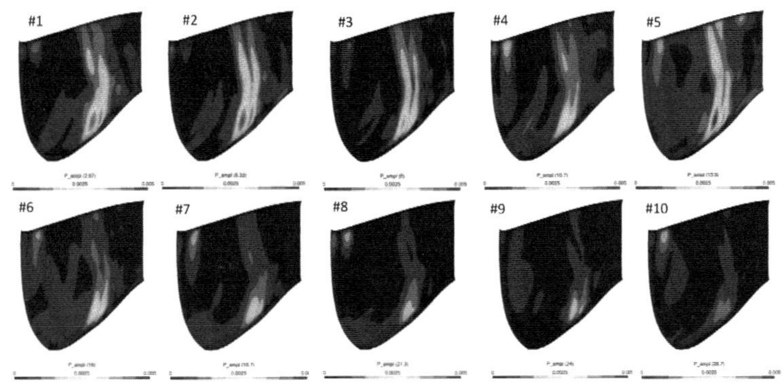

plot	#1	#2	#3	#4	#5	#6	#7	#8	#9	#10
f/f_{runner}	0.2	0.4	0.6	0.8	1.0	1.2	1.4	1.6	1.8	2.0

Figure 4: Pressure plots on blade suction side decomposed into distinct frequencies.

Measuring position C1N is located in the draft tube cone as marked in the sketch of figure 5. Next to it the recorded time signals and the corresponding FFTs are presented. A very good correlation can be stated for $\sigma = 0.11$. For $\sigma = 0.07$ the runner rotational frequency is dominant in the measured signal, which is less pronounced in the simulation result. Four pictures in figure 5 illustrate the origin of the low frequency fluctuations observed in the draft tube cone. Iso-surfaces visualize low pressure structures rotating underneath the runner. The formation and collapse result in a varying amount of these structures which yields the broad band of frequencies.

4. Conclusion

Two unsteady two-phase simulations at deep part load with different Thoma-numbers are performed. Deviations up to 10% are found for the simulated head whereas the torque is in good agreement with the measurements. Deviations can be partly explained by the neglect

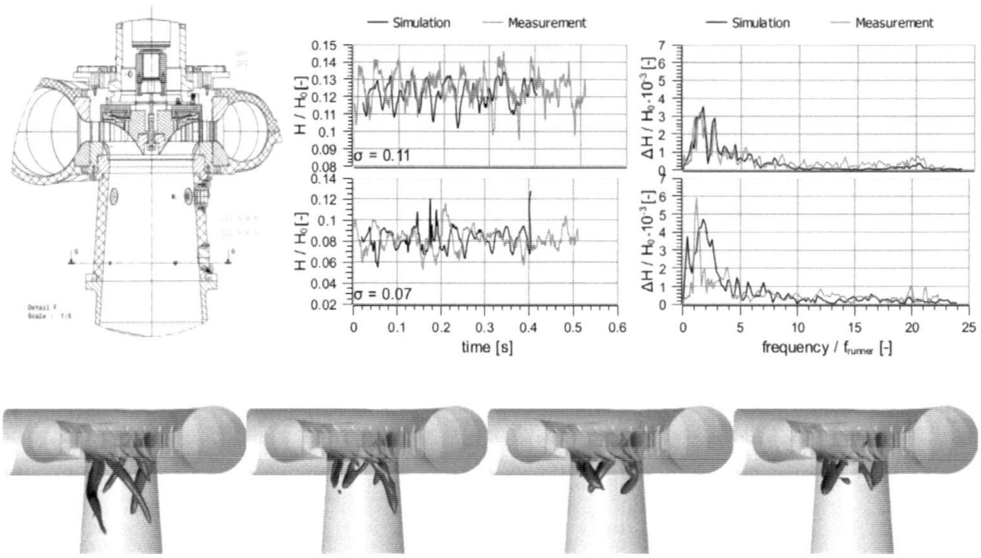

Figure 5: From left to right: Measuring positions, time signals and frequency spectra for both conditions. Bottom: Rotating low pressure zones ($\sigma = 0.03$).

of the runner side gaps and the crown bores. Pressure signals are compared as well. For the lower Thoma-number an increase in amplitude was observed. An acceptable correlation for all measuring locations was found as the low frequencies and the rotor-stator interaction were captured. Multiphase modeling reveals size varying vapor bubbles in the center of the inner blade vortices over time. The simulation approach proves its applicability to low discharge conditions including cavitation. For future work, advanced numerical methods for turbulence treatment like scale resolving simulations could be used to reduce the uncertainties associated with turbulence modeling. The loads calculated by the simulation can be mapped on a finite element model for fatigue calculations.

5. Acknowledgments

The research leading to the results published in this paper is part of the HYPERBOLE research project, granted by the European Commission (ERC/FP7-ENERGY-2013-1-Grant 608532).

References

[1] Magnoli M. V.: *Numerical simulation of pressure oscillations in large Francis turbines at partial and full load operating conditions and their effects on the runner behaviour and fatigue life*, Ph.D thesis Technische Universität München, 2014.

[2] Wack J.; Riedelbauch S.: *Numerical simulations of the cavitation phenomena in a Francis turbine at deep part load conditions*, 9th International Symposium on Cavitation, Journal of Physics: Conference Series 656, 2015.

[3] Yamamoto K.; Müller A.; Favrel A.; Landry C.; Avellan, F.:*Pressure measurements and high speed visualizations of the caviatation phenomena at deep part load conditions in a Francis turbine* 27th IAHR Symposium on Hydraulic Machinery and Systems, 2014.

[4] Menter F. R.; Garbaruk A.V.; Egorov Y.: *Explicit Algebraic Reynolds Stress Models for Anisotropic Wall-Bounded Flows* EUCASS - 3rd European Conference for Aero-Space Sciences, Versailles, 2009.

[5] Wallin S.: *Engineering turbulence modelling for CFD with a focus on explicit algebraic Reynolds stress models* Ph.D thesis Royal Institude of Technology, Stockholm, 2000.

[6] ANSYS Inc.: *ANSYS CFX-Solver Theory Guide*, Release 16.1, 2015

Simulation model of a variable-speed pumped-storage power plant in unstable operating conditions in pumping mode

G Martínez-Lucas[1], J I Pérez-Díaz[1], J I Sarasúa[1], G Cavazzini[2], G Pavesi[2] and G Ardizzon[2]

[1] Department of Hydraulic, Energy and Environmental Engineering. Technical University of Madrid. C/ Profesor Aranguren 3. 28040 Madrid, Spain.
[2] Department of Industrial Engineering. University of Padova. Via Venezia 1. Padova 35131, Italy.
E-mail: guillermo.martinez@upm.es

Abstract. This paper presents a dynamic simulation model of a laboratory-scale pumped-storage power plant (PSPP) operating in pumping mode with variable speed. The model considers the dynamic behavior of the conduits by means of an elastic water column approach, and synthetically generates both pressure and torque pulsations that reproduce the operation of the hydraulic machine in its instability region. The pressure and torque pulsations are generated each from a different set of sinusoidal functions. These functions were calibrated from the results of a CFD model, which was in turn validated from experimental data. Simulation model results match the numerical results of the CFD model with reasonable accuracy. The pump-turbine model (the functions used to generate pressure and torque pulsations inclusive) was up-scaled by hydraulic similarity according to the design parameters of a real PSPP and included in a dynamic simulation model of the said PSPP. Preliminary conclusions on the impact of unstable operation conditions on the penstock fatigue were obtained by means of a Monte Carlo simulation-based fatigue analysis.

1 Introduction

Variable speed operation offer several advantages for a pump-turbine both in pumping and generating modes such as a) the possibility of an almost instantaneous active power control, b) reactive power control and c) higher efficiency and wider range of operation [1].

However, at part load in pumping mode, reversible pump-turbine machines suffer from unstable behavior related to a strong fluid-dynamical interaction between rotor and stator parts that causes the development of unsteady phenomena, dissipating the fluid energy with a consequent decrease in head. Moreover, if this unstable operating zone is located below the highest operating head, it could prevent the start-up in pumping mode at this head negatively affecting the plant's regulation capacity.

The pressure and torque pulsations associated with these unsteady phenomena might propagate both along the power plant conduits and to the electrical grid. The propagation of pressure pulsations might cause fatigue in the penstock and, as a consequence, reduce the power plant lifetime or require a higher investment. The torque pulsations might in turn have a certain influence on the power quality and, as a consequence, affect the design and cost of the converter filters.

The material fatigue is a phenomenon of extreme importance as regards the reliability and safety of various types of engineering structures and is recognized as one of the major causes of destruction of materials [2]. However, to the authors' knowledge, no previous paper has evaluated whether or not the pressure pulsations that develop at part load in pumping mode cause a significant fatigue in the penstock. In [3], the authors evaluated the risk of fatigue in the penstock of a specific hydropower plant due to the tracking of the secondary load-frequency control set-point signals, and concluded that the secondary load-frequency control do not cause a significant fatigue with a proper set of turbine governor parameters. As the authors stated in the paper, fast hydraulic transient phenomena such as those related to the pump-turbine start-up and shutdown, fast mode change, etc., were not considered in the study.

The objective of this paper is twofold: to propose and experimentally validate a set of functions to model the pressure and torque pulsations that develop in a pump-turbine operating at part load in pumping mode; to preliminary evaluate the impact of the propagation of the pressure pulsations along the power plant conduits on the penstock fatigue. The pursued objective can be of special interest for existing pumped storage power plants (PSPP) which, maintaining the existing pump turbine runner, will be upgraded to variable speed, as is the case in some PSPP recently analysed in the eStorage European project [4]. The main reason for such an upgrade is that, as demonstrated in [5], the variable speed technology allows the PSPP to participate in the restoration reserve services in pumping mode and therefore to increase its revenue.

For the purposes mentioned above, both detailed numerical and experimental analyses have been carried out on a reversible pump-turbine by Dep. Industrial Engineering – University of Padova.

A dynamic simulation model of the laboratory-scale PSPP has been developed. The model considers the dynamic behavior of the power plant conduits taking into account the water hammer effects, as well as the pump-turbine operating as a variable speed pump. The pump-turbine operation curves (head-flow-rotation speed and torque-flow-rotation speed) have been obtained from the CFD model results. The model synthetically generates both pressure and torque pulsations that reproduce the operation of the pump-turbine in its instability region by means of a set of sinusoidal functions. These sinusoidal functions have been properly calibrated from the results of a spectral analysis of the pressure and torque signals provided by the CFD model. The results of the simulation model match the numerical results of the CFD model with reasonable accuracy.

The pump-turbine operation curves and the functions used to generate pressure and torque pulsations have been up-scaled according to the design parameters of a real PSPP, and then incorporated in a dynamic simulation model of the said plant. Preliminary conclusions have been obtained on the impact of the pump-turbine unstable operation conditions on the risk of fatigue of the penstock by means of a Monte Carlo simulation based fatigue analysis.

2 CFD model validation

Numerical analyses were carried out by the commercial software ANSYS 15.0 on the first stage of a two stages reversible pump-turbine in pump-operating mode (Figure 1). In order to validate the accuracy of the numerical simulations, the experimental data obtained according to ISO standards from Dep. Industrial Engineering – University of Padova was compared with steady simulations, carried out for some operating points at a constant rotation speed of 600 rpm. The comparison in terms of head is reported in Figure 2.

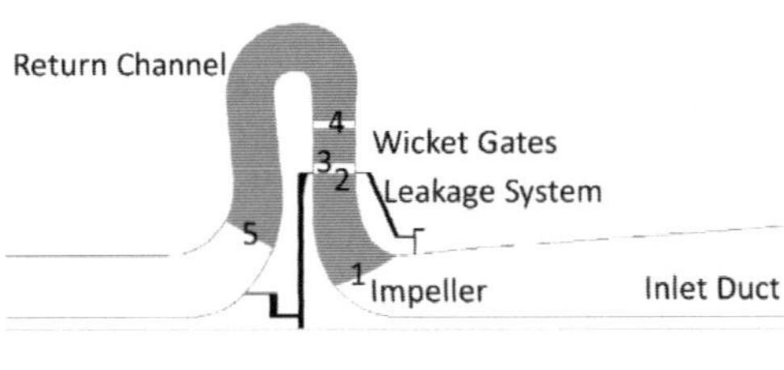

Figure 1. Meridional view of the numerical model. Regions filled in grey refer to the blades; region filled in black refers to the leakage system

The maximum percentage error of -3.84% in head between the experiment and simulation was observed just before the hump-type instability zone. The head showed a tendency to smaller prediction than the experimental data at the BEP (-1.49%) and a tendency to a greater prediction in the hump type instability zone (max +1.24%). The errors may be arising from the practical limitations in numerical model but all the errors between the simulation and experiment are in a reasonable range not only at BEP condition.

3 CFD simulations

To study the evolution of the flow field in a dynamic power reduction scenario the wicket vane was fixed at 18° and kept constant during the simulation. The transient numerical simulation on the entire machine was carried out with a time-varying boundary condition in which the impeller speed was reduced every time step whereas the total pressure at inlet and the average static pressure at outlet were varied according to the experimental values obtained at steady conditions. The initialization was based on the results of the steady simulation carried out for the BEP flow rate and a simple linear variation of the impeller speed from 100% to 88% was fixed (Figure 2). To obtain the variation of pressure in the transient process, monitoring points were created in the whole pump-turbine model.

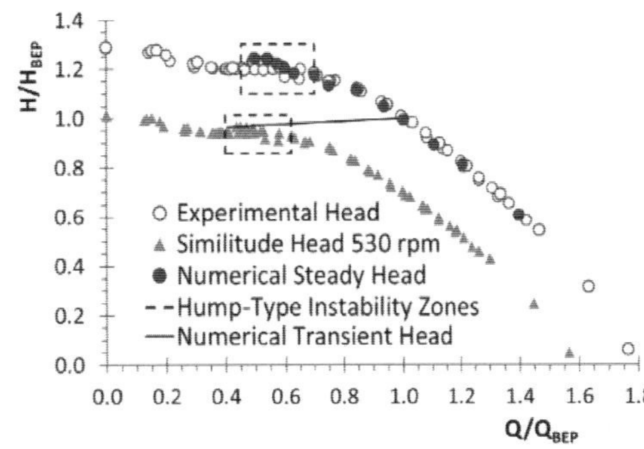

Figure 2. Comparison between numerical and experimental head curves.

The numerical analyses captured the development of unsteady phenomena in the stators affecting the pressure distribution on the vanes, thereby modifying their load distribution and inducing dynamic load on the shear pin or the guide vanes stem.

Spectral analyses in frequency and in time-frequency domain were carried out to characterize the unsteady phenomena and the frequency content of the numerical signals. These signal were acquired at each time step during the whole simulation of power reduction. Time step was determined by using a Hamming window in the Lomb periodogram. More details about the results of the numerical analyses can be found in [6].

4 Laboratory plant simulation model

The main purpose of the laboratory plant simulation model is to reproduce the above described CFD simulation. The model considers the water and pipe elasticity in the suction pipe by means of an approach similar to the one used in [7]. Due to the short length of the impulse pipe, it has been modeled using a rigid water column approach. The flow-head loss function has been obtained by linear regression from the results of the CFD simulations. Upper tank dynamics is considered due to its reduced size.

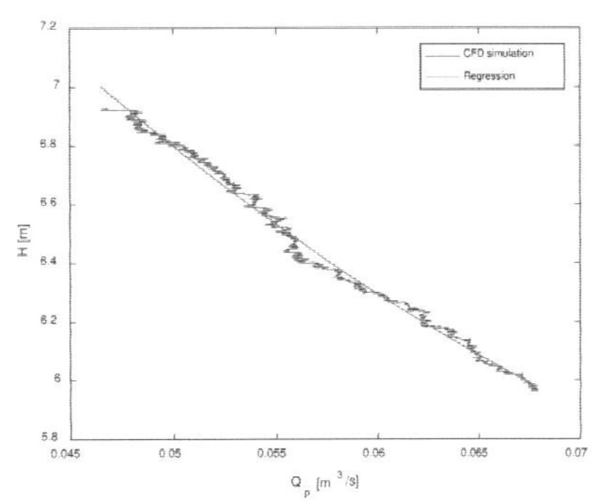

Figure 3. Head-flow operation curve for n=558 rpm.

The pump-turbine has been modeled through its operation curves (head-flow-rotation speed and torque-flow-rotation speed) and a couple of functions that synthetically generate pressure H* and torque C* pulsations (1)-(2). The parameters of the pump-turbine curves corresponding to n=558 rpm were obtained by linear regression from the results of the CFD simulation (see Figure 3). The parameters corresponding to other rotation speeds were then obtained by hydraulic similarity from the previous ones.

IOP Conf. Series: Journal of Physics: Conf. Series **813**(2017) 012028 doi:10.1088/1742-6596/813/1/012028

The amplitude $A_{H,p}$, $A_{C,p}$ and frequency $f_{H,p}$, $f_{C,p}$ parameters in (1) and (2) have been calibrated from the results of the spectral analysis carried out on the results of the CFD simulation. As can be seen from Figure 4 the power spectrum of pressure pulsations corresponding to both the CFD simulation and the sinusoidal function match significantly with each other. In order to harmonize and plot the results of the spectral analysis, frequency values have been formulated in terms of Strouhal number, S_t [6].

$$H^*(n,t) = \sum_{p=1}^{6} A_{H,p}(n) \cdot sin\big(2\pi \cdot f_{H,p}(n) \cdot t\big) \tag{1}$$

$$C^*(n,t) = \sum_{p=1}^{8} A_{C,p}(n) \cdot sin\big(2\pi \cdot f_{C,p}(n) \cdot t\big) \tag{2}$$

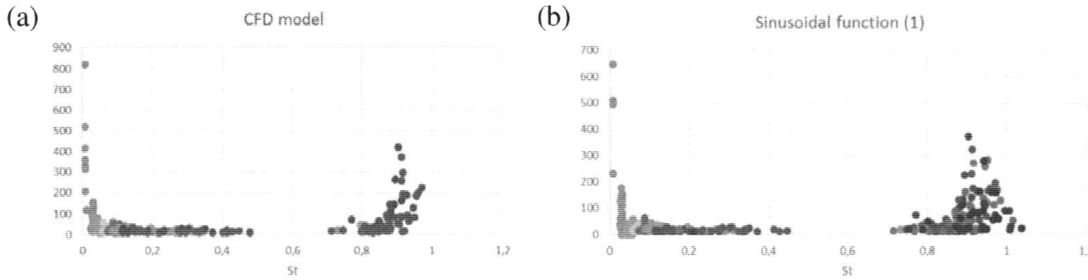

Figure 4. Power spectrum of pressure pulsations corresponding to (a) the CFD simulation and (b) sinusoidal function (1).

The CFD simulation has been reproduced in the simulation model for validation purposes. Figure 5 shows the pump flow and head obtained in both models. As can be seen in the figure, the dynamic simulation model properly reproduces the behavior of the laboratory plant in unstable operating conditions.

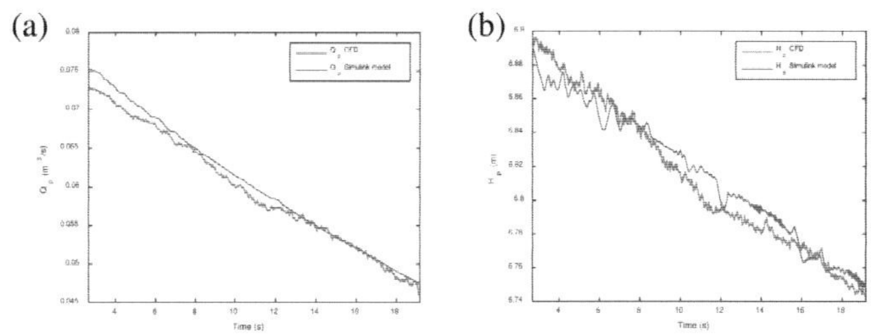

Figure 5. (a) Flow and (b) Head comparison between CFD and Simulink model.

5 Pumped-storage plant simulation model

The pump-turbine operation curves and the sinusoidal functions used to model the pressure and torque pulsations have been up-scaled by hydraulic similarity [8] and incorporated in a dynamic simulation model of a real PSPP. The PSPP considered (Net head=500m, flow = 44,29 m^3/s per unit) consists of two pump-turbines and a single long penstock. Each pump-turbine is connected to the penstock through a short connection pipe. The PSPP upper reservoir is large enough to neglect its level variation.

Because one of the paper objectives is to evaluate the impact of the propagation of pressure pulsations along the PSPP conduits on the risk of penstock fatigue, both the connection pipes and the penstock have been modeled by the characteristic method [9].

In order to study if the propagation of pressure pulsations along the power plant penstock could cause damage due to fatigue phenomenon, a Monte Carlo simulation has been carried out considering a random lag or lead between the positions of the two pump-turbines. For each simulation, a set of tangential stresses is obtained in the most critical penstock section –union between the penstock and the connection pipes- as a function of the pressure amplitude $\Delta\sigma_c = f(\Delta p)$ [3,10]. A single equivalent stress is then obtained for each simulation by using the expression proposed in [2]. Next, a weighted equivalent stress is calculated for all simulations. According to [11], the weighted equivalent stress would cause no lifetime reduction for fatigue reasons (13,89 MPa < 14,57 MPa). However, pressure pulsations, and consequently $\Delta\sigma_c$, are related to PSPP net head. Thus, it would be possible to find fatigue damages in a PSPP with a higher net head.

6 Conclusions

A dynamic simulation model of a variable-speed laboratory pumped-storage power plant operating in unstable operating conditions in pumping mode has been developed. The model has been calibrated from the results of a CFD simulation. The CFD model has been in turn validated from an experiment carried out in the laboratory plant.

The pressure and torque pulsations that develop in the pump-turbine impeller have been modeled through a couple of sinusoidal functions. These functions have been calibrated from the results of a spectral analysis carried out on the results of the CFD simulation.

The dynamic simulation model reproduces the results of the CFD simulation with a reasonable accuracy.

The pump-turbine model has been up-scaled according to the design parameters of a real pumped-storage power plant and has been incorporated in a dynamic simulation model of the said power plant. The simulation model has been used to evaluate the impact of the unstable operating conditions in pumping mode on the risk of penstock fatigue. The results of the fatigue analysis show that the risk of penstock fatigue is negligible. Further research is necessary to evaluate the impact of unstable operating conditions in pumping mode in other pumped-storage power plants with higher heads, where unstable operating conditions might have a greater impact.

7 References

[1] C. Nicolet, Y. Pannatier, B. Kawkabani, J. Simond and F. Avellan, «Hydroelectric interactions with variable speed and fixed speed machines in pumping mode of operation,» de Proc. 2011 HYDRO Conference, 2011.

[2] A. Adamkowski, M. Lewandowski and S. Lewandowski, «Evaluation of the Remaining Lifetime of Steel Penstocks in Hydropower Plants,» Vienna, 2010.

[3] C. Nicolet, R. Berthod, N. Ruchonnet and F. Avellan, «Evaluation of possible penstock fatigue resulting from secondary control for the grid».

[4] Kunz, T., Business case results about potential upgrade of five EU pumped hydro storage plants to variable speed, 2015 eStorage project workshop. Available at (on December 22nd 2016): http://www.estorage-project.eu/documentlibrary

[5] M. Chazarra, JI. Perez-Diaz and J. Garcia-Gonzalez, «Optimal Operation of Variable Speed Pumped Storage Hydropower Plants Participating in Secondary Regulation Reserve Markets,» In *11th International Conference on the European Energy Market (EEM14)* (pp. 1-5). IEEE

[6] G. Pavesi, G. Cavazzini and G. Ardizzon, «Numerical Analysis of the Transient Behaviour of a Variable Speed Pump-Turbine during a Pumping Power Reduction Scenario,» Energies, vol. 9, nº 7, pp. 1-15, 2016.

[7] IEEE Working Group, «Hydraulic turbine and turbine control models for system dynamic studies,» IEEE Transactions on Power Systems, vol. 7, nº 1, pp. 167-179, 1992.

[8] S. Alligné, P. Maruzewski, T. Dihn, B. Wang, A. Fedorov, J. Iosfin and F. Avellan, «Prediction of a Francis turbine prototype full load instability from investigations on the reduced scale mode,» de 2010 IAHR Symposium on Hydraulic Machinery and Systems, 2010.

[9] A. Lohrasbi and R. Attarnejad, «Water Hammer Analysis by Characteristic Method,» American J. of Engineering and Applied Sciences, vol. 1, nº 4, pp. 287-294, 2008.

[10] L. Cuesta and E. Vallarino, Aprovechamientos hidroeléctricos, Madrid: Colegio de Ingenieros de Caminos, Canales y Puertos, 2014.

[11] Eurocode EN 1993-1-9:2005, «Eurocode 3: Design of steel structures - Part 1-9: Fatigues».

Flow characteristics and influence associated with inter-blade cavitation vortices at deep part load operations of a Francis turbine

K. Yamamoto[1], A. Müller[1], A. Favrel[1], C. Landry[2], F. Avellan[1]

[1]EPFL Laboratory for Hydraulic Machines, Avenue de Cour 33bis, 1007 Lausanne, Switzerland

[2]Power Vision Engineering Sàrl, Chemin des Champs Courbes 1, 1024 Ecublens, Switzerland

E-mail: keita.yamamoto@epfl.ch

Abstract. At a deep part load operation, Francis turbines are subject to the formation of inter-blade cavitation vortices inside blade channels, however its characteristics are not fully understood yet. The present study aims to investigate the development of the inter-blade vortex as well as the flow characteristics associated with the vortex formation inside the blade channel, by using the unsteady RANS simulation. The velocity survey reveals the appearance of the back flow region in the vicinity of the hub near the trailing edge, which seems closely linked with the inter-blade vortex development. Furthermore, the local influence of the inter-blade vortex is evaluated by the specific energy loss calculated from the rothalpy in the blade channel. The comparison of the different operating conditions evidences the impact of the inter-blade vortex presence on the energy dissipation through the blade channel.

1. Introduction

It is well known that Francis turbines are subject to a various type of cavitation flows depending on its operating condition [1, 2]. In particular, the cavitation vortex rope developing in a draft tube at full load and part load operations prevents a stable operation of hydraulic machines, therefore several researches have been performed in different ways [3, 4, 5, 6]. Recently, Francis turbines are increasingly required to guarantee a secured operation especially at deep part load condition in order to enable a smooth integration of the renewable energy sources. In such operating condition, the cavitation vortex inside a blade-to-blade channel called inter-blade cavitation vortex is observed, however, its characteristics are insufficiently understood thus far.

In the framework of the HYPERBOLE research project, experiments and numerical simulations for the inter-blade cavitation vortex at deep part load operation have been conducted and reported, contributing to a better comprehension of its characteristics [7, 8, 9]. The present study aims to demonstrate the numerical investigations about the development and the characteristics of the inter-blade vortex by the unsteady RANS simulation. The simulated vortex structure is compared to the visualization results for the validation, then the structure of the flow inside the blade channel associated with the inter-blade vortex development is investigated. Furthermore, the specific energy loss is evaluated by the integration of the specific rothalpy through the blade channel, and the influence of the inter-blade vortex on the energy dissipation is investigated by comparison of different operating conditions.

2. Simulation setups

The numerical simulation is conducted by the unsteady RANS simulation using the commercial flow solver ANSYS CFX 17.1. The calculation domain used for the performed simulation is shown in Fig.1, which includes a spiral case, stay vanes, guide vanes, runner, and draft tube. The total number of the nodes are about 16 million nodes.

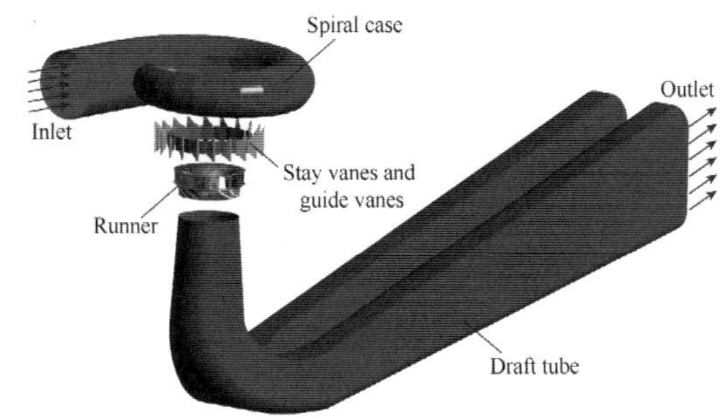

Figure 1. Calculation domain for unsteady RANS simulations

In the performed simulation, a two phase flow is treated as a homogeneous fluid assuming that the relative motion between phases can be neglected. Thus, the fluids in all the phases share a common flow field such as pressure and velocity. Cavitation is taken into account by solving the transport equation for the vapor volume fraction, based on the simplified Rayleigh-Plesset equation [10]. For a time step of the simulation, 0.0001 s corresponding to about 0.5 degree of the runner revolution is selected.

For a turbulence model, SST-SAS model is adopted. This model provides results similar to LES (Large Eddy Simulation), particularly in the unsteady flow region [11]. It has been shown that the model is used for various industrial applications with satisfactory results in recent years [12], including the simulation of the inter-blade cavitation vortex [9].

In the performed simulation, the following operating conditions (see Table 1) at the constant guide vane opening 5 degree are targeted for the comparison of the results. The n_{ED} value of OP2 is equivalent to the rated n_{ED} condition.

Table 1. Targeted operating conditions

OP	α	Runner rotation n	Specific energy E	n_{ED}	Q/Q_{BEP}	σ
1	5 (deg)	14.67 (Hz)	263 (J kg^{-1})	0.317 (-)	0.265 (-)	0.11 (-)
2	5 (deg)	13.33 (Hz)	263 (J kg^{-1})	0.288 (-)	0.270 (-)	0.11 (-)
3	5 (deg)	12.90 (Hz)	263 (J kg^{-1})	0.268 (-)	0.273 (-)	0.11 (-)

3. Simulation result

In Fig.2, the instantaneous view of the vortex structure and cavitation as well as the averaged vortex structure over four runner revolutions are shown. The vortex structure is detected by the non-dimensional Q-criterion which is calculated by the invariant of the velocity gradient tensor $\mathbf{L} = \nabla \mathbf{W}$, as in;

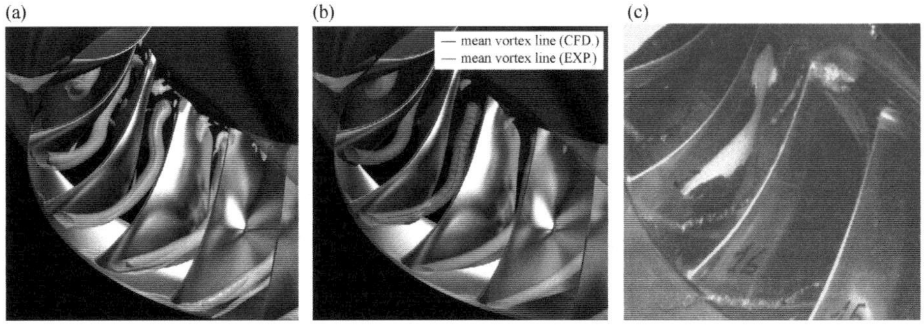

Figure 2. Instantaneous view of the simulated inter-blade vortex structure (white surface) and cavitation (blue surface, iso-surface of vapor volume fraction 10 %), averaged vortex structure together with the mean vortex lines (b) and the experimental visualization of the inter-blade cavitation vortex (c)

$$Q^* = \left(\frac{D_{ref}}{W_1}\right)^2 Q = \frac{1}{2}\left(\frac{D_{ref}}{W_1}\right)^2 [(tr(\mathbf{L}))^2 - tr(\mathbf{L}^2)] \quad (1)$$

where D_{ref} represents a reference diameter of the runner, W_1 a relative velocity at the runner inlet, and $tr(\mathbf{L})$ a first invariant of the velocity gradient tensor. In the presented case, the value of $Q^* = 3.0 \times 10^4$ is chosen to highlight the inter-blade vortex structure. In Fig.2(b), the comparison of the simulated location of the mean vortex center (see blue line in Fig.2(b)) with the vortex location of the experiment estimated based on the visualization (see red line in Fig.2(b)) is also presented. In the simulation result, the maximum value of the non-dimensional Q-criterion at each constant span-wise location is taken as the location of the vortex center. It can be confirmed that the both cavitation and inter-blade vortex structure are well captured by the performed simulation, moreover the simulated mean vortex location has a very good agreement with the experimental result.

In Fig.3, the distribution of the averaged non-dimensional meridional velocity C_m^* at the constant span-wise location in the vicinity of the hub ($s^* = 0.99$) as well as the meridional distribution of C_m^* at OP1 are shown. C_m^* is normalized by using the mean meridional velocity at each streamwise section and then averaged over four runner revolutions. At $s^* = 0.99$, it is shown that the high meridional velocity region appears at the inlet of the blade channel, whereas the velocity becomes remarkably low at the outlet of the runner. In particular, near the trailing edge on the suction side the flow is dominated by the back flow region. The development of the back flow region near the hub is also confirmed by the meridional contour of C_m^* (see Fig.3(b)). It demonstrates that the high velocity flow is consolidated in the vicinity of the shroud especially at the outlet of the blade, on the other hand the flow near the hub is dominated by the back flow region close to the hub. The development of the back flow region on the hub starts at the streamwise location $m^* = 0.60$, which corresponds to the inception of the inter-blade vortex. This evidently confirms that the occurrence of the inter-blade vortex development is closely relevant to the formation of the back-flow region in the vicinity of the hub near the exit of the blade channel. It should be noted that the formation of the back flow region as well as the strength of the inter-blade vortex are drastically reduced at OP2 and OP3.

4. Local influence of the inter-blade vortex

In the present study, the local influence of the inter-blade vortex is evaluated by the specific rothalpy which is used for the quantitative evaluation of the specific energy loss inside the

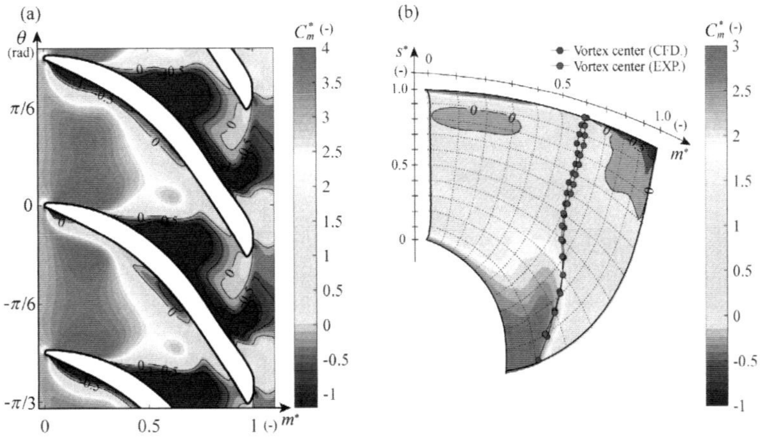

Figure 3. Meridional velocity contour at $s^* = 0.99$ (a) ($m^* = 0$ at the leading edge and 1 at the trailing edge) and at the meridional plane together with the projected inter-blade vortex location(b) at OP1

runner. The variation of an energy flux defined by the specific rothalpy I through a control volume rotating with the runner is expressed as;

$$\Delta F_I = \int_{\partial V} \left(\frac{p}{\rho} + \frac{1}{2}\vec{W}^2 - \frac{1}{2}(r\omega)^2 + gz \right) \rho \vec{W} \cdot \vec{n} dA \equiv \int_{\partial V} I \, \vec{W} \cdot \vec{n} dA \qquad (2)$$

where p represents the static pressure, r the radius, ω the angular rotational frequency, and z the elevation. In the previous study, it was proven that the energy flux ΔF_I is conserved under specific conditions [13], however in the real flow, ΔF_I is dissipated due to the viscosity and the unsteadiness of the flow. By setting the boundaries of the control volume on a plane of a constant streamwise location m^* and the blade channel wall boundaries, the specific energy loss E_r from the blade inlet to a given streamwise location m^* can be calculated as;

$$\begin{aligned} E_{r_{1\div m^*}} &= \frac{1}{\rho Q} \left(\int_A I \, \rho \vec{W} \cdot \vec{n} dA \Big|_{m^*} - \int_A I \, \rho \vec{W} \cdot \vec{n} dA \Big|_{m^*=0} \right) \\ &= \frac{1}{\rho Q} \left(\int_{z_{sh}}^{z_{hb}} \int_{\theta_s}^{\theta_p} \rho C_m I r sin\gamma \, d\theta dz \Big|_{m^*} - \int_{z_{sh}}^{z_{hb}} \int_{\theta_s}^{\theta_p} \rho C_m I r sin\gamma \, d\theta dz \Big|_{m^*=0} \right) \end{aligned} \qquad (3)$$

where z_{hb} and z_{sh} are the axial position of the hub and shroud at a given streamwise location, θ_s and θ_p are the angular position of the suction side and the pressure side of the blade, γ the deflection angle between the span-wise plane and the z-axis. The specific energy dissipation rate is then calculated using the specific energy loss E_r, as in;

$$e = 1 - \frac{E_{r_{1\div m^*}}}{E_{blade}} \qquad (4)$$

where E_{blade} is the specific energy of the evaluated blade channel. In Fig.4, the evolutions of the specific energy dissipation rate e as well as the derivative of e over the streamwise location m^* for three different operating conditions are shown. It demonstrates that the energy dissipation rate is almost constantly decreased near the inlet, however the slope of the decrease is clearly modified around $m^* = 0.55$ which almost corresponds to the region where the inter-blade vortex

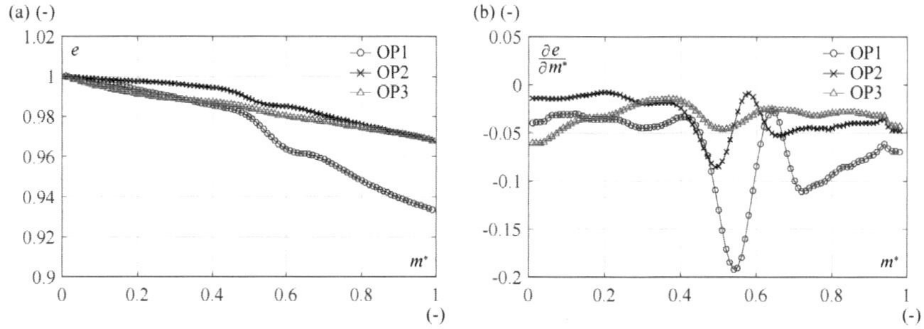

Figure 4. Evolution of the specific energy dissipation rate (a) and the gradient of the energy dissipation rate over the streamwise location (b) for three operating conditions

starts to develop. This significant change is most obvious at OP1, then it is reduced at OP2 and OP3 in accordance with the reduction of the inter-blade vortex intensity. The derivative of e is also significantly amplified at OP1, which evidently shows that the presence of the inter-blade vortex induces the energy dissipation through the blade channel.

5. Conclusion

It is shown that the inter-blade cavitation vortex is successfully captured by the performed unsteady RANS simulation, which is validated by the visualization results. The meridional velocity survey demonstrates the highly non-uniform flow distribution inside the blade channel especially in the vicinity of the hub. Moreover, it evidences the formation of the back flow region around the hub close to the trailing edge, which plays a significant role for the inter-blade vortex development. Furthermore, the negative impact of the inter-blade vortex on the specific energy is confirmed by the energy dissipation evaluated by the specific rothalpy through the blade channel.

Acknowledgments

The results presented in the paper is part of the HYPERBOLE research project, granted by the European Commission (ERC/FP7-ENERGY-2013-1-Grant 608532).

References

[1] Avellan F 2004 *Proceedings of the 6th International Conference on Hydraulic Machinery and Hydrodynamics, Timisoara, Romania*
[2] Escaler X, Egusquiza E, Farhat M, Avellan F and Coussirat M 2006 *Mechanical Systems and Signal Processing* **20** 983–1007
[3] Favrel A, Müller A, Landry C, Yamamoto K and Avellan F 2015 *Experiments in Fluids* **56** 1–15
[4] Favrel A, Mller A, Landry C, Yamamoto K and Avellan F 2016 *Experiments in Fluids* **57**
[5] Müller A, Dreyer M, Andreini N and Avellan F 2013 *Experiments in Fluids* **54**
[6] Landry C, Favrel A, Mller A, Nicolet C and Avellan F 2016 *Journal of Hydraulic Research* **54** 185–196
[7] Yamamoto K, Müller A, Favrel A, Landry C and Avellan F 2013 *IOP Conference Series: Earth and Environmental Science* **22**
[8] Yamamoto K, Müller A, Favrel A, Landry C and Avellan F 2015 *Proceedings of the 6th IAHR International Meeting of the Workgroup on Cavitation and Dynamic Problems in Hydraulic Machinery and Systems, Ljubljana, Slovenia*
[9] Wack J and Riedelbauch S 2015 *Journal of Physics: Conference Series* **656**
[10] Zwart P, Gerber A G and Belamri T 2004
[11] Egorov Y and Menter F 2008 *Notes on Numerical Fluid Mechanics and Multidisciplinary Design* **97** 261–270
[12] Menter F and Egorov Y 2010 *Flow, Turbulence and Combustion* **85** 113–138
[13] Lyman F 1993 *Journal of Turbomachinery* **115** 520–526

IOP Conf. Series: Journal of Physics: Conf. Series **813** (2017) 012030 doi:10.1088/1742-6596/813/1/012030

Transposition of Francis turbine cavitation compliance at partial load to different operating conditions

J. Gomes[1], **A. Favrel**[1], **C. Landry**[2], **C. Nicolet**[2], **F. Avellan**[1]
[1]EPFL Laboratory for Hydraulic Machines, Avenue de Cour 33bis, 1007 Lausanne,Switzerland
[2]Power Vision Engineering Sarl, Ecublens, Switzerland
E-mail: joao.gomes@epfl.ch

Abstract. Francis turbines operating in part load conditions experience a swirling flow at the runner outlet leading to the development of a precessing cavitation vortex rope in the draft tube. This cavitation vortex rope changes drastically the velocity of pressure waves traveling in the draft tube and may lead to resonance conditions in the hydraulic circuit. The wave speed being strongly related to the cavitation compliance, this research work presents a simple model to explain how it is affected by variations of operating conditions and proposes a method to transpose its values. Even though the focus of this paper is on transpositions within the same turbine scale, the methodology is also expected to be tested for the model to prototype transposition in the future. Comparisons between measurements and calculations are in good agreement.

1. Introduction

Hydropower plants are often required to operate in off-design conditions in order to compensate for variations in power consumption and generation. While operating at part load or full load, Francis turbines present a swirling flow in the draft tube that can result in the development of a cavitation vortex. The presence of cavitation changes the propagation characteristic of the pressure wave through the draft tube, drastically reducing the wave speed and therefore the system eigenfrequencies.

As reported by Landry et al. [1], a series of measurements were performed in a reduced scale model of a Francis turbine of specific speed $\nu = 0.27$. These measurements highlighted the influence of the operating parameters, *i.e.* the speed factor n_{ED}, discharge factor Q_{ED} and Froude number Fr on the cavitation compliance C_c. The main objective of this paper is to create a simplified analytical model that enables a better understanding of the physics determining the C_c values. The model presented here assumes an axisymmetric cylindrical cavitating vortex. It does not try to fully describe the kinematics behind the non-axisymmetric vortex rope, but aims at identifying in a simplified way the influence of n_{ED}, Q_{ED}, Fr and Thoma number σ in the resulting cavitation compliance. Based on this model, a method is developed to transpose values of the cavitation compliance and the wave speed to other operating conditions and from the model to the prototype scale.

2. Methodology

2.1. *Simplified equation for the wave speed*

For a straight pipe with 2-phase flow assuming no mass exchange between phases, the continuity and the momentum conservation equations for the liquid phase only can be written as:

$$\begin{cases} \frac{C_m}{\rho}\frac{\partial \rho}{\partial x} + \frac{C_m}{A_l}\frac{\partial A_l}{\partial x} + \frac{\partial C_m}{\partial x} + \frac{1}{\rho}\frac{\partial \rho}{\partial t} + \frac{1}{A_l}\frac{\partial A_l}{\partial t} = 0 \\ \frac{1}{\rho}\frac{\partial p}{\partial x} + \frac{\partial C_m}{\partial t} + C_m\frac{\partial C_m}{\partial x} + g\sin(\alpha) + \frac{\lambda C_m |C_m|}{2D} = 0 \end{cases} \tag{1}$$

where A_l is the cross-section of the water in the liquid state and C_m is the flow velocity in the axial direction. The mass in the vapor phase is neglected. Assuming that C_m is small and neglecting the gravity and viscous effects, these equations can be simplified as:

$$\begin{cases} \frac{\partial C_m}{\partial x} + \frac{1}{\rho}\frac{\partial \rho}{\partial t} + \frac{1}{A_l}\frac{\partial A_l}{\partial t} = \frac{\partial C_m}{\partial x} + \left(\frac{1}{\rho}\frac{\partial \rho}{\partial p} + \frac{1}{A_l}\frac{\partial A_l}{\partial p}\right)\frac{\partial p}{\partial t} = 0 \\ \frac{1}{\rho}\frac{\partial p}{\partial x} + \frac{\partial C_m}{\partial t} = 0 \end{cases} \tag{2}$$

From the system of equations above, the following equation for the wave speed is found:

$$a = \pm\left(\frac{\partial \rho}{\partial p} + \frac{\rho}{A_l}\frac{\partial A_l}{\partial p}\right)^{-1/2} = \pm\left(\frac{\rho}{E_{water}} + \frac{C_c}{A_l g L_p} + \frac{A_p \rho D_p}{A_l E_p e}\right)^{-1/2} \tag{3}$$

where E_{water} is the liquid bulk viscosity, E_p is the pipe elasticity modulus and e is the pipe thickness [2], C_c is the cavitation compliance defined as $C_c = -{}^{\partial V_v}/_{\partial h} = \rho g L_p \left({}^{\partial A_l}/_{\partial p} - {}^{\partial A_p}/_{\partial p}\right)$ [3], h is the piezometric pressure in meters of water column, L_p is the pipe length and A_p is the pipe cross-section which is equal the sum of the liquid and vapor cross-sections (A_l and A_v). Considering the geometry and the mechanical properties of the turbine and test rig where the measurements were performed, it has been noticed that the terms containing the water bulk modulus and the pipe elasticity are orders of magnitude smaller than expected values for the term containing the cavitation compliance. Therefore, these terms can be neglected and the wave speed calculated as:

$$a = \pm\left(\frac{C_c}{A_l g L_p}\right)^{-1/2} = \pm\left[\frac{C_c}{g L_p A_p (1-\beta)}\right]^{-1/2} \tag{4}$$

where β is the void fraction $\beta = V_v/V_p = A_v/A_p = 1 - A_l/Ap$.

2.2. *Simplified analytical model for the calculation of β and C_c*

The model described here considers the following assumptions:

- the pressure inside the cavitation vortex is assumed constant and equal to the vapour pressure;
- the water surface tension is negligible, making the pressure at the interface liquid-vapour equal to the vapor pressure, *i.e.* $p(R_v) = p_v$.
- the cavitation volume and the pipe are axisymmetric with respect to the axial direction;
- the cavitation volume in the draft tube is assumed to have a constant length and a circular cross-section with constant radius;
- the axial velocity C_m is uniform throughout the liquid cross-section A_l.

If the tangential speed C_u is assumed to be constant, the integration of the Navier-Stokes equation in the radial direction from the liquid-vapor interface R_v to the pipe radius R_p gives:

$$C_u{}^2 \ln\left(\frac{R_p}{R_v}\right) = \frac{p_p - p_v}{\rho} \tag{5}$$

where p_p is the pressure at the pipe wall. The term on the right side of Equation 5 can be easily related to the Thoma number $\sigma = NPSH/H$, where H is the net head and $NPSH$ is the Net Positive Suction Head. By ignoring the losses in the draft tube, Equation 5 can be written in terms of β and σ as:

$$\beta = e^{-\frac{2g}{C_u{}^2}(\sigma H + Z_{ref} - Z_{cone} - \frac{Q^2}{2gA_l{}^2})]} \tag{6}$$

where Q is the discharge, Z_{ref} is the reference level for the σ calculation and Z_{cone} is considered here as the level of the draft tube cone center. By deriving Equation 5 with respect to p_p and neglecting the variations of the pipe radius and water density, an expression for the cavitation compliance can be found:

$$C_c = \frac{2gL_pA_l}{C_u{}^2}\beta \tag{7}$$

As the real non-axisymmetric cavitation vortex rope is approximated by an axisymmetric model, the tangential velocity C_u can be seen as an indication of the swirl intensity of the vortex.

In order to assess the validity of this model, the first eigenfrequency of the whole test rig was determined while the reduced scale model of the turbine operated at different operating conditions, *i.e.* with different values of n_{ED}, Q_{ED}, Fr and Thoma number σ [4]. By knowing the properties of all the other parts of the test rig, a SIMSEN model is then used to identify the cavitation compliance at the location of the vortex rope [5]. For these measured points, once C_c is known, C_u and β can also be calculated using Equations 7 and 6.

For all the transpositions presented in this article, the polynomial curve fit based on the measurements presented in Figure 1(a) are used as a reference from which all the values are transposed. The x axis of this figure is $Q_{\mathrm{ED}}/Q_{\mathrm{ED}}{}^{S=0}$, where $Q_{\mathrm{ED}}{}^{S=0}$ corresponds to the value in the swirl-free condition for the same speed factor n_{ED}. Measurements performed at different n_{ED} values have shown a very similar curve as long as they keep the same $NPSH$ and Fr (Figure 1(b)). It can then be assumed that variations of n_{ED} have a small impact on the Cc values, as long as $Q_{\mathrm{ED}}/Q_{\mathrm{ED}}{}^{S=0}$, $NPSH$ and Fr are kept the same.

To transpose the C_c values from the reference presented in Figure 1(a) to another chosen operating condition, the steps below are followed:

- by knowing the ratio $Q_{\mathrm{ED}}/Q_{\mathrm{ED}}{}^{S=0}$ of the point of interest, the reference C_c value from Figure 1(a) is found. As discussed earlier, n_{ED} is expected to have a minor influence on the C_c value;
- from the reference C_c value, C_u reference is obtained by combining Equations 6 and 7;
- C_u reference is then transposed to the new chosen operating condition assuming that the dimensionless speed ratio C_u/nD remains constant;
- as all the terms in Equation 6 are now known for the new operating condition, β can be calculated;
- the new C_c is calculated using Equation 7, using the new C_u and β.

3. Results and discussions

In Figure 2, the measured and calculated values of C_c are shown for two σ-values different from that presented in Figure 1. A very good agreement between measurements and calculations is

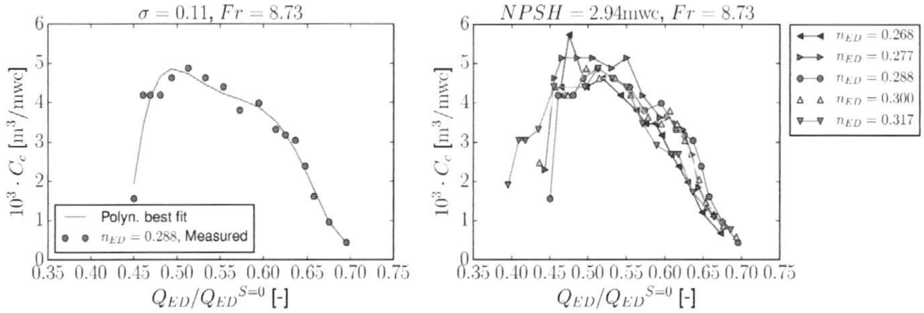

(a) C_c from measurements and the polynomial best fit used as a reference.

(b) C_c for five different n_{ED} values

Figure 1: Cavitation compliance obtained from measurements. (a) $n_{\mathrm{ED}} = 0.288$, used as a reference. (b) Comparison for different n_{ED} values with the same $NPSH$ and Fr.

obtained for $\sigma = 0.10$. For $\sigma = 0.13$, the calculated curve seems to follow a slightly different tendency.

To formulate a hypothesis explaining this deviation, a better understanding of the physics responsible for the shape of the reference C_c curve in Figure 1 is necessary. It is possible to notice three main tendencies: firstly, for low values of Q_{ED} where the flow is very chaotic and turbulent, C_c values increase up to a maximum value where the cavitating vortex rope is fully structured [6]; secondly, as Q_{ED} increases, a small reduction of C_c is noticed as the swirl intensity at the runner outlet also decreases; finally, as the swirl intensity continues to decrease, the vortex rope gets smaller and no longer reaches the draft tube elbow, resulting in a steeper descent in the value of C_c as the model presented here does not take into account this length reduction, but calculates an equivalent cavitation compliance for a constant length.

As the σ-value increases, not only does the vortex rope become smaller, but the range of Q_{ED} values where the vortex rope is fully structured, well developed and reaches the draft tube elbow seems to become shorter as well.

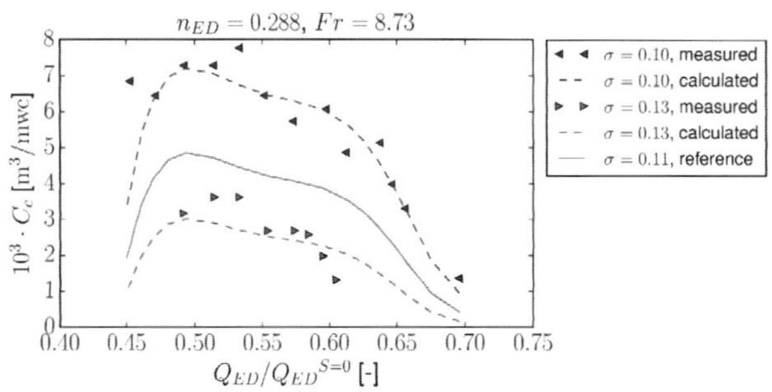

Figure 2: Comparison between the calculated and measured cavitation compliance while varying only the σ-value from the reference operating condition.

To verify the influence of both σ and Fr for constant values of Q_{ED} and n_{ED}, another comparison between measured and calculated C_c values is presented in Figure 3. In this figure, the reference values obtained from the polynomial best fit shown in Figure 1 is represented by the red dots. From this two reference values at two different Q_{ED}, the values represented by the blue lines are calculated. The calculated values show a small variation of the C_c values for the different Fr conditions that were tested, but these variations could not be properly assessed through these results due to the uncertainty of measurements. Except for very low σ-values, good agreement between measurements and calculations are obtained.

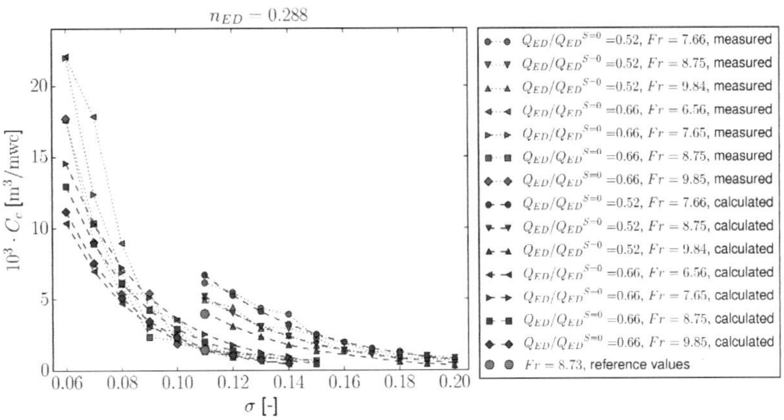

Figure 3: Comparison between calculated and measured cavitation compliance for two different Q_{ED}, while varying both σ and Fr values from the reference.

4. Conclusions

This paper presents a simple model representing the physics of the cavitating vortex rope. Based on this model, a method allowing the transposition of cavitation compliance values from a given operating condition to another is proposed. The method has been tested only for partial load conditions so far.

To be used as a reference, the model requires the determination of the cavitation compliance C_c for at least one n_{ED} value, the Q_{ED} value varying through the partial load condition. From this reference, C_c can be calculated for another n_{ED}, σ and Fr conditions. Although it has not been tried in this paper, this method is expected to be used for the estimation of the C_c at the prototype scale based on the measurements made on the reduced scale model. As the Fr similarity is not always respected and the value of σ cannot be chosen during tests on the prototype, transposition of C_c might become necessary.

Further investigations will verify the applicability of this model to full load conditions. Different flow velocity profiles might be tested. Comparisons between model and prototype measurements are also expected in the future.

References

[1] Landry C, Favrel A, Müller A, Nicolet C and Avellan F 2016 *Journal of Hydraulic Research* **54** 185–196
[2] Nicolet C 2007 *Hydroacoustic modelling and numerical simulation of unsteady operation of hydroelectric systems* Ph.D. thesis STI Lausanne
[3] Alligne S, Nicolet C, Tsujimoto Y and Avellan F 2014 *Journal of Hydraulic Research* **52** 399–411
[4] Favrel A, Müller A, Landry C, Yamamoto K and Avellan F 2016 *Experiments in Fluids* **57** 168
[5] Nicolet C, Avellan F, Allenbach P, Sapin A and Simond J J *Proceedings of the Hydraulic Machinery and Systems 21st IAHR Symposium, Lausanne, Switzerland 2002*
[6] Favrel A, Müller A, Landry C, Yamamoto K and Avellan F 2015 *Experiments in Fluids* **56** 1–15

The comparison between the acquisition vibration data obtained by different types of transducers for hydraulic turbine head cover

LI Youping[1], LU Jinsong[2], CHENG Jian[1],YIN Yongzhen[1] and WANG Jianlan[1]

[1] Technology and Research Center of China Yangtze Power Corporation, Yichang Hubei,443002, P.R.China

[2] XiangJiaBa Hydraulic Power Plant, Yibin, Sichuan 644000,P.R.China

Abstract. Based on the summaries of the rules about the vibration measurement for hydro-generator sets with respect to relevant standards, the key issues of the vibration measurement, such as measurement modes, the transducer selection are illustrated. In addition, the problems existing in vibration measurement are pointed out. The actual acquisition data of head cover vertical vibration respectively obtained by seismic transducer and eddy current transducer in site hydraulic turbine performance tests during the rising of the reservoir upstream level in a certain hydraulic power plant are compared. The difference of the data obtained by the two types of transducers and the potential reasons are presented. The application conditions of seismic transducer and eddy current transducer for hydro-generator set vibration measurement are given based on the analysis. Research subjects that should be focused on about the topic discussed in this paper are suggested.

1. Introduction

1. 1. Evaluation of vibration

The level of mechanical vibration can be expressed by displacement , velocity or acceleration, in which the displacement reflects the amplitude, the velocity reflects the energy and the acceleration reflects the impact force. It can also be considered that the level of vibration is in proportion to the displacement while the frequency of vibration is low, to the velocity while the frequency is medium and to the acceleration while the frequency is high[1]. Therefore, the peak-peak value of displacement is measured for low speed (less than 300 rpm) machines, the root mean square value of speed is measured for medium and high speed (300 rpm ~ 1800 rpm) machines[2].

1. 2. Character of hydro-generator vibration

Generally, the rotational speed of hydro-generators is relatively low, which is about 1 ~ 2 Hz for large and medium capacity hydro-generators. The vibration caused by the vortex in draft tube cone is lower, which is about 1/6 ~ 1/2 times of the rotational frequency.Because the hydraulic factor has a great influence on the head cover, the frequencies of head cover vibration are mostly low components.It is widely believed that the measurement results of head cover vibration are not accurate enough[2-3].

1. 3. Measurement mode for hydro-generator vibration

There are two types of transducers to measure hydro-generator vibration. Contacting transducers are usually used to measure the absolute vibration for non-rotational components, such as bearing seats and brackets. Seismic transducers are generally used as contacting transducers.Non-contacting transducers can be fixed to rigid mounting brackets and applied to measure directly the vibration displacement when the vibration of the mounting brackets can be ignored. Eddy current transducers are commonly used as non-contacting transducers.

1. 4. Demands of hydro-generator vibration measurement devices

The frequency response range of vibration measurement devices for Francis is specified by IEC 994 as following[2]. lower frequency of measurement device frequency response f_l and the upper frequency of measurement device frequency response range f_u should be according to equation (1) and equation (2).

$$f_l = 0.1 \times f_{sp} \tag{1}$$

$$f_u = \max\{z_0 \times z_1 \times f_{sp}, Sh \times V_w \times \delta^{-1}\} \tag{2}$$

Where, f_{sp} is rotational frequency of hydro-generator at specified speed, z_0 is number of guide vanes, V_w is flow velocity, δ is thickness of outlet edge of a hydraulic profile (guide vane, runner blade,etc), Sh is Strouhal number, z_1 is number of runner blades.

1. 5. Transducers for hydro-generator vibration measurement

The rated rotational speed of some large capacity hydro-generators in the world are listed in Table 1. Because of the limitation of technology and craftsmanship, the bandwidth of the seismic transducers at present is difficult to meet the demands of the lower limit mentioned above.

Table 1 rated speed of some large hydro-generators

Plant	Manufacturers	Rated Output (MW)	Rated Speed (rpm)	0.1× Rotational Frequency (Hz)
Right Bank and Underground of Three Gorges Plant	DEC	700	75.0	0.13
	ALSTOM	700	71.4	0.12
	HEC	700	75.0	0.13
XiLuoDu Plant	-	770	125.0	0.21
XiangJiaBa Plant	HEC	800	75.0	0.13
	ALSTOM	800	71.4	0.12
Grand Coulee IIIPlant	-	700	85.7	0.14
Itaipu Plant	-	700	91.6	0.15
Raul Leoni II Plant	-	600	112.5	0.19

The key feature of the eddy current transducer is that it can accurately measure the static and dynamic distances between the measured metal and the probe of the transducer[3]. Additionally, it is provided with the performance of zero frequency response and can be used to measure the hydro-generator vibration with low frequency components. The main problem lies in the erection difficulty of the mounting brackets of the transducers. Furthermore, the vibration of the mounting bracket is inevitable, which will adversely affect the measure results when the eddy current transducer is used to measure the vibration.

In a stability and performance test for a certain hydro-generator, contacting seismic transducers and non-contacting eddy current transducers were used to measure the head cover vibration. The acquisition vibration data respectively obtained by the two types of transducers in the tests are compared and the influence of the transducer type on measurement result is analyzed in this paper.

2. Parameters of the hydro-generator and the test conditions

2. 1. Hydro-generator parameters

The hydro-generator's rated power is 770 MW, its rated speed is 125 rpm, its rated water head is 197 m.The maximum diameter of the head cover is 11 m, and the runner has 15 blades.

2. 2. Test conditions

Stability and performance tests were carried out under 23 water heads while the gross water heads changed from175 m to 215 m.

2. 3. Transducer features

The nominal frequency response range of the seismic transducer is 0.35 Hz to 200 Hz, the nominal span range is ±1000 μm and the nominal sensitivity is 8 mv/μm. The characteristic curve calibrated in factory acceptance test is shown as figure 1. It can be seen that the vibration response of the transducer is stable while the vibration frequency is exceeding 0.5 Hz, and is degenerative while the vibration frequency of is less than 0.5 Hz.

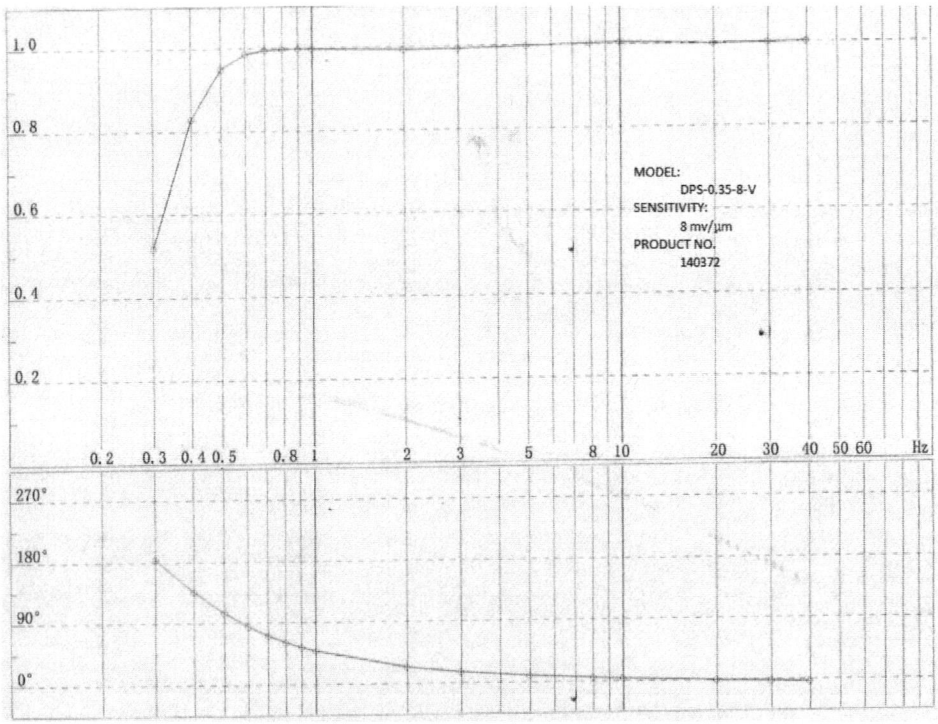

Figure 1 characteristic curve of seismic transducer

The nominal span range of the eddy current transducer is 1.5 mm to 3.5 mm and the nominal sensitivity is 8 mv/μm. The characteristic curve of the transducer is shown as figure 2.

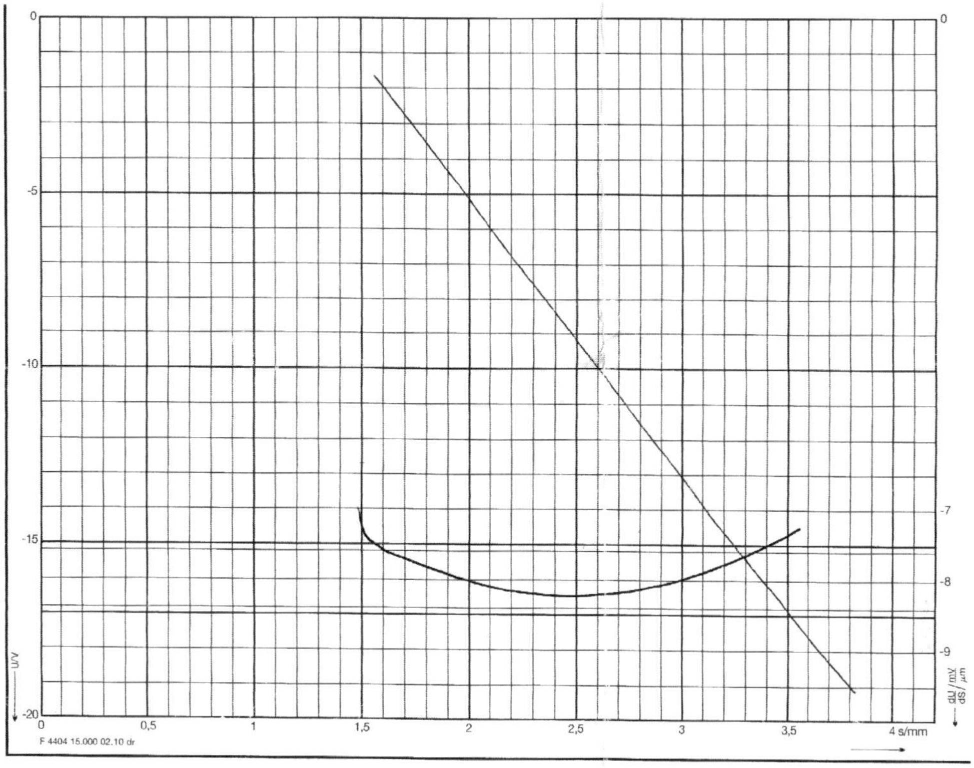

Figure 2 characteristic curve of eddy current transducer

2. 4. Transducer installations

Seismic transducers were installed in the direction of +X and +Y of the head cover inner side, the eddy current transducer is installed close to the direction of +X(shown as figure 3). The mounting bracket of the eddy current transducer consisted of several steel pipes. One end of the pipes were welt to the pit liner and the other ends were hanged in the air.The impending ends of the pipes were connected together by a short steel pipe(shown as figure 4).

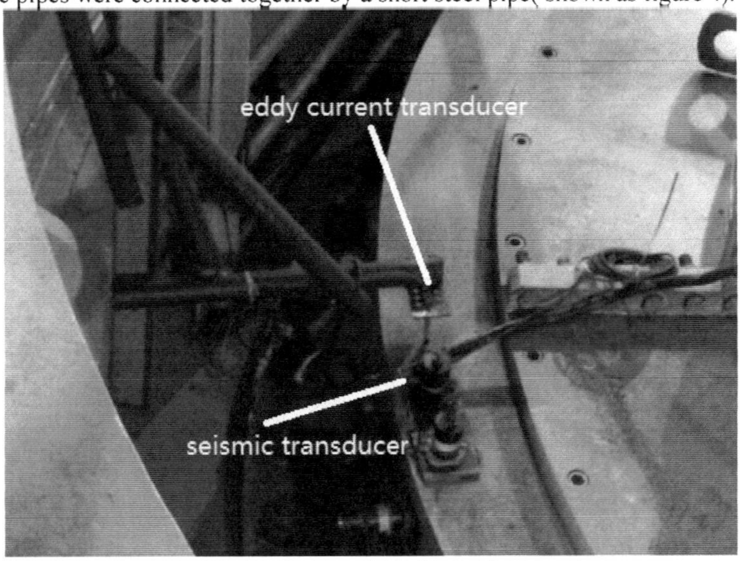

Figure 3 erection of transducers

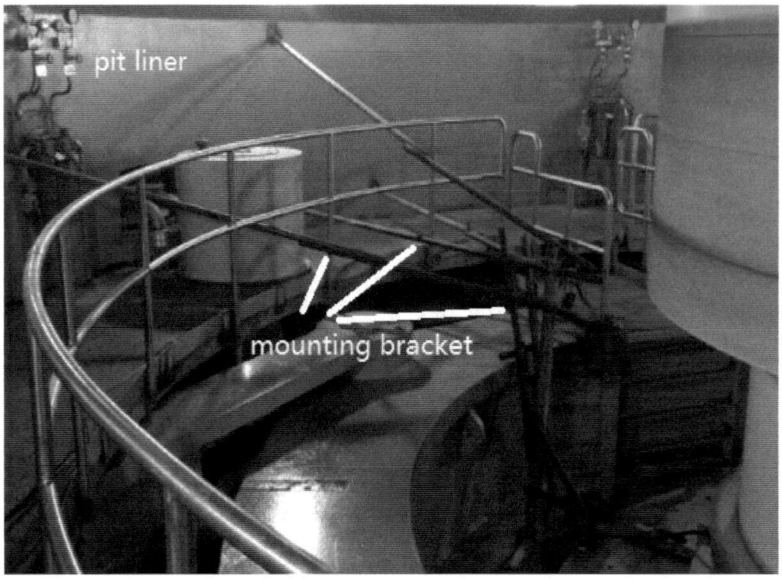

Figure 4 erection of eddy current transducer and mounting bracket

3. Test data analysis

3. 1. Acquisition vibration data

The data obtained by seismic transducers and eddy current transducers are shown in figure 5 and figure 6 respectively.

The vibration data of head cover obtained under different output conditions with typical gross water heads of 214.7 m,198.2 m and 183.3 m are selected to be compared. The vibration peak-peak values are listed in Table2, the 1st to 4th main frequencies and the corresponding amplitudes of the head cover vertical vibration are listed in Table 3 to Table 5 respectively.

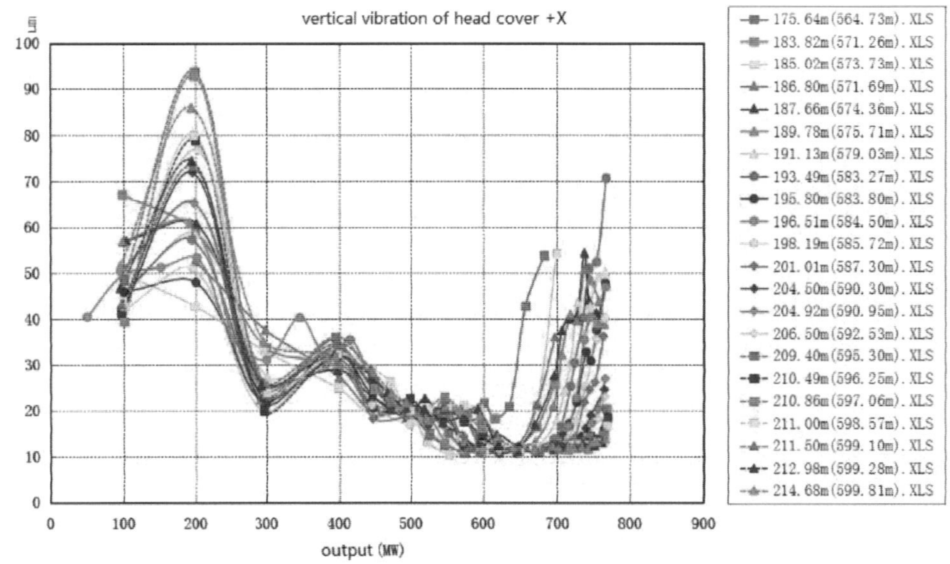

Figure 5 vertical vibration of head cover (obtained by seismic transducer)

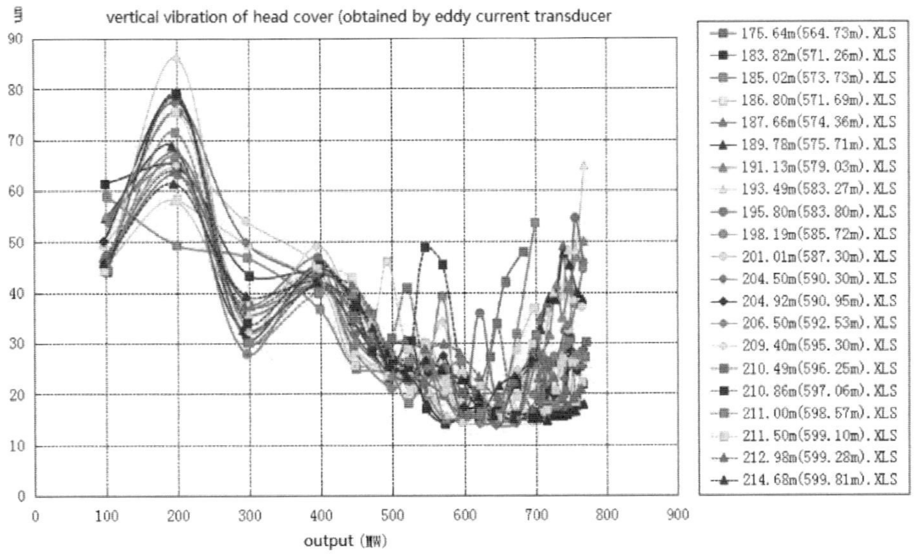

Figure 6 vertical vibration of head cover (obtained by eddy current transducer)

Table 2 vertical vibration peak-peak value of head cover under representative heads

Vibration (μm) H=214.7 m			Vibration (μm)H=198.2 m			Vibration (μm)H=183.8 m		
Output (MW)	Seismic Transducer	Eddy Current Transducer	Output (MW)	Seismic Transducer	Eddy Current Transducer	Output (MW)	Seismic Transducer	Eddy Current Transducer
767	14	18	766	40	45	696	26	26
755	13	16	756	36	38	673	16	22
747	13	16	740	25	29	645	12	15
743	12	16	726	18	27	621	12	15
735	12	16	715	14	18	595	11	15
724	12	16	706	12	19	573	11	14
715	12	15	697	12	20	546	13	17
703	12	16	669	12	22	526	15	30
694	12	15	646	11	14	496	19	25
669	12	15	621	11	16	449	20	39
641	12	16	594	11	15	398	30	44
619	14	19	570	14	39	299	34	43
595	19	23	546	20	25	200	59	65
569	20	25	520	18	23	99	67	61
547	21	27	499	17	21			
521	18	23	446	21	29			
493	22	28	394	33	45			
473	24	33	296	20	28			
447	28	38	196	51	66			
394	32	42	99	42	45			
296	35	39						
195	86	61						
97	52	46						

Table 3 1st～4th main frequencies and the corresponding amplitudes as H=214.7 m

Output(MW)	Vibration +X(Seismic Transducer)								Vibration(eddy current transducer)							
	1st Fre* (Hz)	Amp 1st (μm)	2nd Fre (Hz)	Amp 2nd (μm)	3rd Fre (Hz)	Amp 3rd (μm)	4th Fre (Hz)	Amp 4th (μm)	1st Fre (Hz)	Amp 1st (μm)	2nd Fre (Hz)	Amp 2nd (μm)	3rd Fre (Hz)	Amp 3rd (μm)	4th Fre (Hz)	Amp 4th (μm)
767	2.08	3	31.23	3	0.39	1	0.52	1	2.08	4	31.23	3	28.61	3	28.65	2

IOP Conf. Series: Journal of Physics: Conf. Series **813** (2017) 012031 doi:10.1088/1742-6596/813/1/012031

755	2.08	3	31.26	2	0.20	1	0.24	1	2.08	4	28.61	3	31.26	3	14.69	2
747	2.08	3	31.24	2	0.22	2	18.65	1	2.08	4	28.59	2	31.24	2	14.70	2
743	2.08	3	31.25	2	18.65	1	14.91	1	28.62	4	2.08	3	31.25	3	28.72	2
735	2.08	3	31.23	2	15.08	1	18.66	1	2.08	4	31.23	3	28.63	2	77.04	2
724	2.08	3	31.23	2	18.66	1	14.99	1	2.08	4	14.69	3	31.23	3	14.71	2
715	2.08	3	31.25	2	0.47	1	18.69	1	2.08	3	31.25	2	28.64	2	0.02	2
703	2.08	3	31.25	2	18.69	1	1.50	1	2.08	3	13.16	2	31.25	2	28.65	2
694	2.08	3	31.25	2	18.71	2	0.14	1	2.08	3	31.25	2	28.64	2	18.71	2
669	2.08	3	31.24	2	18.72	1	15.11	1	2.08	3	28.63	3	31.24	2	28.69	2
641	2.08	3	18.76	2	31.24	1	0.40	1	2.08	3	14.68	2	18.76	2	28.69	2
619	0.45	3	2.08	3	0.42	2	0.51	2	2.08	3	0.45	3	0.42	3	28.59	2
595	0.65	8	2.08	3	1.29	3	1.27	2	0.65	9	2.08	3	1.29	3	0.63	2
569	0.51	8	1.03	5	1.54	4	0.53	4	0.51	9	1.03	5	1.54	5	0.53	5
547	0.50	9	1.00	6	1.50	4	2.08	2	0.50	11	1.00	7	1.50	4	0.02	3
521	0.52	4	0.54	4	1.05	4	2.08	2	0.52	6	0.54	5	1.05	5	2.08	3
493	0.51	7	0.53	4	0.49	3	2.08	3	0.51	9	0.53	4	0.49	4	2.08	3
473	0.51	9	0.30	3	0.99	2	2.08	2	0.51	11	0.30	5	0.20	5	0.22	4
447	0.48	14	0.46	4	0.96	3	0.50	3	0.48	17	0.46	5	0.22	4	0.40	4
394	0.47	13	0.36	6	0.83	5	0.79	4	0.47	15	0.36	8	0.32	6	0.49	6
296	0.32	6	0.38	6	0.45	5	0.36	5	0.02	7	0.20	4	0.36	4	1.32	4
195	0.26	18	0.28	14	0.29	13	0.18	13	28.60	10	0.29	6	0.33	5	76.96	5
97	0.34	11	0.30	10	0.26	10	0.24	7	28.59	7	0.30	5	0.81	5	0.93	4

*

1st Fre: 1st main Frequency
Amp 1st : Amplitude of 1st main frequency
2nd Fre: 2nd main frequency
Amp 2nd : Amplitude of 2nd main frequency
3rd Fre: 3rd main frequency
Amp 3rd : Amplitude of 3rd main frequency
4th Fre: 4th main frequency
Amp 4th : Amplitude of 4th main frequency

Table 4 1st～4th main frequencies and the corresponding amplitudes as H=198.2 m

Output(MW)	Vibration +X(Seismic Transducer)								Vibration(eddy current transducer)							
	1st Fre (Hz)	Amp 1st (μm)	2nd Fre (Hz)	Amp 2nd (μm)	3rd Fre (Hz)	Amp 3rd (μm)	4th Fre (Hz)	Amp 4th (μm)	1st Fre (Hz)	Amp 1st (μm)	2nd Fre (Hz)	Amp 2nd (μm)	3rd Fre (Hz)	Amp 3rd (μm)	4th Fre (Hz)	Amp 4th (μm)
766	0.97	12	1.09	9	0.85	7	1.01	6	0.02	13	0.97	10	1.09	8	1.01	6
756	1.04	7	1.18	7	0.96	5	0.98	5	0.02	8	1.04	6	1.18	6	0.96	5
740	2.36	4	1.11	4	1.27	4	1.19	3	13.20	6	1.11	3	13.22	3	1.27	3
726	2.08	3	0.95	2	31.22	2	1.23	2	28.63	5	17.64	5	13.20	5	17.62	4
715	2.08	3	31.23	2	18.78	1	1.13	1	2.08	3	14.69	2	31.23	2	13.22	2
706	2.08	3	31.23	2	0.32	1	0.53	1	0.02	7	2.08	3	0.04	3	31.23	2
697	2.08	3	31.23	2	18.80	1	0.22	1	0.02	6	28.62	4	2.08	3	0.04	3
669	2.08	3	31.23	2	18.84	1	0.40	1	0.02	13	2.08	3	0.04	2	31.23	2
646	2.08	3	31.22	2	18.86	1	14.98	1	2.08	3	28.62	2	31.22	2	14.69	2
621	2.08	2	31.22	2	18.89	1	0.32	1	0.02	3	2.08	3	0.04	3	14.69	2
594	2.08	2	18.73	2	31.22	2	15.01	1	28.61	3	2.08	3	18.73	2	14.69	2
570	0.46	4	2.08	2	18.94	2	0.44	2	0.02	28	0.04	5	0.46	5	0.08	4
546	0.44	7	0.48	2	2.08	2	1.09	2	0.44	7	0.24	4	0.20	3	0.48	3
520	0.53	8	1.05	4	1.03	3	0.51	2	0.53	10	1.05	4	1.03	3	1.56	3
499	0.53	7	1.05	5	1.58	2	0.42	1	0.53	9	1.05	6	1.58	2	13.22	2
446	0.53	4	1.37	3	1.11	2	0.51	2	0.53	5	0.28	5	0.34	4	0.02	3
394	0.46	15	0.49	7	0.44	6	0.95	5	0.46	18	0.49	9	0.44	7	0.95	5
296	0.73	3	0.42	3	0.79	2	0.36	2	0.06	5	0.32	4	0.28	3	0.73	3
196	0.34	12	0.32	6	0.55	6	0.53	5	0.02	16	72.88	8	17.65	6	0.06	6
99	0.36	7	0.28	6	0.20	6	0.53	6	0.10	5	0.32	5	0.04	5	0.26	5

Table 5 1st～4th main frequencies and the corresponding amplitudes as H=183.3 m

Output(MW)	Vibration +X(Seismic Transducer)								Vibration(eddy current transducer)							
	1st Fre (Hz)	Amp 1st (μm)	2nd Fre (Hz)	Amp 2nd (μm)	3rd Fre (Hz)	Amp 3rd (μm)	4th Fre (Hz)	Amp 4th (μm)	1st Fre (Hz)	Amp 1st (μm)	2nd Fre (Hz)	Amp 2nd (μm)	3rd Fre (Hz)	Amp 3rd (μm)	4th Fre (Hz)	Amp 4th (μm)
696	1.04	4	1.11	4	1.24	4	1.06	4	13.19	4	1.04	4	1.24	3	1.06	3
673	2.08	3	18.65	2	31.22	2	1.29	1	28.58	4	28.62	4	77.02	3	2.08	3
645	2.08	2	0.30	1	31.27	1	0.38	1	28.61	3	2.08	3	14.68	2	14.70	2
621	2.08	2	18.74	1	18.76	1	31.24	1	2.08	3	28.60	2	77.06	2	14.69	2
595	2.08	2	18.79	1	31.23	1	15.06	1	14.69	3	2.08	2	77.03	2	14.68	2
573	2.08	2	18.84	1	18.99	1	31.27	1	28.63	3	2.08	2	13.19	2	14.67	1
546	0.48	2	2.08	2	0.45	2	18.86	2	0.48	2	0.02	2	0.45	2	2.08	2
526	0.43	2	0.45	2	18.87	2	0.50	2	0.02	7	0.03	6	0.05	5	0.08	4
496	0.55	8	1.07	4	0.53	4	1.09	2	0.55	10	1.07	5	0.53	4	0.02	4
449	1.09	3	0.79	3	0.53	3	1.32	2	0.02	19	0.03	7	0.30	5	0.07	4
398	0.49	7	0.51	6	0.81	5	0.30	4	0.49	9	0.30	9	0.81	6	0.32	6
299	0.49	9	0.99	5	0.64	4	0.53	4	0.49	12	0.16	7	0.34	6	0.02	5
200	0.36	10	0.33	9	0.61	9	0.28	7	0.02	11	28.62	8	28.55	8	28.60	7
99	0.22	12	0.24	10	0.49	9	0.29	9	0.02	15	28.57	7	28.55	7	0.20	6

3. 2. Data analyses

1) Based on the comparison between figure 5 and figure 6, it can be seen,

① For the head cover vibrations obtained by the two types of transducers, there both exist two peaks when the output is near 200 MW and 400 MW respectively under different water heads. The main frequencies of the vibration at the two peaks are 1/6～1/2 times of the rotational frequency (2.08 Hz) and are typical vortex frequency components[4].

② Low extreme point of the head cover vibration occurs when the output is within 600 MW～700 MW. The head cover vibration increases sharply with the increasing of the output after the low extreme point, the output at the low extreme point increases with the increasing of water head.

The reasons causing the above-mentioned phenomena are detailed as follow. The circumferential component in the runner outlet is minimum at the low extreme point and the runner is under zero circumferential component condition. At this point, the pressure fluctuation reaches the minimum and hence the vibration of the head cover reaches the minimum. The circumferential component in the runner outlet increases with the increasing of the guide vane opening or the output. In this case, the pressure fluctuation increases, which results in the dramatic increasing of the head cover vibration. On the other hand, the output at the zero circumferential component condition increases with the increasing of the water head. Consequently, the output at the low extreme point of the head cover vibration increases with the increasing of the water head[5-6].

According to ① and ②, the peak-peak values of head cover vibration obtained by seismic transducer and eddy current transducer can both correctly reflect the vibration variation with the water head and output.

2) It can be seen from Table 2,

③ The peak-peak values of head cover vibration obtained by the two types of transducers are well-matched in general under high output (when the output exceeds 500 MW) conditions. The differences between the data obtained by the two types of transducers are less than 10 μm and the data obtained by seismic transducers are generally less than that obtained by eddy current transducers.

The possible reasons are as following. The main frequency components of head cover vibration are no longer low vortex frequency under large output conditions as the large output conditions are far away from vortex conditions. The head cover vibration frequency components under these conditions are within the stable measurement frequency range of the seismic transducer. As for the eddy current transducer, the vibration of the mounting bracket itself is inconspicuous under such condition and has less influence on the obtained data of the head cover vibration. Therefore, the two types of transducers can both measure the head cover vibration stably under large output conditions. At the same time, as the vibration of the mounting bracket is unavoidable, the data obtained by eddy current transducers are slightly greater than that obtained by seismic transducers.

④ The data obtained by the two types of transducers have larger differences under low output conditions (when output is less than 400 MW). The difference is 25 μm under the condition of H=214.7 m and P=195 MW. There is no obvious regularity of the quantity relationship between the data obtained by the two types of transducers.

The possible reasons are as following. The main frequency components of the head cover vibration are relatively small under the low output conditions. The attenuation of the vibration with different low frequencies obtained by the seismic transducer are not the same, which may result in the reduction of the measurement stability of the seismic transducer. For the eddy current transducer, as the vibration of the hydro-generator is large under the low output conditions, the vibration transmitted from the pit liner to the mounting bracket is relatively large and the data obtained by eddy current transducers are of poor stability.

3) It can be seen from Table 3 to Table 5,

⑤ The main frequency components and the corresponding amplitudes of the head cover vibration obtained by the two types of transducers are quite same under medium and large output conditions (when output is larger than 400 MW), and they can both reflect the characteristic frequencies of Francis turbine, such as the vortex frequency components, rotational frequency component, product of rotational frequency and number of runner blades.

⑥ The similarity of the main frequency components of the head cover vibration obtained by the two types of transducers is small under the low output conditions (when output is less than 400 MW), the data obtained by seismic transducer can reflect the vortex frequency components. But the data obtained by eddy current transducer can not. The correlation between the vibration frequencies obtained by the eddy current transducer and known hydraulic turbine characteristic frequencies is not strong. However, the main frequency components of the vibration obtained by the eddy current transducer under different water heads are similar in general, which probably results from the inherent frequency of the mounting bracket or the vibration transmitted from the pit liner.

3. 3. Summary

The above-mentioned analyses indicate that,

1) The measure results obtained by seismic transducers are comparatively accurate when the main frequency components of the hydro-generators vibration are within the stable frequency response range of the transducer. Therefore, seismic transducers can be applied in the vibration measurement when the rotational frequency exceeds 4～6 times of the low limit of the stable frequency response range.

2) The measurement results obtained by eddy current transducers are comparatively accurate when the vibration of the mounting bracket is small. The deviation between the measurement result and the actual vibration is significant when the vibration of the mounting bracket is obvious.

4. Conclusions

The actual head cover vibration data obtained by seismic transducers and eddy current transducers are compared in this paper. It is pointed out that the hydro-generator vibration can be correctly measured by both the two types of transducers under certain conditions, and the suitable conditions for the two types of transducer are pointed out also.

Because of some limitations of the instance in the paper, the influence research of the transducer type on the measurement results of hydro-generator vibration is not sufficiently comprehensive. The following aspects on this subject can be further studied:

1) As the runner rotational speed measured in this paper is quite high, the effect of the seismic transducer being applied in the vibration measurement for lower rotational speed of

the hydro-generator has not been verified. Therefore, the subsequent research can be carried out aiming at the hydro-generators with lower rotational speed .

2) The inherent frequency of the eddy current mounting bracket and the vibration transmitted from the pit liner are not measured in the instance while using eddy current transducer to measure the vibration. As a result, the vibration of the mounting bracket are not deducted from the measurement results, which should be taken into account in the subsequent tests and researches.

3) Some other types of transducers, such as fiber optic transducers, are presently being used to measure the vibration of rotational machines[7]. The application of such transducers in the vibration measurement of hydro-generators, as well as the influences of different types of transducers on the hydro-generator vibration measurement, can be further studied.

Acknowledgments
Supported by the Open Research Subject of Key Lab of Hydraulic Machinery Transients,Ministry of Education of China.

References

[1] Qu Weide.*Manual of mechinal vibration[M]*.Beijing:Chine Machine Press,1992 [in Chinese]

[2] CEI/IEC 994 : 1991,Guide for field measurement of vibrations and pulsation in hydraulic machines (turbines, storage pumps and pump-turbines)[s]:59

[3] HU Huan,Xu Zhaoxi.The principle of eddy current sensor and its application on turbine supervisory instrumentation in power plant[J].*Bao steel technology*.**2015(5)**:62-66

[4] Shi Zhensheng. Hydraulic turbine [M]. Beijing:Hydraulic and electric power press.1992:155

[5] Youping Li,Hanli Weng, Jian Cheng,FeiXiang.Performance Analysis for Francis Hydraulic Turbine based on Normalized Operating Condition and its Application[J]. *CHEMICAL ENGINEERING TRANSACTIONS*.**2015(46)**:1135-1140

[6] Qin Daqing,Zhao Hongtian,Zhaoyang.Some New Views on Francis Turbine Stability[J]. *Large Electric Machine and Hydraulic Turbine* .**1998(3)**:43-50

[7] Zhang Xiaodong,Xie Siying,Niu Hang,Zhang Ping,Jia Binghui.Research on Dynamic MeasurementTechnology of Fiber Optic Sensors and Their Development[J]. *Journal of Vibration,Measurement & Diagnosis*.**2015,35(3):**409-416

RANS computations for identification of 1-D cavitation model parameters: application to full load cavitation vortex rope

S Alligné[1], J Decaix[2], A Müller[3], C Nicolet[1], F Avellan[3] and C Münch[2]

[1] Power Vision Engineering Sàrl, Chemin des Champs Courbes 1, 1024 Ecublens, Switzerland
[2] HES-SO University of Applied Sciences and Arts Western Switzerland, Route du Rawyl 47, Sion, Switzerland
[3] Laboratory for Hydraulic Machines, Avenue de Cour 33 bis, 1007 Lausanne, Switzerland

E-mail: jean.decaix@hevs.ch

Abstract. Due to the massive penetration of alternative renewable energies, hydropower is a key energy conversion technology for stabilizing the electrical power network by using hydraulic machines at off design operating conditions. At full load, the axisymmetric cavitation vortex rope developing in Francis turbines acts as an internal source of energy, leading to an instability commonly referred to as self-excited surge. 1-D models are developed to predict this phenomenon and to define the range of safe operating points for a hydropower plant. These models require a calibration of several parameters. The present work aims at identifying these parameters by using CFD results as objective functions for an optimization process. A 2-D Venturi and 3-D Francis turbine are considered.

1. Introduction

The extensive development of new renewable energy sources provokes electrical power flow fluctuations. In order to manage these fluctuations, hydraulic power plants, taming also green energy, are used due to their ability to respond quickly to a variation of the load. However, in order to inject the suitable amount of power in the grid, the hydraulic turbines have to operate far from their best efficient point. In the case of Francis turbines, the turbine off design operating conditions yield to the development of flow instabilities. One of them, that develops at full load, is the cavitation vortex rope leading to self-excited pressure and discharge oscillations [1]. One-dimensional (1-D) unsteady cavitation models are used to investigate such instabilities [2]. However, the 1-D cavitation model requires a calibration of the physical parameters. The calibration can be achieved by experimental measurements [3] or Computational Fluid Dynamics (CFD) simulations ([4],[5]). However, the calibration of the second viscosity by CFD is still challenging [6].

In the present paper, an approach is described to compute the aforementioned parameters by comparing the system dynamics response of both the CFD and 1-D unsteady models. The calibration is achieved by using the CFD results as objective functions for the 1-D model.

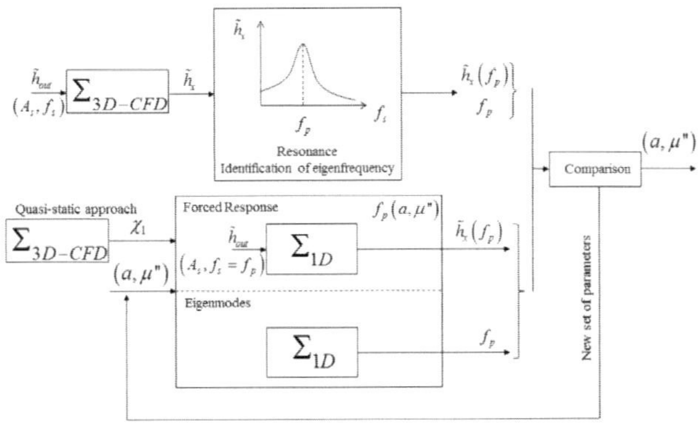

Figure 1. Methodology for identification of 1-D cavitation model parameters.

2. 1-D cavitation model

The modelling of the draft tube cavitation flow is based on both continuity and momentum equations (see Equations 1 and 2) including the convective terms and the divergent geometry [2]:

$$\mathrm{d}Q = -\frac{\mathrm{d}V_c}{\mathrm{d}t} = C_c\frac{\mathrm{d}h}{\mathrm{d}t} + \chi\frac{\mathrm{d}Q}{\mathrm{d}t} \tag{1}$$

$$\frac{1}{gA}\frac{\partial Q}{\partial t} + \frac{Q}{gA^2}\frac{\partial Q}{\partial x} - \frac{Q^2}{gA^3}K_x + \frac{\partial h}{\partial x} + \frac{\tau_0 \pi D}{\rho g A} - \frac{\mu''}{\rho g A}\frac{\partial^2 Q}{\partial x^2} + S_h = 0 \tag{2}$$

This set of equations involve four cavitation vortex rope parameters to be assessed:

- the local wave speed a in a control volume of length dx, which yields the value of the cavitation compliance $C_c = \frac{gA\,dx}{a^2} = -\left.\frac{\partial V_c}{\partial h}\right|_{Q=cste}$;
- the mass flow gain factor $\chi = -\left.\frac{\partial V_c}{\partial Q}\right|_{h=cste}$ corresponding to a variation of the cavitation volume as function of the inlet discharge by keeping the pressure constant;
- the second viscosity μ'' introducing dissipation due to the phase change;
- the momentum excitation source S_h. This parameter is not considered since the phenomenon of interest at full load conditions is of self-excited nature.

3. Methodology

The proposed methodology to find the 1-D cavitation model parameters is presented in Figure 1. Three parameters need to be found: the mass flow gain factor χ_1 , the wave speed a and the second viscosity μ''. The mass flow gain factor is determined with a quasi-static approach: several CFD simulations are performed by keeping constant the pressure downstream the cavitation volume for different inlet discharges. The variation of the cavitation volume as a function of the inlet discharge yields the mass flow gain factor. For identification of the wave speed and the second viscosity, unsteady simulations are performed with an imposed outlet pressure fluctuation. The resonance frequency and the pressure fluctuations at a given location are then used as the objective functions for the optimization process. The optimization process allows the determination of the set of 1-D cavitation parameters that minimizes the errors for the two objectives between the 1-D and the CFD models.

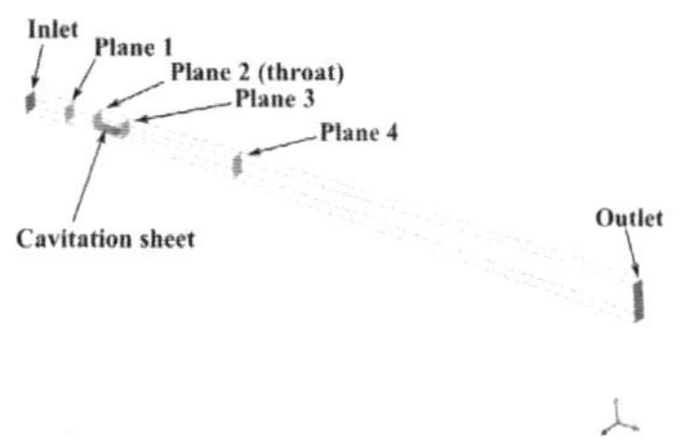

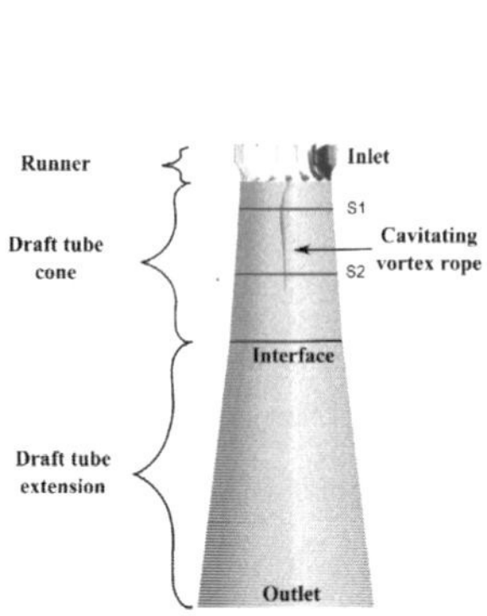

Figure 2. Computational domain of the 2-D Venturi geometry (left) and the 3-D Francis turbine (right).

4. Cases studies

Two test cases are considered: a 2-D Venturi geometry and a simplified 3-D Francis turbine for which only the runner and the axisymmetric simplified draft tube cone are considered, see Figure 2.

5. Numerical set up

5.1. CFD modelling

The CFD simulations are carried out with the Ansys CFX 15.0 software. The flow is modelled using the Unsteady Reynolds-Averaged Navier-Stokes (URANS) equations write for a homogeneous two-phase mixture. The cavitation model is the one proposed by Zwart [8] with the default coefficient values. For the 2-D Venturi, a structured mesh is set up with 15'000 nodes. For the 3D Francis turbine, a structured mesh, with 4 million nodes, is set up in the runner and the cone domains, whereas an unstructured coarse mesh is used for the draft tube extension. This extension avoids that the vortex rope reaches the outlet section. In both cases, the velocity field is imposed at the inlet and a sinusoidal static pressure is specified at the outlet to excite the system.

5.2. Hydro-acoustic modelling

The Venturi and Francis turbine test cases are modelled with the SIMSEN software which includes the cavitation model described previously [2]. The cavitation zone, where cavitation parameters must be identified, is spatially discretized with a lumped model in the case of the Venturi geometry, whereas a distributed model is used for the Francis turbine draft tube. In the cavitation free region, the wave speed is set to 1'000 m/s. Finally, the boundary conditions of the 3-D model are reproduced in the 1-D model: a constant discharge source at the inlet and a sinusoidal static pressure at the outlet.

6. Cavitation model parameters

6.1. Mass flow gain factor

To derive the mass flow gain factor, several 3-D computations are performed by keeping the pressure, as close as possible to the cavitation volume, constant for different inlet discharges. Using a linear regression, the value of the mass flow gain factor χ_1 is determined: $\chi_1 = -0.001$ s for the Venturi test case and $\chi_1 = -0.034$ s for the Francis turbine test case.

6.2. Wave speed and second viscosity

The CFD simulation of the Venturi test case features a resonance frequency value of $f_n = 2Hz$ where a maximum of pressure and cavitation volume fluctuations are experienced [6]. Time

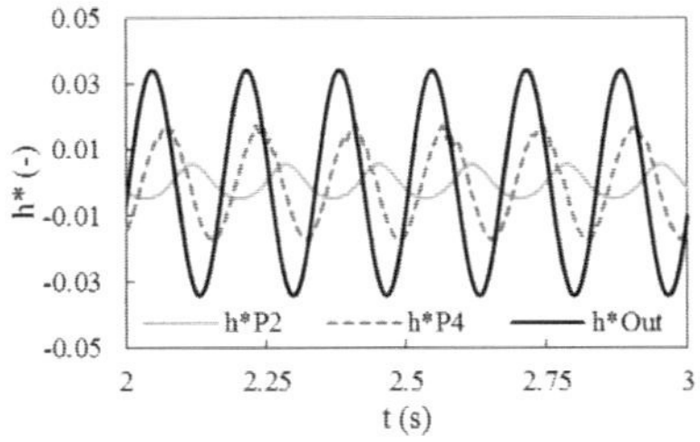

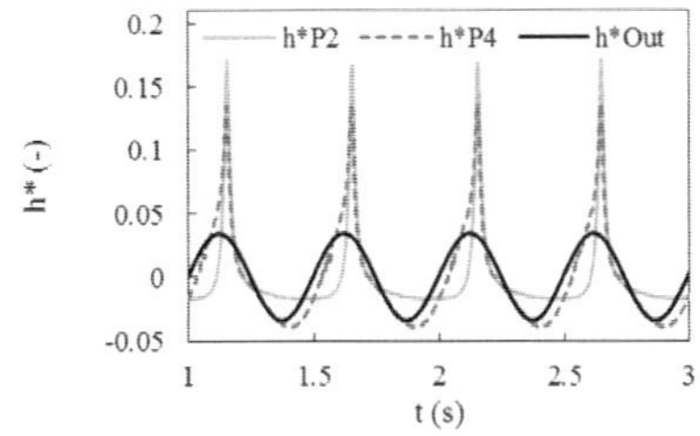

Figure 3. Pressure fluctuations in out of resonance conditions $f_s = 6Hz$ (left) and in resonance conditions $f_s = 2Hz$ (right). CFD results, Venturi test case.

Table 1. 1-D cavitation model parameters for the Venturi and the Francis turbine test cases.

		Venturi	Francis Turbine
With χ_1	a	4.3 $m \cdot s^{-1}$	85.8 $m \cdot s^{-1}$
	μ''	800 $Pa \cdot s$	17'723 $Pa \cdot s$
	χ	-0.001 s	-0.034 s
Without χ_1	a	4.3 $m \cdot s^{-1}$	81 $m \cdot s^{-1}$
	μ''	790 $Pa \cdot s$	16'280 $Pa \cdot s$

history of dimensionless pressure fluctuations in Plane2 and in Plane4 are plotted and compared to the excitation outlet pressure set to a frequency value of $f_s = 6Hz$ and to the resonance frequency $f_s = 2Hz$ in Figure 3. For out of resonance conditions, the amplitude of pressure fluctuations decreases from the outlet to the cavitation sheet. For resonance conditions the amplitude near the cavitation sheet is higher than the outlet excitation source. Moreover, pressure fluctuations do not feature sinusoidal shape anymore. This different behaviour between resonance and non-resonance conditions are also observed for the Francis turbine test case. In Figure 4, pressure fluctuations at the interface of the draft tube are compared to the excitation outlet pressure set respectively to f_s = 30 Hz and f_s = 20 Hz. In resonance conditions, the non-linearity of the system response is observed.

The resonance frequency f_n and the amplitude of pressure fluctuations h^* at the location of interest are set as the objective functions of the global optimization process to derive the wave speed and the second viscosity parameters. The set of 1-D cavitation model parameters that match the eigenfrequency and the amplitude of pressure fluctuations are given in Table 1 for the two test cases. For each test case, two identifications have been performed with and without mass flow gain factor to assess its influence on the wave speed and the second viscosity values. In the case of the Venturi, the low value of the mass flow gain factor barely influences the parameters. In the case of the Francis turbine, the parameters are slightly modified. To compensate the stabilizing effect of the mass flow gain factor and to keep the same eigenfrequency, the wave speed parameter is increased with the second viscosity.

In Figure 5, the forced response of the 1-D model is plotted along the system abscissa. The CFD pressure fluctuation amplitudes at planes Plane2 and Plane4 for the Venturi and at the interface for the Francis turbine are compared for the two investigated frequencies of each case study. At the resonance conditions (surrounded symbol), for which the identification process has been carried out, the 1-D model matches the CFD results. Out of resonance, a difference is observed between the 1-D model and the CFD results. This difference observed for out of resonance conditions could be explained by either frequency dependent cavitation parameters or non-linearities of the system at resonance conditions.

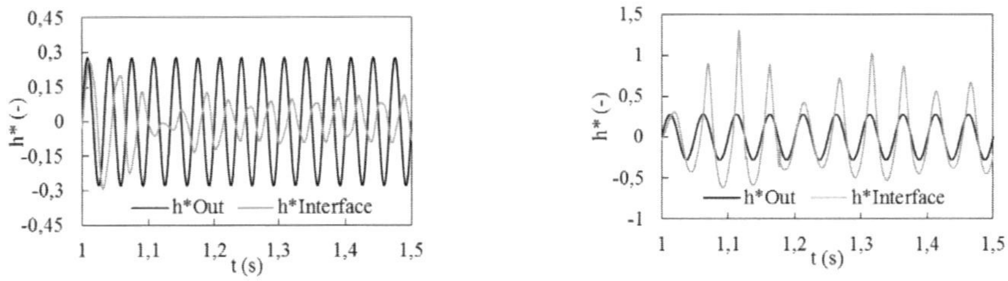

Figure 4. Pressure fluctuations in out of resonance conditions $f_s = 30Hz$ (left) and in resonance conditions $f_s = 20Hz$ (right). CFD results, Francis turbine test case.

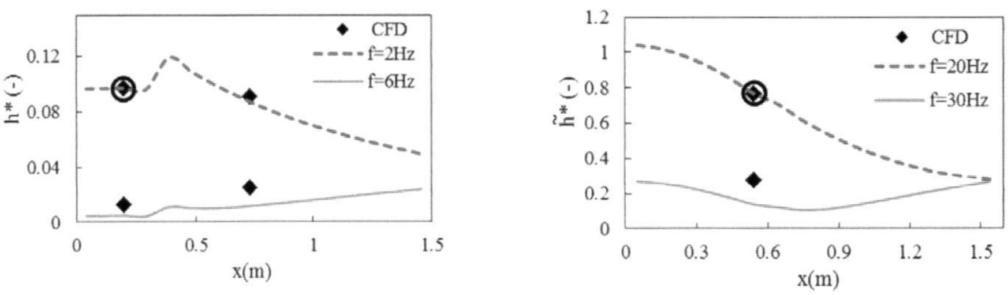

Figure 5. Forced response of the system for different excitation frequency obtained with 1-D model. Venturi test case (left) and Francis turbine test case (right).

7. Conclusion

A calibration process of the parameters involved in a 1-D model designed for stability analysis of power plants has been described. This process is based on an optimization procedure that uses CFD results as objective functions for the 1-D model. Firstly, a 2-D test case has been considered to prove that the second viscosity can be identified from URANS computations. Then, the methodology has been applied to a Francis turbine. The obtained values are close to the mentioned parameters available in the literature. The identified parameters are not intrinsic to the cavitation model dynamics of the RANS model. An investigation of the influence of the grid and the cavitation model would be also valuable in the future.

References

[1] Müller A, Dreyer M, Andreini N and Avellan F 2013 Draft tube discharge fluctuation during self-sustained pressure surge: fluorescent particle image velocimetry in two-phase flow *Experiments in fluids* **54** 4 pp 1-11
[2] Alligné S, Nicolet C, Tsujimoto Y and Avellan F 2014 Cavitation surge modelling in Francis turbine draft tube *J. of Hydraulic Research.* **52** 3 pp 399-411
[3] Landry C, Favrel A, Müller A, Nicolet C and Avellan F 2016 Local wave speed and bulk flow viscosity in Francis turbines at part load operation *J. of Hydraulic Research.* **54** 2.
[4] Koutnik J, Nicolet C, Schohl GA and Avellan F 2006 Overload surge event in a pumped storage power plant *Proc. of the 23rd IAHR Symp. on Hydraulic Machinery and Systems*
[5] Dörfler PK, Keller M and Braun O 2010, Francis full load surge mechanism identified by unsteady 2-phase CFD *Proc. of the 25th IAHR Symp. on Hydraulic Machinery and Systems*
[6] Alligné S, Nicolet C and Avellan F 2011, Identification of Francis turbine helical vortex rope excitation by CFD and resonance simulation with the hydraulic system *Proc. of ASME-JSME-KSME Joint Fluids Eng. Conf.* pp 481-493
[7] Müller A, Yamamoto K, Alligné S, Yonezawa K, Tsujimoto Y and Avellan F 2016 Measurement of the self-oscillating vortex rope dynamics for hydroacoustic stability analysis *ASME J. of Fluids Eng.* **138** 2 021206
[8] Zwart P, Gerber A and Belamri, T 2004, A two-phase flow model for predicting cavitation dynamics. *Fifth Int. Conf. on Multiphase Flow*, Yokohama, Japan, May 30 - June 3, **152**.

IOP Conf. Series: Journal of Physics: Conf. Series **813** (2017) 012033 doi:10.1088/1742-6596/813/1/012033

Set-up of a pump as turbine use in micro-pumped hydro energy storage: a case of study in Froyennes Belgium

A Morabito[1], J Steimes[1], O Bontems[2], G Al Zohbi[1] and P Hendrick[1]

[1]Aero-Thermo-Mechanics Dept. (ATM), École Polytechnique, Université Libre de Bruxelles (ULB), Avenue F.D Roosvelt 50, 1050 Brussels, Belgium
[2]Agence Intercommunale IDETA scrl, Quai Saint Brice 35, 7500 Tournai, Belgium

corresponding author: alessandro.morabito@ulb.ac.be

Abstract. Its maturity makes pumped hydro energy storage (PHES) the most used technology in energy storage. Micro-hydro plants (<100 kW) are globally emerging due to further increases in the share of renewable electricity production such as wind and solar power. This paper presents the design of a micro-PHES developed in Froyennes, Belgium, using a pump as turbine (PaT) coupled with a variable frequency driver (VFD). The methods adopted for the selection of the most suitable pump for pumping and reverse mode are compared and discussed. Controlling and monitoring the PaT performances represent a compulsory design phase in the analysis feasibility of PaT coupled with VFD in micro PHES plant. This study aims at answering technical research aspects of μ-PHES site used with reversible pumps.

1. Introduction

Hydropower on a micro-scale is frequently used to bring electricity supply in remote or rural areas. Unlike large-scale hydropower plants, these applications are mostly developed without dams and wide reservoirs. A local river simply feeds them with a weir crossing the river and little or no water stored. However, an important drawback emerges: the energy production is dependent on spot water availability and thus on seasonally weather. Likewise, solar and wind energy have to deal with intermittency, even if the power plant is properly located and designed.

This usually leads to install fossil-fuel energy sources, not environmentally helpful, to supply base-load demand, e.g. diesel engines. A frequent mismatch between renewable energy production and consumption then might occur. This is why energy storage systems (ESSs) are strong allies [1].

Among energy storage technologies currently available on a small and micro scale, pumped hydro energy storage (PHES) is recognized as one of the most cost-effective because of its predictable energy characteristics, its long-term reliability and its reduced global environmental effects [2]. During off-peak demand, PHES pumps back water to an upper reservoir from a lower reservoir and stores potential energy exploitable by hydraulic turbines when it is needed.

The objective of this paper is to confirm the relevance of the complete chain of tools and methodologies to design a μ-PHES site in a concrete application by means modelling and integrating the impact of μ-PHES applications. The studied site is located in Negundo Innovation Center in Froyennes, Belgium. The micro PHES is connected to a mini-grid system, which counts already 5.2 kW peak of photovoltaic panels and 4.8 kW peak of wind turbines.

1.1. Micro PHES technology

Like hydropower, micro PHES plants require specific topologies, such as a geodetic head (H), defined by the different altitude of the water level between upper and lower reservoirs, and the possibility of water harnessing, defined by the exploitable flow rate (Q).

These systems are very site-specific and a peculiar analysis is needed. Unlike micro hydropower, which has been successfully implemented in remote regions, there are few references in the literature regarding the technical and economic feasibility of PHES at micro scale. Manolakos et al. [3] describe a stand-alone photovoltaic (PV) plant coupled with micro-PHES in Donoussa Island, Greece. A hybrid system with PV and cogenerative internal combustion engine (ICE) is coupled with lead-acid batteries, thermal and water storage and a PHES system [4]. Silva and Hendrick underline the technical feasibility in PHES in buildings but show that it may be not economically competitive when compared to other in micro scale storage technologies [5].

The hydraulic turbine used in generating mode (GM) is selected based on its operating range and its overall performance characteristics. According to H, Q and the rotation speed at best efficiency point (BEP) it is possible to calculate the specific speed of the turbo-machine. This is a relevant parameter to define the type and geometry of the machine with the best performance. During the pumping mode (PM) a single pump or a set of pumps are used to refill the emptied upper reservoir when it is economically profitable.

1.2. Pump as Turbine

The economics of a PHES system is strongly affected by the chosen set of turbo-machines in GM and PM. Relevant differences in capital costs and in efficiency emerge from this choice [6]. Using a centrifugal or diagonal pump as turbine (PaT) in GM is a valid trade-off between capital cost and performances. With respect to maintenance it has many advantages compared to custom-made turbines. Pumps are relatively simple machines, are easy to maintain and are readily available in most developing countries. From an economical point of view, it is often stated that PaTs, in the range of 1 to 500 kW, are profitable and allow capital payback periods of two years or less (which is considerably less than that of a conventional hydraulic turbine [7]).

2. The micro-PHES site in Froyennes

To design PHES, it is essential to define the capacity of the site and the choices of management of the plant itself. This imposes the available head, the capacity and the pumping and generating time frames.

2.1. Topology

In Froyennes (figure 1), the upper basin of 1500 m^3 already exists. The lower reservoir can be loaded to maximum 625 m^3: it limits the plant capacity to about 17 kWh. Due to the small size of the reservoir it may register a large variation (in percentage) of the available head during the working sessions from 10 m to 6 m.

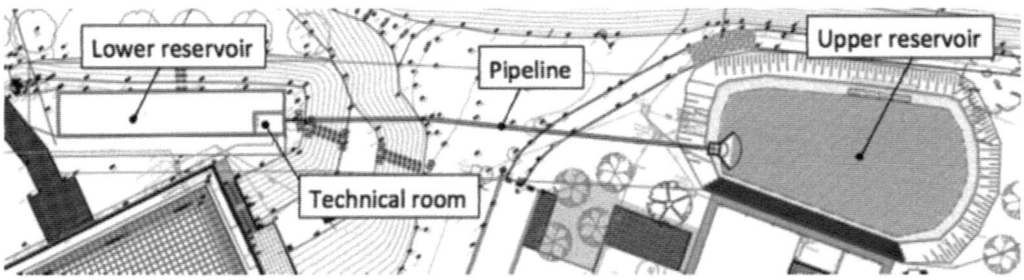

Figure 1. Overview in Negundo Innovation Center

2.2. *Pipeline and technical room*

A 60 m underground pipe connects the lower reservoir and the upper reservoir. The performance of a pump in a pumping system is determined by pipeline characteristics and size. The pressure losses occur due to friction of the liquid particles against each other (shear forces) and pipe walls. Bernoulli's equation is used to obtain the characteristic curve of a system. In general, the pressure losses appear along the pipeline (linear pressure drop) and in valves, bends and other piping fittings, which are called local pressure losses.

The rugosity of contact surface of the selected pipeline is approximately equal to 0,02 mm (for polyethylene) and the pipe internal diameter out of the technical room is 0.300 m. In the Froyennes site, according to the Coolebrok-Prandtl relation, the head loss due to the losses in straight pipe will be 0,48 m (value at the maximum predicted flow rate). Local pressure drops have been estimated under 0,52 m for the diffusers, curved pipefittings and valves. Moreover a straight pipe of length $L \geq 2D_{pipe}$ should be fitted before the impeller to avoid non-uniform condition entry. The formation of swirling eddies on the water surface at the lower limit of submerged is caused by asymmetric flow into the suction pipe. The velocity distribution in the sump depends on many factors: shape of the sump, disposition of the suction pipes, vertical or axial pump and the way water is fed into the sump [8].

The two pipe inlets are equipped with fine grid to protect the pump, located in the technical room next to the lower reservoir. Three valves are needed to maintain accessibility to the devices and thus allow maintenance. Due to the presence of large and heavy equipment units, the powerhouse stability must be completely secured. The excavation work eliminates superficial weathered layer, leaving a sound rock foundation. The minimum distance between the water level in the lower reservoir and the end of the suction pipe is also important to avoid any undesired pre-swirl cavitation phenomena.

2.3. *PaT Selection*

A centrifugal pump takes care of the regular refill of the upper reservoir and of the generating process. A variable frequency driver (VFD) will be installed in order to allow the machine to rotate in both direction and at different speeds while preserving high efficiency.

Usually pump manufactures do not normally provide the characteristic curves of pumps in reverse operation. Therefore, establishing a correlation enabling the passage from the pump characteristics to the turbine characteristics is the main challenge in selecting PaT. The hydraulic behaviour of a pump when rotates as a turbine changes. Many researchers have presented theoretical and empirical relations to predict PaT characteristics at the BEP from the performance of the pump but the results predicted by these methods are not reliable for all pumps with different specific speeds and capacities [9].

3. Experimental setup

In order to measure the real behaviour and validate the efficiency prediction computations in turbine mode, the pump has to be monitored. Figure 2 shows a schematic layout of the machines and sensors located in the powerhouse. Static pressure and temperature transmitters are installed at the inlet and outlet of the pump. Other sensors record the water levels in the reservoirs to avoid flooding. An ultrasonic flow-meter will be installed on the penstock among 10 diameters upstream and downstream of straight pipe run. In order to calculate the mechanical and electrical efficiency separately a torque-meter is equipped on the motor-pump coupling. These outputs signals at 4-20 mA will be connected to a Labview control for real-time measurement acquisition. The set up is designed for flexibility, meaning that the outputs of the system can be regulated/adjusted, since Labview also controls the VFD and an electro-mechanic valve. Further signals are linked to the monitoring system to detach water leaks, overheating of the instrumentation and danger of ice formation in the pump volute.

Attention has to be paid to the starting and stopping of the pump to safeguard the motor against unintentional rotation in reverse direction. Moreover, such monitoring shall be carried out to avoid water hammer, which could have dangerous consequences on the integrity of the whole pipeline.

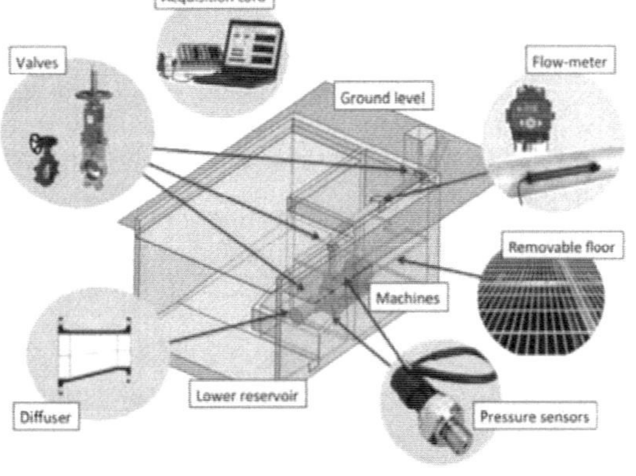

Figure 2. Schematic layout of the technical room: pressure values are measured before and after the pump. The ultrasonic flowmeter can be directly clipped on the pipe and it is able to capture the flow rate in both directions. Fittings and diffuser are designed to avoid air pockets and to reduce local pressure losses.

4. Performance prediction of the turbo-machinery

From the desired working conditions in GM, the methods are applied backwards to find the best suitable pump to use in reverse mode. The predicted ratio of the PaT head and the pump head is presented by the dimensionless parameter h, while q represents the discharge ratio [10-11]. A comparison between different methods depicts a weak reliability in the pump selection (figure 3). However, the subgroups defined by the impeller diameter lead to a better match with the experimental tests available in literature [12]. For pumps between 0.250 and 0.300 m impeller diameter, which likely suit the requirements in Froyennes site, the following equations depending on the specific speed n are found [11]:

$$h = 5.196 \cdot n^{-0.323}, \quad q = 3.1276 \cdot n^{-0.219}$$

According to the site data (Section 2) and linking to manufacturers catalogue, the following pump has been chosen: ICN200-250F, a centrifugal pump manufactured by Ensival-Moret.

A VFD is installed to regulate the rotation speed in pumping and in reverse mode: in general, pumps operating in reverse mode work with higher head and discharge at the same rotational speed. In order to use the same pump in turbine mode, its rotation speed has to been reduced properly. Moreover an adjustment of the speed is required due to the variation of the available head in PM and GM. Based on the affinity laws, a hill diagram can be estimated and allow the machine to work as close as possible to high efficiency at different working conditions (figure 4). Its predicted working conditions as pump and turbine are showed in table 1. In design, the methods provide a range of possible PaT performances and the VFD will assist us for the last tuning on site.

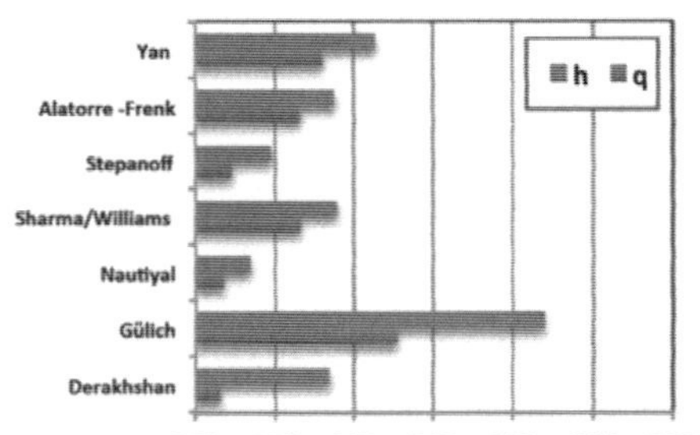

Figure 3. Comparison of the ratio h and q estimated by several methods [9].

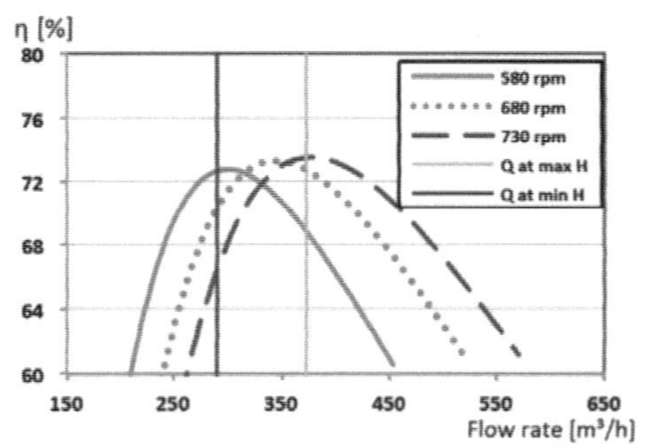

Figure 4. Predicted PaT efficiency curve at variable rotation speed.

Table 1. Pump and PaT coupled with VFD working condition data

	Rated pump	Rated PaT	
		Maximum head	Minimum head
Selected speed [rpm]	1000	730	580
Electric power [kW]	10.5	7.5	3.4
Predicted efficiency [%]	82.6	73.7	73.2
Flow rate [m^3/h]	289.9	372.9	287

5. Conclusion and future work

A reliable storage capacity can provide an opportunity to exploit renewable resources in collaboration with power grids, meeting the requirements for a more sustainable future. To design reliable power grids, which include in both conventional and renewable energy plants, it is essential to propose proper storage systems with adequate capacities. This paper presents such a pilot project, for μ-PHES, located in Froyennes (Belgium). The model results of the simulator will be compared with the experimental results confirming the method adopted. A methodology presenting the different components required to install and characterize such set-up are presented. It allows to record the characteristic curves for the selected pump and provides the needed. This measurement database will be helpful to determine the accuracy of theoretical model for PaT. The results will validate performance modelling of a micro configuration carrying out optimization and relevant consideration on transient state on PaT coupled with VFD. This will be useful for the community and to help in design of μ-PHES sites.

Acknowledgments
Authors wish to acknowledge assistance from Marc Lejoly and Claude Mockels of Ensival Moret for technical support. This research was supported by Region Wallonia in the SmartWater project.

References

[1] Ferreira H, Garde R, Fulli G, Kling W and Lopes J 2013 Characterisation of electrical energy storage technologies *Energy* **53** p 288-298
[2] Ardizzon G, Cavazzini G and Pavesi G 2014 A new generation of small hydro and pumped-hydro power plants: Advances and future challenges *Renew. Sust. Energ. Rev.* **31** p 746-761
[3] Manolakos D, Papadakis G, and Kyritsis S 2004 A stand-alone photovoltaic power system for remote villages using pumped water energy storage *Energy* **29** p 57-69
[4] Stoppato A, Benato A and Mirandola A 2016 A model for the optimal design and management of a cogeneration system with energy storage *Energ Buildings* **124** p 241-7
[5] Silva G and Hendrick P 2016 Pumped hydro energy storage in buildings *Appl. Energ.* **179**
[6] Steimes J and Hendrick P 2016 Cost and revenue breakdown for a pumped hydroelectric energy storage installation in Belgium *Sustainable Hydraulics in the Era of Global Change* 269–276
[7] Paish O 2002 Micro-hydropower: Status and prospects *PI. Mech. Eng. A-J Pow* **216** p 31-40
[8] Nourbakhsh A, Jaumotte A and Hirsch C 2008 *Turbopumps & pump system* Springer
[9] Sanjay V, Rajesh N. Patel, 2014 Investigations on pump running in turbine mode: A review of the state-of-the-art *Renew. Sust. Energ. Rev.* **30** p 841-868
[10] Gülich J F 2010 *Centrifugal pumps* Springer-Verlag Berlin
[11] Morabito A, Steimes J and Hendrick P 2016 Pumped hydroelectric energy storage: A comparison of turbomachinery configurations *Sustainable Hydraulics in the Era of Global Change* p 261–268
[12] Derakhshan S 2008 Experimental study of characteristic curves of centrifugal pumps working as turbines in different specific speeds *Exp. Therm. Fluid. Sci.* **32** p 800-807

IOP Conf. Series: Journal of Physics: Conf. Series **813** (2017) 012034 doi:10.1088/1742-6596/813/1/012034

On the sloshing free surface in the draft tube cone of a Francis turbine operating in synchronous condenser mode.

Vagnoni E., Andolfatto L., Avellan F.

EPFL-LMH, Avenue de Cour 33 bis, 1004, Lausanne (VD), Switzerland.
elena.vagnoni@epfl.ch

Abstract. Hydropower plants may be required to operate in synchronous condenser mode in order to supply reactive power to the grid for compensating the fluctuations introduced by the intermittent renewable energies such wind and solar. When operating in this mode, the tail water in the Francis turbine or pump-turbine is depressed below the runner by injecting pressurized air in order to spin in air to reduce the power consumption. Many air-water interaction phenomena occur in the machine causing air losses and a consequent power consumption to recover the air lost. In this paper, the experimental investigation of the sloshing motion in the cone of a dewatered Francis turbine performed by image visualization and pressure measurements is presented. The developed image post processing method for identifying the amplitude and frequency of the oscillation of the free surface is described and the results obtained are illustrated and discussed.

1. Introduction

Hydropower is the most exploited renewable energy source at a global scale. Thanks to its flexibility, it has the capability to compensate the variations of the energy production of the intermittent renewable energies such solar and wind. Moreover, since these fluctuations of energy production cause grid instability, a supply of reactive power is required for the grid stabilization. Hydropower plants can provide reactive power to the grid by operating in synchronous condenser mode. When operating in synchronous condenser mode, Francis turbine and pump-turbine are operated in dewatered conditions in order to minimize the power consumption [1][2]. The guide vanes are closed and the tail water level is depressed below the runner by injecting pressurized air in the cone of the draft tube.

Several problems occur to run the synchronous condenser mode operation in a hydropower plant since significant power losses are recorded due to the pressurized air lost in the machines. Previous study, such as Tanaka et al. [3] and Ceravola et al. [4], investigated the air-water interaction phenomena in a Francis turbine or pump turbine operating in synchronous condenser mode focusing on the sloshing motion of the free surface below the runner.

Tanaka et al. [3] studied the behavior of the water free surfaces and the related air losses changing the operation condition of the turbine by applying the densimetric Froude similitude to reach the similarity with the prototype. Air-water interaction phenomena such as sloshing motion of a free surface, turbulent waves causing the formation and entrainment of bubbles diffusing in the water volume and formation of droplets interfering with the runner were observed. The main influencing parameters identified were the densimetric Froude number defined in Eq.1, the air density and the water level in the cone of the draft tube.

$$Fr_d = \sqrt{\frac{\rho_a}{\rho_w}}\frac{N\sqrt{D_e}}{\sqrt{g}} \tag{1}$$

ρ_a and ρ_w are respectively the air and water density, N is the rotational speed and D_e the low pressure diameter of the runner. Several studies were conducted to understand the behavior of an oscillating free surface such Royon-Lebeaud et al. [5] and Komori et al. [6]. These studies showed the dependence of the mass transfer of a gas into water on the wave frequencies and amplitude on the surface caused by a wind shear. The mass transfer plays an important role for understanding the air diffusion in water which is one of the causes of the air losses. As such, there has been a shortage of research into Francis turbine operating in condenser mode related to the determination of air losses and the description of the two-phase phenomena involved in the machine.

In the present paper a successful set-up and post-processing method for the experimental investigation of the sloshing motion of the air-water free surface below a Francis runner operating in synchronous condenser mode is presented. Image acquisition and pressure measurements are performed and analyzed in both time and frequency domain. The amplitude and frequency of the sloshing motion are illustrated and followed by a discussion of the results achieved.

2. Experimental set-up

The experiments are conducted in the reduced scaled physical model of a Francis turbine shown in Fig.1. The scaled model - composed by the upstream inlet pipe, the spiral case, the runner and the draft tube - is installed in a closed loop test-rig which is equipped with two axial double-volute pumps to generate the specific head. A generator connected to the model runner regulates the rotational speed. To operate in condenser mode, the guide vanes are closed and a blind plate was mounted at the end of the draft tube. A cooling discharge is injected through the runner and pressurized air is injected in the cone of the draft tube. A transparent draft tube cone with a flat water box window is installed to perform undistorted images acquisition. A camera is mounted as illustrated in Fig.1a) and a 90% uniform LED screen is employed to have uniform white light on the background to enhance the contrast between the liquid and gaseous phase in the cone of the draft tube. Image acquisition is performed at 30 frames per second for 60 s. Measurements of the wall static pressure are performed at $1.02D_e$ downstream the runner outlet with one pressure sensor installed at 35° from the investigated focus plane of the camera. Measurements are performed for 60 s at 3'000 Hz and synchronized with the image acquisition by an external trigger. The Froude similarity with the prototype is respected to choose the operational condition to investigate. Two gauge pressures and six densimetric Froude numbers are tested by varying the pressure in the cone and the rotational speed of the runner. The tested operational conditions of the machine are resumed in Fig.1b).

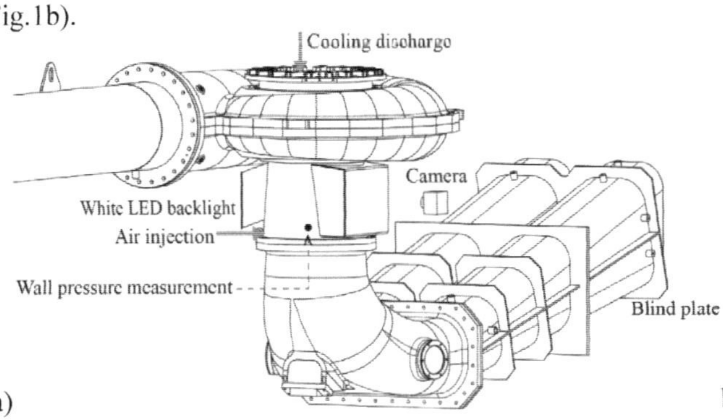

a)

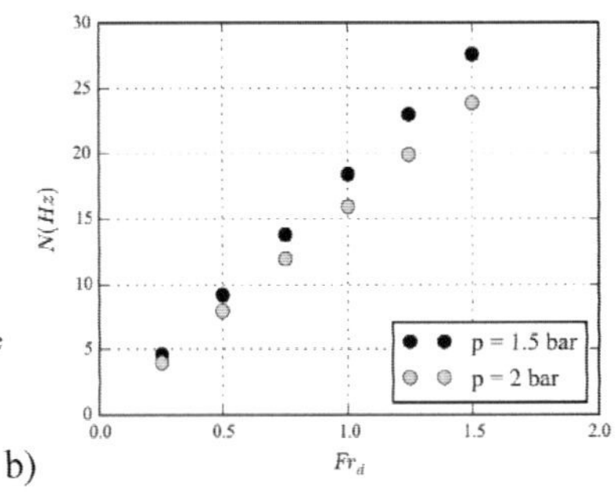

b)

Figure 1. a) Reduced scale model of the Francis turbine on the test rig together with the experimental instrumentation. b) Investigated operation condition of the Francis turbine operating in condenser mode.

3. Methodology

A method for the images post-processing is developed to measure the amplitude and frequency of the sloshing motion of the air-water free surface below the runner. A calibration of the camera is performed with a 200×200 dotted target mounted in the nose of the runner for guaranteeing the alignment of the camera on the horizontal and vertical slit. The focus plane is centered on the central plane of the draft tube cone and a conversion ratio of 2 pixel mm^{-1} is computed. As illustrated in Fig.2, two fixed matrices of pixels of 15 mm width are extracted from each frame and an adaptive threshold filter [7] is applied to identify the illuminated liquid phase in contrast with the dark background. The pixel discretization and the presence of bubbles and droplets induce errors in the identification of the sloshing amplitude since they can locally change the position of the free surface. To limit this problem, the method is based on the identification of the linear equations describing the segments AB and CD in order to identify the position of the points B and C. As a first step, the position of the free surface on the y axis, introduced in the reference system in Fig.2, is identified for each column i of the matrices of pixels by solving Eq.2:

$$y_{\mathrm{AB},i} = \sum_j \left(\mathrm{E}(j,i) = 255\right) \quad \text{and} \quad y_{\mathrm{CD},i} = \sum_j \left(\mathrm{F}(j,i) = 255\right) \tag{2}$$

As second step, the linear regression to best fit a linear equation to the data set of the computed positions of the free surface is performed by using the least squares method. This leads to identify the linear equation describing the segments AB and CD and the positions of the points B and C by solving Eq.3.

$$y_{\mathrm{B}} = \mathrm{a}x_{\mathrm{B}} + \mathrm{b} \quad \text{and} \quad y_{\mathrm{C}} = \mathrm{c}x_{\mathrm{C}} + \mathrm{d} \tag{3}$$

The constants a and c are the slopes and b and d the intercepts on the y axis of the linear equations computed by the linear regression. Finally, the sloshing amplitude h is computed as follows in Eq. 4:

$$h = y_{\mathrm{B}} - y_{\mathrm{C}} \tag{4}$$

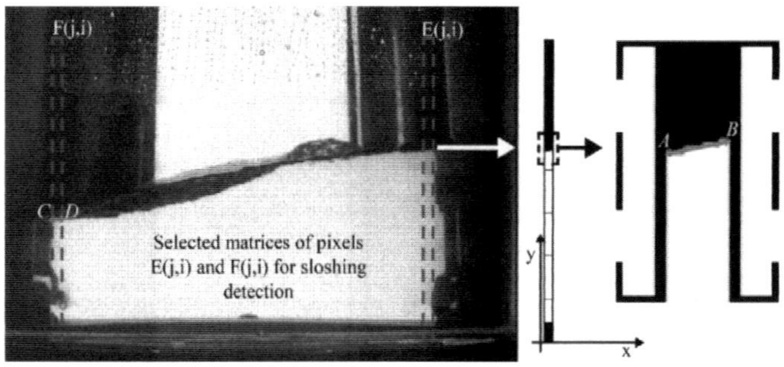

Figure 2. Extracted matrices of pixels and filtering for sloshing detection.

4. Result and discussion

4.1. Amplitude and frequency of the sloshing motion

The image post-processing is applied to the visualization of the sloshing motion, enabling a quantitative description of the oscillation of the free surface for all the investigated values of the densimetric Froude number. A sequence of images of the oscillation of the air-water free surface below the runner at p = 2 bar and Fr_d = 0.75 is presented in Fig.3a). The extracted sloshing amplitude by the method presented in Sec.3 is illustrated in Fig.3b).

A discrete Fourier transform analysis is applied on the sloshing amplitude and pressure measurements to compute the spectrum of the signals and to identify the frequency of the oscillation of the free surface for the investigated operation condition of the Francis turbine. The magnitude of the spectrum computed by the Fourier analysis is presented in Fig.3c).

4.2. Influence of the densimetric Froude number

The behavior of the air-water free surface below the runner in dependency on the Fr_d is illustrated in Fig.4. For $Fr_d < 0.5$ no sloshing motion is observed while for $Fr_d > 0.5$ the oscillation of the free surface is recorded. At $Fr_d = 0.5$ the behavior of the free surface has a transient condition. Starting from a steady flat condition, the free-surface develops an oscillation until the maximum amplitude as shown in Fig.5

at a gauge pressure of 1.5 bar. The pressure data are normalized by defining a pressure coefficient as follows in Eq.5.

$$C_p = \frac{p-\bar{p}}{0.5\rho_w U_{tip}^2} \tag{5}$$

U_{tip} is the peripheral velocity of the blade tip.

Both amplitude and pressure fluctuations measurements confirm the transient condition of the sloshing motion of the free surface at this operation condition of the Francis turbine. A peak on the sloshing amplitude coincides with a peak in pressure. The phase shift is due to the physical angular shift of the pressure sensor with respect to the focus plane of the camera. Moreover, for $Fr_d > 0.5$ the amplitude of the sloshing motion decreases by increasing the densimetric Froude number as it is noticed by observing the images of the sloshing motion for the investigated Fr_d in Fig.4. The sloshing amplitude in function of the Fr_d is illustrated in Fig.6 (left) in dimensionless term by dividing the measured amplitude by the exit diameter of the Francis runner. On the other hand, the frequency of the oscillation slightly increases with the Fr_d, as shown in Fig.6 (right) where the frequency is represented in dimensionless term by dividing by the exit diameter and the gravity.

4.3. Influence of the gauge pressure

The amplitude of the oscillation of the free surface depends also on the gauge pressure as illustrated in Fig.6. By increasing the gauge pressure a higher amplitude is recorded while the frequency is not affected. The maximum difference in frequency which is measured at the two investigated gauge pressures counts the 0.8% of the computed frequency at 2 bars at $Fr_d = 0.75$.

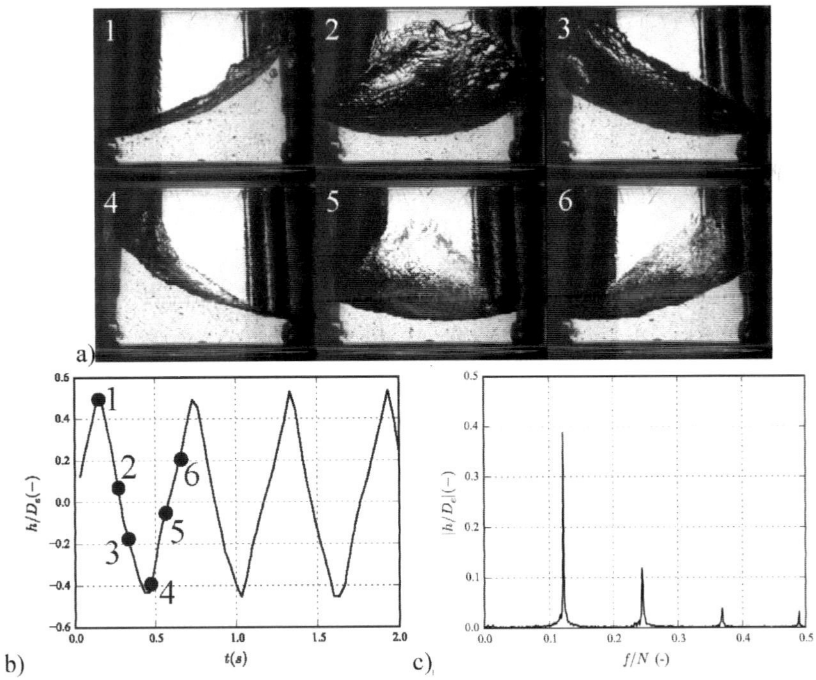

Figure 3. a) Sequence of images of the air-water free surface oscillating during the condenser mode operation of the Francis runner at p = 2 bar and $Fr_d = 0.75$. b) Time resolved signal of the sloshing amplitude measured with the image post-processing method. c) Corresponding spectrum magnitude of the signal in b).

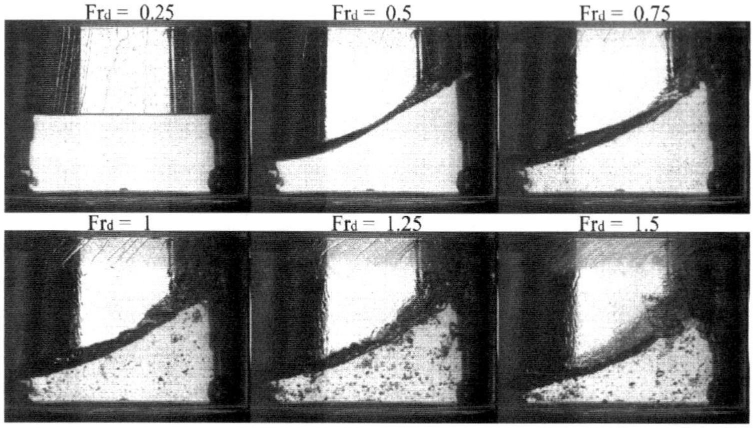

Figure 4. Example images of the sloshing motion of the free surface in the draft tube cone of a Francis turbine operating in condenser mode at p = 1.5 bar for six different densimetric Froude numbers.

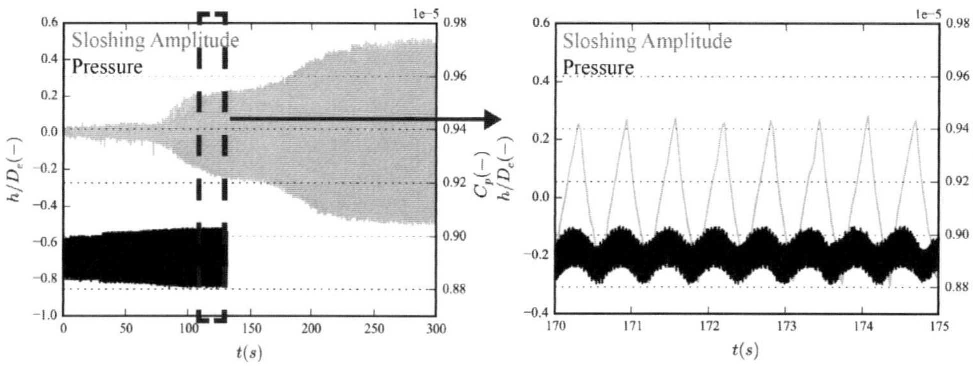

Figure 5. Simultaneous time history of the sloshing motion relative amplitude extracted by the images post-processing and pressure coefficients derived from pressure measurements in the lower section of the draft tube cone at p = 1.5 bar and $Fr_d = 0.5$.

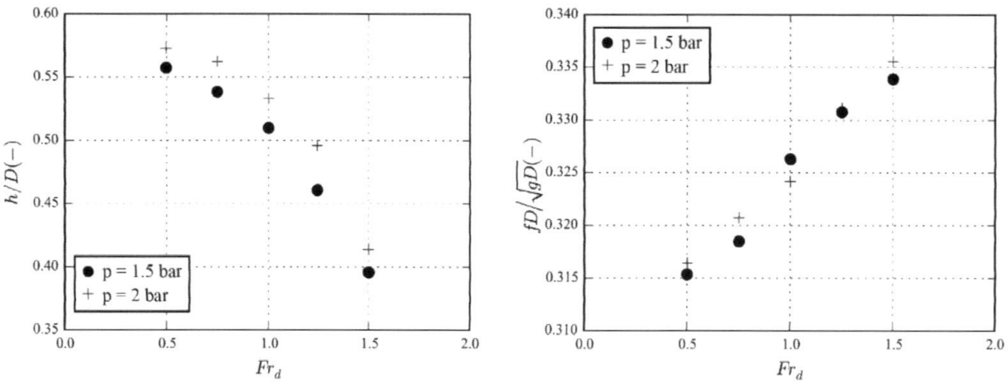

Figure 6. Relative amplitude computed as peak to peak height of the sloshing wave divided by the exit diameter of the Francis runner (left). Frequency of the sloshing motion divided by the gravity g times the exit diameter of the Francis runner (right).

5. Conclusion

Images acquisition of the sloshing motion of the air-water free surface below the runner in the draft tube cone of a reduced scale physical model of a Francis turbine operating in condenser mode was performed. A method for the image post-processing was developed for measuring the amplitude of the sloshing motion. Pressure measurements were also performed in the lower section of the draft tube cone. The pressure signals and the sloshing amplitude data were analyzed and compared in both time and frequency domain and the results are elucidated in the present paper.

Twelve operating conditions were investigated by varying the densimetric Froude number between 0.25 and 1.5 and the gauge pressure at 1.5 and 2 bar in order to evaluate their influence on the amplitude and frequency of the sloshing motion, respectively. Higher pressure value could not be reached because of the pressure limitation of the reduced scale physical test.

The results highlight the dependency of the sloshing motion dynamic behaviour on the Fr_d. The free surface has a steady flat condition for $Fr_d < 0.5$. The transition between the steady flat condition and the oscillation is recorded at $Fr_d = 0.5$ and for $Fr_d > 0.5$ the oscillation is observed. The amplitude of the sloshing motion decreases by increasing the Fr_d, while the frequency slightly increases.

The gauge pressure influences also the amplitude of the sloshing motion: the higher the gauge pressure, the higher is the amplitude. No influence has been recorded on the frequency of the sloshing motion.

In light of the presented results, the key parameters which influence the dynamic behaviour of the air-water free surface below the runner are identified and a quantitative description of the phenomenon is provided.

Acknowledgements

The research leading to the results published in this paper is part of the HYPERBOLE Research Project, granted by the European Commission (ERC/FP7-ENERGY-2013-1-Grant 608532). The authors would also like to thank BC Hydro for making available the reduced-scale model. Moreover, the authors would like to acknowledge the commitment of the Laboratory for Hydraulic Machines' technical staff, especially Georges Crittin, Alain Renaud, and Vincent Berruex.

References

[1] Rychkov I. G., Reactive power control services based on a generator operating as a synchronous condenser, Power Technology and Engineering, 46, pp. 405-409 (2013).

[2] Rossi G. and Zanetti V., Starting in air and synchronous condenser operation of pump-turbines - Model research, 9th IAHR-SHMEC Symposium, vol. 2, pp. 337-352,Fort Collins (USA), (1978).

[3] Tanaka H., Matsumoto K., Yamamoto K., Sloshing motion of the depressed water in the draft tube in dewatered operation of high head pump-turbines, XVII IAHR Symposium on hydraulic machines and cavitation, Beijing, Chine, paper A-8, pages 121-130 (1994).

[4] Ceravola O.,Fanelli M., Lazzaro B., The behaviour of the free level below the runner of Francis turbine and pump-turbines in operation as synchronous condenser, 10th IAHRSHMEC Symposium, vol. 1, pp. 765-775, Tokyo (Japan), (1980).

[5] Royon A., Hopnger E. J. and Cartellier A., Liquid sloshing and wave breaking in circular and square-base cylindrical containers, Journal of Fluid Mechanics, 577, pp. 477-494 (2007).

[6] Komori S., Nagaosa R. and Murakami Y., Turbulence structure and mass transfer across a sheared air-water interface in wind-driven turbulence, Journal of Fluid Mechanics, 249, pp. 161-183 (1993).

[7] Gonzales R. C. and Woods R. E., Digital image processing, pp. 352-357, Pearson Education International,(2008)

Mechanical impact of dynamic phenomena in Francis turbines at off design conditions

F Duparchy[1], J Brammer[1], M Thibaud[1], A Favrel[2], P Y Lowys[1], F Avellan[2]

[1]GE Renewable Energy, Hydro Solutions, 82 Avenue Léon Blum – BP75, 38041 Grenoble Cedex 9, France

[2]EPFL Laboratory for Hydraulic Machines, Avenue de Cour 33 bis, 1007 Lausanne, Switzerland

E-mail: florian.duparchy@ge.com

Abstract. At partial load and overload conditions, Francis turbines are subjected to hydraulic instabilities that can potentially result in high dynamic solicitations of the turbine components and significantly reduce their lifetime. This study presents both experimental data and numerical simulations that were used as complementary approaches to study these dynamic solicitations. Measurements performed on a reduced scale physical model, including a special runner instrumented with on-board strain gauges and pressure sensors, were used to investigate the dynamic phenomena experienced by the runner. They were also taken as reference to validate the numerical simulation results. After validation, advantage was taken from the numerical simulations to highlight the mechanical response of the structure to the unsteady hydraulic phenomena, as well as their impact on the fatigue damage of the runner.

1. Introduction

The increasing need of flexibility for single regulated hydro turbines, such as Francis units, tends to increase the number of hours in operation at partial load and overload conditions, outside of the best operating range for which the turbines were originally designed. In these off-design conditions, hydraulic instabilities can occur. They are the source of high dynamic solicitations that can potentially reduce the lifetime of the components of the turbine, and especially the runner.

Today, numerical simulations can help to understand the dynamic loading applied on the structure as well as its mechanical response. However, modelling accurately such complex phenomena remains challenging. In some cases, it is necessary to further validate the results of the simulations with experimental data, from both a hydraulic and mechanical point of view. In this study, an experimental investigation was performed on a reduced scale physical model of a Francis turbine including an instrumented runner equipped with on-board strain gauges and pressure sensors. In parallel, Computational Fluid Dynamics (CFD) and Finite Element Analysis (FEA) simulations were performed for three operating points corresponding to deep partial load, partial load and overload conditions. The results were then compared with the experimental data. Once consistency between them was achieved, advantage was taken from the numerical simulations to highlight the differences in terms of mechanical behaviour of the runner when subjected to the identified hydraulic phenomena.

IOP Conf. Series: Journal of Physics: Conf. Series **813** (2017) 012035 doi:10.1088/1742-6596/813/1/012035

2. Experimental setup

The test case consisted of a reduced scale physical model of a Francis turbine with 16 blades and 20 guide vanes. It was installed on the test rig PF3 of the Laboratory for Hydraulic Machines (EPFL-LMH) operating in closed loop. The tests were performed in cavitation conditions corresponding to the full-scale turbine conditions.

The instrumentation included a set of pressure sensors distributed along the reduced scale model. In particular, pressure fluctuations were measured in the draft tube cone, where two sections were equipped with 4 flush-mounted pressure sensors (figure 1).

In addition to this standard set of sensors, an instrumented runner was used. It was instrumented with pressure sensors and strain gauges (figure 1). It was specifically designed to measure both the hydraulic pressure fluctuations applied on the blades as well as the mechanical response of the structure. The signals were transmitted to the stationary parts through a telemetry system.

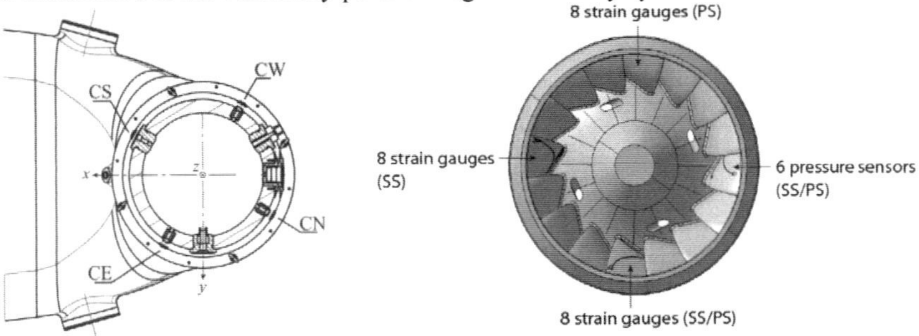

Figure 1: Cut view of the upper section of the draft tube cone with the pressure sensors (left) and instrumented runner with the location and number of the strain gauges (SS: suction side, PS: pressure side) (right)

3. Selection of the calculated points

To identify the relevant operating points to use in the numerical simulations, the experimental data was first analysed. For a fixed n_{ED} value, corresponding to the nominal n_{ED} value, the evolution of the dynamic blade strain over the discharge can be divided into different zones, as shown in Figure 2. n_{ED} and Q_{ED} coefficients are defined according to IEC standard 60193:1999.

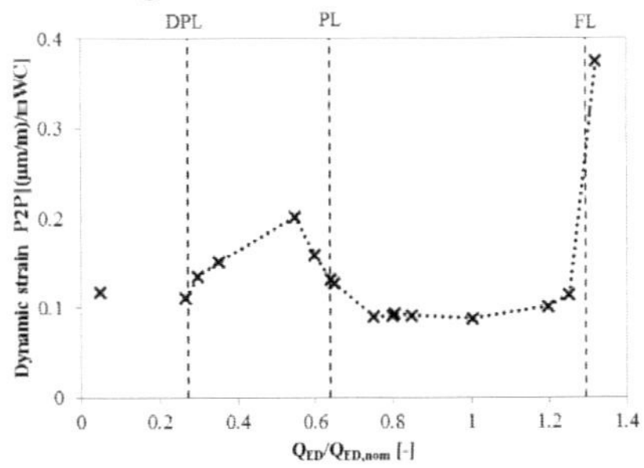

Figure 2: Dynamic strain measurement on a load variation for a strain gauge located close to the trailing edge of the pressure side of the blade, close to the hub.

Around the nominal Q_{ED} value ($0.8<Q_{ED}/Q_{ED,nom}<1.2$), the dynamic strain was low. For this zone, only the rotor-stator interaction (RSI) was identified in the frequency spectra, at $20*f_0$, where f_0 is the rotational frequency of the runner.

At full load ($Q_{ED}/Q_{ED,nom}>1.2$), a significant increase of dynamic strain was observed. It was caused by a self-excited pressure surge due to the presence of the full load axial vortex rope [1]. Its fundamental frequency was $0.23*f_0$ and several harmonics were observed (figure 3c).

At partial load ($0.3<Q_{ED}/Q_{ED,nom}<0.8$), a second dynamic strain elevation was identified. It corresponded to the part load helical vortex zone, with a typical frequency of $0.70*f_0$ in the rotating frame of the runner [3] (figure 3b). This frequency reached its maximum intensity at around $Q_{ED}/Q_{ED,nom}=0.55$ [2].

At deep partial load, ($Q_{ED}/Q_{ED,nom}$<0.3), a global elevation of the spectrum was observed, without clear emergence, except for the RSI (figure 3a). It means that the loading is of a stochastic nature. However, the peak-to-peak dynamic strain was low. It should be noted that for this operating point no inter blade cavitation vortex was observed.
Finally, three operating points were selected for the numerical study (figure 2): at full load $Q_{ED}/Q_{ED,nom}$=1.30 (FL), partial load $Q_{ED}/Q_{ED,nom}$=0.64 (PL) and deep partial load $Q_{ED}/Q_{ED,nom}$=0.27 (DPL).

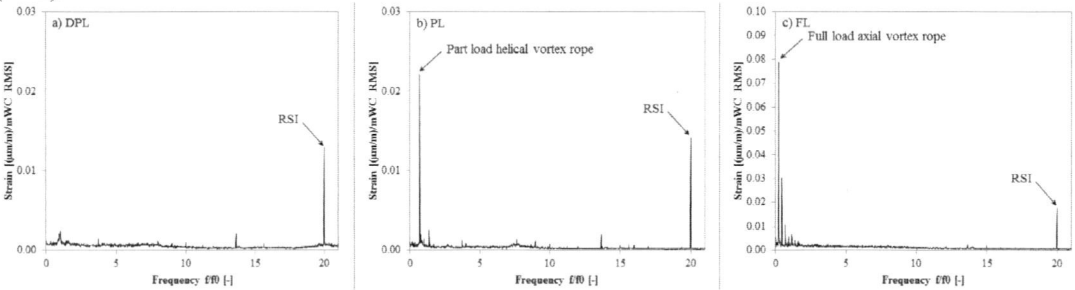

Figure 3: Frequency spectra for the three operating points selected for the numerical study. Different scale for FL.

For the selected operating points, CFD and FEA simulations were performed with ANSYS software. The unsteady CFD simulations generated pressure fields that were then used as an input for the FEA study.

4. Numerical modelling

4.1. CFD simulation

Each selected operating point exhibited different hydraulic and mechanical characteristics, and therefore the complexity of each CFD model was adapted to calculate them as accurately as possible. Each Hyperbole project partner was allocated an operating point, and such differences include the mesh refinement, the turbulence model and the time step. Also, for the full load calculation it was necessary to impose a fluctuating pressure level at the spiral case inlet to simplify and represent the hydro-acoustic characteristics. These unsteady calculations had the additional complexity of modelling cavitation, so the choice of cavitation model was also considered, and sensitivity tests were performed to confirm the results.

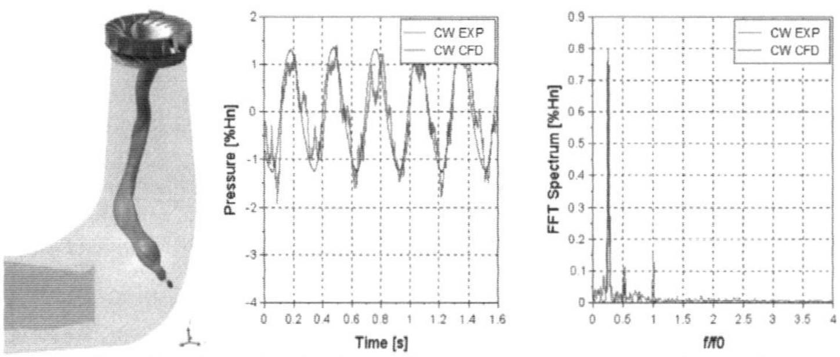

Figure 4: Part load operation: CFD flow visualisation (left) and draft tube pressure fluctuation comparisons between CFD and experimentation (right)

Before using the CFD pressure fields to perform FEA simulations it was crucial to have confidence in the CFD results, otherwise any errors would be passed on to the mechanical analysis. For this purpose, CFD simulations were validated by comparison with the experimental results. Figure 4 shows such an example for the part load vortex rope. As well as confirming visually that the cavitating vortex rope in the CFD corresponded to what was observed on the test rig, pressure sensor measurements were also compared and FFT analysis showed that both frequency and amplitude of the part load vortex rope were well predicted.

4.2. FEA simulation

For the mechanical simulation, the entire runner was considered, including the blade cavities made for the strain gauges and the grooves for the wires. The mesh was refined around the strain gauge locations and a set of two nodes was used to extract the strain values at the exact position and direction of the gauges. This enabled an accurate comparison of the results of the simulation with the measurement. A total of 2.3M nodes were used (figure 5).

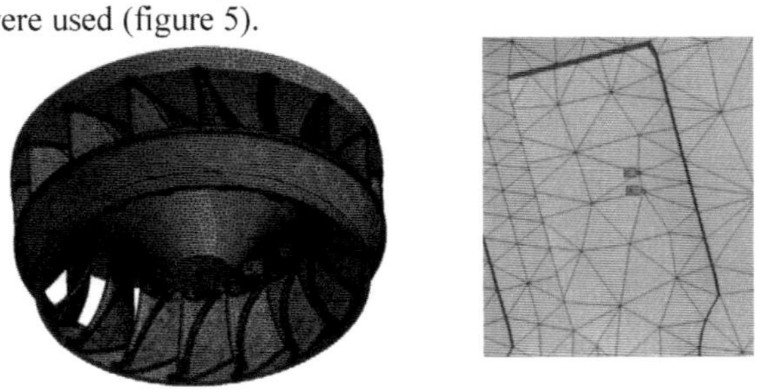

Figure 5: FEA model and meshing (left) and detail of the strain gauge imprint mesh (right)

Considering that the natural frequencies of the runner were much higher than the excitation frequencies contained in the CFD pressure fields, a quasi-static approach was used, i.e. successive static calculations were performed. The hydraulic load was applied in several time steps in order to capture the maximum stress variation. The time step was selected to be consistent with the frequencies of the cavitation phenomena observed in off-design conditions.

Results of FEA analysis were compared to the experimental data. First, the time signals were superimposed to qualitatively check that they had a similar evolution (figure 6). At this stage, it was observed that the time step was sufficient to simulate low frequency phenomena such as the vortex rope, while high frequency solicitations such as RSI were roughly described. It was also checked that RSI could also be modelled accurately by repeating the simulation at DPL with a refined angular step.

Finally, the peak-to-peak strain values were compared with the measurements (figure 6) and they showed a good agreement with the experimental results, except for some strain gauges for which measurement issues were suspected.

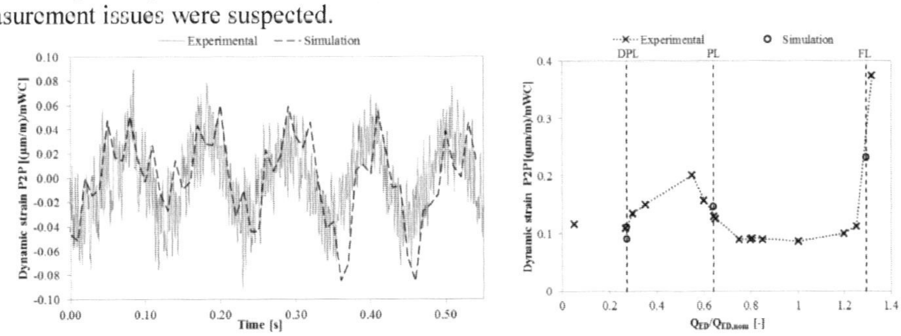

Figure 6: Comparison between simulations and measurements. Time signal of a strain gauge at PL (left) and comparison of peak-to-peak values for the three calculated points (right)

5. Impact on fatigue damage

The good correlation between simulation and measurement gave confidence to the fact that the physics of the phenomena, as well as the reaction of the structure, was well predicted. The simulations were used to understand how the structure behaved in the presence of the different hydraulic phenomena. Contrary to the measurements that gave information only for a limited number of sensor locations, the numerical simulations provided results at each node of the model. Numerical results were therefore used to highlight the differences of behaviour for each operating point, in terms of structure displacement, stress distribution and stress concentration, as shown in Figure 7.

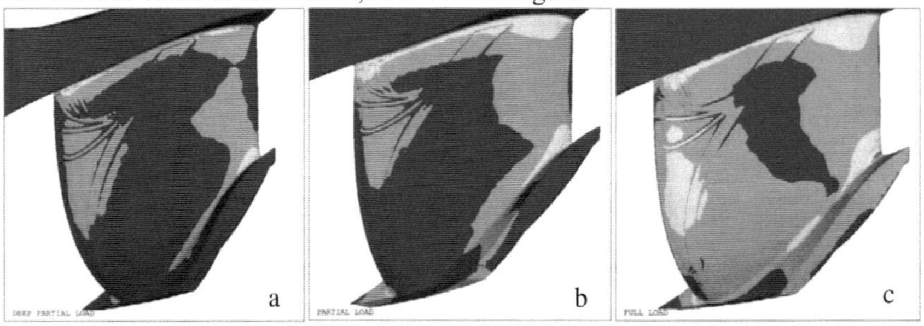

Figure 7: Dynamic stress distribution for deep part load (a), part load (b) and full load (c). Same scale (red color corresponds to highest stress)

Moreover, hot spot factors that link the dynamic strain at the strain gauge locations to the maximum stressed area of the runner were extracted from the simulation and used to extrapolate the measurement. A Rainflow cycle counting algorithm was then applied to the extrapolated signals and fatigue damage was compared for each operating point (Figure 8).

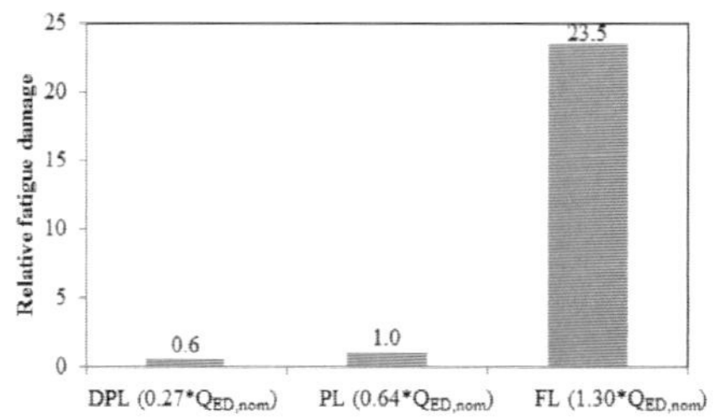

Figure 8: Relative fatigue damage calculated at the trailing edge of the blade-to-hub junction

6. Conclusion

Coupled unsteady CFD and FEA simulations showed good capability to capture hydraulic phenomena that occur at off-design conditions and their mechanical impact on the runner. The results were validated by comparison with measurements performed on a reduced scale physical model that included a special runner equipped with on-board sensors.

These two complementary approaches were used to investigate the impact of each dynamic phenomenon on the mechanical behaviour and the fatigue damage of the runner. This method can be used not only for existing designs but also for new projects [4].

An additional measurement campaign on the full scale industrial turbine, including runner strain gauges and pressure sensors has also been carried out. Therefore, the transposition of the dynamic solicitations from the reduced scale model to full-scale runner will be studied.

Acknowledgments

The research leading to the results published in this paper is part of the HYPERBOLE research project, granted by the European Commission (ERC/FP7- ENERGY-2013-1-Grant 608532).

References

[1] Müller A, Dreyer M, Andreini N and Avellan F 2013 Experiments in Fluids 54(4):1-11
[2] Favrel A, Müller A, Landry C, Yamamoto K and Avellan F 2015 *Experiments in Fluids* **56:215**
[3] Duparchy F, Favrel A, Lowys P Y, Landry C, Müller A, Yamamoto K and Avellan F 2016 Journal of Physics: Conference Series **656** 012061 (CAV2015, Lausanne, Switzerland)
[4] Brammer J, Duparchy F, Lowys P Y, Thibaud M, Wheeler K and De Colombel T 2016 *Increasing the operating range of Francis turbine by considering dynamic phenomena at partial load*, Hydro 2016, Montreux, Switzerland

Dynamic response of the MICA runner. Experiment and simulation

David Valentín, Alexandre Presas, Eduard Egusquiza, Carme Valero, Matias Bossio

Center for Industrial Diagnostics and Fluid Dynamics (CDIF), Polytechnic University of Catalonia (UPC). Av. Diagonal, 647, ETSEIB, 08028 Barcelona.

david.valentin@mf.upc.edu

Abstract. Studying the dynamic response of Francis turbine runners is of paramount importance in order to avoid resonance problems during operating conditions. For this purpose, the natural frequencies as well as their associated mode-shapes and damping ratios of the runner have to be determined. In this paper, an Experimental Modal Analysis (EMA) of the runner of a Francis turbine prototype has been performed. By means of this experimental technique, natural frequencies, mode-shapes and damping ratios have been estimated in air. Results obtained have been compared with a Finite Element Method (FEM) model in order to check the accuracy of the simulation.

1. Introduction

The structural dynamic response of Francis turbines has to be studied during the stage of design of the machine in order to avoid resonance and fatigue problems during operating conditions [1, 2]. One of the components that could be strongly affected by fatigue problems is the runner. The runner is a rotating structure submerged and confined in water with small gaps to the stationary parts. The effect of the water and the confinement on the natural frequencies of the runner is well-known and it is demonstrated with the added mass theory [3-5]. This added mass effect could be accurately estimated by using acoustical-numerical Fluid Structure Interaction (FSI) simulations [5-8]. However, the damping factor associated to each natural frequency and mode-shape plays also an important role on the dynamic response of structures, since it is the mechanism of the structure to dissipate energy due to inertia and elastic forces.

The damping ratio associated to each natural frequency and mode-shape can be understood as a sum of the material damping ratio and the hydrodynamic damping ratio or added damping due to the surrounding water [3, 9, 10]. This hydrodynamic damping has been demonstrated to be higher when the submerged structure is near a rigid wall [3, 10] and even higher when the surrounding water flows through the submerged structure [9]. The estimation of this damping ratio using simulation is difficult and challenging and therefore experimental results are usually necessary.

The mode-shapes associated to each natural frequency of the structure has to be also studied in detail to understand the dynamic response of the machine under operating conditions. In case of resonance during operation, not only the frequency of the excitation has to coincide with a natural frequency of the runner, but also the shape of the excitation has to match with the mode-shape of the natural frequency [11]. According to many studies, confining the structure in still water does not change its mode-shapes [3, 5]. However, when the runner rotates in water every mode-shape became

IOP Conf. Series: Journal of Physics: Conf. Series **813** (2017) 012036 doi:10.1088/1742-6596/813/1/012036

into two travelling waves rotating in the same and in the opposite direction than the runner but maintaining the same shape than without rotation [12].

In this paper, an Experimental Modal Analysis (EMA) of a medium-head Francis turbine has been carried out. The Francis turbine selected for the study is located in British Columbia, Canada, and it has 444 MW of rated power. The study of the dynamic response of this prototype is included in the research carried out in the Hyperbole project. Natural frequencies, damping ratios and mode-shapes of the runner in air have been determined and compared with simulation results.

2. Experiment

Taking profit of an outage of the prototype in 2016, an EMA could be performed in the runner. The runner was accessible from the draft tube mandoor as well as from the spiral casing. Impacts were carried out with an instrumented hammer (Dytran 5802A, sensitivity 220 uV/N) in different points of the runner and the response was measured with accelerometers located always in the same point. This method is called rowing hammer method. The runner was impacted at 16 different equidistant points in the band outlet and in every blade outlet. One accelerometer (Kistler 8750A50, sensitivity 100 mV/g) was located in the band outlet and another two in the crown and band inlet (see Figure 1).

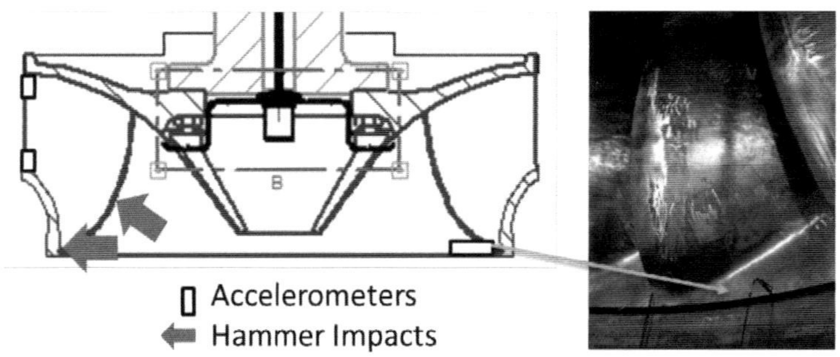

Figure 1. Instrumentation and impacts location.

Five impacts were always done in every point to use their average during the analysis. An exponential window (τ=400 ms) was applied to the accelerometers time signal and a transient window (100ms) to the hammer signal for the FRF (Frequency Response Function) computation. Coherence between the accelerometers signals and the hammer was computed always to ensure that the response is due to the impact. Figure 2 shows and example of one FRF computed at the same time than the coherence. It is seen that for all the peaks found in the FRF, which are the natural frequencies of the runner, the coherence is maximum.

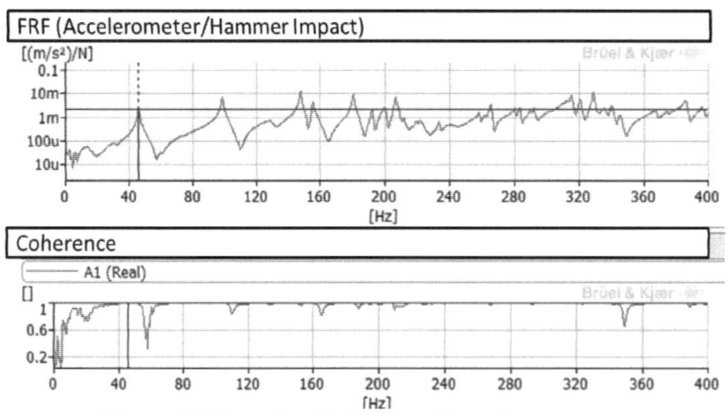

Figure 2. Example of a FRF and its coherence.

This procedure was repeated for all the different points (32) where the runner was impacted. To know how the mode-shapes of the structure are, all these points were assigned to a simplified structure which represents the runner performing ODS (Operation Deflection Shape). Natural frequencies and damping ratios were extracted for every mode using Complex Mode Indicators Functions (CMIF) of the commercial software Pulse® Reflex [13].

3. Numerical model

The numerical model considered the runner, the shaft and the generator. However, as only impacts in the runner were done experimentally, only the runner mode-shapes are shown in this paper. The details of the optimal mesh and the different parameters fixed in the FEM simulation were previously published in [14]. The FEM simulation was solved using the commercial code Ansys® v16.2. In this case, the surrounding water of the turbine is not considered in order to compare the simulation with the experimental case. Moreover, this simulation does not consider any damping in the numerical model, therefore this parameter will not be compared with experimental results.

4. Results

As a first view, comparing the response of different points of the runner (band inlet, band outlet and blades outlet, Figure 3), it is clearly seen that the response of the band and the blades at the outlet is rather higher than the response in the inlet. This means that maximum displacement of the mode-shapes is basically concentrated in the band and blades outlet.

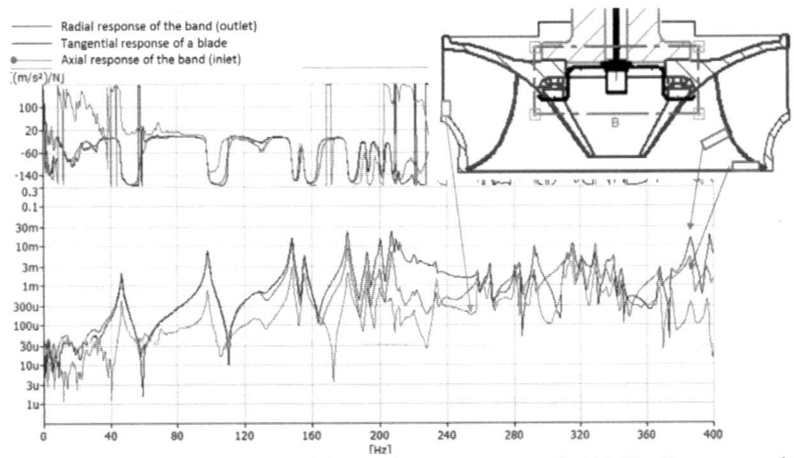

Figure 3. Runner natural frequencies within the frequency range 0-400 Hz. Response of the structure in different points (Green=response in the band inlet, Blue=response in the blade outlet, red=response in the band outlet).

The mode-shapes of Francis turbine runners are normally classified by the number of nodal diameters (iND) [14], but these different mode-shapes with different number of nodal diameters have also another common characteristics that can be used to recognise them. Figure 4 shows which are these common characteristics between the mode-shapes of the Francis turbine of study. There is a first zone in frequency (from 0 to 160 Hz) where the amplitude of the band and blades outlet is the same, these modes have been named as Global Modes (G). However, from 160 Hz onwards, the blade deformation is higher than the band deformation and therefore these modes have been named as Blade Dominant (BlD) modes. Moreover, analysing the deformation and the phases of two consecutive blades (Blades number 15 and 16), there are some modes where both are in phase (IPh), and another in counter phase (CPh) (as seen in Figure 4). Therefore, different combinations of nodal diameters (iND), Global Modes or BlD, or IPh or CPh modes can be found for this Francis turbine runner.

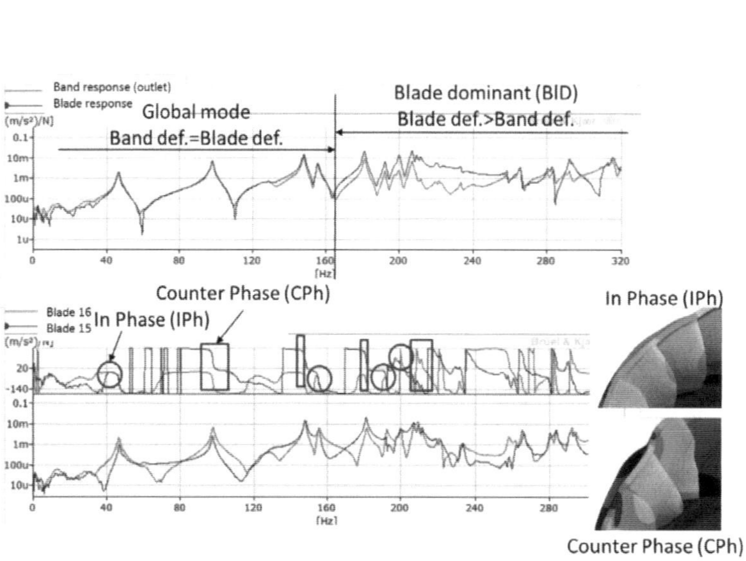

Figure 4. Classification of the mode-shapes.

Table 1. First mode-shapes found within the frequency band 0-210 Hz.

Mode-Shape	f_n (Hz) Experiment	f_n (Hz) Simulation	Difference Exp-Sim (%)	Damping ratio (%)
2ND-G-IPh	46.44	44.27	4.66	1.0272
3ND-G-IPh	98.24	93.6	4.73	0.5498
1ND-G-IPh	129.14	126.28	2.22	2.4669
4ND-G-CPh	148.08	140.65	5.02	0.3094
2ND-BlD-IPh	155.58	152.44	2.02	0.3469
5ND-BlD-CPh	181.03	171.34	5.35	0.2713
3ND-BlD-IPh	192.48	189.06	1.77	0.2394
2ND-BlD-CPh	199.70	188.54	5.59	0.2894
0ND-BlD-IPh	206.85	202.49	2.11	0.2490
1ND-BlD-CPh	209.38	196.86	5.98	0.2413

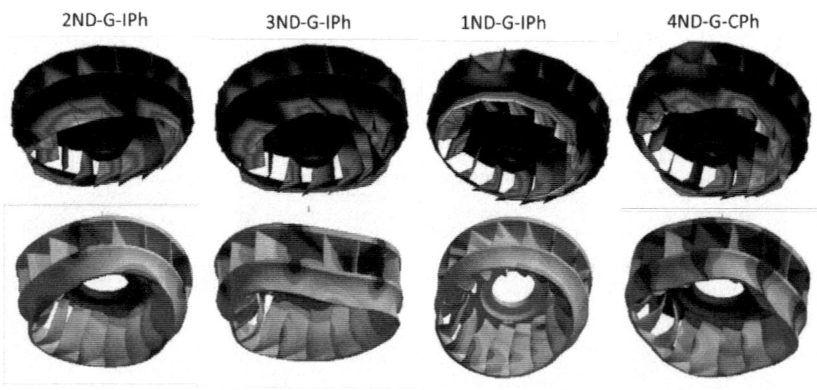

Figure 5. ODS of the first global mode-shapes.

Table 1 shows the value of the natural frequencies and damping ratio for the first ten mode-shapes found for the runner of study. Good agreement between experimentation and simulation is found for the natural frequencies values as well as for the mode-shapes. A comparison between the ODS obtained by experimentation and simulation is also shown in Figure 5.

5. Conclusions

Natural frequencies, damping ratios and mode-shapes of a Francis turbine runner have been obtained by means of an Experimental Modal Analysis. The runner was impacted in 32 different points and the response was measured with accelerometers in different positions. After the FRF computation an ODS analysis was performed and the mode-shapes, natural frequencies and damping ratios were obtained. These results have been compared with simulation having a good agreement.

This experimental analysis validates the numerical simulation and therefore the numerical model can be used to obtain the natural frequencies when the runner is surrounded by water.

References

[1] Egusquiza, E., Valero, C., Huang, X., Jou, E., Guardo, A., and Rodriguez, C., 2012, "Failure investigation of a large pump-turbine runner," Engineering Failure Analysis, 23(0), pp. 27-34.
[2] Ohashi, H., 1994, "Case Study of Pump Failure Due to Rotor-Stator Interaction," International Journal of Rotating Machinery, 1(1), pp. 53-60.
[3] Valentín, D., Presas, A., Egusquiza, E., and Valero, C., 2014, "Experimental study on the added mass and damping of a disk submerged in a partially fluid-filled tank with small radial confinement," Journal of Fluids and Structures, 50(0), pp. 1-17.
[4] Askari, E., Jeong, K.-H., and Amabili, M., 2013, "Hydroelastic vibration of circular plates immersed in a liquid-filled container with free surface," Journal of Sound and Vibration, 332(12), pp. 3064-3085.
[5] Liang, Q. W., Rodríguez, C. G., Egusquiza, E., Escaler, X., Farhat, M., and Avellan, F., 2007, "Numerical simulation of fluid added mass effect on a francis turbine runner," Computers & Fluids, 36(6), pp. 1106-1118.
[6] Rodriguez, C. G., Flores, P., Pierart, F. G., Contzen, L. R., and Egusquiza, E., 2012, "Capability of structural–acoustical FSI numerical model to predict natural frequencies of submerged structures with nearby rigid surfaces," Computers & Fluids, 64(0), pp. 117-126.
[7] Jacquet-Richardet, G., and Dal-Ferro, C., 1996, "Reduction method for finite element dynamic analysis of submerged turbomachinery wheels," Computers & Structures, 61(6), pp. 1025-1036.
[8] Valentín, D., Presas, A., Egusquiza, E., and Valero, C., 2016, "On the Capability of Structural–Acoustical Fluid–Structure Interaction Simulations to Predict Natural Frequencies of Rotating Disklike Structures Submerged in a Heavy Fluid," Journal of Vibration and Acoustics, 138(3), pp. 034502-034502.
[9] Monette, C., Nennemann, B., Seeley, C., Coutu, A., and Marmont, H., 2014, "Hydro-dynamic damping theory in flowing water," IOP Conference Series: Earth and Environmental Science, 22(3), p. 032044.
[10] Naik, T., Longmire, E. K., and Mantell, S. C., 2003, "Dynamic response of a cantilever in liquid near a solid wall," Sensors and Actuators A: Physical, 102(3), pp. 240-254.
[11] Presas, A., Egusquiza, E., Valero, C., Valentin, D., and Seidel, U., 2014, "Feasibility of Using PZT Actuators to Study the Dynamic Behavior of a Rotating Disk due to Rotor-Stator Interaction," Sensors, 14(7), pp. 11919-11942.
[12] Presas, A., Valentin, D., Egusquiza, E., Valero, C., and Seidel, U., 2015, "Influence of the rotation on the natural frequencies of a submerged-confined disk in water," Journal of Sound and Vibration, 337(0), pp. 161-180.
[13] Binda, M., Harčarik, T., and Šimčák, F., "Modal parameters estimation by using PULSE Reflex modal analysis," Proc. 4th International Conference of Modelling of Mechanical and Mechatronic Systems, pp. 17-21.
[14] Valentín, D., Ramos, D., Bossio, M., Presas, A., Egusquiza, E., and Valero, C., "Influence of the boundary conditions on the natural frequencies of a Francis turbine," Proc. IOP Conference Series: Earth and Environmental Science, IOP Publishing, p. 072004.

IOP Conf. Series: Journal of Physics: Conf. Series **813** (2017) 012037 doi:10.1088/1742-6596/813/1/012037

Overview of the experimental tests in prototype

Eduard Egusquiza, David Valentín, Alexandre Presas, Carme Valero

Center for Industrial Diagnostics and Fluid Dynamics (CDIF), Polytechnic University of Catalonia (UPC). Av. Diagonal, 647, ETSEIB, 08028 Barcelona.

egusquiza@mf.upc.edu

Abstract. Experimental tests in prototype are necessary to understand the dynamic behaviour of the machine during different operating points. Hydraulic phenomena as well as its effect on the structure need to be studied in order to avoid instabilities during operation and to extend the life-time of the different components. For this purpose, a complete experimental study of a large Francis turbine prototype has been performed installing several sensors along the machine. Pressure sensors were installed in the penstock, spiral case, runner and draft tube, strain gauges were installed in the runner, vibration sensors were used in the stationary parts and different electrical and operational parameters were also measured. All these signals were acquired simultaneously for different operating points of the turbine.

1. Introduction

The dynamic behavior of Francis turbines is strongly dependent on their operating point [1]. There are certain points were instabilities could appear, affecting the hydraulic field, the structural parts but also the electrical output. Hydraulic resonances produced at part load [2] or full load [3] can affect the structure by oscillating the mechanical torque and resulting in oscillating output power, which in fact, is an important problem for the electrical grid stability.

The hydraulic field is usually studied in models in laboratory and its behavior could be transposed to the prototype with rather good agreement [3]. However, its application to the structure is not the same in models than in prototype since the structural characteristics are rather different. As a consequence, some dynamic problems that appear in prototypes cannot be studied in models and therefore they need to be addressed on-site. However, on-site measurements are usually difficult to be performed since the installation of sensors inside the machine needs time, preparation and availability and the economic cost is high.

In this paper an overview of experimental tests carried out in a large Francis turbine is presented. The turbine was selected for study under the European Project Hyperbole [4], where different companies, universities and research centers are involved. This turbine is located in British Columbia, Canada, and it has a rated power of 444 MW. The tests were performed during three weeks (November 7th to November 27th 2016), using the first two weeks for isolation, installation of sensors and deisolation of the machine, one day for testing and the last week for dismantling and removing the sensors.

2. Participants

Four different teams participated in the on-site measurements. A group of three people from Voith® participated installing strain gauges and pressure sensors in the runner, as well as a telemetry system to transfer the signals from the rotating to the stationary part. Two people from GE Renewable® installed pressure sensors in the penstock, spiral casing and draft tube. A group of three people from UPC

(Universitat Politècnica de Catalunya, Barcelona) installed vibration sensors in the stationary parts and coordinated all the measurements acquiring simultaneously all the sensors installed by the other partners. Finally, a group of three people from EPFL (École Polytechnique Fédérale de Lausanne) calibrated the wicket gate opening angle with the servomotor stroke signal in order to compare results obtained with the model.

3. Description of the equipment

3.1. General sketch of all sensors installed

Due to the big number of sensors and dimensions of the machine, a distributed measurement system was envisaged. The system from Bruel&Kjaer LAN XI Type 3053 was used for the tests. Figure 1 shows a sketch off the sensors installed in every location of the machine. Each group of signals was sent to acquisition modules Pulse (7 modules of 12 channels). From each module an Ethernet cable was routed to a switch and then to the main computer with the software for acquisition and synchronization of all modules. In that way the length of the cables and installation time was minimized. Sensors were located in the hydraulic system, on-board on the turbine (on the runner and on the shaft), on the stationary system (bearings and shaft) and in the electrical system.

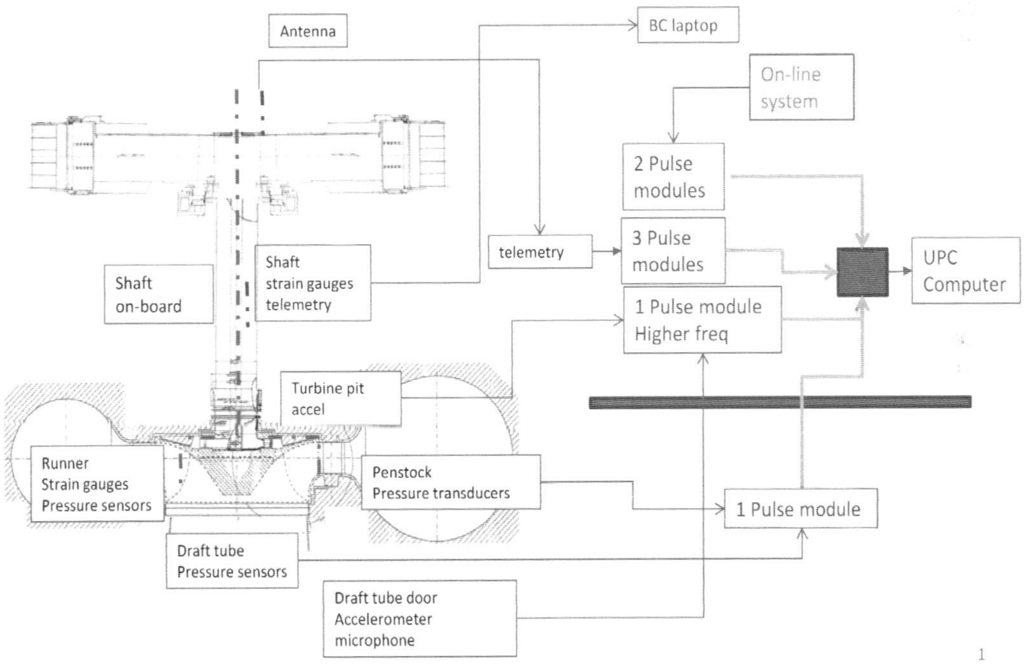

Figure 1. Sketch of the measurement equipment.

3.2. Runner instrumentation

24 strain gauges were located in two different blades of the runner. 8 pressure sensors were also installed in the same blades (Figure 2). The telemetry system was located in the cone of the runner and the rotating antenna was routed until the top of the shaft in the generator side. There, a stationary antenna was located and the signals were connected to the general acquisition system.

IOP Conf. Series: Journal of Physics: Conf. Series **813** (2017) 012037 doi:10.1088/1742-6596/813/1/012037

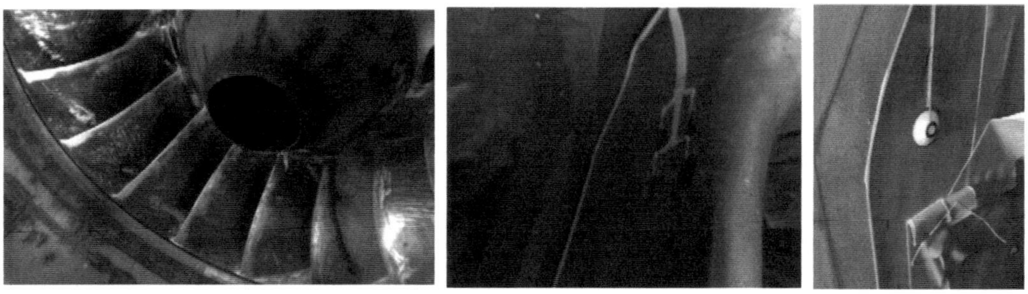

Figure 2. View of the cone of the runner where the telemetry system was installed and view of some strain gauges and pressure sensors located in the runner.

3.3. Penstock and spiral casing instrumentation
Pressure sensors were located in the penstock and spiral casing (see Figure 3).

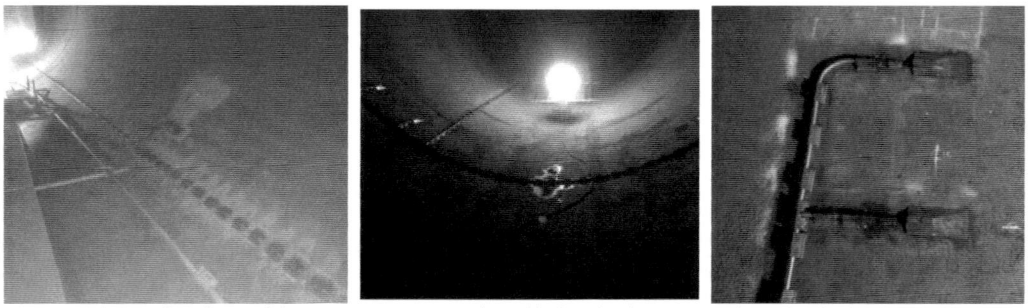

Figure 3: View of the pressure sensors installation in the penstock and spiral casing.

3.4. Draft tube instrumentation
Four pressure sensors have been installed in the draft tube cone (see Figure 4).

Figure 4. General view of the pressure sensors in the draft tube.

3.5. Torque measurement
For measuring the torque an on-board system with telemetry using strain gauges was installed on the shaft (see Figure 5).

Figure 5. Telemetry system installed on the shaft for torque measurement.

3.6. Vibration measurements

Several accelerometers were placed in the stationary parts of the machine: accelerometers in the bearings in radial and axial direction, on the head cover, on two different wicket gates, draft tube wall and casing. Moreover, four existing proximity probes (two in the turbine bearing and two in the generator bearing) were also used to measure shaft displacement.

3.7. Electrical measurements

Active power and voltage and intensity of one phase were also acquired at the same time than the other signals. These signals were taken from existing transducers in the power plant.

3.8. Operating signals

Rotational speed and wicket gate position were also taken from the existing Scada system and acquired simultaneously with the other signals. The upper and lower reservoir levels were also recorded during the tests.

4. Measurements

4.1. Measurements program

The tests followed the following sequence that can be observed in Figure 6. Several ramp-ups and ramp-downs, staying at different loads to identify the instability regions were carried out. Condenser mode and a load rejection were also tested. Operating conditions tested on a reduced scale model were reproduced in the prototype.

4.2. Overview of the results

As a first view, with the time signals of the different sensors, some instabilities can be observed. Figure 7 shows the time signals for different operating points. For a certain range of power, at part load, the fluctuation of the signals as well as the power is of relevant importance.

5. Conclusions

Experimental tests using several sensors on the rotating and stationary parts on a large Francis turbine prototype were successfully done. Pressure sensors were installed in the penstock, spiral case, runner and draft tube, strain gauges were installed in the runner, vibration sensors were used in the stationary parts and different electrical and operational parameters were also measured. All these signals were acquired simultaneously for different operating points of the turbine.

Hydraulic resonances and instabilities were identified for different operating points with different sensors. Further analysis is necessary to understand the phenomena.

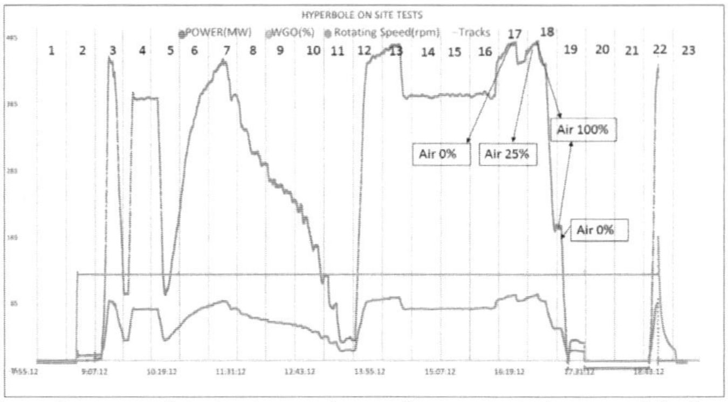

Figure 6. Evolution of power, wicket gate opening and runner rotating speed during the measurements.

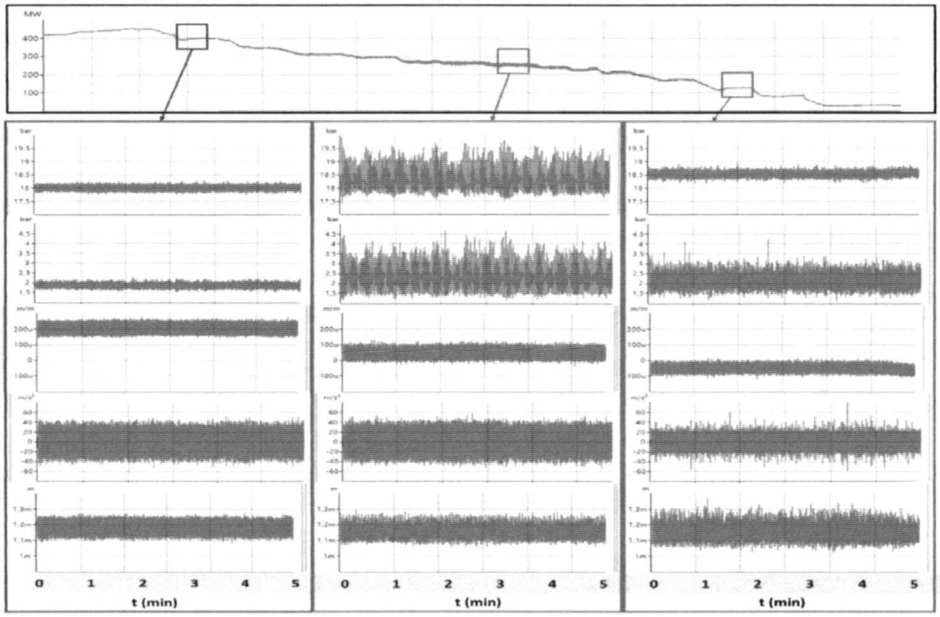

Figure 7. Time signals of different sensors for different operating points. From left to right, 400 MW, 260 MW and 180 MW.

References

[1] Trivedi, C., Cervantes, M. J., Gandhi, B. K., and Dahlhaug, O. G., 2013, "Experimental and Numerical Studies for a High Head Francis Turbine at Several Operating Points," Journal of Fluids Engineering, 135(11), pp. 111102-111102.

[2] Nicolet, C., Arpe, J., and Avellan, F., "Identification and modeling of pressure fluctuations of a Francis turbine scale model at part load operation," Proc. Proceedings of the 22nd IAHR Symposium on Hydraulic Machinery and Systems, Stockholm, Sweden, June 29 – July 2, 2004.

[3] Alligné, S., Maruzewski, P., Dinh, T., Wang, B., Fedorov, A., Iosfin, J., and Avellan, F., "Prediction of a Francis turbine prototype full load instability from investigations on the reduced scale model," Proc. IOP Conference Series: Earth and Environmental Science, IOP Publishing, p. 012025 (2010).

[4] https://hyperbole.epfl.ch, 2013, "HYdropower plants PERformance and flexiBle Operation towards Lean integration of new renewable Energies."

IOP Conf. Series: Journal of Physics: Conf. Series **813** (2017) 012038 doi:10.1088/1742-6596/813/1/012038

Detection and analysis of part load and full load instabilities in a real Francis turbine prototype

Alexandre Presas, David Valentin, Eduard Egusquiza, Carme Valero

Center for Industrial Diagnostics and Fluid Dynamics (CDIF), Polytechnic University of Catalonia (UPC). Av. Diagonal, 647, ETSEIB, 08028 Barcelona.

alex.presas@mf.upc.edu

Abstract. Francis turbines operate in many cases out of its best efficiency point, in order to regulate their output power according to the instantaneous energy demand of the grid. Therefore, it is of paramount importance to analyse and determine the unstable operating points for these kind of units. In the framework of the HYPERBOLE project (FP7-ENERGY-2013-1; Project number 608532) a large Francis unit was investigated numerically, experimentally in a reduced scale model and also experimentally and numerically in the real prototype. This paper shows the unstable operating points identified during the experimental tests on the real Francis unit and the analysis of the main characteristics of these instabilities. Finally, it is shown that similar phenomena have been identified on previous research in the LMH (Laboratory for Hydraulic Machines, Lausanne) with the reduced scale model.

1. Introduction

One of the advantages of using large hydraulic turbines for power generation is that they output power can be adjusted depending on the instantaneous energy demand of the grid. In order to regulate the output power, these units have to operate out of its design conditions. Particularly, many Francis units, which are nowadays the most installed hydraulic turbines in power plants, have to work out of its best efficiency point for long periods of time. Nevertheless, some operating points can be unstable, limiting the effective operating range of the unit. For this reason, it is of paramount importance to accurately predict (during the design phase of the unit) and to determine (during the operation of the installed unit) these particular points.

Inside the HYPERBOLE project (FP7-ENERGY-2013-1; Project number 608532) a large Francis unit has been investigated by means of numerical models [1, 2] and experimentally on both the reduced scale model of this unit [3-6] and the real prototype [7].The real unit has a rated power of 444MW and it is located on the British Columbia in Canada. The tests on the reduced scale model (complete model with scale 1/16) were performed in the Laboratory of Hydraulic Machines (EPFL) in Switzerland.

The unstable points of this unit were firstly predicted based on the numerical and experimental studies made on the reduced scale model. As a final step of the HYPERBOLE project, a measurement campaign on the real unit has been performed. One of the goals of this campaign was to determine and to investigate the unstable operating points of the real prototype of the investigated unit.

In this paper, the experimental data obtained during the prototype tests are analyzed. The main focus of the analysis is firstly to determine and to analyze the main unstable operating points of the unit. Finally, it is observed that similar phenomena was found in the tests made on the reduced scale model.

2. Experimental tests

The on-site tests in the real prototype planned inside the HYPERBOLE project took place in November 2016. The objective was to measure different parameters of the unit such as pressure fluctuations, stresses on the runner, vibrations on the bearings and power fluctuations for many operating conditions. Schematically, the sensors where installed according to Figure 1.

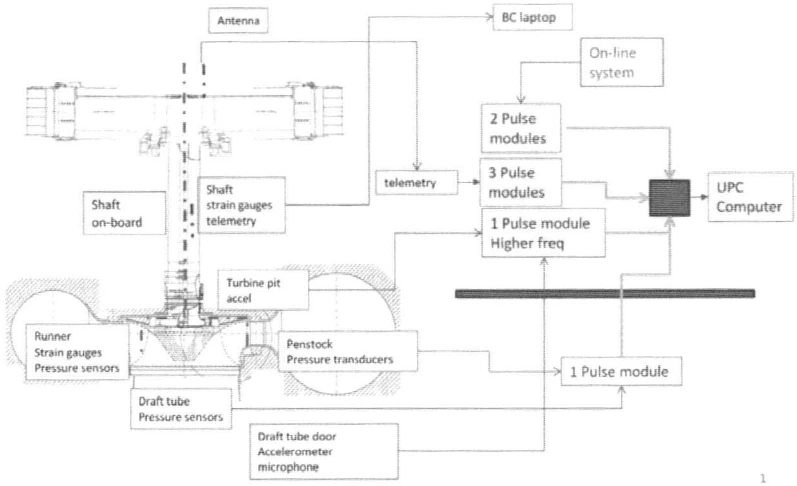

Figure 1. Overview of the installed acquisition system

As seen in Figure 1, the sensors installed were: strain gauges and pressure sensors on the runner, pressure sensors on the draft tube and spiral case& penstock, accelerometers on the bearings and other stationary parts and strain gauges on the shaft. Furthermore operating signals such as Power or Wicket Gate Opening and Gross Head were also monitored.

All the signals were simultaneously acquired with a B&K LAN XI module (Pulse module). The sampling rate was selected at 4096Hz. All the data were stored in the central computer. Figure 2a shows the active power of the unit and gives an overview of the operating points measured. For the analyses presented in this paper, firstly the slow ramp-up (approximately 1 hour) was analyzed. To have such a characteristic with the signal of Power, the operating gross head was approximately constant and the wicket gates were opened very slowly. The second part analyses the stabilized points, for which the operating parameters of the unit were maintained constant for approximately 5 minutes. This permits to analyze each particular condition with more details.

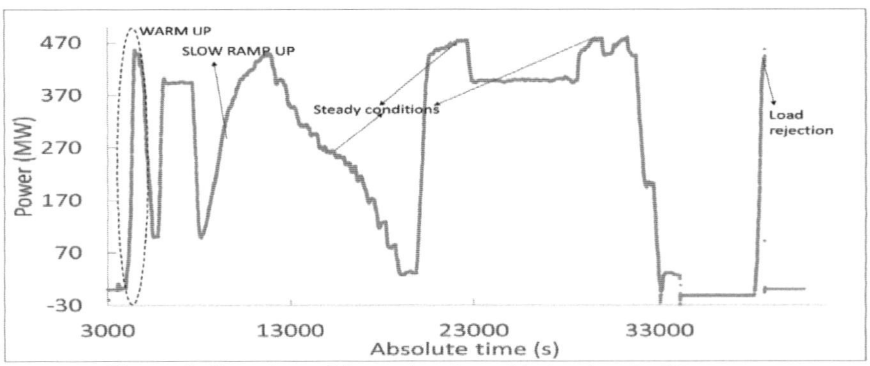

Figure 2. Overview of the testing day with the signal of Power

3. Results

3.1. Detection of unstable points during the slow ramp-up

Analysing one of the pressure sensors located on the draft tube (Figure 3) one can identify possible unstable points. In Figure 3, the part load instability corresponding to the resonance condition between the vortex rope precession and the natural frequency of the unit can be identified for P = 255 MW. In a similar way the full load instability has been investigated. The main operating characteristics of these points and the coefficients n_{ED}, Q_{ED} are summarized in Table 1.

In order to analyse the main characteristics of these two unstable conditions, these are studied with more detail during the steady conditions (see Figure 2).

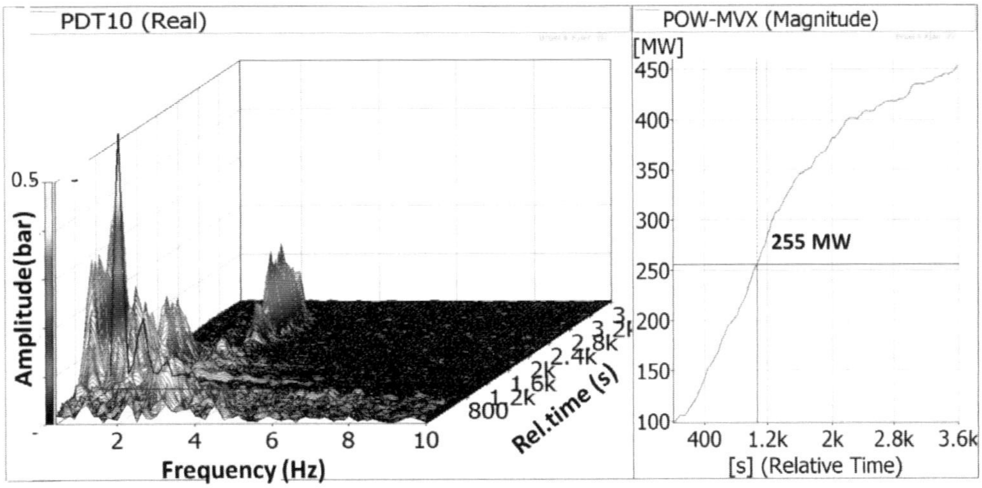

Figure 3. Time-Frequency analysis during the slow ramp up of the tests. Unstable point with maximal pressure fluctuation detected at 255MW

Table 1. Main frequency (Pressure sensor), n_{ED}, Q_{ED} and power and for the unstable points found

Unstable point	Main frequency	Power	Q_{ED}	n_{ED}
Part Load	0,625 Hz	255,6 MW	0.136	0.276
Full Load	0,8125 Hz	471 MW	0.2533	0.279

3.2. Analysis of the unstable points during the steady conditions

For the operating points shown in Table 1, a FFT (Fast Fourier Transform) analysis of the time signals has been performed. The steady conditions (see Figure 2) are now considered. For the part load and full load instability a signal of approximately 180 seconds is taken for the analysis. The analysis is performed with a resolution of 1/16Hz (window of 16 seconds weighted by a Hanning) and by averaging all the signals in the frequency domain.

For these two unstable points (Figure 4a), a large power fluctuations of the unit are observed. These large fluctuations of the output power are not convenient for the grid. The period of oscillation of these time signals for these two unstable conditions corresponds to the period detected with the pressure sensor at the respective unstable points for the slow ramp up (see Table 1).

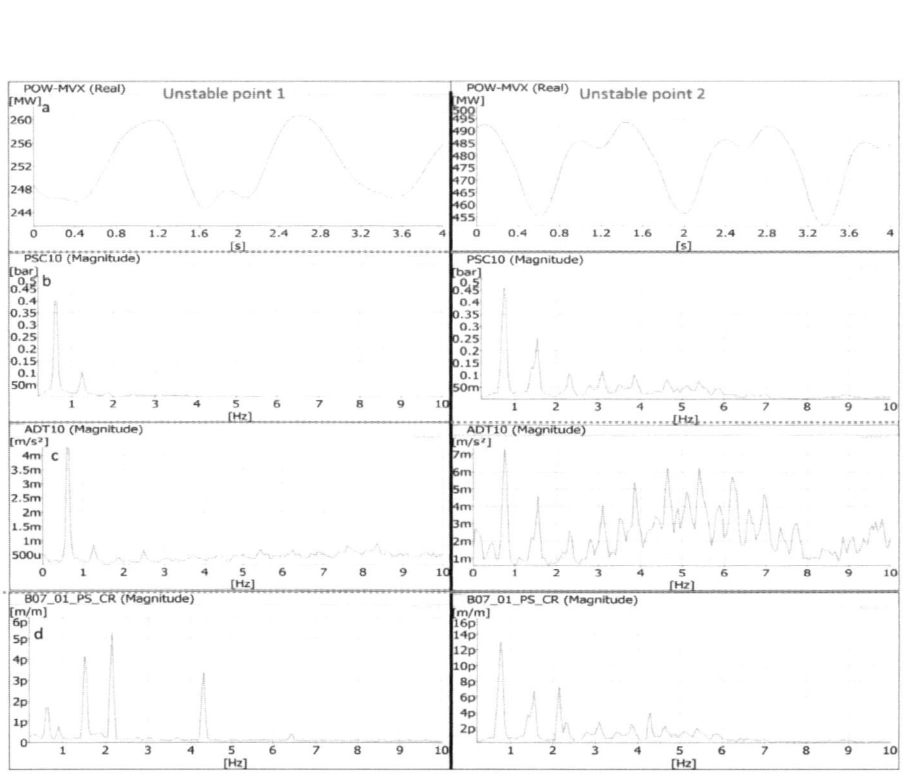

Figure 4. Frequency analysis of the instabilities. a) Power, b) Pressure, c)Accelerometer and d) Strain gauge on the runner

One of the pressure sensor installed on the spiral case, detects an oscillation in pressure of approximately 0.5 bar (RMS value) for both unstable points (Figure 4b). The same frequencies are clearly detected with an accelerometer installed on the draft tube wall (Figure 4c). Regarding the strain gauge sensor installed on the runner, there is a difference in the autospectrum when comparing the two instabilities (Figure 4d). For the full load instability, the frequency detected with this sensor on the rotating frame is the same as the frequency detected with the stationary sensors. For the part load instability the peak of 1.5 Hz dominates instead of the peak of 0.625Hz. According to the part load [5] and full load analyses [6, 8] on the reduced scale model, the first unstable point is excited by a vortex rope with a precession motion while for the full load instability, the excitation source is an axial vortex rope. This precession produces differences in the frequency content of the pressure field on the rotating and stationary frame. The main frequency on the rotating frame can be calculated according to $f_{vortex,rot} = f_{runner} - f_{vortex,stat}$, where $f_{vortex,rot}, f_{vortex,stat}$ are the vortex rope frequency from respectively the rotating and stationary frame and f_{runner} is the rotating speed of the machine. Similar phenomena was found in [9].

For the unstable point 1 (Figure 4), the ratio between the vortex frequency and the rotational frequency of the runner is $\frac{f_{vortex,stat}}{f_{runner}} \approx 0{,}29$. Favrel et al. in [4] found approximately the same ratio for the same range of Q/Q_0. According to [4, 5] (tests in the reduced scale model) this unstable point could be explained by a resonance caused by a coincidence of $f_{vortex,stat}$ with one natural frequency of the hydraulic system.

For the full load unstable point, the study made by Müller et al. in [8] is taken into account. In that case, $\mathbf{Q_{ED}}$ was approximately the same as the $\mathbf{Q_{ED}}$ for the full load instability point identified on the prototype and $\mathbf{n_{ED}}$ was slightly larger. Same as in the prototype tests, a large pressure fluctuations, due to an axial oscillating vortex rope were detected with a pressure sensor installed on the wall of the draft tube. The similarity in the shape of the pressure signal can be clearly seen in Figure 5.

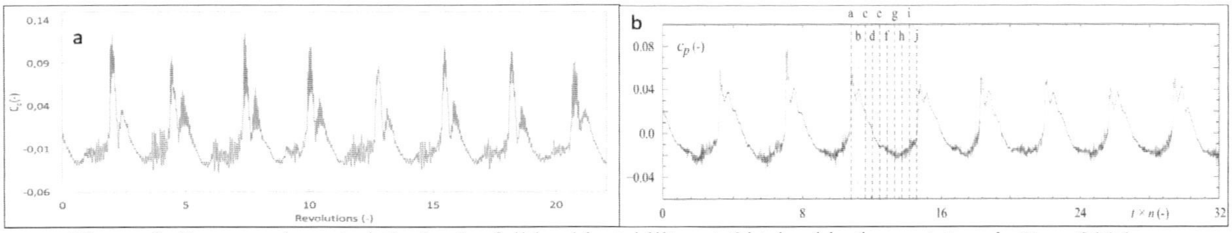

Figure 5: Pressure characteristic for the full load instability. a) Obtained in the prototype in Nov. 2016 and b) Obtained in the model test (picture taken from [8])

4. Conclusions

The data obtained after the prototype tests of the HYPERBOLE project, has been analysed. The analyses were focused on the detection and the study of two unstable operating points found during the tests. For the tests, the operating head of the unit was approximately constant and the wicket gates were moved from its minimum opening to almost its maximum.

These unstable points have been confirmed analysing many types of sensors located on the stationary frame and on the rotating frame (runner). Furthermore, the results obtained in the prototype tests seems to be in good qualitative agreement with the ones from the reduced scale model test.

Acknowledgments

The research leading to the results published in this paper is part of the HYPERBOLE research project, granted by the European Commission (ERC/FP7- ENERGY-2013-1-Grant 608532). The authors would like to acknowledge BC HYDRO for making the prototype tests possible.

References

[1] K. Yamamoto, A. Müller, A. Favrel, C. Landry, F. Avellan, Numerical and experimental evidence of the inter-blade cavitation vortex development at deep part load operation of a Francis turbine, IOP Conference Series: Earth and Environmental Science, IOP Publishing, 2016, pp. 082005.

[2] C. Landry, A. Favrel, A. Müller, C. Nicolet, F. Avellan, Local wave speed and bulk flow viscosity in Francis turbines at part load operation, Journal of Hydraulic Research, 54 (2016) 185-196.

[3] A. Favrel, C. Landry, A. Müller, K. Yamamoto, F. Avellan, Hydro-acoustic resonance behavior in presence of a precessing vortex rope: observation of a lock-in phenomenon at part load Francis turbine operation, IOP conference series: earth and environmental science, IOP Publishing, 2014, pp. 032035.

[4] A. Favrel, A. Müller, C. Landry, K. Yamamoto, F. Avellan, Study of the vortex-induced pressure excitation source in a Francis turbine draft tube by particle image velocimetry, Experiments in Fluids, 56 (2015) 1-15.

[5] A. Favrel, A. Müller, C. Landry, K. Yamamoto, F. Avellan, LDV survey of cavitation and resonance effect on the precessing vortex rope dynamics in the draft tube of Francis turbines, Experiments in Fluids, 57 (2016) 168.

[6] A. Müller, M. Dreyer, N. Andreini, F. Avellan, Draft tube discharge fluctuation during self-sustained pressure surge: fluorescent particle image velocimetry in two-phase flow, Experiments in fluids, 54 (2013) 1-11.

[7] D. Valentín, D. Ramos, M. Bossio, A. Presas, E. Egusquiza, C. Valero, Influence of the boundary conditions on the natural frequencies of a Francis turbine, IOP Conference Series: Earth and Environmental Science, IOP Publishing, 2016, pp. 072004.

[8] A. Müller, A. Favrel, C. Landry, K. Yamamoto, F. Avellan, On the physical mechanisms governing self-excited pressure surge in Francis turbines, IOP Conference Series: Earth and Environmental Science, IOP Publishing, 2014, pp. 032034.

[9] F. Duparchy, A. Favrel, P. Lowys, C. Landry, A. Müller, K. Yamamoto, F. Avellan, Analysis of the part load helical vortex rope of a Francis turbine using on-board sensors, Journal of Physics: Conference Series, IOP Publishing, 2015, pp. 012061.

IOP Conf. Series: Journal of Physics: Conf. Series **813** (2017) 012039 doi:10.1088/1742-6596/813/1/012039

1D simulation of pump-turbine transition

N Ruchonnet[1] **and O Braun**[2]
[1]Andritz Hydro SA, turbine physics, Vevey, Switzerland
[2]Andritz Hydro AG, pumps, Graz, Austria

Abstract. Variable speed motor-generators based on full size converter offer new options for operation of reversible pump-turbines. To explore the challenges related with this new technology a specific test rig configuration is developed to perform such transition with a PT reduced scale model. Transition from pump to turbine and vice versa are performed with various guide vanes opening and transition time. In order to extrapolate the results obtained at model scale to prototype, it is important to understand the system behaviour including the waterways and its influence on the flow in the pump-turbine. For this purpose, a 1D model of the test rig is realised. After validation with measurement results, the 1D model has been used to test the influence of the waterway on the pump-turbine transition.

1. Introduction

Variable speed motor-generators based on full converter solutions offer new options for operation of reversible pump-turbines (PT) [1], including fast transition from pump to turbine mode, where the pump is operating in pump-brake mode for a while. To explore the challenges related with this new technology a specific test rig configuration is developed to perform such transition with a PT reduced scale model. Transition from pump to turbine and vice versa are performed with various guide vanes opening and transition time. Dynamic pressure sensors have been installed to capture transient flow phenomena and various mechanical components have been instrumented to evaluate their loadings. In order to extrapolate the results obtained at model scale to prototype, it is important to understand the system behaviour and its influence on the results. For this purpose, a 1D model of the test rig is realised. After validation with measurement results, the 1D model can be used to verify the influence of various parameters on the transition [3].

2. Experimental setup and results

The measurements have been carried out on the universal test rig at ANDRITZ Hydro GmbH in May 2015. A PT model of specific speed NQE=0.17 (nSQ=207) with Zs=24 guide vanes and Zr=7 runner blades was selected for the test.

At prototype scale, the machine head is determined by the level difference between upstream and downstream reservoir, the losses in the hydraulic circuit and the fluid acceleration during transitions. For the reduced scale model, a pump is used to deliver the head. The available head is therefore strongly dependant on the discharge through the pump-turbine. For this reason, a specific test rig configuration, see Figure 1, is used in order to perform the full transition with realistic head.

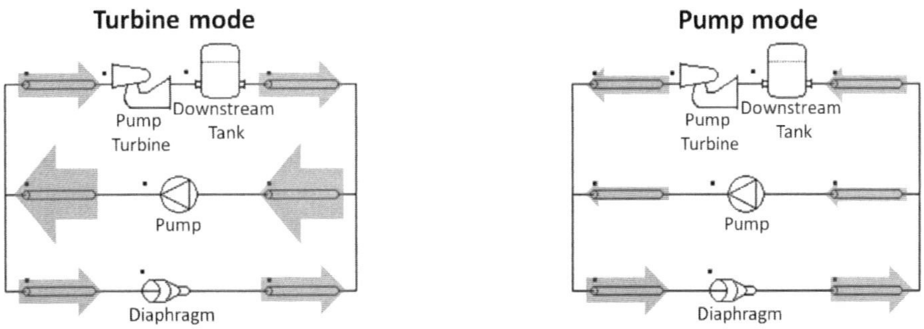

Figure 1. Test rig configuration for pump-turbine transition

The pump-turbine reduced scale model is mounted in parallel with a pump and a diaphragm made of a perforated plate. The head losses through the diaphragm are proportional to the discharge. In pump mode, the discharge of both pump and pump-turbine flows through the diaphragm. In turbine mode, part of pump discharge flows through the pump-turbine, the other part through the diaphragm. The head is adjusted by varying the pump rotating speed. During transitions, linear speed ramps are imposed to both the pump-turbine and the pump using variable speed electric motor-generator. With this alternative test rig configuration, the transition as seen from the pump-turbine is close to the transition at prototype scale. The head variation is mainly driven by the fluid acceleration within the pump turbine branch of the circuit.

In addition to the standard model test instrumentation, dynamic pressure transducers are installed in the rotating and stationary parts of the turbine. Strain gauges are installed on guide vanes stems and turbine shaft for torque measurement. A dynamic flow meter is used to record the discharge variation. Measurements have been performed at constant guide vane opening for various transition time, guide vane opening, cavitation level and head. More details regarding the experimental setup and results can be found in [2].

The transition tests have been performed for the following guide vanes and transition times:

- Guide vanes openings: 5°, 15° and 25°;
- Transition time: 4 s, 8 s, 12 s, 20 s.

In this paper the focus is placed on the integral quantities which are considered in the 1D model.

The head is computed using the pressure difference between turbine inlet and outlet. Dynamic pressure sensors are used to capture the transient pressure as well as the fluctuation related to turbulence.

The mechanical torque is measured using an instrumented shaft equipped with strain gauges. Due to instrumentation failure during the measurement campaign, torque records are not available for the final measurements. Fortunately torque records from preliminary measurements are available for most of the operating conditions.

The discharge is measured using a MID flow meter. The nominal time constant of the instrument is 0.07s; preliminary measurements shown that the actual time constant is much larger. Additionally the high frequency flow meter shows drop-outs from a few tenths of seconds to seconds, it was found that this was related to small quantities of air being convected in the test rig. The presence of those bubbles in the flow meter stopped the acquisition of valuable values. In this case, the analogue output of the flow meter that was connected to the acquisition system remains at a constant value until the flow meter goes to normal operation again.

Using pressure difference between sensors located in the discharge measurement section, it is possible to estimate the time constant of flow meter. According to the momentum equation, the time derivate of discharge is proportional with the pressure difference along a pipe section. The best fit between measured discharge (using flow meter, red line) and computed discharge (using pressure sensors, blue line) is obtained assuming a delay of the flow meter of 0.5-0.6s. Example is given in Figure 2 for a 4s transition.

To facilitate the comparison of the simulations with the measurements, all measured signals have been resampled at 100Hz.

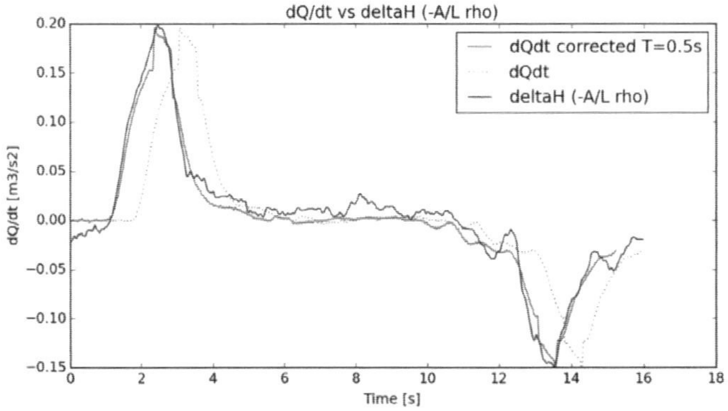

Figure 2. Evaluation of flowmeter delay based on pressure difference

3. 1D simulation model

An equivalent 1D model of the test rig has been prepared using the simulation software SIMSEN, see Figure 3. The model includes:

- The water inertia of the entire circuit;
- The PT (FTURB);
- The pump (PUMP02) characteristics.

Due to the complexity of the junctions, the head losses have been determined a posteriori to fit with the measurements. All head losses are aggregated per branch; LOSS_PT for the PT branch, LOSS_P for the pump branch; DLOSS for the perforated plate branch. Analysis of the measurements shows that the losses are influenced by the cavitation level in the circuit. Most probably, the cavitation level in the wake of the perforated plate is the major driver of this phenomenon.

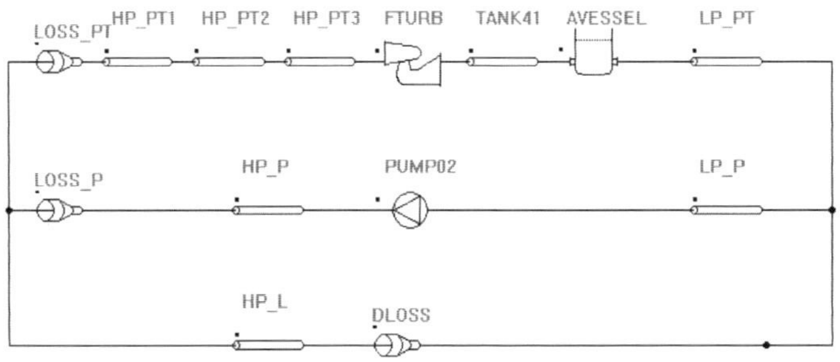

Figure 3. 1D equivalent model of the test rig

3.1. Validation of 1D model

Simulation is compared with measurement for 8s transition in Figure 4. Fair agreement is found for all variables. The pressure peak amplitude and timing is well predicted, as shown later the amplitude depends mainly on the water inertia. Considering a delay of 0.5s for the measured discharge, good agreement is observed between simulation and measurement. The flow meter drop-out is clearly visible at 21-22s, the measurement is not relevant during this period of time. The predicted torque is also in good agreement with the measurement except at the end of the transition from turbine to pump. From ~22s when the nominal rotating speed is reached the torque is over predicted in the simulation. This effect is observed for all transition time. The strong torque fluctuation observed during the transition is due to the stepwise rotating speed induced by the frequency converter. This effect is not considered in the simulation. The imposed rotating speed is used to synchronize the simulations with the measurements.

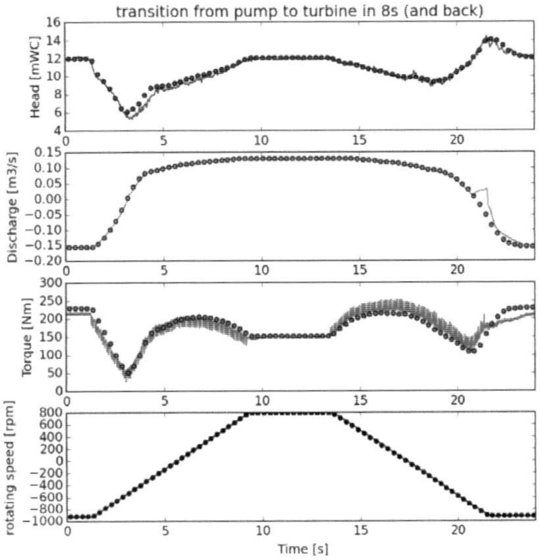

Figure 4. Comparison of simulation (dots) with measurement (continuous line) for 8s transition

3.2. Parameter study on water inertia

Various circuit configurations have been simulated in order to assess the influence of the different elements of the circuit on the pump-turbine transition.

To isolate the effect of the pump, a simulation is performed assuming constant pressure at both end of the pump-turbine branch, see Figure 5. This situation corresponds to an ideal pump providing constant pressure independently of the discharge.

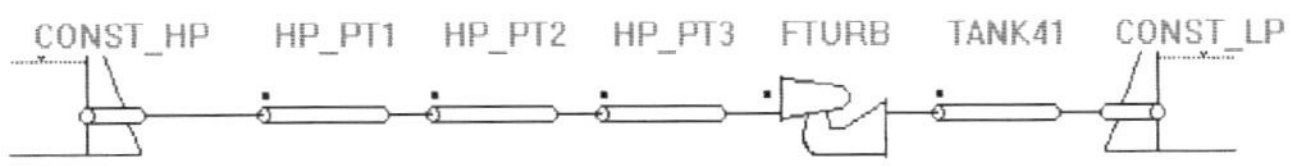

Figure 5. Setup 2: Circuit with constant upstream and downstream pressure

To evaluate the influence of the water inertia in the circuit, Setup3 corresponds to Setup 2 as shown in Figure 5, but the lengths of all pipes are divided by 2; the inertia of water inside the turbine remains unchanged. Finally during transient, the pressure difference between inlet and outlet of the pump-turbine or head is due to the quasi-static head on one hand and to the fluid inertia inside the turbine on the other hand. This inertia component induces a deviation from the characteristic curve especially during fast transient, the quasi-static head being the head measured during steady state operation.

To evaluate the influence of the different parameters, the results of the 3 setups are compared in Figure 6 for the head, discharge, torque and trajectory in the n11-Q11 diagram.

The influence of the pump is clearly visible in the head. In the pump turbine, the flow inversion takes place at the beginning of the transition from pump to turbine. Inversely for the transition from turbine to pump, the flow inversion takes place at the end of the transition. On the pump side, the rotating speed and resulting head changes continuously during the 12s transition leading to low head at the end of the transition from pump to turbine and at the beginning of the transition from turbine to pump.

The water inertia has a direct influence on the transition speed; this is visible in the discharge. The flow inversion for shorter circuit (green line) is faster than for longer circuit (blue line).

The influence of water inertia in the turbine is visible in the n11-Q11 diagram. For fast transition, deviation from the quasi-static characteristic curve (black line) is observed; this deviation is due the influence of water inertia inside the turbine on the turbine head. In this study, this inertia is estimated based on the turbine geometry.

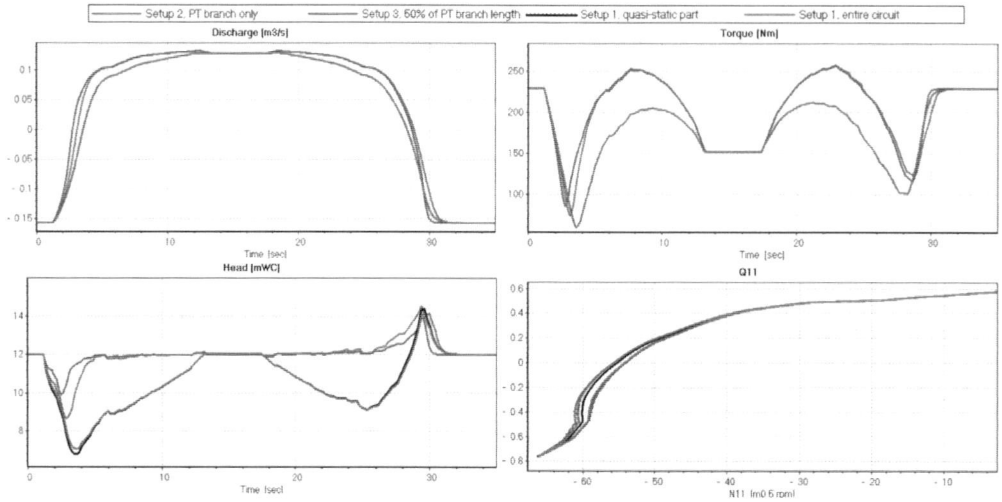

Figure 6. Discharge, Torque, Head and trajectory in n11-Q11 diagram during 12s transitions, influence of water inertia

4. Conclusion and outlook

1D transient simulation has been performed for a speed controlled transition from turbine to pump and vice-versa. The results are compared with experimental results obtained on reduced scale model. The 1D model help to better understand the relation between experimental results obtained at model scale and actual prototype behaviour. The influence of the specific test rig configuration on the transient head is clearly shown. The influence of the water inertia in the pump-turbine branch is analysed. Its effect on the transition speed is enhanced. The water inertia limits the transition speed and increases the pressure spike during transition.

In the present experiment, the head varies quasi-linearly with the pump speed during the entire transition. This leads to some deviation with respect to the theoretical case with constant head at circuit boundaries. For future work an interesting improvement would be to control the pump speed with the measured pump-turbine head in order to achieve a more realistic transition such as load rejection with surge tank oscillations expected on the prototype.

5. References

[1] Hell J., Egretzberger M., Lechner A., Schürhuber R., Vaillant Y. Full size converter solutions for pumpedstorage plants – a promising new technology. *Hydro2012, Bilbao*. 2012.

[2] Ruchonnet, N., Braun, O., "Reduced scale model test of pump-turbine transition". 6th IAHR International Meeting of the Workgroup on Cavitation and Dynamic Problems in Hydraulic Machinery and Systems, Ljubljana. 2015.

[3] Nicolet, C., Béguin, A., Bollaert, E., Boulicaut, B., Gros, G, "Hydro-Clone: a Real-time simulation monitoring system for hydropower plant transients survey". International Journal on Hydropower and Dams. Issue 5 2015, p. 62-69

Acknowledgments

The research leading to the results published in this paper is part of the HYPERBOLE research project, granted by the European Commission (ERC/FP7-ENERGY2013-1-Grant 608532). The authors would like to thank their colleagues in Linz for performing the test rig measurement.

IOP Conf. Series: Journal of Physics: Conf. Series **813** (2017) 012040 doi:10.1088/1742-6596/813/1/012040

Simulation of pump-turbine prototype fast mode transition for grid stability support

C Nicolet[1], O Braun[2], N Ruchonnet[2], J. Hell[3], A Béguin[1], F Avellan[4]

[1] Power Vision Engineering Sàrl, Ch. des Champs Courbes 1, CH 1024 Ecublens, Switzerland
[2] ANDRITZ HYDRO SA, Rue des Deux Gares 6, CH-1800 Vevey, Switzerland
[3] ANDRITZ HYDRO GmbH, Eibesbrunnergasse 20, A-1120 Vienna, Austria
[4] Ecole Polytechnique Fédérale de Lausanne, Laboratory for Hydraulic Machines, CH 1007 Lausanne, Switzerland

Abstract. The paper explores the additional services that Full Size Frequency Converter, FSFC, solution can provide for the case of an existing pumped storage power plant of 2x210 MW, for which conversion from fixed speed to variable speed is investigated with a focus on fast mode transition. First, reduced scale model tests experiments of fast transition of Francis pump-turbine which have been performed at the ANDRITZ HYDRO Hydraulic Laboratory in Linz Austria are presented. The tests consist of linear speed transition from pump to turbine and vice versa performed with constant guide vane opening. Then existing pumped storage power plant with pump-turbine quasi homologous to the reduced scale model is modelled using the simulation software SIMSEN considering the reservoirs, penstocks, the two Francis pump-turbines, the two downstream surge tanks, and the tailrace tunnel. For the electrical part, an FSFC configuration is considered with a detailed electrical model. The transitions from turbine to pump and vice versa are simulated, and similarities between prototype simulation results and reduced scale model experiments are highlighted.

1. Introduction

The integration of the constantly growing capacity of New Renewable Energies, NRE, mainly composed by wind and solar energies, is a challenging task as far as the power network stability is concerned, due to the intermittent nature of these energy sources. Beside storage and substitution production capabilities, pumped storage power plants can significantly contribute to improve the stability of power network due to their production flexibility, fast response time and large energy storage capability. Variable speed motor-generator solutions enable to provide a variety of new control services to the electrical grid to be carefully considered, [2]. In particular, solution based on synchronous motor-generator with Full Size Frequency Converter, FSFC, offer the possibility to achieve turbine to pump fast transition mode and vice versa [1], [2], [4]. The feasibility of this new ancillary service was deeply investigated in the framework of the HYPERBOLE European Research Project, by means of reduced scale model and numerical simulation [3]. This paper presents experimental and numerical results of fast mode transition from pump to turbine and vice versa performed on a reduced scale model tests as well as simulation results for an existing 2x210 MW pumped storage power plant equipped with a pump-turbine homologous to the reduced scale model, where conversion from fixed speed to FSFC variable speed solution is assumed. The plausibility of fast transition mode is investigated by means of full hydroelectric power plant simulation including both hydraulic system as well as electrical installations.

2. Investigation at reduced scale model

The feasibility of fast mode transition from pump to turbine and vice versa is investigated first at the reduced model scale by means of both experimental and numerical approach enabling a comprehensive survey of the dynamic behavior of the pump-turbine unit and its interaction with the related hydraulic circuit. Figure 1 left presents the reduced scale model of the pump-turbine installed at the ANDRITZ Hydro Hydraulic Laboratory in Linz, Austria. This reduced scale model is quasi-homologous to the prototype unit of the 2x210 MW test case, as both feature the same specific speed of NQE=0.17 (nSQ=207) and the same blade number and main dimensions, [3].

A 1D model of the test rig has been setup using the SIMSEN simulation software developed by the Ecole polytechnique fédérale de Lausanne, EPFL. The model includes the pipes dimensions of the entire test rig; the downstream reservoir; the pump and pump-turbine characteristics. Transitions from pump to turbine and return where carried out on the test rig for transition time of 4s, 8s, 12s and 20s where the rotational speed is varied linearly using specific motor-generator drive. Figure 1 right presents the comparison between the 1-D simulation results and test rig measurements for transition time of 8s and guide vane opening Y=15°. The transition starts at t=3s in 8s. Once in turbine mode, the speed is kept constant during 4s before returning to pump mode with the same transition time. The resulting time evolutions of head and discharge are in good agreement with the measurements, see Figure 1 right.

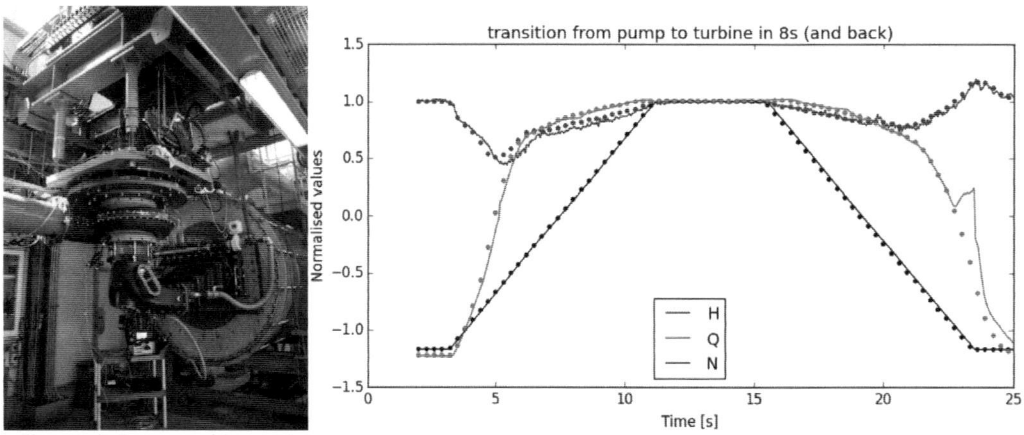

Figure 1. Pump-turbine reduced scale model (left) and comparison between simulation results (dots) and measurements (solid line) of the head, discharge and rotational speed during pump to turbine transitions and vice-versa carried out in 8s and GVO Y=15°.

3. Modelling of prototype pumped storage power plant

The layout of the pump-turbine unit of the 2x210 MW test case is presented in Figure 2 left, while the corresponding main characteristics are provided in Table 1.The SIMSEN model of the pumped storage power plant, see Figure 2 right, includes all hydraulic components and takes into account [3]:

- Water hammer phenomena in piping systems (pipe head losses, water inertia and pipe elastic behavior (fluid compressibility and pipe wall deformation));
- The 4 quadrants transient behavior of the pump-turbine including so-called S-shape unstable characteristics corresponding to the one considered for the reduced scale model tests, and the link with rotating inertias;
- The surge tank mass oscillations phenomena between the lower surge tanks and the lower reservoir taking into account variable cross section area of the surge tanks and asymmetric head losses at the inlet of the surge tank;
- The generating unit rotating train torsion dynamics with turbine and motor-generator rotor rotating inertia linked through coupling shaft with given torsional stiffness corresponding to the first natural torsional frequency of the unit.

IOP Conf. Series: Journal of Physics: Conf. Series **813** (2017) 012040 doi:10.1088/1742-6596/813/1/012040

Moreover, the SIMSEN model of Figure 1 right included a detailed simulation model of the FSFC set, considering the following elements: a synchronous machine, the machine side and grid side converters and the related control, the unit step-up transformer, a connection to the grid with a given finite short-circuit power and the unit rotating inertia. The model of the synchronous machine is of high order and takes into account the excitation currents and damper currents, see [3].

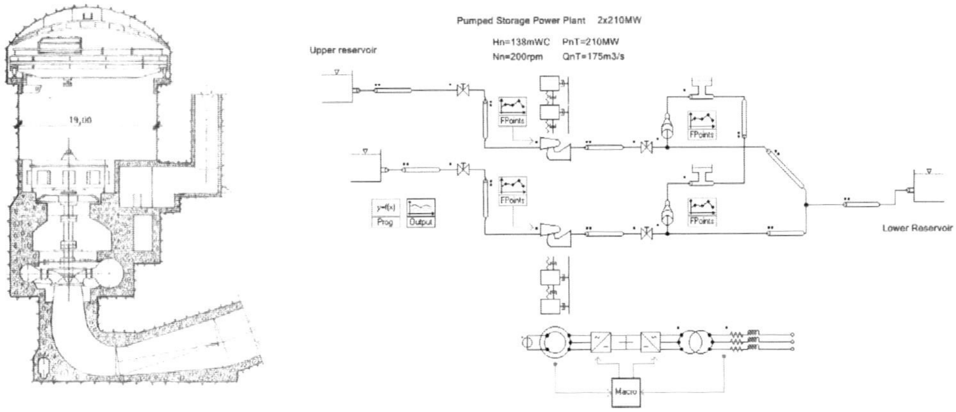

Figure 2. Layout of the 210 MW pump-turbine unit (left) and the SIMSEN model of the pumped storage power plant test case (right).

Table 1. Main characteristics of the pumped-storage power plant test case, see Figure 2 right.

Francis pump-turbine:	
Nominal rotational speed	200 rpm
Turbine nominal power	210 MW
Turbine nominal discharge	175.1 m^3/s
Turbine nominal net head	133.5 mWC
Turbine nominal torque	9.74 MNm
Low pressure side diameter	3.86 m
Total unit inertia including generator and turbine (J=MR2)	4106 $t*m^2$
Mechanical time constant ($\tau_m = J_{tot} \cdot \omega_n^2 / P_n$)	8.6 s

4. Simulations of fast transition mode of prototype unit

The Froude similitude has been considered to transpose the transition time from reduced scale model to the prototype scale, by expressing the reference velocity as function of the unit net head H, and then expressing the head H as function of the runner peripheral velocity $U = \omega \cdot R$ with $H = U^2/(2 \cdot g)$ as follows:

$$Fr = \frac{C_{ref}}{\sqrt{g \cdot D_{ref}}} = \sqrt{\frac{g \cdot H}{g \cdot D_{ref}}} = \sqrt{\frac{U^2}{2 \cdot g \cdot D_{ref}}} = \frac{\omega}{2}\sqrt{\frac{D_{ref}}{2 \cdot g}} \quad (5)$$

According to the dimension ratio of $D^P/D^M = 15.82$ between the prototype and the model, the transition time ratio is given equal to:

$$\frac{T^P}{T^M} = \sqrt{\frac{D^P}{D^M}} = \sqrt{\frac{3.86}{0.244}} = 3.98 \cong 4$$

According to the above transposition, the transition time considered during reduced scale model tests of 4s, 8s, 12s, and 20s, respectively leads to transition time of 16s, 32s, 48s and 80s.

The numerical simulation of the fast transition from pump to turbine mode and vice versa is simulated at the prototype scale with transition time of 32s from pump nominal rotational speed N=-200 rpm to the turbine nominal rotational speed N=+200 rpm, and considering constant guide vane of Y=15° (y=0.375 pu). Figure 3 left presents the time evolution of the dimensionless characteristic quantities of the Prototype pump-turbine resulting from the fast transition and the related transient operating point, see Figure 3 right, to be compared with the simulation and measurement results at the reduced scale of Figure 1 (right). It could be highlighted that despite completely different hydraulic circuits between model and prototype, the qualitative evolution of the dimensionless net head h and discharge q shows remarkable similarities with sudden and sharp net head drop during the pump to turbine transition and a net head rise during the turbine to pump transition. These net head variations result respectively from the discharge acceleration and deceleration. But of course, direct transposition of the transient behavior of the pump-turbine unit from reduced scale model test to prototype is not possible since the related hydraulic systems are not in similitude. Hence transposition can only be investigated by means of numerical simulation. Figure 4 left presents the time evolution of the pump-turbine spiral case and draft tube pressure resulting from the fast transitions which remains within acceptable values. Figure 4 right presents the time evolution of the electromagnetic torque and the related active power. As expected, the transition from pump to turbine does not lead to large values of the electromagnetic quantities as the transition induces a discharge reversal supported by the gravity forces. However, during the transition from turbine to pump, extreme values of electromagnetic torque and active power are reached due to discharge reversal against gravity forces leading to 1.7 times the nominal torque at t=115s and a maximum active power of 240 MW at t=132s representing 114% of the nominal power of the unit. The maximum torque occurs during the zero rotational speed cross over, while the maximum active power occurs at the end of the fast transition.

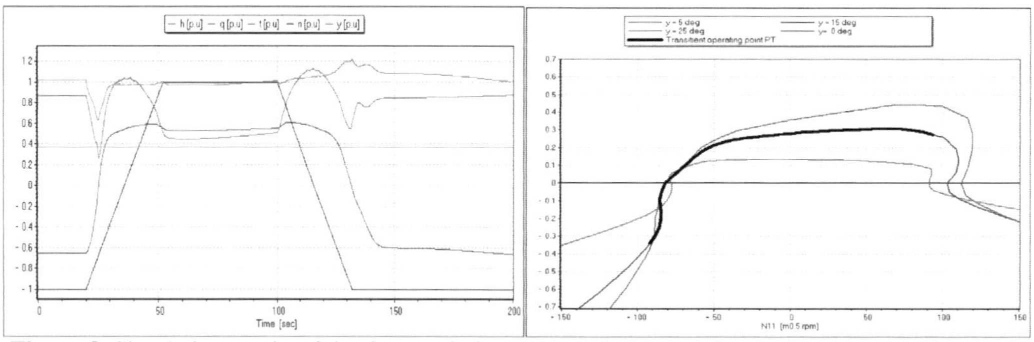

Figure 3. Simulation results of the time evolution of the pump-turbine key quantities (left; h: head, q: discharge, t: torque, n: rotational speed, y: guide vane opening) and related transient operating point (right) in case of fast transition mode from pump to turbine achieved in 32s and vice-versa.

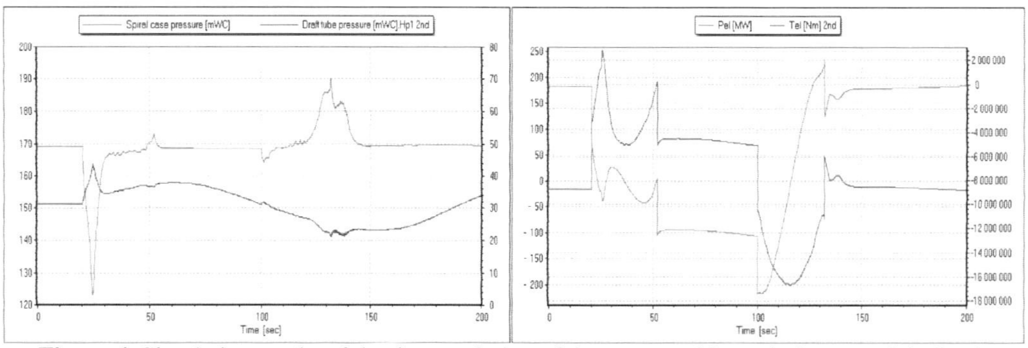

Figure 4. Simulation results of the time evolution of the pump-turbine spiral case and draft tube pressure (left) and active power and air-gap torque (right) in case of fast transition mode from pump to turbine achieved in 32s and vice-versa.

IOP Conf. Series: Journal of Physics: Conf. Series **813** (2017) 012040 doi:10.1088/1742-6596/813/1/012040

Figure 5 presents simulation results of the time evolution of the motor-generator air gap torque obtained for different transition time varying from 16s to 80s considering a specific guide vane closing and reopening sequence defined to mitigate extreme values of electrical quantities, pointing out the potential for fast transition optimization.

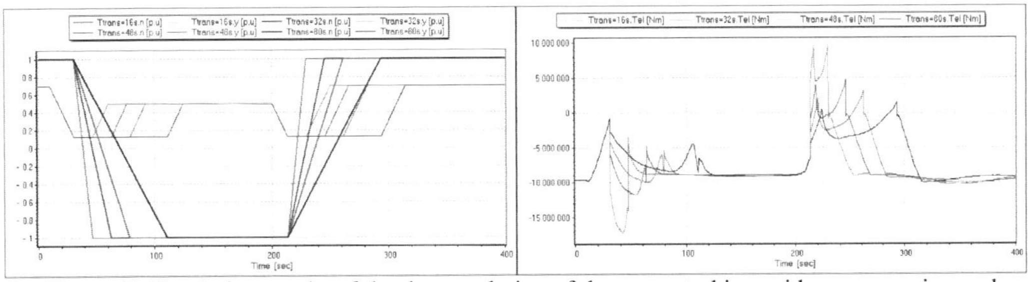

Figure 5. Simulation results of the time evolution of the pump-turbine guide vane opening and rotational speed (left) and resulting air-gap torque (right) in case of fast transition mode from pump to turbine and vice-versa achieved for different transition time.

5. Conclusions

The lean integration of intermittent New Renewable Energies requires additional power control services capabilities to ensure electrical grid stability. In this context, pumped storage power plants operating flexibility combined with recent development of power electronics enable to envisage new ancillary services such as fast active power injection or absorption and inertia emulation which can significantly contribute to secure power network operation. Full Size Frequency Converter synchronous machine unit offer new perspectives such as fast transition from pump to turbine and vice versa enabling to take full advantage of the entire pumped storage capacity from consumption to generation mode. Such opportunity was deeply investigated in the framework of the European Research Project HYPERBOLE in order to assess the feasibility of fast mode transitions. Therefore, reduced scale model test investigations were carried out by means of both experimental and numerical approaches while the transposition to prototype was investigated by means of numerical simulation for an existing power plant whose pump-turbine is quasi homologous to the reduced scale model, demonstrating the feasibility of fast mode transition at the prototype scale.

6. References

[1] Hell J., Egretzberger M., Lechner A., Schürhuber R., Vaillant Y., "Full size converter solutions for pumped storage plants – a promising new technology", Hydro2012, Euskalduna Congress Centre Bilbao, Spain, 29-31 October 2012.

[2] Kuwabara, T., Shibuya, A., Furuta, H., Kita, E., Mitsuhashi, K., "Design and Dynamic Response Characteristics of 400 MW Adjustable Speed Pumped Storage Unit for Ohkawachi Power Station", IEEE Transactions on Energy Conversion, vol. 11, issue 2, pp. 376-384, June 1996.

[3] Nicolet, C., Braun, O., Ruchonnet, N., Beguin, A., Avellan, F., Full Size Frequency Converter for Fast Francis Pump-Turbine Operating Mode Transition, Proceedings of HYDROVISION International 2016, July 26 - 29, 2016, Minneapolis Convention Center, Minneapolis, MN, USA.

[4] Schlunegger, H., Thöni, A., "100MW full-site converter in the Grimsel 2 pumped-storage plant", Innsbruck, Hydro 2013.

Acknowledgments

The research leading to the results published in this paper is part of the HYPERBOLE research project, granted by the European Commission (ERC/FP7- ENERGY-2013-1-Grant 608532) [32]. The authors are grateful to the contribution of the team of the ANDRITZ Hydro Hydraulic Laboratory in Linz Austria. The authors also would like to thank Iberdrola Generación, S.A.U. in Madrid for their collaboration in the modelling of the prototype power plant test case considered in this paper.

Condition monitoring of a prototype turbine. Description of the system and main results.

C Valero[1], E Egusquiza[1], A Presas[1], D Valentin[1], M.Egusquiza[1], M. Bossio[1]
[1]Center of Industrial Diagnosis. Avda. Diagonal 647. 08028 Barcelona. Spain.
E-mail: valero@mf.upc.edu

Abstract. The fast change in new renewable energy is affecting directly the required operating range of hydropower plants. According to the present demand of electricity, it is necessary to generate different levels of power. Because of its ease to regulate and its huge storage capacity of energy, hydropower is the unique energy source that can adapt to the demand.
Today, the required operating range of turbine units is expected to extend from part load to overload. These extreme operations points can cause several pressure pulsations, cavitation and vibrations in different parts of the machine.
To determine the effects on the machine, vibration measurements are necessary in actual machines. Vibrations can be used for machinery protection and to identify problems in the machine (diagnosis). In this paper, some results obtained in a hydropower plant are presented. The variation of global levels and vibratory signatures has been analysed as function as gross head, transducer location and operating points.

1. Introduction
Hydropower is the most flexible source of power generation available and it is capable of responding to demand fluctuations in minutes. Hydroelectric generating units are able to start up quickly and operate efficiently almost instantly, even when used only for one or two hours [1]. As a result of this flexibility, hydropower is an ideal complement to new renewables.
The required operating range of turbine units today has increased dramatically. In this new scenario, more cases of damage have been reported and a more advanced and effective vibration monitoring is necessary [2]
Vibration monitoring is the process of determining the condition of machinery while in operation. The state of a machine is determined by the acquisition and processing of information and data acquired. Depending on the criticality of the machine, condition monitoring can be ongoing or intermittent. The former reacts more quickly to a sudden change and, therefore, better protects machines from sudden faults that cannot be predicted; the latter is cheaper and avoids the permanent installation of sensors, often a difficult task if not considered in the design stage.
In this paper the vibration monitoring in a hydraulic turbine is analysed. The study has been done from data obtained by on-line monitoring system working from January 2016.

2. Monitoring system
The turbine unit studied belongs to hydropower Mica located in British Columbia, Canada. It is a Francis turbine with 500MW of maximum power, rotating at 125rpm. It is a vertical machine with two radial bearings and one axial bearing. The runner has 16 blades and the distributor 20 vanes.

Vibration sensors, pressure transducers and an OneProd MVX monitoring system were installed in the unit.
For vibration monitoring two accelerometers located at 90° were installed in each bearing. Pressure transducers were situated in draft tube and spiral casing. Signals from proximity probes from Bently-Nevada system were also acquired, see Figure 1.

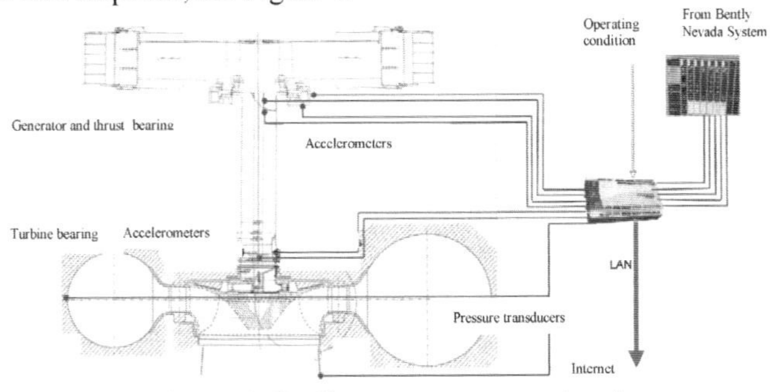

Figure 1. On-line measurements sketch

The system enables databases to store the vibrations from sensors connected to the MVX. Each database values can be stored in the time period desired (every hour, every two hours, once a day…). The databases of the same group can be classified according to the machine operation conditions. The acquisition system will be located inside a cabinet near the turbine floor. From the cabinet an Ethernet cable will connect the acquisition system with a computer.

3. Analysis of turbine operation

Gross head (difference of levels between higher and lower reservoir) changes significantly with season. There is a minimum level in spring (month of May) and a maximum level in autumn (month of October). The turbine studied has limited their operation between 300MW and 430MW see Figure 2.
Several problems appear if the machine works outside of these limits.

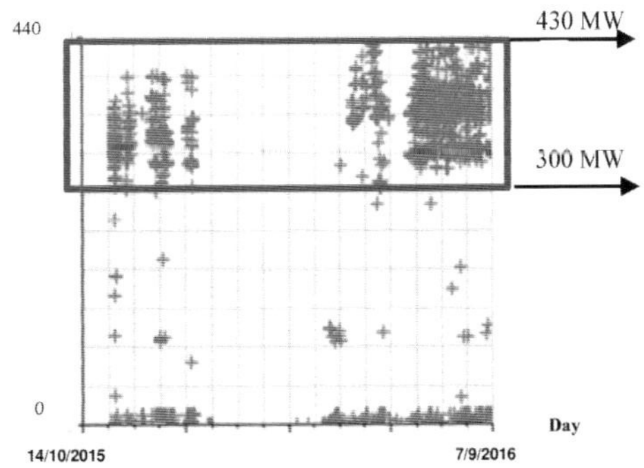

Figure 2. Power of turbine from January 2016

Anyway, vibration levels change with operating conditions in normal operation (between 300MW and 400MW). The variation of global levels and vibratory signatures has been analysed as function of the gross head. The study has been done from data obtained by on-line monitoring system working from January 2016 to September 2016.
In Figure 3 the change in the overall vibration levels (RMS values from 2 to 500Hz) with load and with head in normal operation (300MW to 400MW) has been represented.
It can be seen that vibrations are more important in the turbine bearing and in the draft tube and reaches maximum values at 300 MW. In addition, overall levels increase when head increases for all accelerometers.

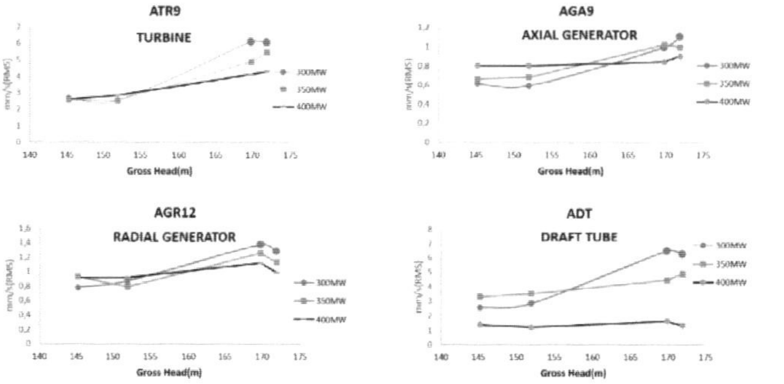

Figure 3. Overall vibration levels (RMS value of mm/s) of function as function of the head: A-TR9 Accelerometer Turbine bearing radial, A-GA9 Accelerometer Generator bearing Axial, A-GR12 Accelerometer Generator bearing Radial, A-DT Accelerometer Draft Tube.

To know the origin for this increase, the vibration signatures (from 0 to 500Hz) has been analyzed for different gross heads and loads for turbine and generator bearing see Figure 4.

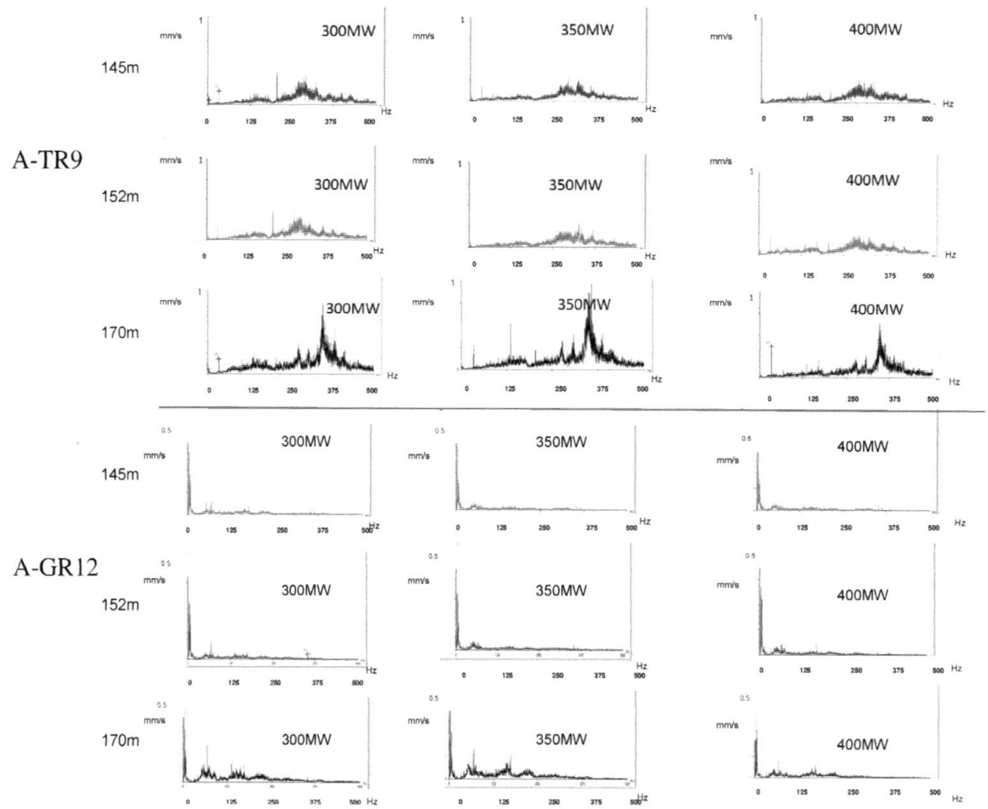

Figure 4. Vibrations signatures (RMS mm/s) for different gross heads and loads for A-TR9 (up) and for A-GR9 (down)

Vibrations can be generated by different type of excitation forces of hydraulic, mechanical and electromagnetic origin. Hydraulic excitations can have periodic characteristics like rotor/stator interaction (RSI) or random characteristics like turbulence and cavitation. The final vibration amplitudes depend on the excitation type and on the response (hydraulic and mechanical) of the system. The main periodic excitation in hydraulic turbines are generated by RSI which is generated by the interference between the rotating blades of the runner and the stationary guide of the distributor. (RSI) [3,4]. Its frequency is:

$$f_b = n \cdot Z_b \cdot f_f \qquad (1)$$

Where f_f is the rotating frequency of the runner, Z_b is the number of blades and n is the order of harmonic. Taking a look at the vibration spectra in the turbine and generator bearings (Figure 4), it can be seen that for a constant load (for example 300MW) vibrations increase when gross head increases especially in the band between 250Hz and 350 Hz. For a constant head, vibration signatures don't change significantly; minimum values of vibrations are observed at 400MW which is near the design point where flow turbulence should be at minimum. Vibrations in generator bearing are lower than in the turbine bearing because flow excitations act basically on the turbine bearing.

In Figure 5 the vibration signal measured in the turbine bearing at 300 MW with 152m of head gross is shown. Some periodic peaks can be identified at 34.21Hz and its harmonics generated by RSI and a broad band vibration almost in all the frequency range indicating that turbulence/cavitation is the predominant excitation. The system response enhances vibrations in some of the frequency bands.

In Figure 6 vibration signals from generator bearing at 300 MW with 152m of head gross are shown. For this position, the main peaks are detected at the rotating frequency of 2.18Hz (f_f) and its harmonics are detected. Also, natural frequencies at around 4,8 and 7.81Hz are a little bit excited. Based in a FEM model, this are rotor natural frequencies.

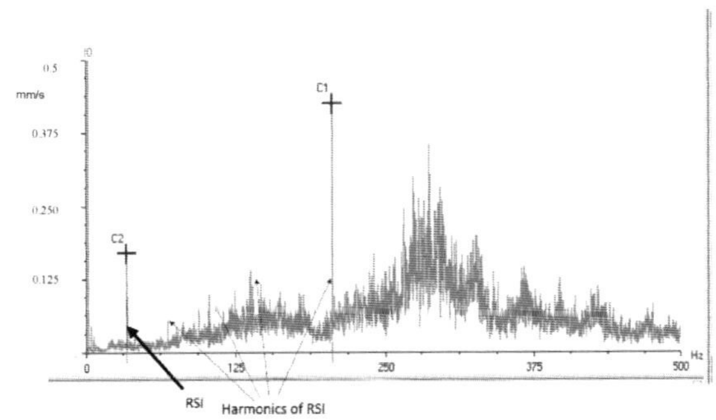

Figure 5. Spectral signatures for A-TR12

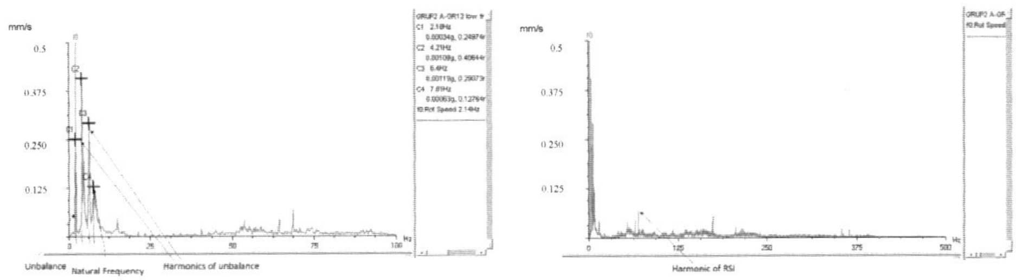

Figure 6. Spectral signatures for A-GR12 bearing from 0 to 100 Hz (left) and from 0 to 500Hz (right)

When the machine operates at very low load (for example at 100MW), vibrations increase in the radial and axial accelerometer located in generator as shown in Figure 7. The natural frequencies are more excited than before because when operating at very low loads the hydraulic excitation increases.

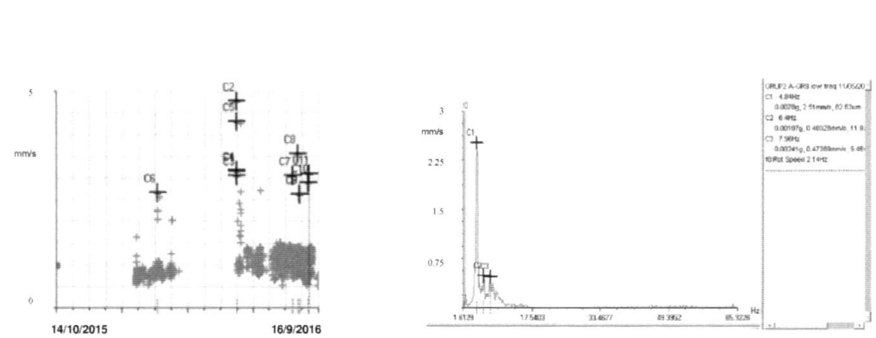

Figure 7. Overall levels RMS (mm/s) (right) and power spectrum (left) for AGR-9

In Figure 8 a waterfall spectra is shown for the machine working from start up to maximum load. When the load reaches 200MW and 250MW a part load instability appears and a power instability is produced. Also, an overload instability can be detected at 470 MW when gross head is maximum.

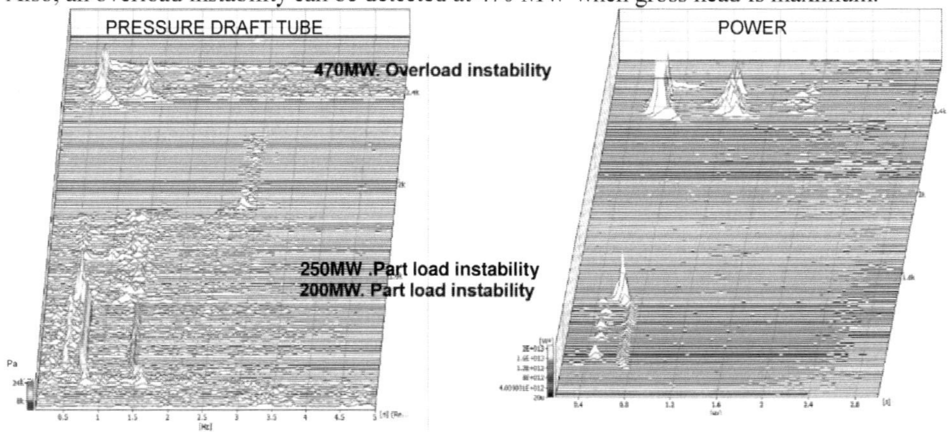

Figure 8. Waterfall spectra working machine from start –up to maximum load

4. Conclusions

This paper presents a preliminary analysis of the measurements obtained from an on-line monitoring system installed in a Francis turbine located in British Columbia (Canada). When the machine is working at normal operating conditions (300-430MW) overall vibration levels increase when head gross increases. Maximum values of vibrations are reached at 300MW with maximum gross head. Most important amplitudes are measured in the turbine bearing and in the draft tube. Predominant excitation is turbulence for all cases. Vibration amplitudes also increase when the machine works at very low loads. At part load (200-250MW) and at overload (470MW) instabilities appear inducing power fluctuations.

Acknowledgements

The research leading to the results published in this paper is part of the HYPERBOLE research project, granted by the European Commission (ERC/FP7- ENERGY-2013-1-Grant 608532).

References

[1] Renewable energy technologies: cost analysis Volume 1: Power Sector Issue 3/5 *Hydropower*

[2] E.Egusquiza et al. 2015 Condition monitoring of pump-turbines. New challenges.*Measurement*, 67 pp 151-163.

[3] G. Jacquet-Richardet et al., 2013 Rotor to stator contacts in turbomachines. Review and application. *Mech. Syst. Signal Process*. 40 (2) pp 401–420.

[4] Tanaka, H., 2011 Vibration Behavior and Dynamic Stress of Runners of Very High Head Reversible Pump-turbines. *Int. Journal of Fluid Machinery and Systems*,. 4(2): pp 289-306.

Dynamics of large-diameter water pipes in hydroelectric power plants

G. Pavić[1], F. Chevillotte[2], J. Heraud[3]

[1]LVA, INSA de Lyon, 69621 Villeurbanne, France
[2]Matelys Research Lab, 69120 Vaulx en Velin, France
[3]MPSH, EDF-DTG, 38040 Grenoble, France

goran.pavic@insa-lyon.fr

Abstract. An outline is made of physical behaviour of water - filled large pipes. The fluid-wall coupling, the key factor governing the pipe dynamics, is discussed in some detail. Different circumferential pipe modes and the associated cut-on frequencies are addressed from a theoretical as well as practical point of view. Major attention is paid to the breathing mode in view of its importance regarding main dynamic phenomena, such as water hammer. Selected measurement results done at EDF are presented to demonstrate how an external, non-intrusive sensor can detect pressure pulsations of the breathing mode in a pressure pipe. Differences in the pressure measurement using intrusive and non-intrusive sensors reveal the full complexity of large-diameter pipe dynamics.

1. Introduction

The effect of fluid-structure coupling (FSC) was known since more than a century ago[1]. Studies on transient pipe dynamics have already started at the turn of 20th century[2]. Various models have been established so far to simulate effects of non-rigid pipe walls. Simplified formulae of sound speed subjected to pipe external mechanical conditions were published in papers[3] and books[4,5]. Further enhancement of pipe modelling was done by allowing the pipe to move via junction coupling which is not included in the classical pipe models[6]. A broad account of pipe dynamics under varying external boundary conditions aimed at diagnostics applications was recently presented in thesis work[7].

Based on thin-wall assumptions a comprehensive FSC model of a free fluid-filled cylindrical shell has been produced by Lin and Morgan[8]. It has been shown that the sound speed is not constant but changes with frequency. Using a frequency-dependent multi-mode model the excitation by a point source located in a free elastic cylindrical shell has been worked out[9]. Similar modelling has been extended to the propagation of pulsation and vibration energy along fluid-filled cylindrical shells[10].

This paper focuses at the dynamics of a large fluid-filled pipe which, free at the outer surface. The ultimate goal is to show how an external strain sensor can be employed to measure pressure pulsations using a pipe model which allows for pressure variation in axial, circumferential and radial directions.

2. Model of a free, thin-walled fluid-filled pipe

Fig. 1 shows the coordinate system used, shell geometry and wall displacements. The coupled fluid-wall motion can be expressed using the equations of motion of the fluid and the wall and the equation of continuity of radial motions at the fluid-wall interface, Eq. (1). The symbols stand for: t – time, p – dynamic pressure, E,ν,ρ_s – Young's modulus, Poisson ratio and mass density of the wall, B,ρ_f – bulk modulus and mass density of the fluid, Ξ – 3×3 differential operator of shell elasticity. In what follows Flügge's operator Ξ will be employed[11].

Assuming harmonic motion with angular frequency $\omega = 2\pi f$ Eq. (1) can be resolved by factorization. This leads to a radial pressure distribution which is governed by Bessel functions $J_n(k_r r)$, $n \in \mathbb{Z}$. In the axial and circumferential sense both the fluid and the wall satisfy $\exp(\pm j k_x x)$ and $\cos(n\theta + \vartheta)$ laws. The symbols k_x and k_r denote the axial and radial components of the wavenumber $k = \omega / c_{f\infty}$, $c_{f\infty}$ being the speed of sound in the unbounded fluid.

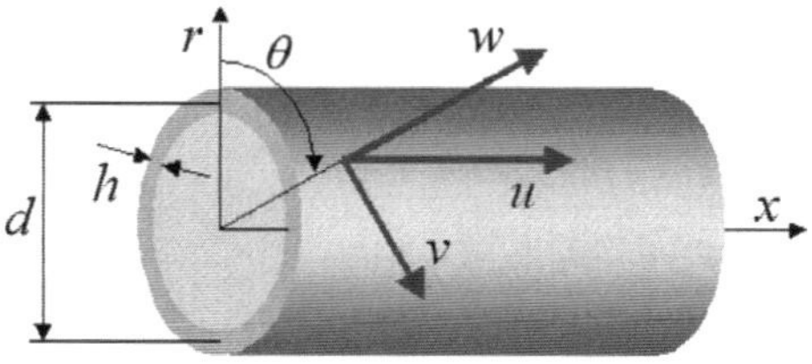

Figure 1. Shell geometry

$$B\Delta p - \rho_f \frac{\partial^2 p}{\partial t^2} = 0, \quad \left(\frac{4Eh}{d^2(1-\nu^2)} \Xi - h\rho_s \frac{\partial^2}{\partial t^2} \right) \begin{bmatrix} u \\ v \\ w \end{bmatrix} = p_{r=d/2} \begin{bmatrix} 0 \\ 0 \\ 1 \end{bmatrix}, \quad \frac{\partial p}{\partial r} + \rho_f \frac{\partial^2 w}{\partial t^2} = 0 \tag{1}$$

To resolve the equations (1) a power series development is used[12]. For any particular solution k_r / k_x the relationship between the radial displacement amplitude of the wall W and the fluid pressure amplitude P can then be obtained in the following way:

$$\frac{P}{W} = \omega^2 \frac{\rho_f d}{2} \frac{J_n(\kappa)}{\kappa J_n'(\kappa)}, \quad \kappa = \frac{k_r d}{2}, \quad J_n'(\kappa) = \frac{\mathrm{d}J_n(\kappa)}{\mathrm{d}\kappa} \tag{2}$$

The $n = 0$ shell order represents the axially symmetric case, called breathing motion. In this case the hoop (circumferential) strain due to breathing is simply $\varepsilon_{\theta\theta} = 2w/d$, which allows direct use of Eq. (2) in providing the strain-pressure relationship. This relationship will be used as a basis of the method of non-intrusive pressure pulsation measurement, further discussed in §4.

3. Wave motion

Any wall deformation of a straight shell can be decomposed into circumferential orders $n = 0, 1\ldots$ Fig. 2. To each n is associated a number (theoretically infinite) of radial pressure distributions indexed by $s = 1,2\ldots.$ as defined in §2. Each pair n,s physically describes one particular wave entirely defined by wavenumber pairs: radial $k_{r,n,s}$ and axial $k_{x,n,s}$. A given wave can be either propagating or evanescent in dependence of whether its axial wavenumber $k_{x,ns}$ is purely real (propagating), or not (evanescent).

At low frequencies only a few of these waves can propagate. These are *pulsation wave* ($n = 0$, $s = 1$), *torsional wave* ($n = 0$, $s = 2$), *extensional wave* ($n = 0$, $s = 3$), and *flexural wave* ($n = 1$, $s = 1$). Evanescent waves may exist at low frequencies too; these waves do not propagate but instead decay exponentially. As the frequency increases, more and more evanescent waves turn into propagating ones. The frequency at which the conversion takes place is called *cut-on frequency*. Table 1 shows the first cut-on frequencies for waves of different orders relative to two water-filled steel pipes of 10.5 mm thickness used at EDF hydroelectric power plants. These frequencies are very low which means that in the frequency range of a few tens of Hz a large number of different waves may propagate.

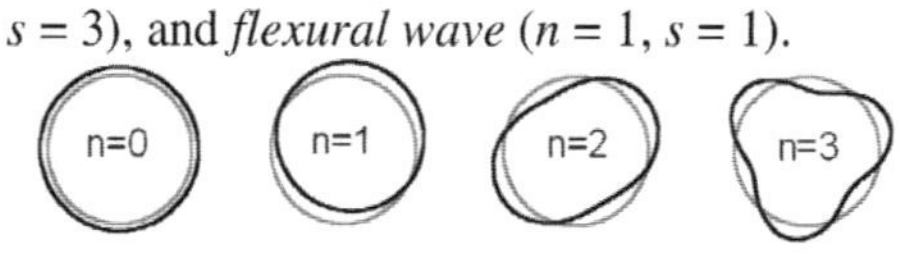

Figure 2. Lowest orders of shell motion.

Table 1: First cut-on frequencies in Hz of water-filled 10.5mm thick steel pipes

order n	2	3	4	5	6	7	8	9	10	11	12
dia 2.4 m	1.81	5.78	12.2	21.3	33.2	48.1	66.0	87.0	111	139	169
dia 5.5m	0.241	0.777	1.66	2.94	4.63	6.78	9.40	12.5	16.1	20.3	25.0

With the exception of torsional wave any given type of wave occurs simultaneously in the fluid and in the wall due to the fluid-wall coupling. Each fluid-wall wave pair has the same axial wavenumber, while the radial pressure distribution in the fluid is governed by the matching radial wavenumber.

The wave defined by $n = 0$, $s = 1$, called the pulsating wave, is of a particular importance: this wave typically generates the largest part of pressure pulsation within the pipe. However other $n = 0$ waves superpose to the pulsation wave, each of these having its own amplitude and its own speed. This is also true for any order other than $n = 0$, so the total wave motion will consist of a superposition of groups of waves of different orders with each group consisting of different wave types: some propagating, the others evanescent. If the flow in the fluid is non-homogeneous or of non-negligible speed the physical picture of wave motion becomes even more complex.

4. The non-intrusive measurement of pressure pulsations

A PVDF wire connected to a charge amplifier has been proposed as a sensor for the detection of pressure pulsations in a pipe[13]. It responds to the hoop surface strain relative to the line along which it is wound to the pipe. Where the breathing mode (n=0) is concerned the radial displacement and the integrated circumferential strain are proportional to each other, while for $n > 0$ modes the integrated circumferential strain is zero. The objective is thus to measure the integrated circumferential strain and to deduce from it the internal pressure originating from the pulsation waves using Eq. (2).

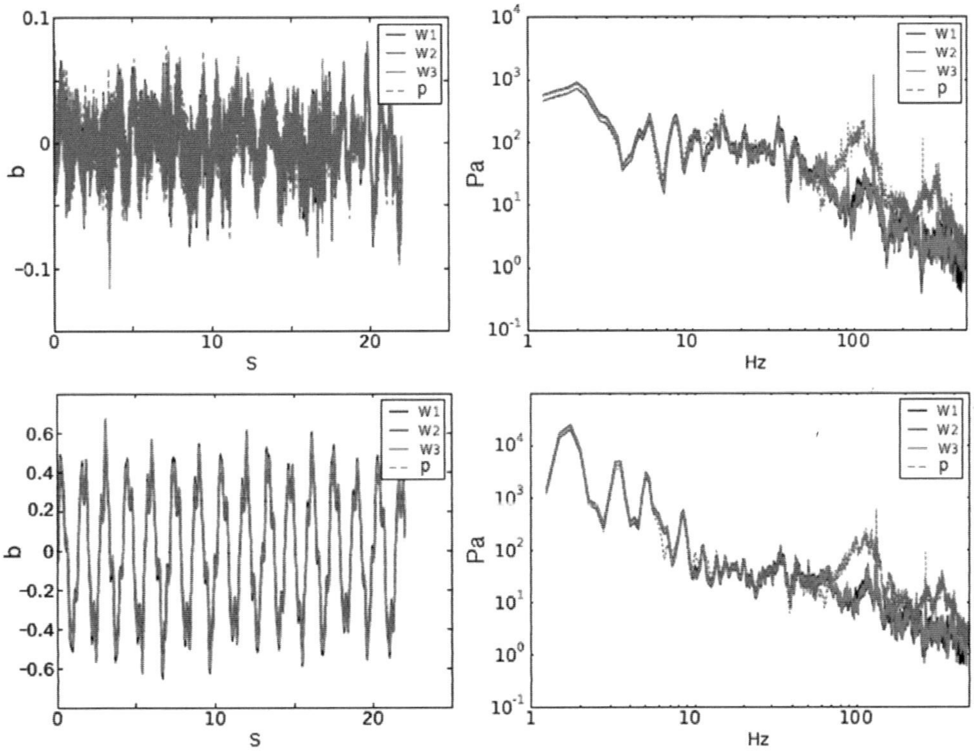

Figure 3. Pressure pulsation in a 2.4m dia 10.5mm thick steel pressure pipe of a small hydroelectric power plant. Top: normal regime; bottom: regime of strong oscillations. Left: time history; right: RMS pressure spectrum. Symbols: w – external wire sensor, p – pressure sensor.

On Fig. 3 are shown pressure pulsations measured in a pressure pipe using three closely spaced PVDF sensors and a classical intrusive piezo sensor placed nearby. The measurements are shown for two regimes of operation in both time and frequency domains. Good overall matching can be seen in both cases. The mismatch in the spectra above 50 Hz is believed to originate from the difference in the character of the measured pressures: the intrusive sensor refers to a single point while the wire outputs the pressure averaged around the circumference. The difference between the two increases with frequency as a consequence of increase in number of propagating higher order waves.

The mounting of the external wire sensor can be simply done using a double-face adhesive tape. In some cases it is sufficient to install the wire directly on the pipe as shown on Fig. 4. The wire is connected to a charge amplifier. The signal from the wire is converted into frequency domain and the relationship (2) is applied to obtain the pressure spectrum. Alternatively, Eq.(2) can be brought in a form of a filter and the conversion of strain to pressure can then be done directly in time domain.

Figure 4. PVDF wire sensor installed on the pressure tube of a hydropower plant.

5. The wave speed in a pipe

The speed of sound in a free (unbound) fluid $c_{f\infty}$ is given by Newton-Laplace equation:

$$c_{f\infty} = \sqrt{B/\rho_f} \tag{3}$$

The elasticity of pipe wall makes the apparent fluid bulk modulus, and thus the sound speed, drop. In the case of heavy liquids, i.e. water, and small wall thickness the decrease of sound speed can be rather substantial. The term "sound speed" is usually used for the speed of pulsation waves. Waves other than pulsation waves have own speed of propagation which varies with frequency in a rather complex fashion. For any type of wave the speed is obtained from the wavenumber by $c = \omega/k_x$,

The sound speed in function of frequency was computed for two EDF water-filled steel pipes using the solution for wavenumber k_x based on Eq. 1, Fig. 5 left. The nominal speed $c_{f\infty}$was taken to be 1450 m/s. The drop of speed is seen to be significant and increases with frequency. Fig. 5, right, shows a particular cost function G^{14} used to obtain the actual speed in the 2.4m dia pipe from pulsation measurements. The function G, calculated for a range of assumed speeds from the signals measured by 3 PVDF sensors, indicates the true speed by the position of its minimum.

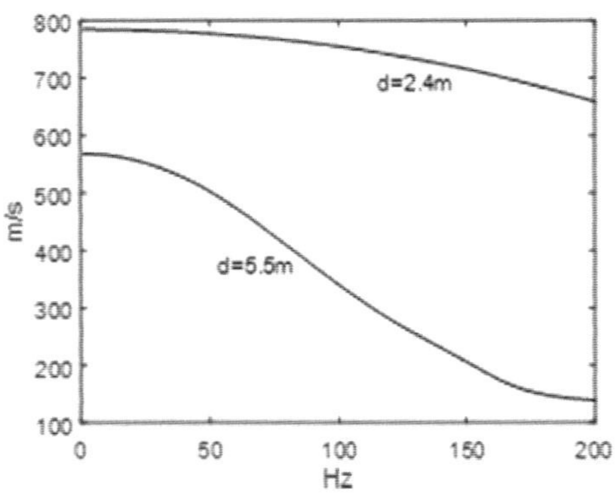

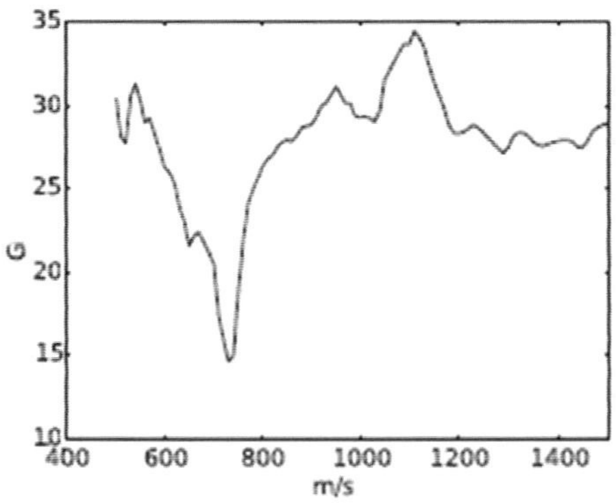

Figure 5. Sound speed of a water-filled 10.5mm steel pipe. Left: computed; right: measured on a 2.4m dia pipe using the cost function approach.

6. Vibration of pipe wall

In order to monitor wall vibration six accelerometers were mounted around the circumference of the dia 2.4m pipe. The instantaneous radial displacement of the wall is shown in Fig. 6 at some selected frequency bands. The plots clearly show the complexity of pipe motion which is a consequence of the superposition of a multitude of waves having different circumferential orders. In view of Eq. 1 such complex vibration patterns indicate a complex distribution of internal pressure pulsations. It can be thus concluded that point measurement of pressure pulsations using classical intrusive sensors may not be representative where global pulsations averaged along the circumference are concerned.

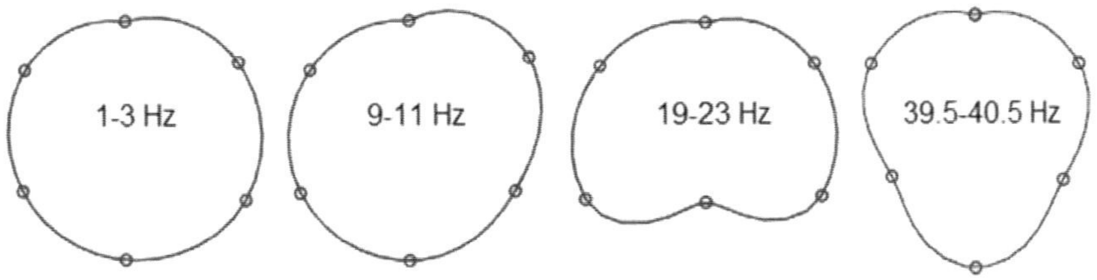

Figure 6. Instantaneous displacement of the 2.4m pressure pipe.

7. Conclusions

It has been demonstrated that an external wire strain sensor wound around a large pipe can be used for the measurement of pressure pulsations within the pipe. This requires use of a mathematical model of fluid-wall coupling. The measurements done on a large pipe using the wire sensor and a classical intrusive sensor done in situ have shown satisfactory global matching. An array of 3 wire sensors was shown to be capable of measuring the true speed of sound in the pipe.

References

[1] Korteweg D.J. On the velocity of propagating sound in elastic tubes. Physics Annals, 174, 525-542, 1878 (in German).
[2] Zhoukovsky N.E. On hydraulic shock in water piping. Mem. Acad. Sci., St Petersburg, series. 8. Physics-mathém. Class, 9 (5), 1900 (in German).
[3] Halliwell A.R. Velocity of a water hammer wave in an elastic pipe. Proc. Amer. Soc. Civil Engrs - J. Hydraulics Division, 89 (4), 1-21, 1963.
[4] Parmakian, J. Waterhammer analysis. Dover Publications, New York, 1963.
[5] Jaeger C. Fluid Transients in Hydro-Electric Engineering Practice. Blackie and Son, Glasgow, 1977.
[6] A.S. Tijsseling, C.S.W. Lavooij, Waterhammer with fluid–structure interaction, Appl. Sci. Res. 47, 273–285, 1990.
[7] F. Hachem Monitoring of steel-lined pressure shafts considering water-hammer wave signals and fluid-structure interaction, PhD thesis N° 5171, Ecole polytechnique fédérale de Lausanne, Octobre 2011.
[8] Lin T.C., Morgan G.W. Wave propagation through fluid contained in a cylindrical elastic shell. Journal of the Acoustical Society of America 28 (6), 1165-1176, 1956.
[9] Merkulov V.N., Prikhodko V.Yu., Tyutekin V.V. Excitation and propagation of normal modes in a thin cylindrical elastic shell filled with fluid. Akusticheskii Zhurnal, 24, 723-730, 1978.
[10] Fuller C.R., Fahy F.J. Characteristics of wave propagation and energy distributions in cylindrical elastic shells filled with fluid. J. Sound Vib. 81 (4), 501-518, 1982.
[11] Flügge W. Stresses in shells, 2nd ed. Springer, New York, 1973.
[12] Pavić G. Vibroacoustical energy flow in elastic circular cylindrical shells. J. Sound Vib. 142 (2), 293-310, 1990.
[13] Pinnington R.J., Briscoe A. Externally applied sensor for axisymmetric waves in a fluid-filled pipe. J. Sound Vib. 173 (4), 503-516, 1994.
[14] Pavić G. Acoustical analysis of pipes with flow using invariant field functions. J. Sound Vib.263 (1), 153-174, 2003.

IOP Conf. Series: Journal of Physics: Conf. Series **813** (2017) 012043 doi:10.1088/1742-6596/813/1/012043

Cavitating vortices in the guide vanes region related to the pump-turbine pumping mode rotating stall

U Ješe[1], A Skotak[2] and J Mikulašek[2]

[1]Litostroj Power d.o.o., Ljubljana, Slovenia
[2]ČKD Blansko Engineering, a.s., Blansko, Czech Republic

uros.jese@litostrojpower.eu

Abstract. Reversible pump-turbines used in Pumped Storage Power Plants are among the most cost-efficient solutions for storing and recovering large amount of energy in short time. Presented paper is focused on the pump-turbine pumping mode part-load instabilities, among them the rotating stall and the cavitating vortex in the distributor region. Rotating stall can be described as a periodic occurrence and decay of the recirculation zones in the distributor with its own rotational characteristics frequency. Unstable behaviour can result in high radial forces, high pressure fluctuations and local velocity fluctuations that can in some cases lead into the occurrence of the cavitating vortex in the distributor region, even though the distributor is located in the high pressure region. Computationally demanding calculations have been performed using commercial CFD code. Analysed results have been compared to the experimental data obtained in the ČKD Blansko Engineering hydraulic laboratory.

1. Introduction

To maintain the stability of the electrical grid, the pump-turbines need to switch between the generating and the pumping mode several times per day what expose them to extended operating under off-design conditions. Main instabilities in the part load pumping mode operating are the rotating stall in the distributor and the appearance of the cavitation in different pump-turbine regions. Rotating stall has been initially investigated in the axial and the centrifugal compressors [1, 2], much later also for the case of the pump-turbines [3, 4, 5]. Sometimes, the rotating stall is related to the positive slope of the performance curve also called hump zone [6] which is unstable and potentially dangerous operating region. If presented, it can lead into uncontrollable changing of the discharge through the machine and consequently strong vibrations. Operating under described conditions should be completely avoided. However, rotating stall can be present even if the slope on the performance curve is negative.

The intensity of the rotating stall in pump-turbines can vary. As shown several times experimentally and numerically [3, 6, 7] changing discharge and guide vane opening angle can lead into different number of the stalled cells and different rotating stall frequency. Various shapes of rotating stall influence pressure fluctuations, radial forces acting on the impeller as well as guide vanes vibrations and can have negative effects on the machine. Sometimes, when the rotating stall is very intensive, appearance of the cavitating vortex is possible in the distributor region as it was mentioned by [3]. However, the phenomenon have not been so far phenomenologically described in the literature.

2. Experimental data

Experimental measurements took place in ČKD Blansko Engineering hydraulic laboratory [7]. Investigations have been performed on the pump-turbine model. Additional to the standard performance measurement instrumentations, 8 pressure sensors have been distributed around vaneless space between the impeller and the guide vanes. One constant guide vane opening a_0 = 20 mm have been analysed. Rotational speed of the machine have been set to n = 1400 min^{-1}. Guide vane channel and vaneless space have been visually observed during the measurements.

The goal of the experimental setup was to measure low frequency pressure pulsations in pumping and generating mode. This paper will focus on the part load pumping mode only, where the phenomenon called rotating stall occurs together with related flow instabilities. Measurements have been done for whole part load regime, however, for the analysis, operating points at best efficiency (BEP) and at Q = 0,65 Q_{BEP} have been chosen.

Figure 1 presents comparison of the pulsations for the operating point at BEP and investigated point at Q = 0.65Q_{BEP}. It can be clearly observed that at the BEP, the flow is stable, therefore without pressure pulsations and instabilities. On the contrary, one can see periodic pressure pulsations at Q = 0.65 Q_{BEP}. Governing frequency of the low frequency pressure pulsations is f = 0.59 Hz, which corresponds to around 2.5% of the pump-turbine rotation frequency. Pressure measurements at mentioned operating point confirm the presence of the rotating stall.

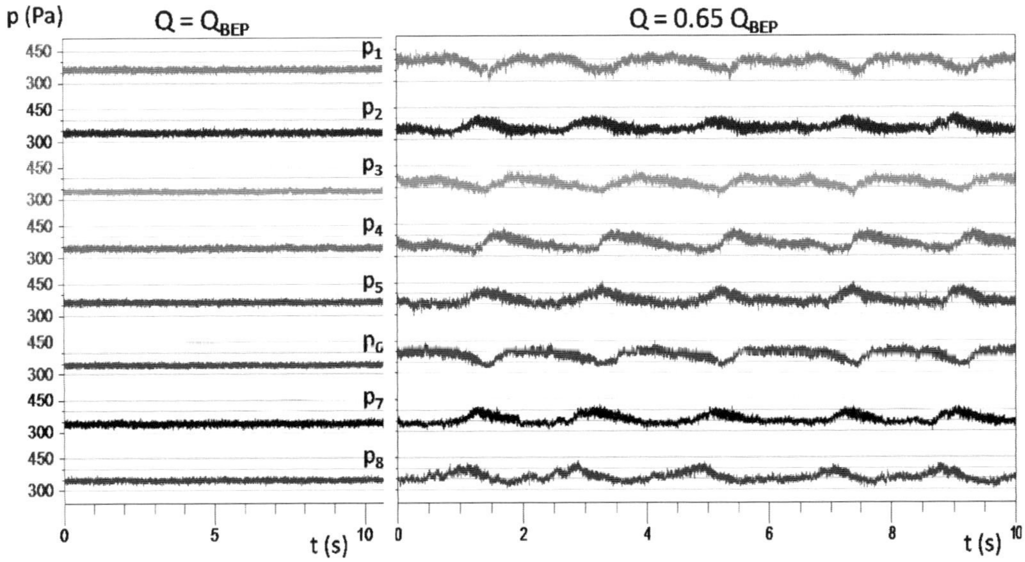

Figure 1. Pressure fluctuations at Q = Q_{BEP} (left) and Q = 0.65 Q_{BEP} (right)

Occasionally, during the pressure measurements at Q = 0.65 Q_{BEP}, the cavitating vortex have been observed in the distributor between the guide vanes, as seen on figure 2. Sometimes, there was one vortex, attached to the suction side of the guide vane (figure 2, left). At some other instants, the phenomenon have been observed as several separate, smaller cavitating vortices, as seen on figure 2, right. In both cases, the vortices occur only for a short time. It should be pointed out that cavitation in the distributor region is highly unusual due to very high pressure in the surrounding. Analysis of the cavitating vortex is the main subject of the presented paper and have been additionally analysed numerically.

Figure 2. Cavitating vortices in the distributor

3. Numerical setup

For the flow analysis, the numerical flow simulation (Computational Fluid Dynamics) software is nowadays the most common tool. It uses a set of Navier-Stokes equations to compute transport of mass and momentum in all parts of the computational domain. Commercial software has been used for the simulation. Transient simulation were performed using URANS equations and turbulence model based on k-ε model. Chosen time step corresponds to 2° of the impeller revolution.

Meshing of the domains has been done by commercial software, using structured and unstructured mesh. Total mesh contains of around 10 million cells, special attention has been put into meshing the distributor region, since this would be the place where the rotating stall occurs. Rotational speed of the impeller has been set to $n = 1400\ min^{-1}$, guide vane opening has been set to $a_0 = 20$ mm. Around 20 revolutions of the impeller have been simulated, each fifth time step has been saved for the analysis. Steady results have been used as an initial solution for transient simulations.

4. Numerical results

Unsteady CFD analysis have been focused on the rotating stall parameters and related phenomena. Operating point at $Q = 0.65\ Q_{BEP}$ have been chosen for the comparison to the experiment. Three regions with high velocity have been found (figure 3) what corresponds to the experimental findings. Numerical rotating stall frequencies have been estimated to 0.5 Hz. Since the frequencies of the rotating stall are very low, more impeller revolutions should be simulated for more accurate frequency prediction. However, we can say that the phenomenon has been accurately described by using CFD and simple k-ε based turbulence model.

Cavitating vortex in the distributor channel have been occasionally observed during operating under rotating stall. The vortex is attached to the guide vanes and reaches the next upstream channel, as seen on figure 4. Constant Q-criterion has been used for the vortices representation. Estimated time of cavitating vortex appearance is around 0.011s what corresponds to around one quarter of the impeller revolution. The vortex appears and disappears several times on different positions around the distributor during the impeller revolution. The appearance time is very short, which means that the phenomenon is visually demanding to observe. The vortex appears when the discharge through the guide vane channel is partially blocked. Firstly, the blockage appears at the hub side and later also at the shroud side. The moment when the flow is still strong at the shroud side and blocked at the hub side is favourable for the occurrence of the cavitating vortex in the guide vane channel. Comparing experimental and numerical cavitating vortex (figure 2, figure 4) show very good agreement.

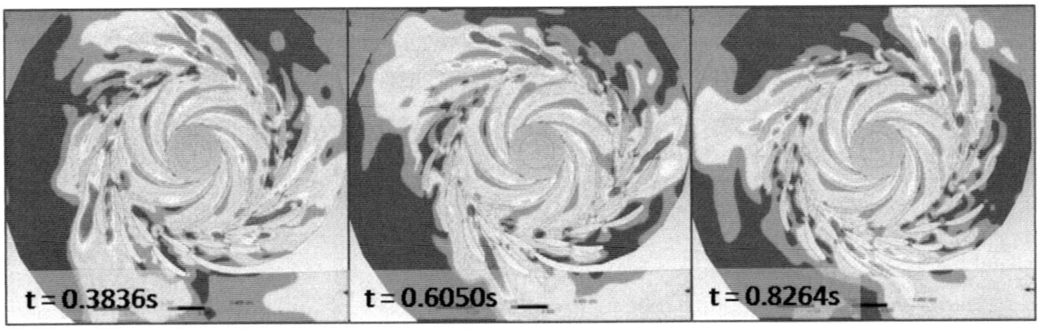

Figure 3. Meridional velocity time sequence on the mid-span

5. Conclusions

Rotating stall conditions have been simulated using CFD at the operating point Q = 0.65 Q_{BEP}. Results have been compared to the experiment. In both cases the same number of rotating stall cells (three) have been found. Similar rotating stall frequencies have been found experimentally and numerically. Cavitating vortices in the distributor region that have been observed during the experiment, have been successfully simulated by the CFD. The analysis explains the governing mechanisms of the cavitating vortex. Methodology can be used for pump-turbine impeller and guide vanes geometry optimisation. The findings will be used for a development of a new modern pump-turbine designed for wide range of operation.

6. References

[1] Frigne P and Van den Braembussche R 1984 Distinction between different types of impeller and diffuser rotating stall in a centrifugal compressor with vaneless diffuser *Trans of ASME Journal of Engineering for Gas Turbines and Power* 106(2):468-474

[2] Haupt U, Seidel U, Abdel-Hamid A and Rautenberg M 1988 Unsteady flow in a centrifugal compressor with different types of vanes diffuser *Trans of ASME Journal of Turbomachinery*, 110(3):293-302

[3] Braun O 2009 Part load flow in radial centrifugal pumps *PhD Thesis* EPFL, Lausanne

[4] Pacot O, Kato C, Guo Y, Yamade Y and Avellan F 2016 Large eddy simulation of the rotating stall in a pump-turbine operated in pumping mode at a part-load condition *Journal of Fluids Engineering*, ASME, doi: 10.1115/1.4033423

[5] Ješe U 2015 Numerical study of pump-turbine instabilities: Pumping mode off-design conditions *PhD Thesis* Univ. Grenoble Alpes, Grenoble

[6] Ješe U and Fortes Patella R 2016 Unsteady numerical analysis of the rotating stall in pump-turbine geometry *Proc. 28th IAHR symposium on Hydraulic Machinery and Systems*, Grenoble

[7] Veselý J, Půlpitel L and Troubil P 2005 Model research of rotating stall on pump-turbines *Proc. HYDRO 2005*, Villach

Figure 4. Cavitating vortex development - CFD analysis

IOP Conf. Series: Journal of Physics: Conf. Series **813** (2017) 012044 doi:10.1088/1742-6596/813/1/012044

Experimental analysis of the onset and development of cavitation in a centrifugal pump

G Mousmoulis[1, 2], N Karlsen-Davies[2], G Aggidis[2], J Anagnostopoulos[1] and D Papantonis[1]

[1]Laboratory of Hydraulic Turbomachines, School of Mechanical Engineering, National Technical University of Athens, Heroon Polytechniou 9, Zografou, Athens, 15780, Greece

[2]Lancaster University Renewable Energy Group, Department of Engineering, Lancaster University Bailrigg, Lancaster, LA1 4YW, United Kingdom

mousmoulisgeorge@gmail.com

Abstract. Centrifugal pumps are considered as machines of high importance in a wide range of industries, hence several strategies and methods have been developed in order to ensure their optimum performance under different operating conditions. An important mechanism that can affect pump's steady and dynamic operation is cavitation, which appears in the low static pressure zone at the inlet of the impeller. Several researchers have studied experimentally the physics of the phenomenon in order to develop different methodologies for its safe detection. The aim of this paper is to study the onset and evolution of cavitation in a centrifugal pump with unshrouded impeller by performing of flow visualization and acoustic emission measurements. The results showed that the acoustic emission technique is able to detect the onset of cavitation, while the leakage backflow through the clearance affects the development of cavitation between the impeller blades.

1. Introduction

Cavitation is a hydrodynamic phenomenon that appears in hydraulic machinery, where water may be vaporized due to local pressure drop in the flow field. In centrifugal pumps water flows from areas of low pressure to areas of high pressure, so bubbles could be possibly created in the inlet of the machine and travel until a point where pressure increases again, where they are reliquified. Reliquification is a very violent procedure, where during the abrupt elimination of vapor volume and bubble implosion, shock waves are generated that could overstress solid boundaries and create unwanted noise and vibration [1]. For this reason engineers investigated different methodologies in order to be able to detect cavitation inception and its development. Mc Nulty [2] used piezoelectric sensors with resonant frequency of 130 kHz and proved that the energy of discrete high frequencies are able to monitor inception of two phase flow inside the machine. In addition, energy of noise was reduced as NPSH decreased mainly because larger cavities damped noise emitted until a point where flow became very unstable and increased again. Similar trend in their results obtained also Alfayez and Mba [3], by measuring with acoustic emission (AE) signals in a large scale pump. According to them, noise increase in low NPSH values was the result of backflow cavitation phenomenon where flow recirculation chocks pump's flowrate. Neil et al. [4] used high frequency acoustic emission (AE) sensors for the detection of incipient cavitation. Their results showed that it is possible to quantify bubbles onset and development with the use of frequency spectrum. In addition, larger bubble areas decrease the AE root mean square (r.m.s.) value, however further NPSH reduction didn't increase the noise. Chudina [5, 6] proposed the use of microphones in the audible frequency range for the monitoring of cavitation. Despite the disadvantages introduced from the low frequency mechanical noise, they showed that there is a discrete frequency in the low frequency range related with the blade passing frequency that it is not affected from surrounding noise and was increased only when bubbles existed. Cernetic et al. [7] measured mechanical vibrations in two pumps; one with a closed impeller and a second with a semi open impeller. In the first, cavitation could be detected in the high frequency range between 1 and 10 kHz where the amplitude of this frequency increased in cavitating conditions. However in the second pump the frequency tone discussed by Chudina [5, 6] increased intensively with cavitation development and proved how geometrical characteristics of the impeller could affect noise emission. Finally, Zhang [8] et al. measured vibration response in different locations and in different frequency ranges. Similarly to Mc Nulty, showed that the onset of bubbles is detected clearer in high frequencies as well as vibration energy decreased with cavitation development until the lowest NPSH value, where noise increased again only for small flow rates. Summarizing the above works, important progress can be observed in the field of cavitation monitoring techniques. From the methods developed so far it is possible to detect the onset of cavitation with safety and predict its development depending on the transducers and frequency ranges used. However, different impellers or flowrates gave differences in the noise measurements irrespective of the methodology used. This depicts the need of deeper understanding of cavitation development and its behavior in different flow or geometrical conditions. In this paper, acoustic emission r.m.s. is measured in different flowrates in a mini centrifugal pump the casing of which is made by plexi-glass. By this way it is possible to visualize the flow with the use of high speed stroboscope and study the onset and development of cavitation as a function of noise characteristics and the NPSH value.

2. Experimental Set Up

The model of centrifugal pump that is used in this study is presented in figure 1. It is a low specific speed pump of 0.58 kW power at 12 m head with 7.5 m^3/sec flow rate at its best efficiency point (BEP). Water is pumped from the tank and passes from the suction valve before it enters the semi open impeller that consists of 12 straight blades. Before impeller inlet, a 90 degrees bending pipe is installed that eradicates the axial symmetry of the flow and creates an extra pressure drop. This resulted in a non-axisymmetric pressure flow field in the inlet of impeller and as a consequence cavitation inception appeared always on the left side of the impeller eye. This is the reason that all the photos taken in this experiment are from this side of the machine, as it presented in figure 1.

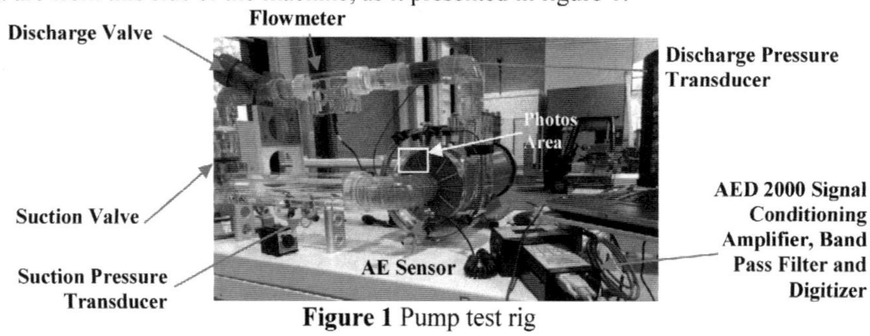

Figure 1 Pump test rig

After the pump, water passes from the flow rotometer and from the discharge valve, which controls the flowrate and returns to the tank. For cavitation experiment, NPSH value is decreasing in order to study its inception and development. This is done by throttling the suction valve in order to increase the losses and reduce the total suction pressure, while properly regulating the discharge valve to retain the flow rate. However, this resulted in cavitation inception far before the pump's inlet, in the suction valve, for the lowest NPSH value irrespective of the flowrate. Four different flowrates have been tested for an impeller with straight blades and rotational speed has been chosen at 1800 rpm. According to McNulty [2], Neil [4] and Zhang [8], cavitation affected more intensively high frequency ranges no matter the physical quantity measured. As a consequence, acoustic emission sensor is used in order to validate the ability of detecting bubbles inception as well as to associate noise emitted with cavitation behaviour by the study of the photos of the flow field obtained. The sensor is connected with AED 2000 in order to filter, amplify and digitize the signal. After signal digitization, conditioner transfers in laptop, every second, the average of 90 volt r.m.s. values. The AE r.m.s. results presented here deal with the average value of AE obtained during two minutes of measurement at each operating point. For flow visualisations a stroboscope has been used and photos were taken with a DSLR camera. Tests were performed three times in order to ensure repeatability both at the measurements of flow quantities but also for cavitation inception and development. Repeatability error for flowrate and total head calculations was 0.2% and 0.5% respectively. Quantitative repeatability is not expected for AE measurements as well as in visual observations. However, the trends of noise as a function of the NPSH were repeatable as well as the photos of cavitation inside the impeller.

3. Experimental Results

Four different flowrates are tested in order to derive the NPSH - H_{TOT} and NPSH - AE r.m.s. graphs that are presented in figure 2. As the flow rate increases, it is more difficult to approach low NPSH values mainly because of the use of the valve as the mean of total suction pressure drop. As a result, total head drop is much clearer in lower flow rates than in higher flow rates such as 159 lt/min. Another limitation introduced by the suction valve is the inability to obtain measurements and photos between the two lowest NPSH values. At the last point pressure dropped heavily in the suction of the pump and cavities were created in the valve and travelled inside the impeller. Measurements at this point present high interest because differences are obtained in noise trends between already published studies [2, 3, 4 &7]. However, flow visualization approach provides interesting conclusions, which are discussed afterwards.

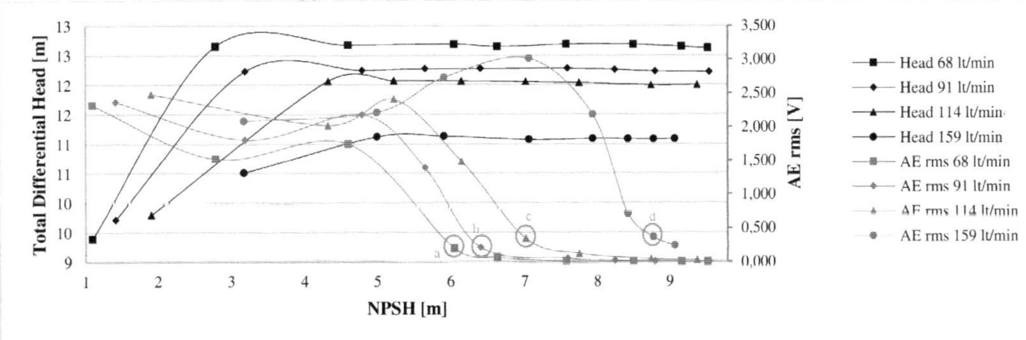

Figure 2 Total head and AE curves in red impeller for four flow rates

The AE analysis confirmed that the increase of flowrate increases the NPSH value where cavitation started due to the higher velocity field inside the impeller. Also it was validated that the increase of noise coincides with cavitation inception by observing the flow with the use of a stroboscope.

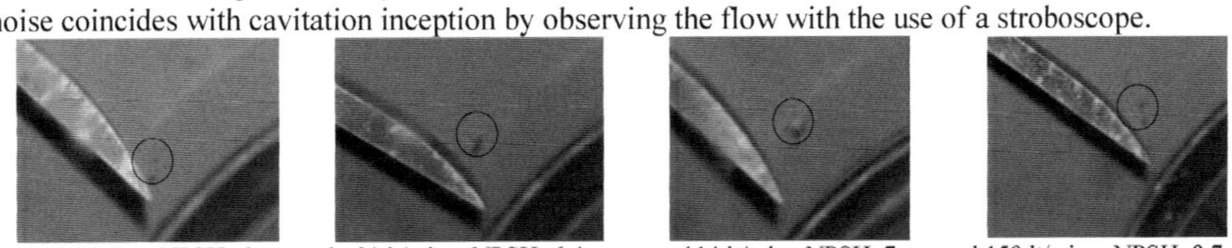

a. 68 lt/min – NPSH: 6 m b. 91 lt/min – NPSH: 6.4 m c. 114 lt/min –NPSH: 7 m d.159 lt/min – NPSH: 8.7 m

Figure 3 Cavitation inception that corresponds to visual inception point in different flow rates

In figure 3 the appearance of the first bubble in the flow field is presented. As it is expected, vapour is formed in the suction side of the leading edge (LE) of the blade, irrespective of the flowrate studied. It is very interesting that AE r.m.s. value increases just after the appearance of the first small cavity, as it can be seen in figure 2, for points a-d, for all the flowrates. This is the result of correct use of band pass filtering that allows only frequencies in the range of 100 to 150 kHz to pass and removes all the surrounding low frequency environmental noise of the motor and the pump. A further decrease of NPSH value results in an increase of the number and magnitude of cavities for all flow rates, as can be observed in figure 4. At this NPSH value, AE r.m.s. reaches a local maximum value, similarly to the majority of the results presented in other works [2-4]. Until this point the shock pressure waves generated from bubble implosion travel without resistance through the liquid flow and they are transferred to pump casing and to the sensor.

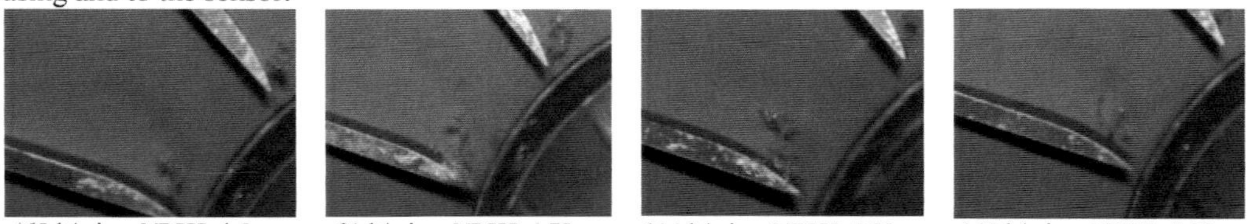

68 lt/min – NPSH: 4.6 m 91 lt/min – NPSH: 4.79 m 114 lt/min – NPSH: 5.23 m 159 lt/min – NPSH: 7.06 m

Figure 4 Photos that correspond to local maximum value of AE r.m.s. in different flow rates

However, further decrease of total suction pressure increases the two phase flow area thus resulting in some decrease of noise emission. Gulich [9] has noticed that NPSH decrease creates large cavities that absorb the pressure waves created. Figure 5 presents the flow field inside the impeller when noise is decreasing. For all flowrates, cavitation area has increased and close to the blades inlet two phase flow extends in a large part of the blade passage. In addition the sharp shape of LE blade accelerates locally the flow and deteriorates vapour creation.

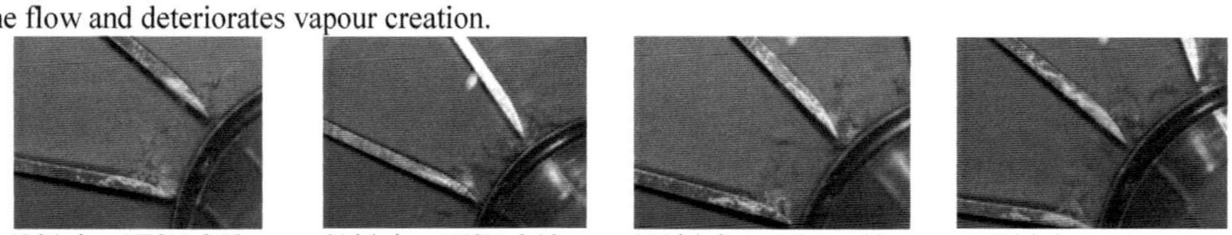

68 lt/min – NPSH: 2.79 m 91 lt/min – NPSH: 3.19 m 114 lt/min – NPSH: 4.32 m 159 lt/min – NPSH: 5 m

Figure 5 Photos from impeller that correspond to AE r.m.s. decrease in different flowrates

Finally, figure 6 shows the photos taken for the lowest NPSH value of point 4 of figure 2, where the discharge valve was almost fully opened and the suction valve was throttled. For the majority of the flowrates this was the operating point with the most intensive noise and vibration. At this point the flow becomes very unstable and exhibits intensive recirculation and separation regions. In addition, cavitation due to leakage flow recirculation can be observed at the yellow brackets marked in Fig. 6, where bubbles exist in the space between the impeller and the casing. Alfayez [3] and Brennen [1] discussed the same phenomenon and noted it as backflow cavitation. Also, the thickness of the clearance of this impeller ranges from 1 mm in the trailing edge (TE) to 1.3 mm in the LE, introducing high volumetric losses and reducing pump efficiency. The corresponding blade height ranges from 3.2 to 7 mm, respectively. As a consequence, part of the flow that already consists of bubbles enters the clearances, where it is further accelerated, thus increasing cavitation. This flow is driven in the low pressure area of the impeller and it is re-joined with the two phase flow already established in the suction of the pump. At this point, cavitation dominates at the impeller inlet, vapour blocks the flow, pump's head drops, and noise increases heavily. However, this phenomenon becomes less extended as the flow rate takes higher values, as can be observed in figure 6 (green boundary lines), as also in the AE r.m.s. results of figure 2. This happens because, as the pump flow rate is growing, the percentage fraction of the flow that is passing through the clearance becomes lower. As a result, recirculating flow affects more the flow characteristics in low flowrates.

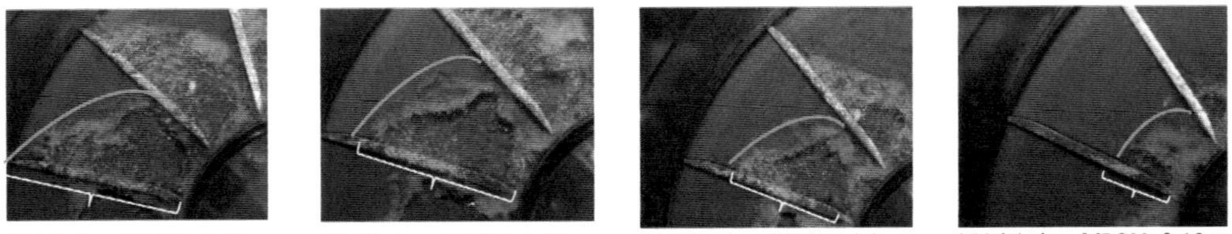

68 lt/min – NPSH: 1.09 m 91 lt/min – NPSH: 1.42 m 114 lt/min – NPSH: 1.9 m 159 lt/min – NPSH: 3.18 m

Figure 6 Photos from impeller that correspond to the lowest NPSH point for different flowrates

4. Conclusions

This study investigates the inception and development of cavitation in a centrifugal pump with unshrouded impeller and straight blades. In order to detect and study the cavitation behaviour, one AE sensor is placed close to the suction side of the pump. Furthermore, the casing of the machine is made from plexi-glass in order to be able to visualize the flow between the blades and validate noise measurements. The obtained total head curves confirmed that intense cavitation in low NPSH values can block the flow rate of the machine and reduce its ability to maintain water head. In addition, the AE signal is found to increase exactly at the point of the onset of cavitation, and this was validated by flow observations. Also, the emitted noise r.m.s. decreases when cavitation is increased, irrespectively of the flow rate. Finally, interesting observations have been made at the lowest NPSH value, where backflow cavitation dominates the flow and creates a very turbulent and noisy environment that was depicted both in photos taken from the impeller, as well as in the AE results. The obtained results aim to contribute to the deeper understanding of cavitation inception and evolution in such pump configurations. In addition, the flow visualisation validates the effective and reliable use of AE measurements as cavitation inception detection methodology. However, more research could be done in order to study the effect on cavitation of other critical geometric characteristics, such as the blade shape and inlet angle, as also the effect of different clearances.

References

[1] Brennen, C. E., 2011, Hydrodynamic of pumps, Cambridge University Press.

[2] McNulty, P.J. and Pearsall I.S., 1982, "Cavitation Inception in Pumps," Journal of Fluids Engineering, Vol. 104(1), pp. 99-104

[3] Alfayez, L., & Mba, D., 2005 "Detection of incipient cavitation and determination of the best efficiency point for centrifugal pumps using acoustic emission," Proceedings of the Institution of Mechanical Engineers, Part E: Journal of process mechanical engineering, 219(4), 327-344.

[4] Neil, G. D., Reuben, R. L., Sandford, P. M., Brown, E. R. and Steel, J. A., 1997. "Detection of incipient cavitation in pumps using acoustic emission," Proceedings of the Institution of Mechanical Engineers, Part E: Journal of process mechanical engineering, Vol. 211, No. 4, pp. 267-277

[5] Chudina, M., 2003, "Noise as an indicator of cavitation in a centrifugal pump," Acoustical Physics, Vol. 49, No. 4, pp. 463-474

[6] Chudina M., and Prezelj J., 2009 "Detection of cavitation in situ operation of kinetic pumps: Effect of cavitation on the characteristic discrete frequency component," Applied Acoustics, 70.9, 1175-1182.

[7] Černetič, J., Prezelj, J. and Čudina, M., 2008, "Use of noise and vibration signal for detection and monitoring of cavitation in kinetic pumps," The Journal of the Acoustical Society of America, Vol. 123, No. 5, pp. 3316-3316.

[8] Zhang, N., Yang, M., Gao, B. and Li, Z., 2015, "Vibration characteristics induced by cavitation in a centrifugal pump with slope volute", Shock and Vibration, Vol. 2015, pp. 1-10

[9] Gülich, J. F., 2008, Centrifugal pumps, Springer.

Numerical Prediction of Cavitating Vortex Rope in a Draft Tube of a Francis Turbine with Standard and Calibrated Cavitation Model

D Jošt[1,3], A Škerlavaj[1,2], M Morgut[2] and E Nobile[2]

[1] Kolektor Turboinštitut, Ljubljana. Slovenia
[2] University of Trieste, Trieste, Italy
[3] corresponding author dragica.jost@kolektor.com

Abstract. Transient simulations of flow in a Francis turbine were performed with a goal to predict pressure pulsation frequencies and amplitudes caused by rotating vortex rope at part load operating regime. Simulations were done with the SAS SST turbulence model with curvature correction on basic and refined computational meshes. Without cavitation modelling too small values of frequency and amplitudes were obtained. With mesh refinement the calculated amplitudes were a bit closer to the measured values, while the accuracy of predicted frequency did not improve at all. Agreement between measured and numerical values was significantly improved when cavitation was included in simulations. In addition, the predicted value of the dominant frequency was slightly more accurate when, in the Zwart et al. cavitation model, the default condensation and evaporation model constants were replaced with previously calibrated ones.

1. Introduction

Pressure fluctuations caused by cavitating vortex rope at part load operating regimes can be a serious problem for Francis and also for single regulated axial turbines. The consequences of the vortex developed in the draft tube are very unpleasant pressure pulsations, axial and radial forces and torque fluctuation as well as turbine structure vibrations. The result of intensive experimental investigations of draft tube pressure pulsation for various turbines of different specific speed proved that the intensity of the vortex depends strongly on the shape of runner blades and channel [1]. Therefore, an accurate numerical prediction of pressure pulsation caused by vortex rope is crucial in a design stage, before the runner with undesirable dynamic characteristics is manufactured.

The first numerically obtained helical vortex was presented by Skotak [2]. In spite of a too coarse mesh for the Large Eddy Simulation (LES) model, a low pressure zone which agreed well with the rotating rope observed in experiment was obtained. In the following years, several papers about this topic were published. Generally, the frequency of pressure pulsation matched the measured values quite well, but the prediction of amplitudes was less accurate. In [3] the detailed comparison of numerical (SAS SST, no cavitation modelling) and experimental results at different part load operating points showed good agreement of vortex rope shape, length, thickness as well as pressure pulsation frequencies, while the predicted amplitudes were too small. The results improved with mesh refinement, but even on the refined mesh with 12 million nodes in the draft tube, the discrepancy was around 14%. Cavitating vortex rope was modelled in [4, 5, 6] and it was reported that a size of the predicted cavitating rope was smaller than the one observed during experimental tests. In [6] the results obtained with three turbulence models (SAS-SST, RSM and LES) were presented. When cavitation was not modeled, there was no significant difference in the accuracy of the results due to different turbulence models. In all cases predicted pressure pulsation amplitudes were significantly too small. When cavitation was included the results obtained with SAS-SST and RSM were less accurate than the results obtained with LES. The results of the RSM model were so irregular, that pressure pulsation frequency could not be obtained. For the simulation with the LES model, the mesh in the draft tube was refined (23.5 million nodes) and it is possible that the mesh refinement contributed even more to the improvement of the results than the model itself. Anyway, also with the LES model on refined computational mesh, the amplitudes of pressure pulsation were considerably too small, discrepancy with measurements was at one location 12.3% and at the other location even 31.5%.

Kolektor Turboinštitut (Slovenia) and University of Trieste (Italy) joined in the ACCUSIM EU project with the aim to develop reliable, high fidelity methods for accurate predictions and optimization of performances of hydro-machinery and marine propellers. One of the main goals of the ACCUSIM project is to improve the prediction of all types of cavitation in water turbines including cavitating vortex rope in the draft tube of a Francis turbine. With this purpose a cavitating vortex rope was modelled with the Zwart mass transfer model using standard and calibrated evaporation and condensation parameters. The first results for part load operating regime are presented in this paper.

2. Mathematical Model

1.1. Turbulence Modelling

Even though flow in water turbines is unsteady, the efficiency and cavitation in Francis turbines can be predicted by a steady state flow analysis and the results are usually enough accurate for design purposes. But for prediction of rotating vortex rope transient simulations with one of the advanced turbulence models (Reynolds Stress Models (RSM), Large Eddy Simulation (LES), Detached Eddy Simulation (DES) or Scale-Adaptive Simulation (SAS)) have to be performed. From our previous experiences [3, 6] it was concluded that the most economic turbulence model for the vortex rope prediction is the SAS SST model with the curvature correction. Therefore this model was used for the case presented in this paper. A detailed description about SAS SST and curvature correction can be found in [7] and [8].

1.2. Multiphase Flow and Cavitation Modelling

Cavitation refers to the process by which vapour forms in low pressure regions of a liquid flow. Cavitation is numerically usually modelled by the homogeneous multiphase model. The homogeneous model assumes that the velocity field for the transported quantities (with the exception of volume fraction) for the process is the same for all phases. Therefore it is sufficient to solve bulk transport equations for shared fields rather than solving individual transport equations. Mass transfer from liquid to vapour is modelled by different cavitation models. In this study, the Zwart et al. model [9], which is based on the simplified Rayleigh-Plesset equation for bubble dynamics, was used. The evaporation and condensation processes are regulated with empirical constants F_e and F_c. The default settings in ANSYS CFX are equal to F_e=50 and F_c=0.01, respectively.

A calibration of empirical constants using an optimization strategy is presented in [10]. The entire calibration process was driven by the modeFRONTIER 4.2 optimization system, which is a general integration and multi-objective optimization platform commonly used for functional and shape optimization of systems and devices. With the aim to reduce computational costs the empirical coefficients were optimized on a two-dimensional sheet cavity flow around the NACA66(MOD) hydrofoil. The evaporation and condensation coefficients of the Zwart model were tuned within the following ranges: $30<=F_e<=500$ and $0.0005<=F_c<=0.08$. The best result was found with F_e = 300 and F_c= 0.03. Details about optimization process can be found in [10]. Calibrated values for F_e and F_c were successfully used for prediction of cavitating flow around model scale propellers in uniform and non-uniform inflow [11, 12] and also for cavitation prediction in a Kaplan turbine [13]. In this study, besides default values, also calibrated constants were used for prediction of cavitating vortex rope.

3. Model Test

Pressure fluctuations on Francis model turbine were measured in accordance with IEC 90193. KISTLER piezoresistive absolute pressure transducers were located at several locations. Signals from the pressure transducers were wired to the multichannel data acquisition system, based on National Instruments multifunction card with additional SCXI signal conditioning modules. Signals were acquired continuously with 5 KHz sampling frequency and 16 bit resolution. Binary data samples stored on the computer hard disk were at least 30 seconds long. LabVIEW software was used for analysis of pressure signals.

With this measurements information about magnitude and dominant frequency of pressure fluctuations at the spiral casing intake, at two locations at the draft tube cone, at one location at the draft tube elbow and at one location at the extension of the draft tube were obtained. For validation of numerical results only the values at the draft tube cone (see figure 1) were used. Dominant frequency was obtained with the Fast Fourier Transformation. For amplitudes 2% of extreme values in the sample were not taken into account, so actually peak-to-peak amplitudes obtained from 98% of amplitude values were compared to the numerical results.

Although the focus of this study is not efficiency prediction, it was interesting to see the effect of cavitation modelling on accuracy of calculated efficiency values. Therefore calculated values of flow rate, torque on the shaft and efficiency were compared to the measured values.

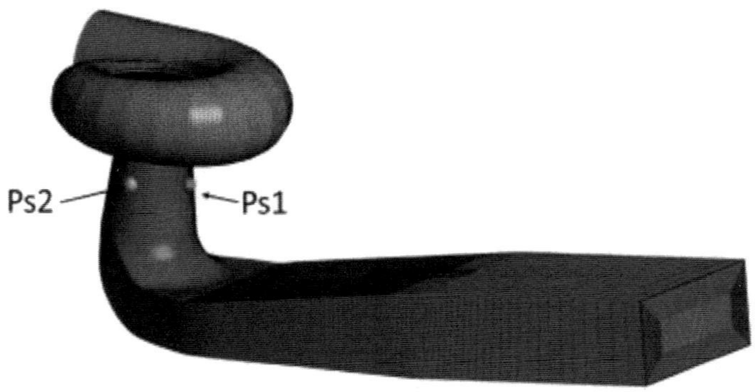

Figure 1: Computational mesh and positions of pressure measurements on conical part of the draft tube

4. Numerical Prediction

Prediction was done for a middle head Francis turbine at one operating point at part load operating regimes (φ/φ_{BEP}=0.8, ψ/ψ_{BEP}=0.97). Transient simulations without cavitation modelling were performed first, then the cavitation was included. The standard and calibrated evaporation and condensation constants were used for cavitation modelling.

4.1. Computational Domain and Meshes

The computational domain for numerical flow analysis consists of a spiral casing with stay vanes, a guide vane cascade, a runner and a draft tube. An extension of the draft tube was added in order to move the outlet boundary condition away from the region of interest. Flow energy losses in the draft tube extension were not taken into account when the turbine efficiency was calculated.

While the mesh in the spiral casing was unstructured, the meshes in all other turbine parts were structured. The meshes are rather coarse, except in the draft tube, where mesh was at first moderately (basic mesh - BM) and later significantly refined (fine mesh - FM). Number of nodes and elements for meshes of all turbine parts are presented in table 1, while the basic computational mesh can be seen in figure 1.

4.2. Boundary and initial conditions

Simulations are more stable, if flow rate is input data and head a result of computation. But the flow rate corresponding to each guide vane opening is not known in advance. The purpose of numerical simulations is to minimize measurements, therefore it is more sensible to check the accuracy of numerical results with prescribed head, while the value of flow rate is a result of numerical simulation. Therefore in this study input data were geometry, rotating speed and head, while the value of flow rate, torque on the shaft and efficiency were the results of numerical simulation. At the inlet of the computational domain, total pressure $P_{tot,in}$ was prescribed. $P_{tot,in}$ was defined by the expression $P_{tot,in} = \rho\ g\ H + P_{tot,out}$, where ρ is water density, g is gravity, H is turbine head and $P_{tot,out}$ is total pressure at the draft tube outlet obtained in the previous iteration of numerical simulation. At the outlet, static pressure calculated from cavitation coefficient σ was prescribed.

Table 1. Number of nodes and elements in various computational meshes

Domain	Nodes	Elements
Spiral casing	167,605	422,336
Stay vane cascade	300,863	270,996
Guide vane cascade	886,200	801,312
Runner	1,000,428	936,832
Draft tube extension	196,353	187,520
Draft tube, BM	2,340,295	2,302,752
Draft tube, FM	12,032,526	11,917,764
Total, BM	4,891,744	4,921,748
Total, FM	15,143,184	15,063,064

In steady state simulations the frozen rotor condition was used between the runner and stationary turbine parts. In transient simulations, a transient rotor stator condition was prescribed. With this condition, the position of runner blades is updated at each time step and all effects of mixing due to runner rotation are taken into account.

At first, a steady state simulation without cavitation modelling was performed on the basic mesh. These results were used as initial condition for transient simulations without cavitation and also as an initial for steady state simulations with cavitation. The results of the latter ones were used as initial conditions for transient simulations with cavitation. The simulation on the refined mesh was performed only without cavitation modelling, it started from the results obtained on the basic mesh.

Time step in transient simulation corresponded to two degrees of runner revolution. Maximal and average values of Courant number in the draft tube were in case of the basic mesh equal to 7.14 and 0.21, respectively. In case of the fine mesh the time step size was not reduced, therefore maximal and averaged values of Currant number increased to 10.1 and 0.55, respectively. Time of transient simulation without cavitation modelling on BM corresponded to 60 runner revolutions, on the refined mesh additional 10 runner revolutions were performed. With cavitation modelling in both cases, with default and calibrated parameters, about 50 runner revolutions were simulated.

4.3. Results and discussion

At the operating point treated in this study, a strong, compact vortex rope was observed on the test rig. Similar shape and thickness of the rope was obtained numerically (see figure 2). For simulations without cavitation modelling the rope is presented with iso-surface of absolute pressure equal to evaporation pressure, while cavitating vortex rope is presented with iso-surface of Vapour Volume Fraction = 0.1. In case of cavitation modelling, the rope has a staircase-like surface due to insufficiently refined mesh in a vertical direction. The cross section of the rope is more elliptical, which is more typical for the operating regimes closer to the BEP.

Pressure pulsations at Ps1 and Ps2 obtained with numerical simulations are presented in figure 3. It is clearly seen that without cavitation modelling the amplitudes are significantly smaller than when cavitation is included. With mesh refinement pressure pulsations become more regular and larger. With cavitation modelling some extra peaks can be observed. Similarly as for experimental results, where 2% of extreme values were not taken into account, also these peaks were ignored when pressure pulsation peak-to-peak amplitudes were determined.

A comparison of calculated flow rate, torque on the shaft and turbine efficiency to the measured values is presented at the top of figure 4. All simulations overestimated values of flow rate and torque on the shaft. When cavitation was not included, the predicted flow rate was about 2.6% larger than the measured one. With cavitation modelling discrepancy increased to about 3.4%. The value of torque was in all simulations overestimated for about 4%. The effect of mesh refinement on calculated flow rate and torque was negligible. An important reason for discrepancy between predicted and measured values of flow rate and torque on the shaft is that in this study volumetric and torque losses in labyrinth seals were not taken into account. From the results for a Francis turbine presented in Workshop Francis'99 [14] it is evident, that these losses are significant at part load, but even at the best efficiency point the volumetric losses and especially the torque losses should not be neglected.

An accuracy of a numerically predicted efficiency value depends on accuracy of predicted flow rate and torque. When cavitation was not modelled, calculated efficiency is about 1.73% larger than the measured one, while with cavitation modelling the discrepancy decreased to about 0.7%. Again, the effect of calibration of cavitation constants is negligible.

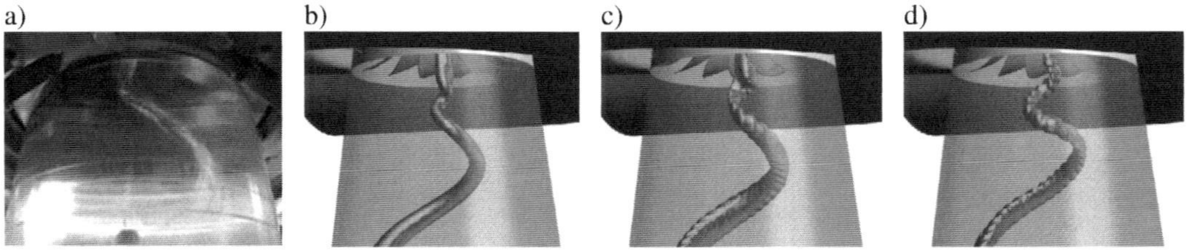

Figure 2. Vortex rope a) on the test rig; b) without cavitation modelling, iso-surface of evaporation pressure, basic mesh; c) default cavitation constants, iso-surface of Vapour Volume Fraction = 0.1 d) calibrated cavitation constants, iso-surface of Vapour Volume Fraction = 0.1.

Due to too large values of flow rate the operating point being treated is in numerical simulations a bit closer to the best efficiency point than in the measurements. This can contribute to the discrepancy between predicted and measured values of pressure pulsation frequencies and amplitudes.

In figure 4 also pressure pulsation frequency and peak-to-peak amplitudes are presented. Pressure pulsation frequencies are divided by frequency of runner rotation. Peak-to-peak amplitudes are presented in relative form: $App = \Delta p / (\rho g H)$, where p is pressure, ρ is water density, g is gravity and H is turbine head. Numerically, without cavitation modelling, significantly too small values of pressure pulsation frequency and peak-to-peak amplitudes were obtained. Predicted pressure pulsation frequency was about 8% smaller than the measured one and no improvement of accuracy was achieved with mesh refinement. On the basic mesh discrepancies between measured and calculated *App* at Ps1 and Ps2 were equal to 56.3% and 29.4%, respectively. Besides, contrary to the experimental results, where pressure pulsations were stronger at Ps1, higher amplitudes were obtained at Ps2. With mesh refinement only very small improvement was achieved for *App* at Ps1, while at Ps2 improvement was larger – the discrepancy decreased to 17.8%.

With cavitation modelling simulations were performed only on the basic mesh. The accuracy of results improved significantly. Dominant frequency was most accurately predicted in case of calibrated evaporation and condensation constants where the discrepancy to the measured value was below 1.4%. Pressure pulsation amplitudes were most accurately predicted with default constants where discrepancies between measured and calculated *App* values at Ps1 and Ps2 were equal to 3.2% and 0.13%, respectively. With calibrated constants the discrepancies for *App* values at Ps1 and Ps2 were equal to 0.3% and 10%, respectively.

The reason for relatively small positive effect of mesh refinement could be that only the mesh in the draft tube was refined, while mesh in the runner, where vortex rope started to form, was rather coarse. On interface between the runner and the draft tube considerably larger elements were on the runner side. Therefore the flow at the inlet of the draft tube was probably not enough accurately calculated. In [3], where positive effect of mesh refinement was clearly seen, the mesh in the runner had 12 million elements and it was much more refined than in this case. Our experiences with cavitation modelling are that mesh should be fine enough to get proper extend of cavitation. Therefore we plan to repeat simulation without cavitation and with cavitation on computational mesh, which will be refined in the runner and in the draft tube.

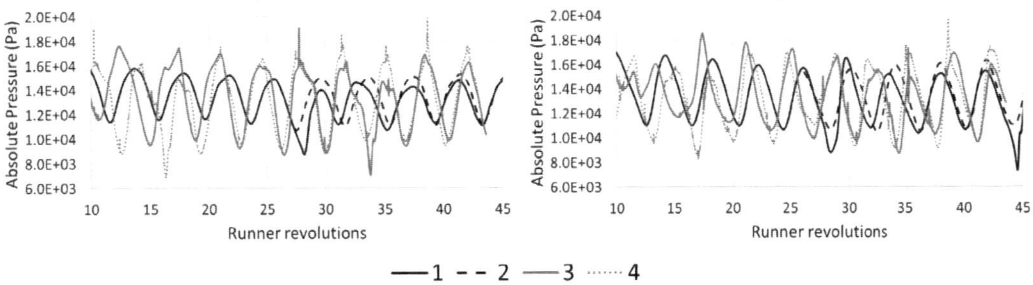

Figure 3. Pressure pulsation at two locations on the cone of the draft tube
1 – no cavitation modelling, BM; 2 – no cavitation modelling, FM; 3 – cavitation modelled with default constants, BM; 4 – cavitation modelled with calibrated constants, BM.

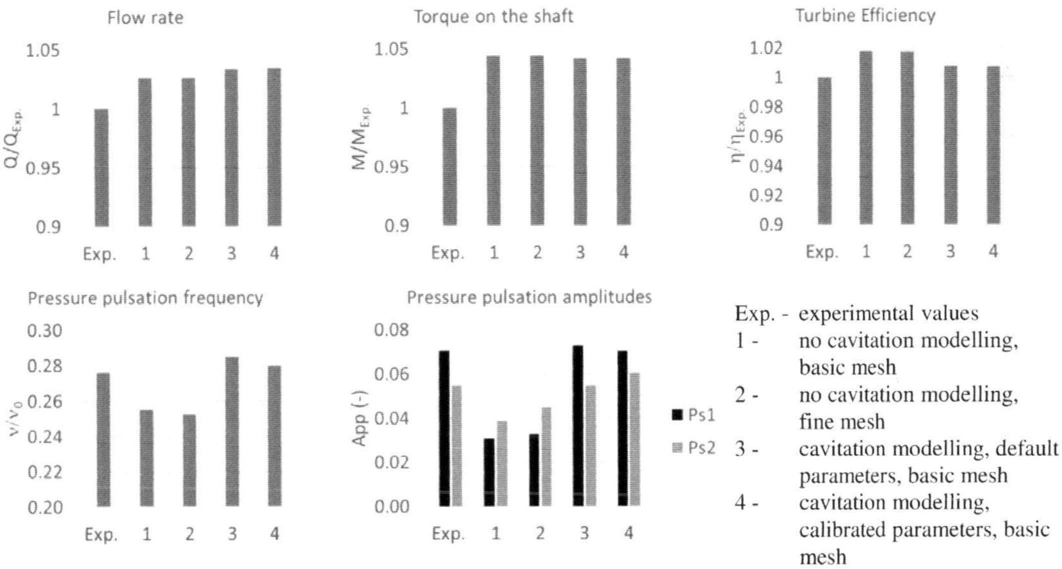

Figure 4. Comparison of numerical and experimental results

5. Conclusions

Transient simulations with SAS SST turbulence model with curvature correction were performed at part load operating regime with and without cavitation modelling. The goal was to predict pressure pulsations caused by rotating vortex rope. Numerical values of dominant frequency and peak-to-peak amplitudes at two locations on the cone of the draft tube were compared to the experimental results. The following conclusions can be deduced:

- Numerical simulations without cavitation modelling significantly underestimated values of pressure pulsation frequency and amplitudes caused by rotating vortex rope. Refinement of the mesh in the draft tube improved only prediction of peak-to-peak amplitude at one location, while generally the results were nearly the same as on the basic mesh.
- Agreement between measured and numerically predicted values of frequency and peak-to-peak amplitudes was significantly improved when cavitation was included in simulations.

- With a previously calibrated evaporation and condensation constants in Zwart at al. cavitation model, a bit more accurate value of dominant frequency was obtained. Accuracy of peak-to-peak amplitudes at one location improved, while at the other location too large value was obtained.

References

[1] Kercan V, Bajd M, Djelić V, Lipej A, Jošt D 1996 Model and prototype draft tube pressure pulsations. *Proceedings of the XVIII. IAHR Symposyum on Hydraulic Machinery and Cavitation, Volume II*, p 994-1003.
[2] Skotak A 1999 Draft tube swirl modeling. *Proceedings of the 9th International Meeting of IAHR Work Group on the Behavior of Hydraulic Machinery under Steady Oscillatory Conditions*, 6 pages, D4.
[3] Lipej A, Jošt D, Mežnar M, Djelić V 2008 Numerical prediction of pressure pulsation amplitude for different operating regimes of Francis turbine draft tube. *24th IAHR Symposium on Hydraulic Machinery and Systems*, 10 pages, (IAHR 93.pdf)
[4] Guo Y, Kato C, Miyagawa K 2006 Large-eddy simulation of non-cavitating and cavitating flows in an elbow draft tube. *23rd IAHR Symposium on Hydraulic Machinery and Systems*, 10 pages, F195.pdf
[5] Zhou L, Wang Z, Tian Y 2006 Numerical simulation of vortex cavitation in draft tube. *23rd IAHR Symposium on Hydraulic Machinery and Systems*, 8 pages, F102.pdf
[6] Jošt D and Lipej A 2001 Numerical Prediction of Non-Cavitating and Cavitating Vortex Rope in a Francis Turbine Draft Tube, *Journal of Mechanical Engineering 57(2011)No. pp.*
[7] Smirnov P E and Menter F 2009 Sensitization of the SST turbulence model to rotation and curvature by applying the Spalart-Shur correction term J. Turbomach. 131 041010
[8] Egorov Y and Menter F 2008 Development and application of SST-SAS turbulence model in the DESIDER project Advances in Hybrid RANS-LES Modelling ed. S-H Peng and W Haase (Heidelberg: Springer) pp 261–70
[9] Zwart P, Gerber A G and Belamri T 2004 A Two-Phase Model for Predicting Cavitation Dynamics, *ICMF 2004 International Conference on Multiphase Flow, Yokohama, Japan*
[10] Morgut M, Nobile E and Biluš I 2010 Comparison of mass transfer models for the numerical prediction of sheet cavitation around a hydrofoil. *Int. Journal of Multiphase Flow*. Vol. 37. No. 6. 2010. pp. 620-626
[11] Morgut M Nobile E 2012 Numerical Predictions of Cavitating Flow around Model Scale Propellers by CFD and Advanced Model Calibration. *International Journal of Rotating Machinery*. Vol. 2012. 2012. Article ID 618180. 11 p.
[12] Morgut M, Jošt D, Nobile E and Škerlavaj A 2015. Numerical investigations of a cavitating propeller in non-uniform inflow. *Fourth International Symposium on Marine Propulsors, smp'15, Austin*, Texas, USA
[13] Jošt D, Škerlavaj A, Morgut M, Nobile E 2015 Cavitation Prediction in a Kaplan Turbine Using Standard and Optimized Model Parameters, *6th IAHR International Meeting of the Workgroup on Cavitation and Dynamic Problems in Hydraulic Machinery and Systems*, September 9-11, 2015, Ljubljana, Slovenia
[14] Jošt D, Škerlavaj A, Morgut M, Mežnar P, Nobile E 2014 Numerical simulation of flow in a high head Francis turbine with prediction of efficiency, rotor stator interaction and vortex structures in the draft tube, *Workshop Francis 99*, Trondheim, Norway.

Acknowledgments

The research leading to these results has received funding from the People Programme (Marie Curie Actions) of the European Union's Seventh Framework Programme FP7/2007-2013/ under REA grant agreement n°612279 and from the Slovenian Research Agency ARRS - Contracts No. 1000-15-0263 and P2-0196.

IOP Conf. Series: Journal of Physics: Conf. Series **813** (2017) 012046 doi:10.1088/1742-6596/813/1/012046

Experimental Investigation of the Unsteady Pressure Field in Decelerated Swirling Flow with 74° Sharp Heel Elbow

D C Moş[1], S Muntean[2,3], A I Bosioc[1], C Tănasă[1] and R. Susan-Resiga[1]

[1] Politehnica University Timişoara, Bv. Mihai Viteazu 1, Timisoara, Romania
[2] Romanian Academy – Timişoara Branch, Bv. Mihai Viteazu 24, Timisoara, Romania

E-mail: seby@acad-tim.tm.edu.ro

Abstract. The unsteady pressure at the wall of the conical diffuser without and with a 74° heel elbow is examined. The self-induced instability is experimentally investigated on the swirl generator test rig. As a result, the asynchronous (rotating) pressure pulsation associated with the vortex rope and it second harmonic are discriminated. The discriminated Fourier spectra with plunging and rotating components of the unsteady pressure signals are analyzed for several geometrical configurations in order to identify the frequency associated to the heel elbow. It is shown that a plunging component of 8 Hz corresponds to a 74° heel elbow in the Fourier spectra.

1. Introduction

The Francis turbines operated at part load conditions develop hydrodynamic instabilities in the draft tube cone due to the residual swirl delivered by the runner. These phenomena are recognized as the main cause of severe pressure fluctuations called draft tube surge by Rheingans [1]. Extensively investigations were conducted during last decades in order to elucidate it mechanism [2-4]. A synopsis outlook about these investigations is presented by Nishi and Liu [5].

The group coordinated by Prof. Nishi has shown that the pressure pulsation consists of a synchronous part (plunging) and an asynchronous part (rotating), respectively. He has conducted experimental investigations to demonstrate that a synchronous component resides in the draft tube elbow [2]. Recently, a low-frequency synchronous component corresponding to the self-induced instability of the swirling flows is measured [6]. In this case, this self-induced low frequency component is one distinct than the frequency of the synchronous component associated to the elbow.

An extensive research program is developed by our group to elucidate the swirling flow interaction with different geometries of the draft tube elbow in order to explain the physical mechanism that produces the plunging pressure pulsations with low-frequencies. The 90° sharp hell elbow geometry corresponds to the old draft tubes available in hydropower plants [7-10]. A low-frequency synchronous component of 7 Hz is measured on our test rig for a 90° sharp heel elbow configuration being associated to swirling flow interaction with elbow. This low-frequency plunging component of 7 Hz is around 40% from the frequency of the asynchronous one [11].

In this paper, our experimental investigations are presented in order to examine the cause of low frequency oscillation in a 74° sharp heel elbow. The 74° heel elbow corresponds to underground draft

[3] To whom any correspondence should be addressed

tube geometry available in a hydropower plant. The experimental setup is described in Section 2 while the Fourier spectra obtained on the wall of the test section is presented in Section 3. The conclusions are drawn in the last section.

2. Experimental setup

The test rig is designed to investigate several swirling flow configurations. Several swirling flow configurations are generated using different guide vane geometries together with a free runner [12-14]. These swirling flow configurations correspond to hydrodynamic conditions associated to Francis turbine operation [14]. The test rig is equipped with an acquisition system in order to record the discharge and runner speed [13]. The most important part of the rig is the test section, Fig. 1. The test section includes the swirl apparatus with two rows of blades (13 guide vanes and 10 runner blades) [13] and the convergent-divergent section [15]. The test section is manufactured from plexiglass in order to allow flow visualization. The throat diameter is 100 mm while the conical diffuser has an 8.5° half-angle and a length of 200 mm. A sharp heel elbow with 74° is added downstream to diffuser providing the geometrical configuration labeled HE74E0 in Fig. 1. Also, several pipe extensions of 100 mm, 200 mm and 300 mm are added between the diffuser and 74° sharp heel elbow leading to the geometrical configuration named HE74E01, HE74E02 and HE74E03, respectively.

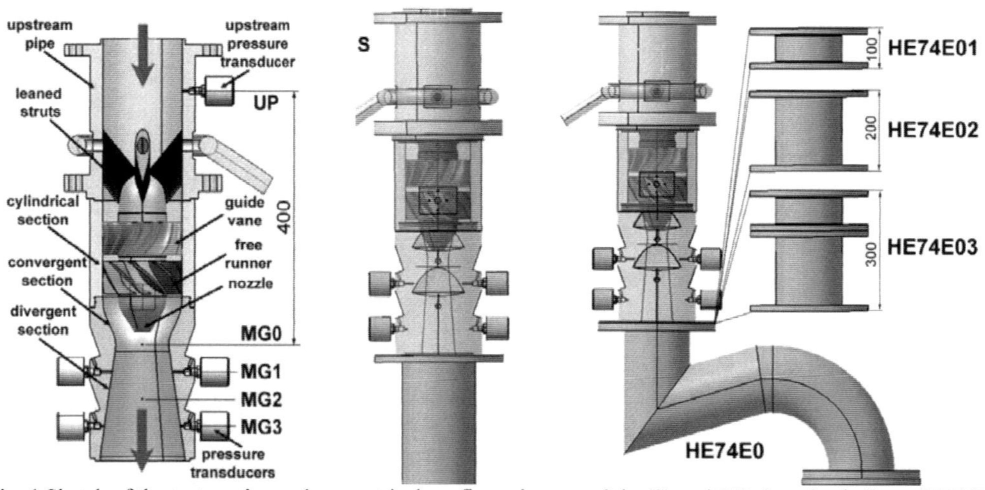

Fig. 1 Sketch of the test section and geometrical configurations: straight (S) and 74° sharp heel elbow (HE74E0) together with three pipe extensions of 0.1 m (HE74E01), 0.2 m (HE74E02) and 0.3 m (HE74E03)

The test section allows us to measure the unsteady pressure on the cone wall at four levels using eight transducers, see Fig. 1. As a result, two pressure transducers are installed on each level being located at 180° each other. This experimental setup allows us to examine the flow structure with one single vortex as it is visualized on our test rig. Moreover, this experimental setup is able to discriminate the type of first harmonic. Several numerical investigations were performed in order to indentify the type of higher harmonics in the Fourier spectra [6, 16].

The first level corresponds to the throat (MG0) and the next levels are displaced at 50 mm (MG1), 100 mm (MG2) and 150 mm (MG3) with respect to the first one. The transducers measurement range was ±1 bar with a precision of ±0.13%. However, the upper limit of frequency for unsteady pressure transducers is beyond to 50 Hz. In our investigations a nominal discharge of 30 l/s is selected. In this case, the swirling flow configuration provided by the stay vanes leads to a free runner speed around 1020 rpm. In this paper only the swirling flows under non-cavitating conditions are examined.

Fig. 2 Photo of the swirling flow test section considering three geometrical configurations: straight (S) and 74° sharp heel elbow together with pipe extensions of 0.1 m (HE74E01) and 0.3 m (HE74E03), respectively

3. Experimental data

Firstly, two unsteady pressure signals acquired on each four level of the cone are used to discriminate the plunging and rotating components. The experimental setup allows us to The Fourier spectra of the decomposed signals recorded from MG0 to MG3 levels on the test section (see Fig. 1) are plotted in Fig. 3 for a straight diffuser (denoted with S) and a straight diffuser with 74° sharp heel elbow (denoted HE74E0), respectively.

The plunging component around 19 Hz can be clearly identified in all power spectra. This frequency was identified in our experimental investigations performed on test rig without any row blades. In this particular case, the flow is axial without any swirling component. This plunging component around 19 Hz is associated to the honeycomb structure installed upstream to the test section in order to uniform the flow at the test section inlet [6].

One can observe that the rotating component frequency associated to the vortex rope is around 18 Hz in power spectra for straight diffuser (denoted with S). The largest amplitude of the rotating component is measured on MG0 level. The frequency around 36 Hz is distinguished on the Fourier spectrum of the rotating component on MG0 and MG1 levels, respectively. This frequency corresponds to second harmonic of the vortex rope (36 Hz = 2 x 18 Hz). The largest amplitude of this frequency is assessed on MG0 level being the same level with rotating component. It is obvious that these two frequencies are connected with the vortex rope. The rotating component is generated by the vortex rope and it remains trapped in the cone [3].

Two frequencies at 8 Hz associated the plunging component (blue) and around 17 Hz and it second harmonic around 34 Hz correspond to rotating one (red) can be clearly identified in power spectra with 74° heel elbow (HE74E0). The difference between both rotating components is coming from a slight deviation of free runner speed. As a result, the synchronous component of 8 Hz is associated to swirling flow interaction with 74° sharp heel elbow. This low-frequency synchronous (plunging) component of 8 Hz is around 47% from the frequency of the asynchronous (rotating) component.

Next, the interaction between decelerated swirling flow developed in the conical diffuser and 74° sharp heel elbow is investigated. This study is performed installing three pipe extensions with different lengths (0.1 m – HE90E01, 0.2 m - HE90E02 and 0.3 m - HE90E03) between the conical diffuser and 74° sharp heel elbow (see Figs. 1 and 2). Consequently, the Fourier spectra with discriminated components (plunging and rotating) on all levels for cases with 74° sharp heel elbow are plotted in Fig. 4.

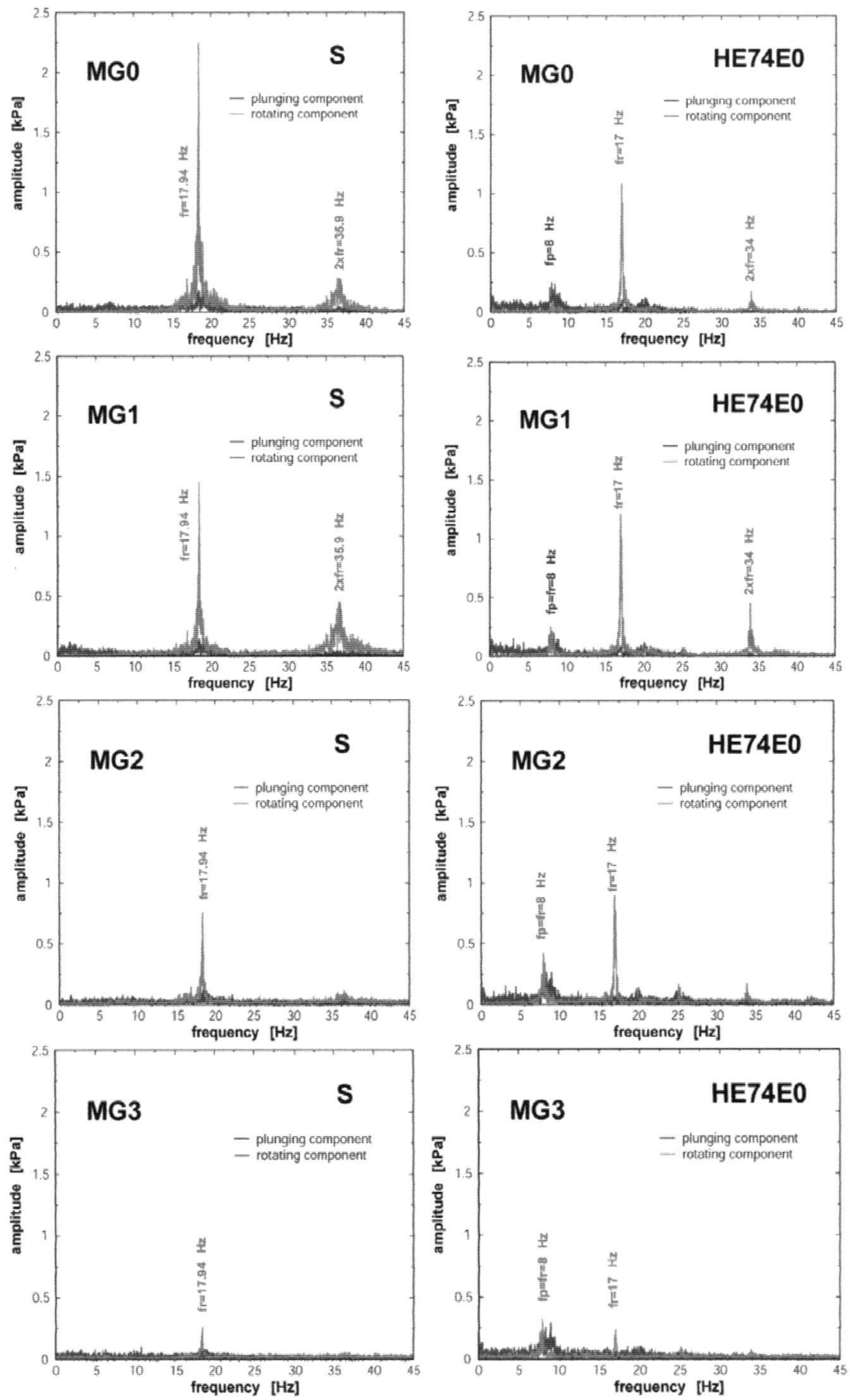

Fig. 3 Fourier spectra of the decomposed unsteady pressure signals on the first three levels at the cone wall (from MG0 to MG3, Fig. 1).

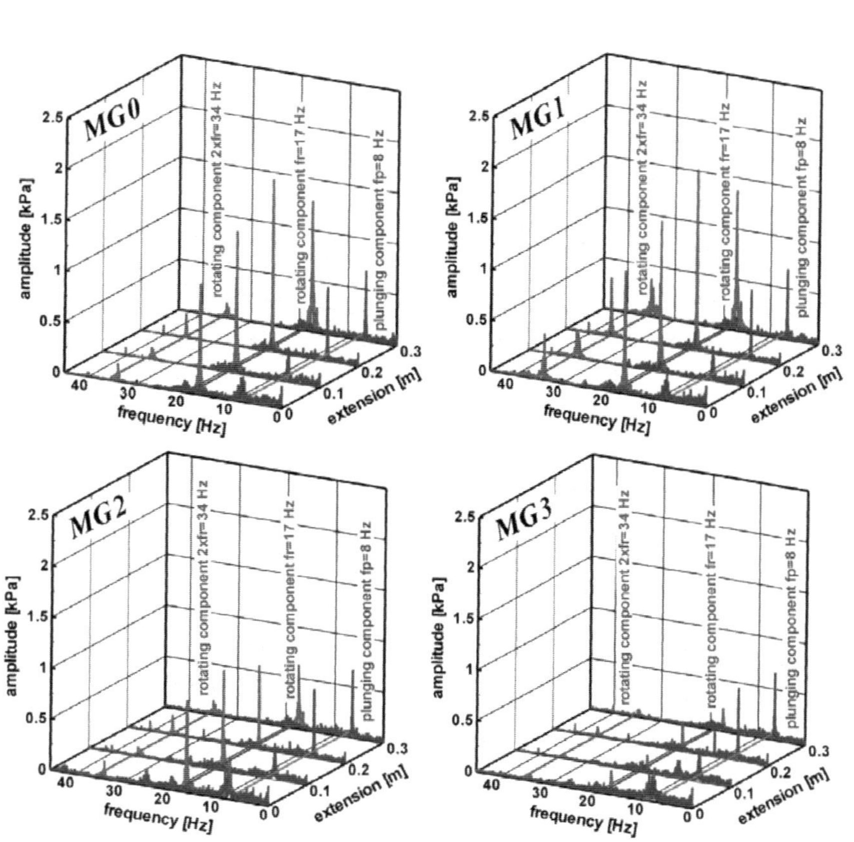

Fig. 4 Fourier spectra with discriminated signals associated to the unsteady pressure levels from MG0 to MG3 for geometrical configurations investigated: straight (S) and 74° heel elbow (HE70E0) together with three pipe length extensions of 0.1 m (HE74E01), 0.2 m (HE74E02) and 0.3 m (HE74E03)

One can observe that all frequencies indentified in the spectra associated to 74° sharp heel elbow case (HE74E0) are recovered for the cases with pipe extensions. However, the amplitude value associated to the plunging component of 8 Hz increases with the pipe length extension. This amplitude grows up over three times for the configuration with the largest pipe extension of 0.3 m (HE74E03 case) considered in our investigation with respect to the configuration without any extension (HE74E0 case), respectively. This observation suggests that a compact geometrical configuration should be selected for the hydraulic turbines to diminish the amplitudes associated to the plunging component.

4. Conclusions

The paper investigates the decelerated swirling flow in a straight draft tube, using a swirl generator that mimics the flow in the discharge cone of hydraulic turbines operated at partial discharge. The unsteady pressure measurements were performed on the cone wall of the test section. Eight points located on four levels were investigated for each geometrical configuration (straight diffuser without and with a 74° sharp heel elbow). The unsteady pressure signal is decomposed in two parts: rotational and plunging components, respectively. The rotational (asynchronous) component of pressure pulsation associated with the self induced instability is identified. The plunging component around 8 Hz is identified for geometrical configuration with 74° sharp heel elbow. This low-frequency synchronous (plunging) component of 8 Hz is around 47% from the frequency of the asynchronous (rotating) one. It amplitude grows up when the distance between conical diffuser and sharp heel elbow is increased. The main conclusion emerging from the analysis of the swirling flow interaction with the sharp hell elbow is responsible for the plunging (synchronous) pressure fluctuations. It is well known that the plunging pulsation propagates into the whole hydraulic system. Therefore, a compact geometrical configuration should be selected for the hydraulic turbines to diminish the plunging component.

References

[1] Rheingans WJ 1940 Power swings in hydroelectric power plants, *Trans ASME* **62** *174* 171-184
[2] Nishi M, Kubota T, Matsunaga S and Senoo Y 1984 Surging characteristics of conical and elbow-type draft tubes, *in Proc. of the 12th IAHR Symp. Section on Hydraulic Machinery, Equipment and Cavitation* (*Stirling, UK*) 272–283
[3] Wu Y-L, Li S, Liu S-H, Dou H-S and Qian Z-D 2013 Vibration of Hydraulic Machinery. Ch. 6 Vibration Induced by Hydraulic Excitation (*Dordrecht: Springer Verlag*)
[4] Dörfler PL, Sick M sand Coutou A 2013 Flow – Induced Pulsation and Vibation in Hydroelectric Machinery. Ch. 2 Low-Frequency Phenomena in Swirling Flow (*London: Springer Verlag*)
[5] Nishi M and Liu S-H 2013 An Outlook on the Draft-Tube-Surge Study *International Journal of Fluid Machinery and Systems* **6** *1* 33-48
[6] Muntean S, Tănasă C, Bosioc AI and Moş D C 2016 Investigation of the Plunging Pressure Pulsations in a swirling flow with precessing Vortex Rope in a straight diffuser *in Proc. 28th IAHR Symp. On Hydraulic Hydraulic Machinery and Systems* (*Grenoble, France*)
[7] Engström TF, Gustavsson LH and Karlsson RI 2001 The second ERCOFTAC Workshop on Draft Tube Flow, *Proc. of Turbine 99 – Workshop 2 on draft tube flow* (Älvkarleby, Sweden)
[8] Cervantes MJ, Engström TF and Gustavsson LH 2005 The third IAHR/ERCOFTAC Workshop on Draft Tube Flow, *Proc. of Turbine 99 – Workshop 3 on draft tube flow* (Porjus, Sweden)
[9] Anderson U 2009 Experimental study of Sharp Heel Draft Tube PhD Thesis, Luleå University of Technology (Luleå, Sweden)
[10] Marjavaara D 2006 CFD Driven Optimization of Hydraulic Turbine Draft Tubes using Surrogate Models PhD thesis, Luleå University of Technology (Luleå, Sweden)
[11] Muntean S, Moş DC, Tănasă C, Bosioc A I and Susan-Resiga R 2016 The Plunging Component with Low Frequency in Swirling Flows through a Straight Diffuser with and without a 90° Heel Elbow, *in Proc. 19th International Seminar on Hydropower Plants* (*Vienna, Austria*)
[12] Bosioc A, Susan-Resiga R, Muntean S and Tanasa C 2012 Unsteady pressure analysis of a swirling flow with vortex rope and axial water injection in a discharge cone, *ASME J Fluid Eng* **134** *8* 081104 1-14
[13] Muntean S, Bosioc A I, Stanciu R, Tănasă C and Resiga R 2011 3D Numerical Analysis of a Swirling Flow Generator, *in Proc. 4th IAHR Int. Meeting of the Workgroup on Cavitation and Dynamic Problems in Hydraulic Machinery and Systems* (*Belgrade, Serbia*) 115-125
[14] Susan-Resiga R and Muntean S 2008 Decelerated Swirling Flow Control in the Discharge Cone of Francis Turbines, *in Proc. 4th Int. Symposium on Fluid Machinery and Fluid Engineering,* (*Beijing, China*) 89-96
[15] Bosioc A, Susan-Resiga R and Muntean S 2008 Design and manufacturing of a convergent-divergent test section for swirling flow apparatus, *in Proc. 4th German-Romanian Workshop on Turbomachinery Hydrodynamics* (*Stuttgart, Germany*) 1 – 15
[16] Javadi A, Bosioc A, Nilsson H, Muntean S and Susan-Resiga R 2016 Experimental and Numerical Investigation of the Precessing Helical Vortex in a Conical Diffuser, With Rotor-Stator Interaction *ASME J Fluid Eng* **138** *8* 081106 1-13

IOP Conf. Series: Journal of Physics: Conf. Series **813** (2017) 012047 doi:10.1088/1742-6596/813/1/012047

Flow field investigation in a bulb turbine diffuser

M Pereira[1], P Duquesne[1], V Aeschlimann[1], C Deschênes[1]

[1] Hydraulic Machine Laboratory (LAMH), Laval University, Department of Mechanical Engineering, Pavillon Adrien-Pouliot, Québec G1V 0A6, Canada

jadid-mauricio.pereira-pabon.1@ulaval.ca

Abstract. An important drop in turbine performances has been measured in a bulb turbine model operated at overload. Previous investigations have correlated the performance drop with diffuser losses, and particularly to the flow separation zone at the diffuser wall. The flow has been investigated in the transition part of the diffuser using two LDV measurement sections. The transition part is a diffuser section that transforms from a circular to a rectangular section. The two measurement sections are at the inlet and outlet of the diffuser transition part. The turbine has been operated at three operating points, which are representative of different flow patterns at the diffuser exit at overload. In addition to the average velocity field, the analysis is conducted based on a backflow occurrence function and on the swirl level. Results reveal a counter-rotating zone in the diffuser, which intensifies with the guide vanes opening. The guide vanes opening induces a modification of the flow phenomena: from a central backflow recirculation zone at the lowest flowrate to a backflow zone induced by flow separation at the wall at the highest flowrate.

1. Introduction

In recent years, increased trading and cost of energy have pushed the hydropower industry to install hydraulic power plants in low head sites and to operate turbines in a more extended range of conditions. Some turbines operated at overload present an efficiency and power sharp drop after the best operating point [1–3]. One important cause of losses, when the turbine is operated in overload conditions, is the flow separation at the wall in the diffuser. This is particularly the case for low head turbines where the pressure recovery produced by the diffuser represents an important part of the total head [4].

Flow separation, described as the ejection of fluid from the boundary layer to the irrotational flow away from the wall, is a numerical and practical challenge. The majority of flow separations found in industrial applications, including in turbine diffusers, are complex because they are turbulent, three-dimensional and unsteady.

One of the main scope of the BulbT project launched in 2011 by the Consortium on Hydraulic Machines and the LAMH (Hydraulic Machine Laboratory of Laval University) is to investigate and document the flow separation zone in a bulb turbine diffuser. Significant experimental works have already been presented at the runner exit [5] and in the diffuser to characterize the flow separation zone [4,6,7]. Despite of these investigations some questions remain in particular because the investigation was not yet extended to the diffuser outlet.

This paper presents two LDV setups performed to obtain the velocity field in two sections of the diffuser transition part. The three selected operating conditions are representative of a bulb turbine operated at overload with important performance decreases after the best operating point. The analysis shows three distinct flow behaviors in the diffuser transition part at overload, including one case with a large flow separation zone.

2. Experimental setup

2.1. Turbine setup

Bulb turbine performance and LDV measurements were performed on the LAMH test bench. The bulb-shaped turbine model is supported by two identical and symmetric profiled piers. The distributor is composed of 16 guide vanes and the runner has four blades. Both sets of blades have adjustable pitch angle. The bulb contains an eddy current brake to control the model rotational velocity and to measure the torque. The diffuser has been specially designed by the Consortium on Hydraulic Machines with one important but realistic divergence in order to produce a sudden drop in efficiency and power after the best efficiency point. As shown in figure 1, the diffuser is composed of two parts. An acrylic cone, located at the runner exit, has an opening half angle of 10.25° and a length of 1.4 times the runner shroud diameter (D_{ref}). The second section is a transition part that transforms the circular section into a rectangular section within a length of 2.3 D_{ref}. The transition section is not symmetric. Right and left sides diverge with an angle of 9.5°, while the top side is near horizontal and the bottom side diverges with an angle of 5°. This section is made of metal sheets but it was fitted with numerous acrylic windows for LDV optical access. The cross-sectional area of the diffuser increases approximately by a factor 4 within the length of the diffuser (L_{dt} = 3.7 D_{ref}). As shown in figure 1, the reference Cartesian coordinate system is defined as follows: Z is the turbine axis pointing downstream, Z = 0 is defined at the runner blade axis, Y is the vertical axis pointing upward and X is set to respect a right-handed coordinate system. The runner rotation direction is from Y+ axis to X+ axis; in this paper the counter-rotating zone refers to the rotations opposite to the runner rotation direction (from X+ axis to Y+ axis).

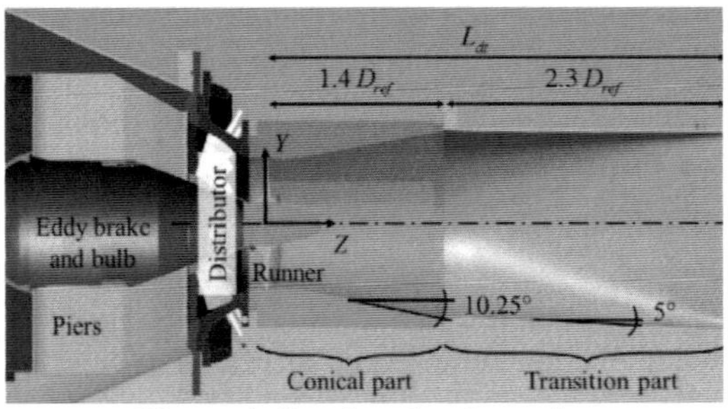

Figure 1. Test bench schematic.

The unit parameters Q_{11}, N_{11}, P_{11} and the efficiency η are used to define the operating points:

$$Q_{11} = Q/\left(H^{1/2} D_{ref}^2\right);\ N_{11} = \left(N D_{ref}\right)/H^{1/2};\ P_{11} = P/\left(D_{ref}^2 H^{3/2}\right);\ \eta = P/\left(\rho g H Q\right) \quad (1)$$

where N is the model rotation speed (rpm), H is the net water head (m), Q is the flow rate (m³/s), P is the mechanical power extracted (W), ρ is the water density (kg/m³) and g is the gravitational acceleration (m/s²).

The best efficiency point measured is at a runner blade angle of 22.5° and N_{11}=150 rpm [4]. To produce significant but representative efficiency drop and flow separation, the runner blade angle has been set at 30.2° and N_{11} = 170 rpm. Five operating points were selected to investigate flow separation in the diffuser (OP 1 to 5). OP 2, corresponds to the best operating point in terms of efficiency for the selected blade angle and unitary speed. For OP 2, tuft visualizations have shown that flow separation occurs in the transition part but remains relatively small and sporadic. At OP 1, where the guide vanes are more closed than at the best operating point (OP 2), the flow separation at the wall was not observed. For OP 3 to 5 the guide vanes are more opened and the flow separation is larger and more frequent.

Figures 2 presents the efficiency curve for these five operating points. In this paper only OP 1, 2 and 5 are presented for brevity reasons.

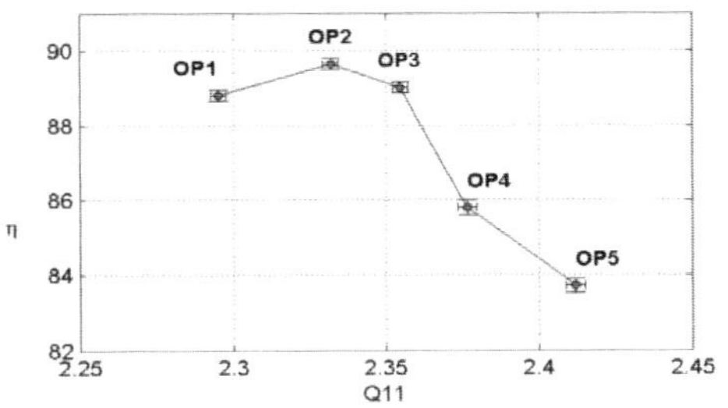

Figures 2. Efficiency curve for selected points of the BulbT project. N_{11}=170 rpm and blade angle at 30.2°.

2.2. LDV setup

Figure 3 shows the two measurement sections. The upstream section, 5a, is represented with blue squares at Z = 1.8 D_{ref}. A double axis displacement system has been used to define a 100 nodes grid with 20 lines and 5 columns, each point being separated by 5.8% D_{ref} in X and Y directions. This section gives information on the core flow at the Z-position where the flow separation front was previously observed [6]. Axial velocity component C_z, and transversal velocity component C_x, have been measured. Because of optical access limitations, the C_x component is not measured for some points of the grid.

Due to the divergence angle of the diffuser lateral sides and the dimension limitation of the optical access, three setups were needed to measure the three velocity components for the downstream plane at Z = 3.12 D_{ref}. The three LDV probe positions are identified in figure 3. The position 5b' used a double axis displacement system and a horizontal window at the diffuser top side. With this position, the axial velocity component C_z and transversal velocity component C_x, have been measured. The two other positions, 5b'' and 5b''', are on each lateral side of the diffuser, a three displacement system has been used to measure the last transversal velocity component C_y. Because of optical access limitations, the last measurement grid has a central region with the three components (position with black dots in figure 3). For the upper and lower regions, only the two components C_z and C_x have been mesured (red dots in figure 3). Also shown in figure 3, the distance between two points is not constant (about 23 × 14% D_{ref} for the central squares).

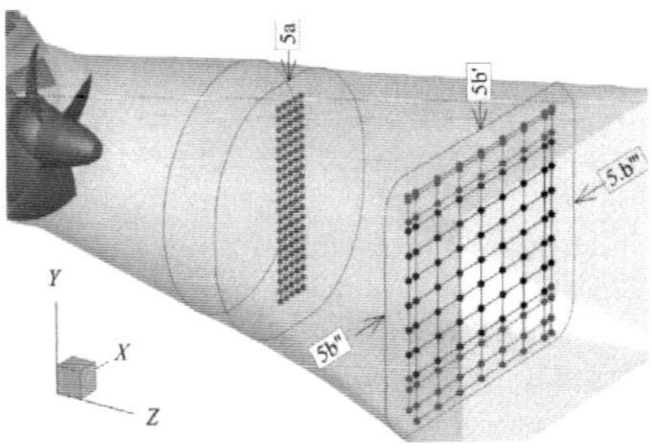

Figure 3. LDV measurement sections. Blue squares represents the measurement grid of the section 5a. Black and red dots represent the grid of the section 5b. Labels represent LDV probe positions and directions. Black lines represent the end of the conical diffuser and measurement sections.

The LDV system is a two-components with four laser beams operating in backward scatter on-axis collection mode. There are two LDA (laser Doppler Anemometer) inside the probe head and each LDA uses two laser beams at the same wavelength to measure one component of velocity. Bragg-cell shifting at 40 MHz is used to resolve directional ambiguity on a 5W Argon-ion laser. The particles used as a reflective medium are silver coated hollow glass spheres. Their diameter is 10 μm and the density 1.4 g/cm^3. The focal length, measurement volume, velocities components, sample size and uncertainty estimation for all measurement campaigns are reported in table 1. The observation time is 90s per node on the measurement grid. The acquisition rate varies from 3 Hz to 5 kHz depending on the distance travelled by the laser in water and the local mean velocity. The given uncertainty corresponds to the quadratic sum of the uncertainty of the LDV measurement considering the probe alignment (B_c) and the random uncertainty (P_c estimated with the standard deviation and the student coefficient with a 95% probability).

Table 1. Focal, measurement volume, measured velocity components, minimum sample size, grid size and total uncertainty.

Position	Focal (mm)	Meas. Volume (mm^3)	Velocity components	Samples (average)	Grid size (nodes)	Uncertainty P=95% (V_{ref}) $\sqrt{B_c^2 + P_c^2}$
5a	402.5	0.0486	C_x, C_z	40 000	100	1%
5b'	1000	0.1204	C_x, C_y, C_z	45 000	116	0.7%
5b'',5b'''	402.5	0.0486	C_x, C_y, C_z	45 000	116	0.7%

In this study, the velocity is normalized using a reference velocity:

$$V_{ref} = 4Q / \pi D_{ref}^2 \tag{2}$$

3. Result and analysis

3.1. *Backflow analysis*

As an indicator of the intermittent backflow, the following averaged backflow occurrence function δ has been defined and calculated at each measured position:

$$\delta_{(X,Y)} = \sum_{i=1}^{nb} TT_{i(X,Y)}\_\text{if}\left(C_{zi} \leq 0\right) \Big/ \sum_{i=1}^{nb} TT_{i(X,Y)} \tag{3}$$

where nb is the sample size and TT_i is the transit time of the particle i inside the LDV measurement volume. δ represents the fraction of the observation time where the axial velocity is negative. Some limitations are associated with this indicator. First, the velocity component C_z is the axial velocity but with the presence of swirl, the flow is not only axial. It is not evident to define the backflow direction as there is not one main flow direction. In the case of a three-dimensional separation zone from the wall, a point in space can be in the separation zone even if there is no axial backflow. Hence, a negative axial velocity is only an approximate indicator of the presence of a separation zone near the point in space considered. Furthermore, the backflow occurrence function is not able to dissociate between backflow zones issued from a separation zone at the wall or issued from a central recirculation.

Figure 4 presents the normalized axial velocity and the averaged backflow occurrence function at the diffuser exit plane (section 5b) for OP 1, 2 and 5.

For OP 1, as shown in blue, an average backflow zone is present at the diffuser center. This average backflow zone is small and intermittent. It covers only around 1% of total section area and the average of the backflow function does not exceed 65%. The contours of the average backflow occurrence function at 1% defines a larger circular zone around. Neither averaged nor intermittent backflows is observed near the wall in the restricted measurement area. Therefore, the average backflow does not seem to be connected to the wall and the zone is associate to a central recirculation area and not to a flow separation at the wall.

At higher flow rate, OP 5, an average backflow zone is located in the (*X*+, *Y*-) corner. The exact size of the zone is not known due to the lack of information near the wall but can be roughly estimated to more than 5% of the total section area. At the opposite corner (*X*-, *Y*+) the flow velocity is increased to balance the flow rate. The difference between the two corners is up to 47% of V_{ref}. The contour values of the average backflow occurrence function increase with the proximity of the wall in the (*X*+, *Y*-) corner, up to 80%. These observations confirm the presence of a large flow separation in the (*X*+, *Y*-) corner already observed in previous experimental works as shows that [6]. This new data set brings complementary information on the flow separation description. The average reattachment line is located downstream the diffuser outlet. In despite of the previous remark and of the wall distance, the averaged backflow occurrence has never reached 100%, therefore unsteady reattachment lines might be present at the end of the diffuser.

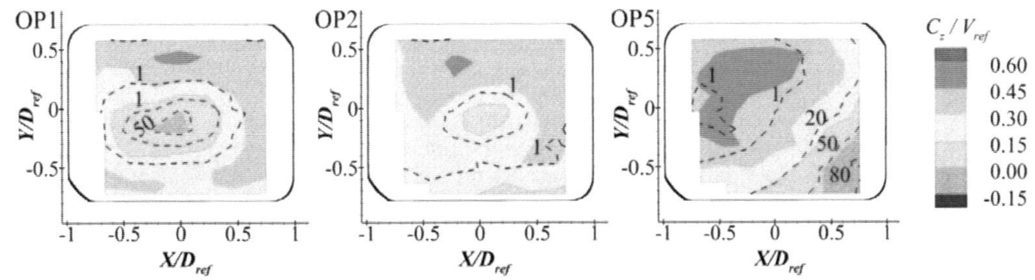

Figure 4. Contour plot of the normalized axial velocity for the section 5b. Dashed lines represent the averaged backflow occurrence function [%].

For the best operating point, OP 2, neither averaged wall separation nor averaged core flow recirculation is present. In the core flow, a recirculation is detected with δ lower than 16%. The averaged backflow occurrence function occasionally points out a wall separation, with δ lower than 17%. The averaged backflow occurrence function might be higher near the wall where no measurement could be performed. If an average flow separation zone exists at OP 2, it would be smaller than the distance between the wall and the measured area.

Backflow zone in the diffuser acts as restrictions of the section, reducing the kinetic energy conversion into pressure. OP 2 presents a better pressure recovery in the diffuser compared to operating points with more closed guide vanes (OP 1) or more open guide vanes (OP 5).

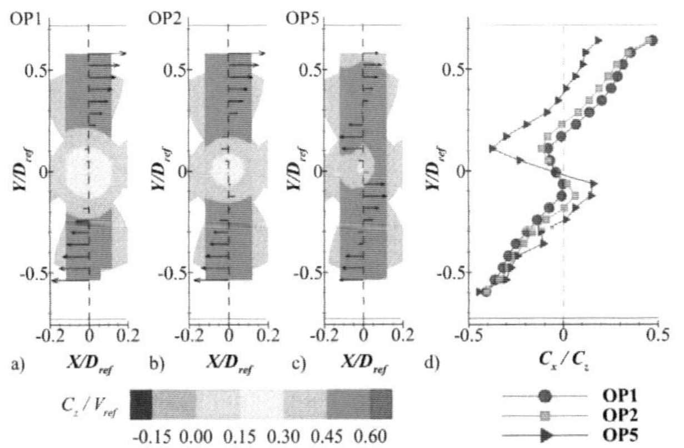

Figure 5. a, b and c, contour plots of the normalized axial velocity, section 5a. Vectors represent the $C_x - \overline{C_x}$. d, swirl values (C_x / C_z) at $X = 0$.

Figure 5 shows the normalized axial velocity at the beginning of the transition part (section 5a) for OP 1, 2 and 5. All operating conditions present a similar global pattern with a central zone of low positive axial velocity. There is no averaged backflow in the core flow, with the lowest velocity of 15% V_{ref} at OP 1. From OP 1 to OP 5, this zone size decreases. The averaged backflow occurrence function has been calculated (not shown in the figure) and its value decreases from 3.5 % at OP 1 to 0.06% at OP 5. This central low velocity zone is probably induce by a combinaison of the hub wake and by the diffuser's diverging shape. At OP 1, the central low velocity zone is larger and the flow rate is lower to the other OPs. But even in this case, no average backflow zone can be noticed. Therefore, the recirculation which has been found at the diffuser exit is not present at the beginning of the transition section. In rare cases, δ < 3.5%, the recirculation at the diffuser exit induces an instantaneous backflow and can be superimposed to the hub wake effect. In OP 2 and OP 5, the δ values are nearly null in the core region, the backflow from the diffuser exit never extends that far upstream. Here, the central low velocity zone can be considered mainly induced by the hub wake and the diffuser diverging shape. With the guide vanes opening, the flow rate increases and tends to fill the velocity deficit in the core flow. At OP 1, the flow rate is insufficient to fill the velocity profile, the hub wake can propagate more downstream and a weaker zone in the center is present at the diffuser exit, which benefits to the central recirculation apparition.

3.2. *Rotating zone analysis*

At the diffuser transition part, section 5a, the flow rotation is analysed with the swirl ratio C_x / C_z and with the C_x vectors along the Y-axis at $X = 0$. Figure 5d, presents the swirl ratio for all operating points. Here, for the positions with $Y/D_{ref} > 0$, the positive values of the swirl ratio correspond to the runner rotation direction. Because swirl ratio is expressed in the cartesian reference frame, for the positions with $Y/D_{ref} < 0$, the runner rotation direction corresponds to the negatives value. Figure 5d shows that all operating conditions present a central zone of rotating flow in the direction opposite to the runner rotation. With the guide vanes opening, from OP 2 to OP 5, the intensity of this counter-rotating zone increases. For example, the maximal swirl value increase at OP 5 is 218% in comparison with the maximum value of OP 1. The position of this maximum does not move with the guide vanes opening, suggesting a stronger vortex but with the same axis.

The spatial averaging magnitude of the C_x velocity along $X = 0$, named here $\overline{C_x}$, changes with the guide vanes opening, from -0.05% V_{ref} at OP 1 to -4% V_{ref} at OP 5. This means that at OP 1 and 2 the flow is not deviated. At OP 5, the negative value of $\overline{C_x}$ underlines the flow deviation due to the flow separation zone in the corner (X+,Y-). To underline the effect of the secondary flow, vectors in figure 5a, b and c represent the difference of the transversale component with the spatial averaging of this component ($C_x - \overline{C_x}$) along $X = 0$. At OP 5, the magnitude of counter-rotating $C_x - \overline{C_x}$ vectors can be consequent, up to 11% V_{ref}, and underlines the high intensity of a counter-rotating region.

Previous LDV measurements at the runner exit exhibited also a counter-rotating zone issued from the runner with the same evolution over the OPs [5]. Therefore the counter-rotating zone from the runner is convected along the diffuser conical section. The flow at the runner exit in this case is composed by three annular rotating zones where the middle ring one is counter-rotating zone [5]. Similar flow parterns have been also observed and well characterized in Francis turbines [8]. So, the counter-rotating zone is annulus, a small backflow zone of co-rotating flow is present at the center, close to the hub and in its wake. It is seems that this small zone vanishes in the diffuser conical section.

As presented in figure 5d, the swirl ratio decreases from OP 2 to OP 5 at the top of the measurement section. This trend is in part due to the global flow deviation in the (X-) direction, previously mentioned. But also, this trend is due to the reduction of tangential velocity from the secondary flow, as shows in figure 5c for OP 5. The global decrease of the swirl in the mesured zone suggests that the swirl decreases also at the wall with the guide vanes opening. A well know effect of the swirl is the prevention of flow separation at the wall [9]. Therefore the swirl reduction has an important contribution to the flow separation apparition in this case.

4. Conclusion

The complex flow in the transition section of a bulb turbine diffuser has been investigated. The turbine model has been operated at three operating conditions in the overload range, including the point with maximal efficiency for the selected blade angle and selected unitary velocity. The two other operating points have respectively a lower and a lager flow rate. Velocity measurements in the transition part have been performed using a 2-component LDV system at two cross-sections. The first section is located near the transition inlet. The second section is at the diffuser outlet.

The velocity field analysis suggests that the counter-rotating zone generated at the runner persists all along the diffuser for all OP. In opposition, the backflow in the wake of the hub seems to disappear in the conical section of the diffuser. The central counter-rotating zone intensity increases with the guide vanes opening but the vortex position remains the same over the OPs.

For the operating condition with the smaller flow rate, in addition to the counter-rotating zone, a backflow central recirculation is present at the diffuser exit. This zone appears disconnected to the flow separation zone at the wall. This central recirculation is induced by the hub wake and the low pressure at the diffuser exit.

At the operating condition with the larger flow rate, a flow separation zone appears at the (*X*+, *Y*-) corner with the help of a decreasing swirl level injected to the wall. In the opposite corner, the flow velocity is increased to balance the flow rate, this induces a zone of strong shear in the diffuser. The three-dimensional envelop of the flow separation zone appears to have an important size. The effects of the flow separation appear in the inlet to the transition part and the average reattachment line is downstream the diffuser. Surprisingly, the central counter-rotating zone, which is stronger for this operating point, seems not to be deviated by the flow separation.

For the best operating point, in terms of efficiency for the selected blade angle and the selected unitary speed, neither the wall separation nor the central recirculation is present in the averaged field. But an instantaneous wall separation or an instantaneous central recirculation might appear from time to time at the diffuser outlet.

This study complements the previous flow investigation in a bulb turbine diffuser operated at overload with flow description in the core flow. In the short term, these observations will be used to validate numerical simulation tools. The velocity measurements will provide a reference at the model outlet. Afterwards, the validated numerical tools will be used to design new diffusers and to predict performances with more confidence.

Acknowledgement

The authors would like to thank the participants of the Consortium on Hydraulic Machines for their support and contribution to this research project: GE Renewable Energy, Andritz Hydro Inc, Hydro-Quebec, Laval University, NRCan, Voith Hydro Inc. Our gratitude goes as well to the Canadian Natural Sciences and Engineering Research Council who provided funding for this research.

References

[1] Mauri S, Kueny J L and Avellan F 2005 Werlé–Legendre Separation in a Hydraulic Machine Draft Tube *J. Fluids Eng.* **126** 976–80

[2] Tridon S, Barre S, Ciocan G D, Leroy P and Ségoufin C 2010 Experimental investigation of draft tube flow instability *IOP Conf. Ser. Earth Environ. Sci.* **12** 12044

[3] Deschênes C, Houde S, Aeschlimann V, Fraser R and Ciocan G D 2014 Modern challenges for flow investigations in model hydraulic turbines on classical test rig *IOP Conf. Ser. Earth Environ. Sci.* **22** 22013

[4] Duquesne P, Fraser R, Maciel Y, Aeschlimann V and Deschênes C 2014 Draft tube flow phenomena across the bulb turbine hill chart *IOP Conf. Ser. Earth Environ. Sci.* **22** 32003

[5] Vuillemard J, Aeschlimann V, Fraser R, Lemay S and Deschênes C 2014 Experimental investigation of the draft tube inlet flow of a bulb turbine *IOP Conf. Ser. Earth Environ. Sci.* **22** 32010

[6] Duquesne P, Maciel Y and Deschênes C 2016 Investigation of Flow Separation in a Diffuser of a Bulb Turbine *J. Fluids Eng.* **138** 011102–011102

[7] Duquesne P, Maciel Y and Deschênes C 2015 Unsteady flow separation in a turbine diffuser *Exp. Fluids* **56** 1–15

[8] Susan-Resiga R, Dan Ciocan G, Anton I and Avellan F 2006 Analysis of the Swirling Flow Downstream a Francis Turbine Runner *J. Fluids Eng.* **128** 177

[9] Clausen P D, Koh S G and Wood D H 1993 Measurements of a swirling turbulent boundary layer developing in a conical diffuser *Exp. Therm. Fluid Sci.* **6** 39–48

IOP Conf. Series: Journal of Physics: Conf. Series **813** (2017) 012048 doi:10.1088/1742-6596/813/1/012048

Decoupled CFD-based optimization of efficiency and cavitation performance of a double-suction pump

A Škerlavaj[1,2], M Morgut[2], D Jošt[1] and E Nobile[2]

[1] Kolektor Turboinštitut, Rovšnikova 7, 1210 Ljubljana, Slovenia
[2] University of Trieste, Department of Engineering and Architecture, Piazzale Europa 1, 34127 Trieste, Italy

E-mail: aljaz.skerlavaj@kolektor.com

Abstract. In this study the impeller geometry of a double-suction pump ensuring the best performances in terms of hydraulic efficiency and reluctance of cavitation is determined using an optimization strategy, which was driven by means of the modeFRONTIER optimization platform. The different impeller shapes (designs) are modified according to the optimization parameters and tested with a computational fluid dynamics (CFD) software, namely ANSYS CFX. The simulations are performed using a decoupled approach, where only the impeller domain region is numerically investigated for computational convenience. The flow losses in the volute are estimated on the base of the velocity distribution at the impeller outlet. The best designs are then validated considering the computationally more expensive full geometry CFD model. The overall results show that the proposed approach is suitable for quick impeller shape optimization.

1. Introduction

Centrifugal pumps are widely used in industrial applications. Double-suction (also called: double-entry) centrifugal pumps allow transportation of greater flow rates than single-entry pumps [1] because they are less prone to cavitation problems (smaller $NPSH_R$, required net positive suction head). Another advantage is counter-balancing of axial hydraulic forces due to double-entry design [1].

The optimization techniques bring many benefits over the traditional "trial-and-error" design process, including less hard-to-spot human-based errors. In turbomachines in general, usually multiple objectives are to be optimized. One of the first multi-objective optimization study was performed by Lipej and Poloni [2]. Due to importance of centrifugal pumps in industrial applications, as well as of pump-turbines, many recent studies aim to optimize their geometry to improve their performance. Shingai et al. [3] performed a multi-objective optimization of pump-turbine, where the objectives were efficiency of turbine, pump efficiency, cavitation characteristics for both types of operation, torque to hold runner blades and total pump head. Xuhe et al. [4] optimized pump turbine for efficiency in both modes of operation whereas Zhang et al. [5] performed optimization of a centrifugal pump for vibration, using fluid-structure interaction (FSI).

The purpose of this work is to propose and validate an economical decoupled simulation method for the optimization of a pump. In the presented decoupled optimization case, only the impeller geometry was allowed to be modified, while the rest of the geometry was fixed. Therefore, initially, flow losses in the volute were estimated from CFD simulations as a function of the velocity distribution at the inlet of the volute. During the optimization only the flow in the impeller passage was simulated, whereas the flow losses in the volute were assessed from impeller outflow velocity distribution. The validation of the method was performed, for the most promising design candidates, by using a full geometry coupled method, considering the suction chambers, full impeller geometry and the double volute. However, given the interest in comparing the two methods, the non-uniformity of the flow at the impeller inlet due to the suction chambers was deliberately disregarded in case of the coupled method, by using the so-called stage (mixing plane) condition at the rotor-stator interface.

Besides the maximization of pump's hydraulic efficiency, also the minimization of the cavitation extent was set as the objective of the optimization. For the latter objective, a simple method for estimation of reluctance to cavitation was used and validated.

In this study, the specific speed nq of a pump was equal to 62 (specified per impeller side).

2. Numerical procedure

The geometry of the double-suction centrifugal pump is presented in figure 1. Instead of using the full-geometry (in this paper also called *coupled*) model (figure 1a), the optimization was performed for a *decoupled* model (figure 1c). The coupled model was used during the validation process, to compare the decoupled model to the coupled one.

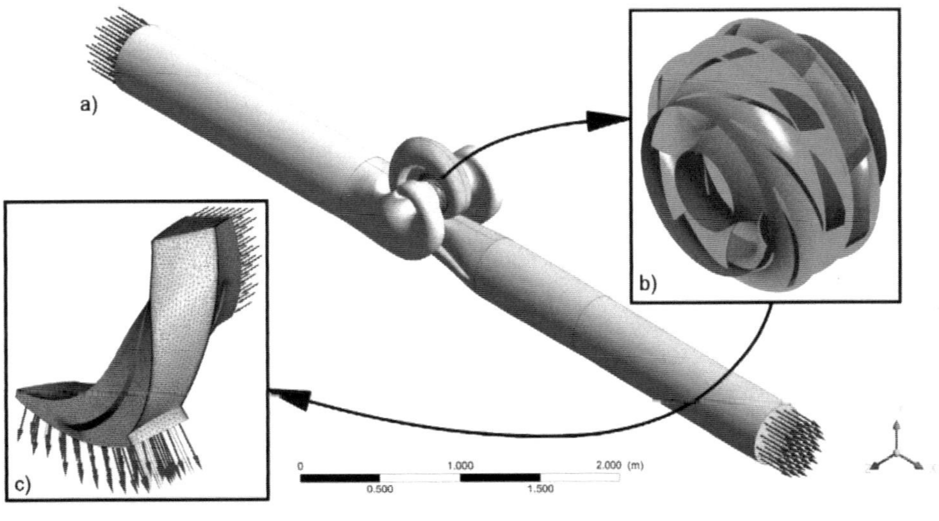

Figure 1. Double-suction centrifugal pump. (a) Geometry. In yellow: inlet pipe with suction chambers. In blue: impeller. In green: double volute with outlet pipe. (b) Double sided impeller. (c) Mesh of the impeller passage. Arrows represent inlet and outlet boundary conditions for the coupled (a) and the decoupled (c) simulation case.

It is well known that the suction chamber delivers a non-uniform flow to the pump impeller, which for instance results in length of the blade-attached cavities, varying upon the theta (circumferential) position of the impeller blade. In [6] it was shown that the growth and reduction of the cavities strongly depend upon the location, strength and rotation direction of the vortices formed in the suction chamber, close to the rib. In [7] the non-uniformity was quantified with the angle $\gamma = \text{arctg}(v_{circ}/v_{axial})$.

Before performing the decoupled optimization, the coupled CFD simulation was performed, to obtain the velocity distribution at the impeller inlet. The angle γ and the velocity distribution at the grid interface between the suction chamber and the impeller is presented in figure 2.

In the optimization phase, which was performed with the decoupled CFD model (figure 1c), the velocity components at the inlet boundary condition were prescribed as a function of radius around the rotation axis (figure 2b). At the outlet, the average pressure boundary condition was set. Periodic boundary conditions (in grey in figure 1c) were used to take into account the periodicity of the impeller passage. For the impeller, the shape of the meridional channel was based on past experience and was not modified during the optimization procedure. On the other hand, the blade passage geometry was generated from a set of 14 input parameters provided by an in-house code that uses an inverse-singularity method. The data was exported to ANSYS DesignModeler commercial tool, where a 3-D geometry of an impeller was automatically created. For each geometry, a tetrahedral-based, unstructured mesh of the impeller was automatically created in ANSYS ICEM, which also provided the generation of 15 prism layers on the walls.

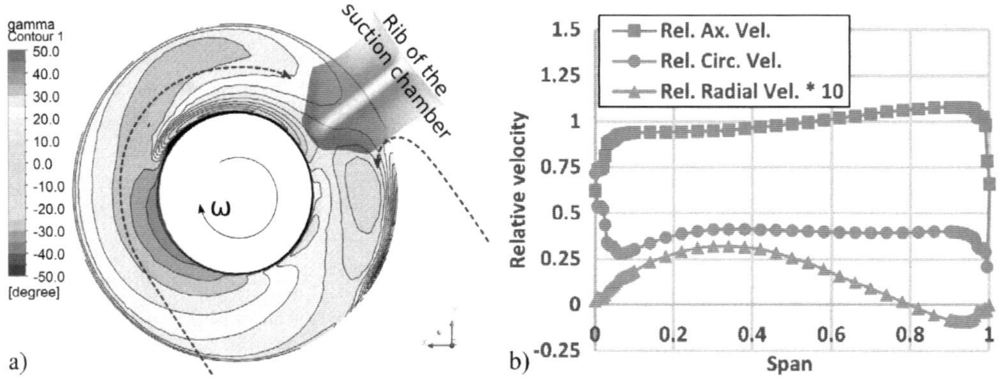

Figure 2. Flow conditions at the impeller entrance. (a) Contour plot of angle γ at the outlet from the asymmetric suction chamber. Blue dashed lines represent flow directions from the suction chamber. (b) Distribution of axial, circumferential and radial velocities at the entrance to the impeller. Velocities are relative to the average axial velocity at the entrance to the impeller. Radial velocity values in graph are multiplied by the factor of 10.

The optimization process was driven by modeFRONTIER 2014 [8] optimization platform. The objectives of the optimization were overall pump efficiency and reluctance to cavitation. The latter was assessed from the pressure difference between the impeller domain inlet boundary surface and the surface on the impeller blade, consisting of the leading edge and a part of the blade suction side. To obtain the overall pump efficiency, the flow losses of the volute ΔH_{volute} were assessed as a function of the angle α_2 between circumferential and absolute velocity at the impeller outlet, at regions near hub, at midspan and near shroud:

$$\Delta H_{volute} = \frac{f(\alpha_{2,hub})+f(\alpha_{2,midspan})+f(\alpha_{2,shroud})}{3}, \quad (1)$$

where the function f, a relationship between the flow losses and the α_2, was obtained prior to optimization from CFD simulations of the volute alone.

Besides the two objectives, two constraints were used: the overall pump head (as an interval), as it also influences the direction of the optimizer towards large heads [9, 10], and the allowed $\Delta\alpha_{2mid\text{-}sh}$, the difference of α_2 angle between the midspan and shroud intervals. Based on the volute simulations, the cases with the value of $\Delta\alpha_{2mid\text{-}sh}$ smaller than two degrees were considered as unfeasible.

Design of experiments (DOE), which fills the design space with the initial set of design variables, was based on the Latin Hypercube [11] samplings, but consisted also of some known high-efficiency designs, obtained in previous optimization cases. The Latin Hypercube sampling guarantees uniform random distribution of points over the variable range. The initial generation thus represented 48 sets of variables. Afterwards, the Multi Objective Genetic Algorithm (MOGA-II) [12], based on generational evolution, was used for 10 generations. Probability of directional cross-over, probability of selection and probability of mutation were set to 0.5, 0.05 and 0.1, respectively.

The validation of the decoupled-simulation-based method was performed for the most promising design candidates, by using a full geometry coupled method. Namely, after performing CFD simulations on 480 cases, each one taking less than six minutes of clock-time to complete, the simulations were ordered by the predicted efficiency. The ones with the largest efficiency value were then validated using the full geometry model (figure 1a), using modeFRONTIER platform as a convenient way to drive the CFD simulations and performing the postprocessing.

Both, coupled and decoupled CFD simulations, were performed with the ANSYS CFX software. These were performed only for the design point, as steady-state single-phase simulations with Shear-Stress-Transport turbulence model [13] with a curvature correction [14] (SST-CC). The 'high-resolution' advection scheme was used, which is an upwind adaptive scheme, based on the Barth and Jespersen's limiter [15]. A local time-scale factor equal to 10 was used. For the *decoupled* case, simulations lasted for up to 550 iterations, or stopped when RMS residuals were smaller or equal to $1 \cdot 10^{-5}$. The result data (head, torque and efficiency) were averaged over the last 20 iterations to produce meaningful result. It was checked that even for simulations with shorter number of iterations the averaging sample size choice did not influence the value of the output variables. For the validation with the *coupled* case, the simulations were performed using the stage condition [16] as a general grid interface between the impeller and the suction chamber on one side, and the volute on the other side. In this case the CFD simulations lasted for 600 iterations. All other settings remained the same. The process of case generation, case simulation and its postprocessing lasted for about 2:45 hours, on 128 computer cores.

3. Results and discussion

Figure 3a presents predicted efficiency by using decoupled and full geometry numerical approach. Results are presented in relative form, ordered by the decreasing value of the efficiency of the decoupled simulations. Case 0 is the case with highest efficiency (relative efficiency of 100%). It can be noticed that the inclination of the trendline of the full geometry model prediction is similar to the inclination of the results of the decoupled approach, which means that in general such method can be used for the optimization. However, the line of predicted efficiency with full geometry approach is jagged, with values oscillating for approximately ±0.5%, so it is worth investigating the reasons for such behaviour.

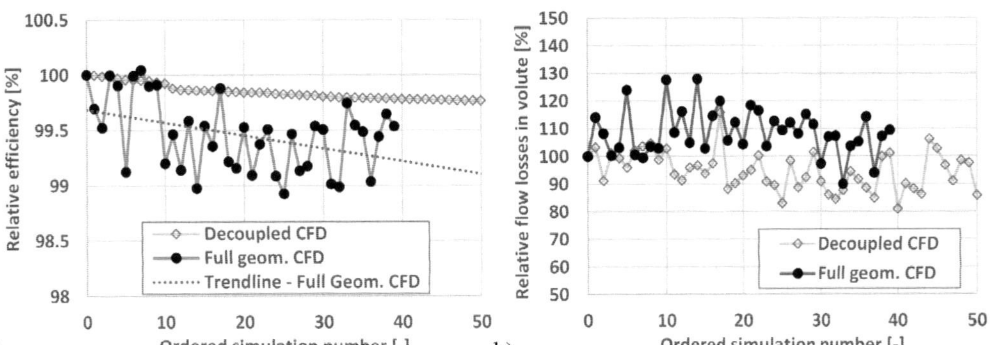

a) b)

Figure 3. Comparison of (a) efficiency and (b) flow losses in the volute for decoupled and full-geometry steady-state simulations. Cases are ordered by efficiency, obtained by the decoupled numerical model. Efficiency and flow losses are relative to the values obtained in case 0.

Since for the decoupled approach the prediction of the flow losses in the volute is based on the α_2 parameter, it is worth comparing predicted values of α_2 at hub, midspan and shroud intervals for the two numerical approaches. In our case, for the decoupled approach, the α_2 at hub was overpredicted by approximately 1.3°, whereas the α_2 at shroud was underpredicted by approximately 1.3° (figure 4). At midspan, the underprediction was negligible, 0.2°. The systematic deviations at hub and at shroud, due to the volute interaction with the impeller, can be taken into account in future optimization processes, to improve the results. When the systematic deviations in figure 4 are eliminated, the discrepancy between the two curves in each graph is small, usually smaller than 1°. Nevertheless, flow losses, predicted by equation (1), do not change much when the corrected value of α_2 is used.

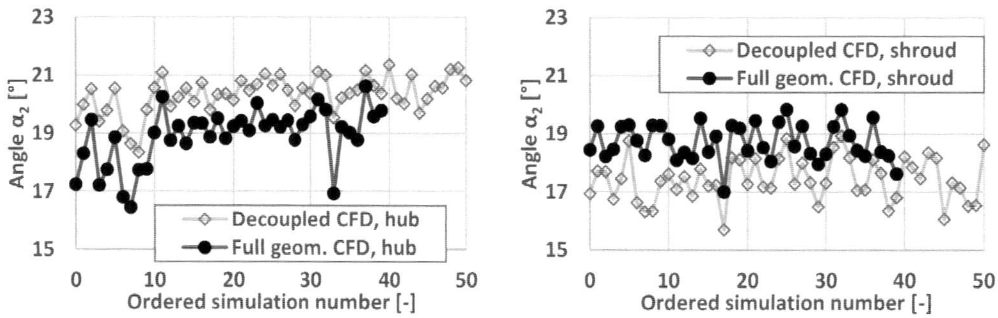

Figure 4. Angle α_2 at hub and shroud, predicted by decoupled and full-geometry steady-state simulations. The order of the cases is the same as in figure 3.

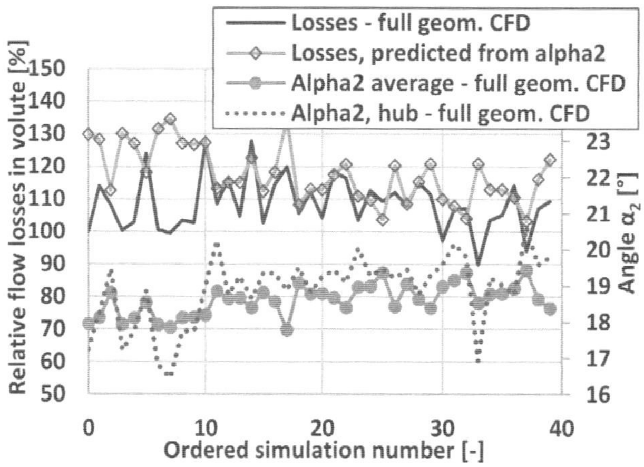

Figure 5. Primary axis: relative (to case 0 baseline, as predicted by the CFD) flow losses in volute for the full geometry model - comparison of CFD and prediction by equation (1). Secondary axis: angle α_2 - average value and value at hub. Order of cases is the same as in figure 3.

For the reason presented above it is necessary to consider a modification of equation (1), to address other reasons influencing the volute flow losses. In figure 5 only the full geometry (coupled) CFD model is considered. For the full model, the flow losses predicted by equation (1), where angles α_2 are obtained from the coupled simulations, are compared to the "real" losses (blue curves). It can be noticed that the approach presented by equation (1) is acceptable for designs from case 10 onwards. However, large discrepancy can be noticed for cases 0, 3, 4, 6, 7, 8, 9 and 33. These cases seem to be the cases where the value of $\alpha_{2,hub}$ is smaller than the average value of α_2 (figure 5). Small value of

$\alpha_{2,hub}$ seems to act beneficially on reducing the flow losses in the volute, so the equation (1) could be modified by using a weight factor, accounting for the low value of $\alpha_{2,hub}$.

Apart from the efficiency, another objective of the optimization was reluctance to cavitation. A very simple variable was considered as indicative of this risk: minimal value of pressure difference Δp between the impeller domain inlet boundary surface and the surface on the impeller blade, consisting of the leading edge and a part of the blade suction side. In figure 6 it is possible to observe that the decoupled and the full geometry CFD approach predicted similar values of Δp. In figure 7 the values of Δp can be observed for cases 14 and 33. Although only the minimal value of such pressure difference is presented in figure 6, it can be observed in figure 7 that low-pressure values are attributed to a part of a surface, rather than just one point. This means that the result is not just a pure coincidence, based on a location of a node of the surface mesh. The proposed variable is therefore a good simple candidate for assessment of reluctance to cavitation.

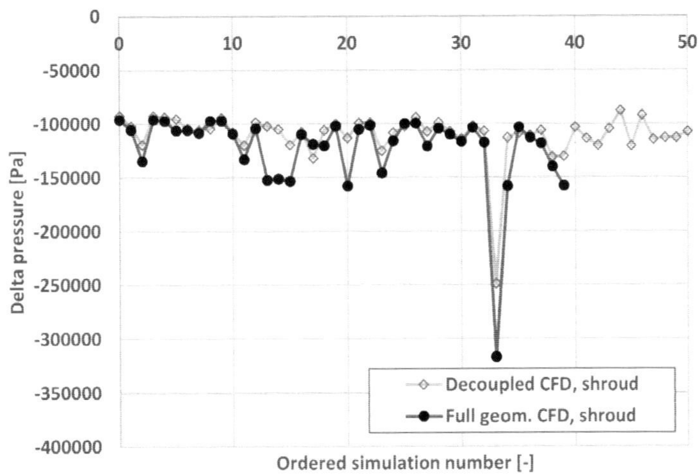

Figure 6. Pressure difference between pressure at inlet to impeller domain and pressure at blade surface (leading edge and suction side). Comparison of decoupled and full-geometry steady-state CFD simulations. Order of cases is the same as in figure 3.

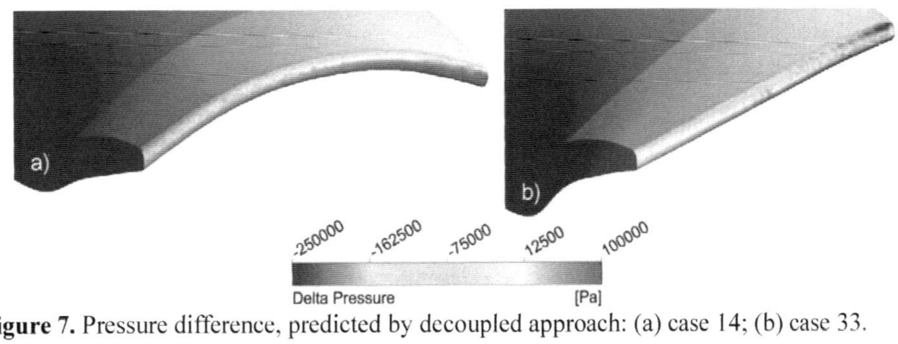

Figure 7. Pressure difference, predicted by decoupled approach: (a) case 14; (b) case 33.

4. Conclusions

A decoupled numerical model for the optimization of a double suction pump's impeller was presented. Results for the most promising candidates for high pump efficiency were validated with the full geometry CFD numerical models. The following conclusions were drawn:

- The presented method with estimation of flow losses in the volute represents a good approach for quick optimization;

- Values of angle α_2 at hub and at shroud predicted with the decoupled approach need small adjustments;
- Low value of angle α_2, when compared to the average α_2 value, acts beneficially on reducing the volute flow losses.
- The value α_2 at hub should be included in the equation for determination of the volute flow losses;
- The presented Δp variable is a good simple candidate for the assessment of reluctance to cavitation.

Acknowledgements
The research leading to these results has received funding from the People Programme (Marie Curie Actions) of the European Union's Seventh Framework Programme FP7/2007-2013/ under REA grant agreement n°612279 and from the Slovenian Research Agency ARRS - Contract No. 1000-15-0263.

References
[1] Gülich JF 2008 *Centrifugal Pumps* (Berlin: Springer-Verlag) p 926
[2] Lipej A and Poloni C 2000 Design of Kaplan runner using multiobjective genetic algorithm optimization *J. Hydraul. Res.* **38**(4) pp 73–9
[3] Shingai K, Omura Y, Tani K, Tominaga H and Ito S 2008 Multi-objective optimization of diagonal flow type reversible pump-turbines *Proc. of 24th Symp. on Hydraulic Machinery and Systems* October 27-31 Foz do Iguassu (Madrid: IAHR)
[4] Xuhe W, Baoshan Z, Lei T, Jie Z and Shuliang C 2014 Development of a pump-turbine runner based on multiobjective optimization *IOP Conf. Ser.: Earth Environ. Sci.* **22** 012028
[5] Zhang Y, Hu S, Zhang Y, Chen L 2014 Optimization and analysis of centrifugal pump considering fluid-structure interaction *Scientific World J.* **2014** 131802
[6] Škerlavaj A and Pavlin R 2014 Effect of vortical structures on cavitation on impeller blades in pumps with suction chambers *IOP Conf. Ser.: Earth Environ. Sci.* **22** 052002
[7] Muntean, S., Škerlavaj, A., Drăghici, I. and Anton, L.E., 2015. Numerical analysis of the flow non-uniformity generated by symmetrical suction elbows of the large storage pumps *Proceedings of 6th IAHR International Meeting of the Workgroup on Cavitation and Dynamic Problems in Hydraulic Machinery and Systems*, Ljubljana, Slovenia.
[8] ESTECO 1999-2016 *modeFRONTIER User Manual* (ESTECO: Trieste)p 504
[9] Kim JH, Oh KT, Pyun KB, Kim CK, Choi YS and Yoon JY 2012 Design optimization of a centrifugal pump impeller and volute using computational fluid dynamics *IOP Conf. Ser.: Earth Environ. Sci.* **15** 032025
[10] Škerlavaj A, Morgut M, Jošt D and Nobile E 2016 Optimization of a single-stage double-suction centrifugal pump *To appear in proc. of 34th UIT Heat Transfer Conference* July 4-6 Ferrara
[11] McKay MD, Conover WJ and Beckman RJ 1979 Latin hypercube sampling: a comparison of three methods for selecting values of input variables in the analysis of output from a computer code *Technometrics* **21** 239–45
[12] Poles S 2003 *MOGA-II An improved Multi-Objective Genetic Algorithm* (Trieste: ESTECO)
[13] Menter FR, Kuntz M and Langtry R 2003 Ten years of industrial experience with the SST turbulence model *Turbulence, Heat and Mass Transfer 4* ed K Hanjalić, Y Nagano et al (New York: Begell House) pp 625-32
[14] Smirnov PE and Menter F 2009 Sensitization of the SST turbulence model to rotation and curvature by applying the Spalart-Shur correction term *J. Turbomach.* **131** 041010
[15] Barth TJ and Jespersen DC 1989 The design and application of upwind schemes on unstructured meshes *Proc. of 27th Aerosp. Sci. Meet.* (Reno: AIAA) 0366 (12pp)
[16] ANSYS 2016 *ANSYS CFX-Solver Theory Guide* (ANSYS: Canonsburg) p 362

IOP Conf. Series: Journal of Physics: Conf. Series 8(2017) 012049 doi:10.1088/1742-6596/813/1/012(

Numerical investigation of draft tube pressure pulsations in a Francis turbine with splitter blades

I Kassanos, J Anagnostopoulos and D Papantonis

Laboratory of Hydraulic Machines, School of Mechanical Engineering, National Technical University of Athens, Heroon Polytechneiou 9, Zografou, 15780 Athens, Greece

E-mail: ikassanos@gmail.com

Abstract. Operation of Francis turbines at part load conditions is related to the appearance of the draft tube helical vortex rope. Splitter blades have been employed in high head Francis turbines in order to improve performance as well as their unsteady characteristics. In this work the draft tube unsteady characteristics of a Francis runner with splitter blades are investigated numerically. Two different splitter designs were analysed, and the performance results were compared to the baseline runner with no splitter blades used. The amplitude of pressure pulsation caused by the precessing vortex rope as well as the related frequency was compared for all cases, for two different operating conditions. From the results a relationship between the pulsation frequency and splitter blade geometry was observed.

1. Introduction

Francis turbines are the most popular reaction turbines due to their high efficiencies, increased reliability and extended lifetime. In order to maximize energy generation, Francis turbines are often required to operate at part load conditions where the precessing vortex rope forms in the draft tube potentially resulting in unwanted pressure fluctuations, power swings and structural vibration [1]. For these reasons, this phenomenon has been the focus of significant research over the past years aiming towards understanding the flow patterns associated, predict the appearance of the phenomenon, understand and quantify its effects on the turbine and at the same time identify corrective measures.

Through experimental measurements the physical mechanism of the precessing vortex rope has been studied at various operating conditions with an effort to understand the underlying flow physics. By obtaining velocity measurements downstream of the runner the evolution of the vortex and its boundary as function of time and cavitation number has been studied [2,3]. These analyses provide useful insight on the behavior of the vortex rope and at the same time can pose a baseline facilitating the development of numerical techniques and prediction models.

With the development of computation fluid dynamics (CFD) tools, it has become possible to perform reliable simulations for the analysis of the of the flow patterns inside the turbine. Several studies have used this tool to simulate the flow in the draft tube in conditions where the vortex rope appears aiming towards the validation of the numerical approach against experimental data [4,5]. As draft tube performance is more important for low head machines, there has been little research on the flow observed in the draft tube of high head turbines. In high head turbines splitter blades have been employed to improve the efficiency of the turbine, minimize secondary flows and reduce pressure fluctuation amplitudes [6,7,8]. Although the use of splitter blades has been popularized in high head machines, there is limited work studying the effect of the splitter blade geometry on the characteristics of the draft tube vortex rope at part load conditions. In this work the effect of the splitter blades on the draft tube pressure pulsation characteristics of two different splitter blade designs was investigated numerically.

2. Numerical Model

In the present study a turbine of specific speed equal to v_o=0.26 as specified by the IEC standard [9] was investigated numerically with a reference runner diameter of D_{2e}=210 mm and 13 runner blades. In order to minimize the computational resources for this study, the distributor and spiral casing were neglected. The numerical model shown in figure 1, consisted of the runner, the draft tube suction cone, as well as a straight extension downstream of the suction cone. The cone angle was 4.2° while all the draft tube dimensions are given in figure 1 non-dimensionalised by the runner reference diameter D_{2e}. The numerical grid used consisted of 2.11 million tetrahedral elements in total, of which 30% were used in the runner while the draft tube was modeled using approximately 1.5 million elements. Two operating conditions were considered corresponding to 0.9 Q_{BEP} and 0.72Q_{BEP} at the design head. The inlet boundary conditions consisted of a constant radial and tangential velocity component corresponding to the guide vane opening for each flow condition, while a zero average static pressure was assumed at the outlet. In table 1 the corresponding boundary conditions and inlet turbulence quantities considered are summarized.

Previously it has been shown that the ke model is associated with a damping effect leading to an under-prediction of the pressure fluctuation amplitudes, while a better agreement with experimental amplitudes and frequencies can be achieved using the SAS model [5]. Recently, the SAS model has been shown to slightly overpredict the pressure amplitudes at higher harmonics compared to experimental values and the RSM model [10]. However, since in this study the effect of different splitter blade geometries on the draft tube vortex is compared numerically and in order to minimise computational requirements, the SAS model was considered adequate and used in all calculations.

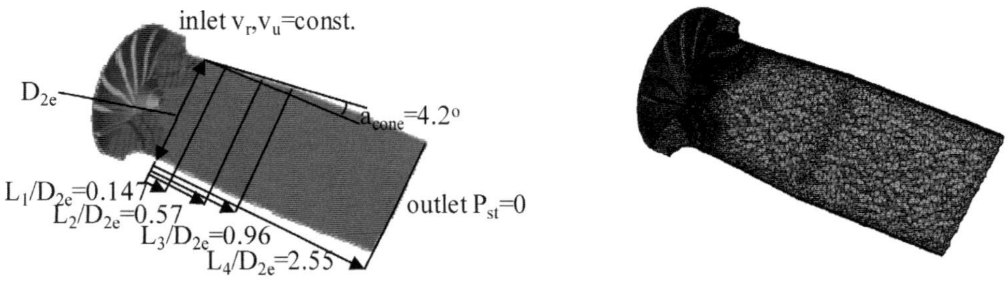

Figure 1. Computational domain and mesh under consideration.

Table 1. Operating points (OP) considered.

Rotational Speed: 750 rpm						
Operating Point	Q/Q_{BEP}	Flow rate parameter Φ	Net Head Parameter Ψ	Radial velocity v_r[m/s]	Tangential velocity v_u[m/s]	Turbulence Intensity [%]
1	0.9 * Q_{BEP}	0.26	2.7	1.88	8.55	6
2	0.72 * Q_{BEP}	0.207	2.7	1.51	9.15	6

For the velocity and pressure coupling, a fully coupled method was assumed, while for all equations a second order discretization scheme in space and time were considered. Finally, a time step corresponding to a 1° rotation per time step was used.

3. Results

3.1. Numerical results of the initial geometry

In all the simulations performed, two monitoring locations at an axial position of L_1=0.147D_{2e} and L_2=0.57D_{2e} downstream of the runner were considered (figure 1). On each section four monitoring points were placed around the circumference of the draft tube wall corresponding to the intersection of the monitoring planes and the x (x0, x1) and y (y0, y1) planes. An initial solution was obtained using a steady state computation and a frozen rotor approach, before the simulation was switched to an unsteady analysis. After switching to the unsteady model approximately 1500 time steps were required before time periodic results were achieved.

In the following figures the results are presented in dimensioless form using the runner rotational speed for the frequencies and the turbine head for the pressure fluctuations. Figure 2 compares the pressure monitor signals in the frequency domain for the two downstream pressure monitors of the initial runner geometry considered (L_1 and L_2) at OP1. Comparing the response between the two downstream locations, it can be seen that the pressure fluctuation is quickly attenuated as the flow moves towards the exit of the suction cone. At the same time, the pressure fluctuations appear at the runner rotation frequency and its harmonics, with the dominant frequency appearing at $2f_n$. At these flow conditions the residual swirl (table 2) entering the suction cone isn't stong enough to initiate the formation of the vortex rope.

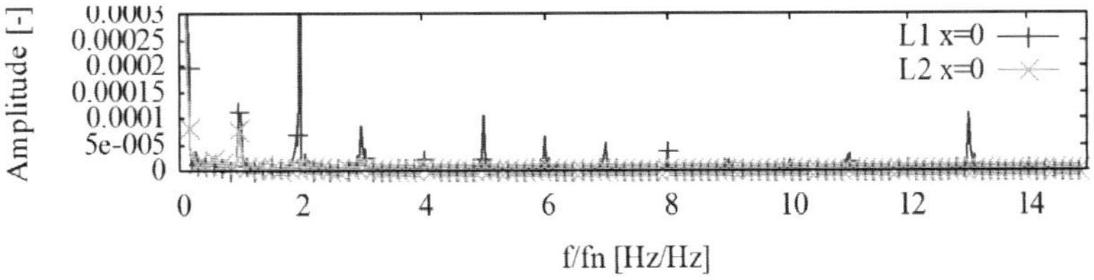

Figure 2. Frequency spectrum of the monitor pressure data for OP1 using the SAS model.

In figure 3 the pressure monitor signal for OP2 in the frequency domain is shown, where the dominant frequency of 0.3fn is identified. Contrary to OP1, the swirl at the suction cone outlet has increased to a point where a recirculation region is formed towards the axis (figure 5), the flow becomes unstable and the vortex rope is formed.

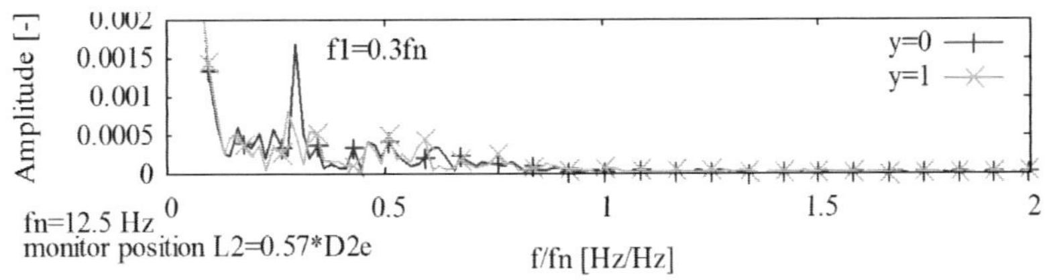

Figure 3. Frequency spectrum of the monitor pressure data for OP2 using the SAS model.

Table 2. Splitter blades geometries and dimensionless flux of moment of momentum (m_2) at L_1 section.

Case #	Area ratio (A_r)	Pitch position (P)	m_2 OP1	m_2 OP2
initial	-	-	0.0189	0.031
1	0.67	0.5	0.0135	0.0263
2	0.64	0.5	0.01396	0.0266

A secondary frequency approximately 0.5 times the rotational frequency can be also identified. This frequency components is believed to be caused due to the self rotation of the vortex rope [3].

IOP Conf. Series: Journal of Physics: Conf. Series 813 (2017) 012049 doi:10.1088/1742-6596/813/1/012

3.2. *Effect of splitter blades on draft tube flow*

Table 2 summarizes the two different splitter blade geometries considered in the present study. The geometries are characterized by the ratio of the surface area of the splitter blade to the surface area of the main blade, as well as the pitch-wise position in the inter-blade passage. In figure 4 the geometries of the two variants are compared with the main runner blade. It can be seen that the main difference of the two splitter blades is that the outlet edge of case 1 lies on a meridional plane, which is not the case for case 2 geometry. In Figure 5 the dimensionless velocity distribution at monitoring location L_1 for all cases considered are compared. These distributions were used to compute the flux of moment of momentum at the runner outlet as a quantitative indicator of the residual swirl associated with a particular design in a similar manner as in [11]. The calculated fluxes are shown in table 2 where it can be seen that at OP2 both splitter blade cases lead to a reduction of the residual swirl of 15.2% and 14.2%, respectively.

Figures 6 and 7 show the frequency spectrum of the pressure monitor signals for the two splitter blade cases at OP1 and OP2. At OP1, comparing the initial runner with case 1, an increase in the pressure fluctuation amplitude can be seen for all harmonics up to $6f_n$. This increase is a result of the increase in number of blades, which have doubled compared to the initial geometry. The dominant frequency in this case is the runner rotation frequency.

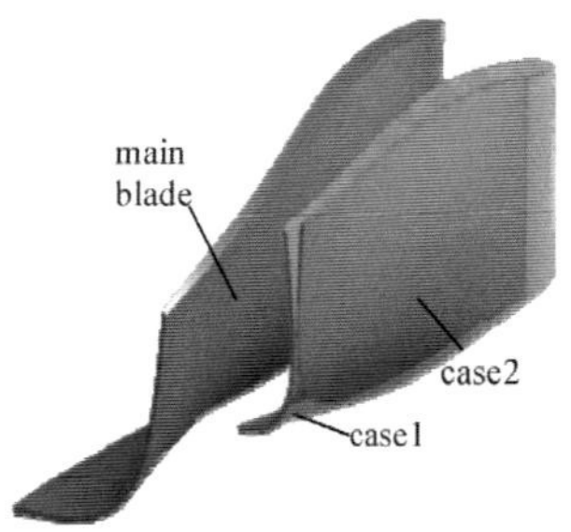

Figure 4. Comparison of splitter blade geometries considered in the present study.

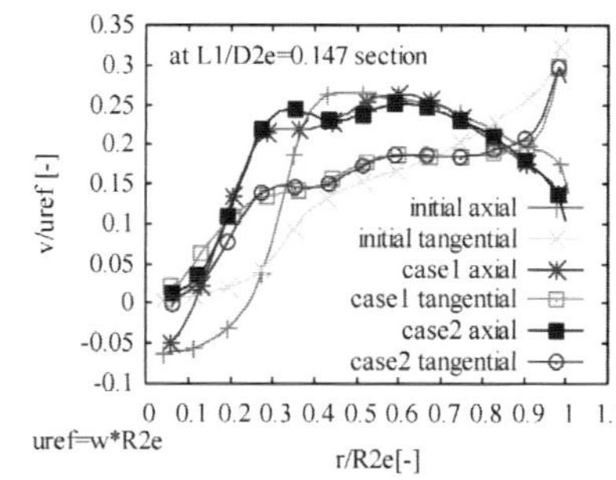

Figure 5. Dimensionless velocity distributions at OP2 at monitor location L_1.

For OP2, an analysis of the pressure signals shows that for case1 the vortex rope frequency has shifted to $0.34f_n$, while three additional frequency components at $0.44f_n$, $0.53f_n$, $0.66f_n$ can be identified, which are simultaneously captured by all four monitors at sections L1 and L2. For case 2, at OP1 the pressure amplitude has further increased in the first three harmonics as a result of the slightly increased residual swirl compared to case 1 (table 2). Similar to case 1, the dominant frequency is the runner rotation frequency. At higher frequency harmonics, similar amplitudes can be observed for all three geometries. At OP2 where the vortex rope has formed, the frequency components of 0.32 fn, 0.54fn, 0.75fn are dominant for case2, with the vortex rope frequency at $0.32f_n$. Similar to case 1 the amplitude of pulsation has been significantly reduced as also suggested by the reduction in the flux of moment of momentum. Compared to case 1, the frequency spectrum of case 2 at OP2 shows more resemblance to the initial geometry spectrum. The above suggest that small changes in the shape of the splitter blade can affect the characteristics of the vortex rope at part load conditions.

4. Conclusions

In this paper, the effect of the splitter blade geometry on the draft tube vortex rope was studied numerically. Two different splitter blade designs were compared against the initial runner with no splitter blades at flow conditions corresponding to $0.9Q_{BEP}$ and $0.72Q_{BEP}$. The results showed that the introduction of the splitter blades alters the frequency response in the draft tube due to a reduction of the residual swirl downstream of the runner. It was shown that small geometry modifications can affect the behaviour of the runner at part load conditions, suggesting that an optimum design could be obtained minimising the effects of the part load vortex rope. The effect of the splitter blades at lower flow conditions in relation to geometry modifications should be further investigated in order to further quantify the effects of the splitter blades on part load performance and the velocity distribution downstream of the runner.

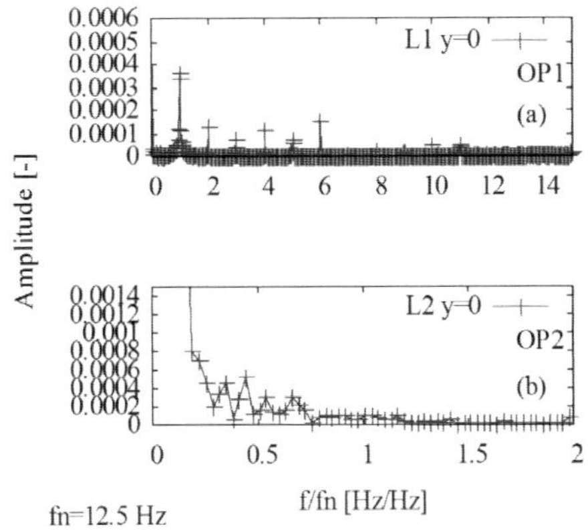

Figure 6. Frequency spectrum of the monitor pressure data for the case2 (a) OP1 (b) OP2.

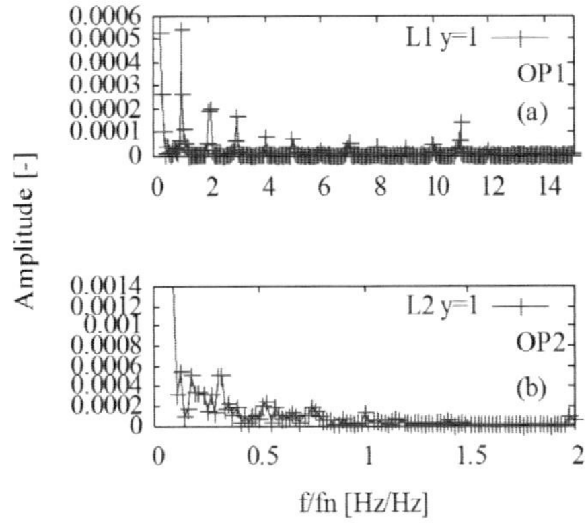

Figure 7. Frequency spectrum of the monitor pressure data for the case2 (a) OP1 (b) OP2.

References

[1] Li S C 2000 *Cavitation of Hydraulic Machinery* (London: Imperial College Press)

[2] Iliescu M, Ciocan G, Avellan F 2008 Analysis of the cavitating draft tube vortex in a Francis turbine using particle image velocimetry measurements in two-phase flow *Journal of Fluids Engineering* **130** 146-157

[3] Kirshner O, Ruprecht A, Gode E 2009 Experimental investigation of pressure pulsation in a simplified draft tube, *Proc. of the 3rd IAHR Int. Meeting of the Work Group on Cavitation and Dynamic Problems in Hydr. Machinery and Systems* (Brno, Czech Republic)

[4] Ciocan G, Iliescu M, Vu T, Nennemann B and Avellan F 2007 Experimental Study and Numerical Simulation of the FLINDT Draft tube Rotating Vortex, *Journal of Fluids Engineering* **129** 146-158

[5] Jost D, Lipej A 2009 Numerical Prediction of the vortex rope in the draft tube, *Proc. of the 3rd IAHR Int. Meeting of the Work Group on Cavitation and Dynamic Problems in Hydr. Machinery and Systems* (Brno, Czech Republic)

[6] Brekke H 2001 State of the art in turbine design. *Proc. Of the XXIX IAHR Congress 21st Century: New Era for Hydraulic Research and its Application* (Beijing, China)

[7] Hou Y 2012 Research on the length ratio of splitter blades for Ultra High Head Francis runners *Procedia Engineering* **31** 92-96

[8] Meng L, Zhang S P, Zhou L J, Wang Z W 2013 Study on the Pressure Pulsation inside Runner with Splitter Blades in Ultra-High Head Turbine, *Proc. of the 27th IAHR Symposium on Hydrulic machinery and system*, (Montréal, Canada)

[9] IEC60193 1999 *Hydraulic turbines, storage pumps and pump-turbines-Model aceptance tests* 2nd ed. (Geneva, Switzerland)

[10] Muntean S, Tănasă C, Bosioc A I, Moş D C 2016 Investigation of the Plunging Pressure Pulsations in a Swirling Flow with Precesing Vortex Rope in a Straigt Diffuser, *Proc. of the 28th IAHR Symposium on Hydrulic machinery and system*, (Grenoble, France)

[11] Susan-Resiga R, Muntean S, Avellan F, Anton I 2011 Mathematical modelling of swirling flow in hydraulic turbines for the full operating range, *Applied Mathematical Modelling* **35** 4759-4773

IOP Conf. Series: Journal of Physics: Conf. Series **813** (2017) 012050 doi:10.1088/1742-6596/813/1/012050

Time-frequency domain identification of dynamic loads on hydropower station

Zhenyue Ma [1], Wanquan Sun[2]

([1] School of Hydraulic Engineering, Dalian University of Technology, Dalian, 116024, China.
[2] Department of Hydraulic Engineering, North China Electric Power University, Beijing, 102206, China.)

Corresponding author's e-mail: dmzy@dlut.edu.cn

Abstract: In this study the wavelet-based analysis technique is proposed for analyzing its time-frequency characteristics, and identifying its maximum dynamic loads with genetic algorithm. The signals extracted by wavelet analysis are furthermore used for estimating the modelling parameters and loads. It is also indicated that the load amplitudes estimated with the extracted data by wavelet analysis can provides more detailed and finer information than those by Fourier analysis technique. This study is a valuable reference for investigating the vibrating source rule and predicting the dynamic response of powerhouse structures.

1. Introduction

The vibration of powerhouse structure is an important problem during its design and operation. The significant dynamic modal parameters include damping ratio, material parameters, boundary conditions and dynamic loads. However, direct measuring of these components will be quite difficult because of their complicated mechanism and the giant structure. Therefore, it is very valuable for finding out an efficient method to set up a rational parametric model utilizing the dynamic response measurements of the structure and the system identification theory [1, 2].

Most of the dynamic analysis for hydropower plant's supporting structure is based on the Fourier Transformation technique (FFT) previously. But the FFT is more appropriate in linear systems than nonlinear systems. Wavelet transform is a fundamentally different approach from Fourier theory [3]. In this novel transformation, a signal is not decomposed into its harmonics, which are global functions that have support on $[-\infty, +\infty]$, but into a series of local basis functions called wavelets, which are of a waveform of effectively limited duration and having an average value of zero. At the finest scale, the wavelets may be very long. By wavelet transform, any particular local features of signals could be detected and identified from the scale and the position of the wavelets.

In this paper, combining the finite element analysis and field test of Xilongchi Pumped-storage Power Station located on the Shanxi province in China, the selection of the wavelet transform for its loads analysis and identification is discussed. The purpose of this paper is to provide the advanced technology that is conducted on vibratory analysis of the powerhouse structure by wavelet analysis.

2. Analysis of field experiments

In this study, we used 8 transducers on different positions of the powerhouse, by radial and vertical directions on different floors, piers and airhoods, as shown in figure 1. The frequency range of vibration-measuring device is 0.5Hz～125Hz. This study identifies chiefly the loads involved in the 50MW partial load operation condition.

Firstly, measured variation signals are analyzed by Fourier transform. The result shows the most important dynamic loads on hydropower plant's supporting structure are 58.08Hz hydraulic pulsation caused by the non-uniform wave of turbine blades; the other loads include hydraulic pulsations as white noise in spiral case and tailrace, the 50Hz or 100Hz unbalanced magnetic pull, and the 8.33Hz rotating unbalanced loads.

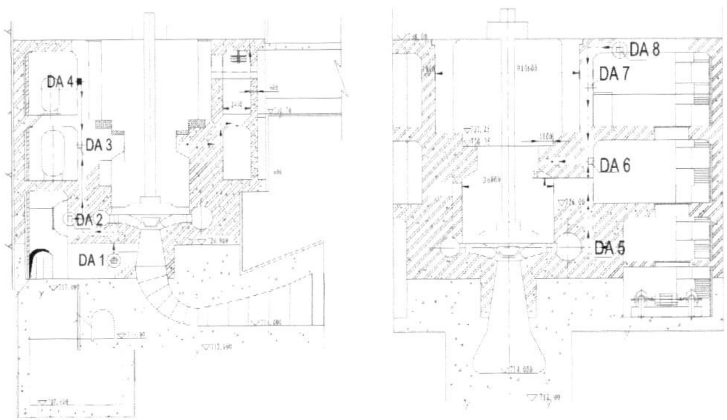

（a）Cross section of unit central line （b）Vertical section of unit central line

Figure 1. Arrangement of transducers

Subsequently, by applying wavelet function of Daubechies of order 2 (db2), all signals are decomposed into level 12. The time-frequency signals of 58.08Hz can be extracted from level 2 detail (D2), with example of the radial signal of DA4 shown in figure 2. Table 1 contrasts its maximum amplitude of wavelet analysis in confidence 97% with the amplitude of the Fourier transform. It is revealed that the amplitude by wavelet analysis is much higher than that by the Fourier transform. This also indicates that the 58.08Hz hydraulic pulsation of turbine blades is quite unstable, and its amplitude varies severely with time.

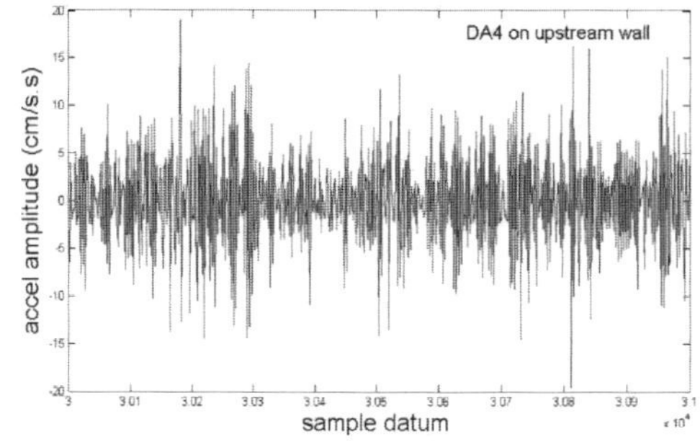

Figure 2. The wavelet signal of 58.08Hz on DA4 radial

Table 1. Contrast of maximum amplitude of wavelet analysis in confidence 97% and of the Fourier transform（cm/s^2）

	radial accel. amplitude (58.08Hz)		vertical accel. amplitude (58.08Hz)	
points	wavelet analysis	FFT analysis	wavelet analysis	FFT analysis
DA8	19.64	5.44	16.18	2.06
DA7	16.53	4.56	15.18	4.93
DA6	8.30	1.59	12.86	4.16
DA5	9.08	2.06	8.26	2.37
DA4	11.02	2.27	9.83	1.27
DA3	6.22	0.78	7.36	1.17
DA2	6.15	1.34	5.63	0.45
DA1	5.56	1.29	2.80	0.25

3. Identification of 58.08Hz hydraulic pulsation

The previous study in Ref.[5] shows that the acting mode of 58.08Hz hydraulic pulsation caused by turbine blades has three ways: the first is hydraulic vibration acting on the spiral case and stay ring etc. directly; the second is concentrated forces acting on lower bracket-arms transmitting from turbine axis; and the another is concentrated forces acting on upper bracket-arms.

3.1 Identifying method

The proposed identifying procedure shown in figure 3 provides a method to transform a parametric identification into an objective function optimization. By iterative calculating and modifying identified parameters to achieve the minimum of the objective function, the optimum load value is obtained. It is implemented by using the software ANSYS and Genetic algorithm. ANSYS is used to generate the FEM model (shown in figure 4), computing structural dynamic response. Genetic algorithm is applied to optimum identifying the loads.

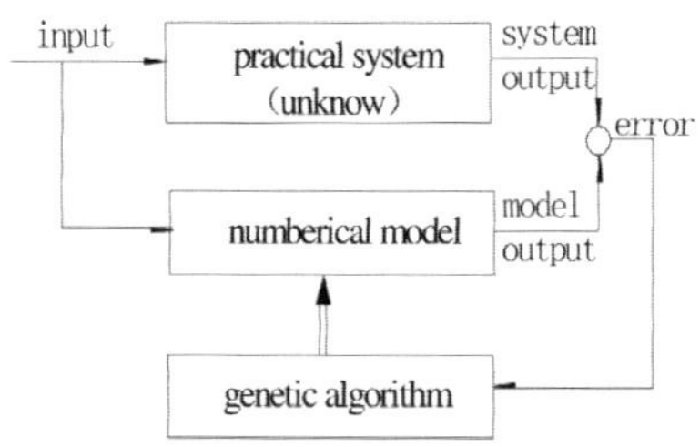

Figure 3. The process of structure system identification

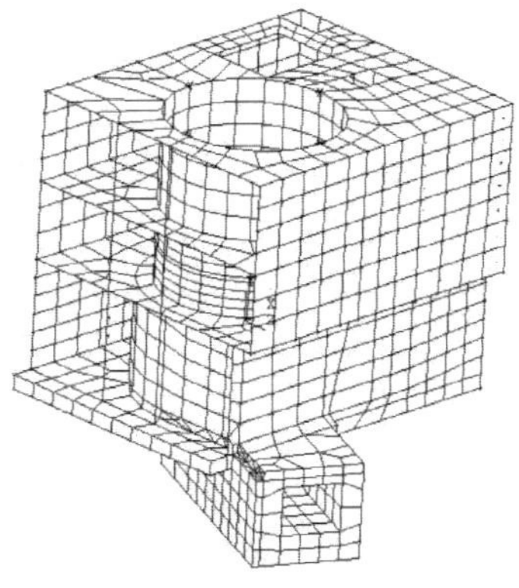

Figure 4. Finite element model of underground powerhouse

The objective function of Genetic algorithm is as following:

$$\varepsilon = \frac{\sqrt{\sum_{i=1}^{M}[(A_m(i) - A_e(i))/A_m(i)]^2}}{M}$$

Where $A_m(i)$ and $A_e(i)$ are the measured amplitude and the calculated amplitude respectively; M is the total of survey points.

3.2 Identified results and analysis

Using the radial datum of all measuring points to identify, the load results are obtained as following:

Based on the wavelet analysis, the single amplitude of 58.08Hz hydraulic vibration in spiral case is 17.6KN/m^2 (equal to 1.8m head pressure); the single amplitude of resultant force on lower bracket-arms is 57KN; the single amplitude of resultant force on upper bracket-arms is 1.12KN.

Otherwise, based on the Fourier analysis, the single amplitude of 58.08Hz hydraulic vibration in spiral case is 4.83KN/m^2 (equal to 0.5m head pressure); the single amplitude of resultant force on lower bracket-arms is 8.28KN; the single amplitude of resultant force on upper bracket-arms is 0.352KN.

The identified results show that the load values of hydraulic vibration caused by turbine blades are smaller during the 50MW partial load operation condition. The primary transmitted load resulting to the powerhouse vibration is hydraulic excitation acting on the spiral case liner, the secondary load excitation source is acting on the lower bracket-arms, and the load acting on the upper bracket-arms is relatively very small.

The identified load values based on the wavelet analysis are much greater than the values based on the Fourier analysis. Respectively, the values of hydraulic load acting on the spiral case and the force on lower bracket-arms identified from wavelet analysis are 3.6 and 6.9 times greater than the values identified from Fourier analysis. So the time-frequency analysis of wavelet transformation can provides more detailed and finer information about non-stationary signals than traditional Fourier analysis in the dynamic analysis of the powerhouse structures.

4. Summary and conclusion

In this paper, with the field test of Xilongchi Pumped-storage Power Station, the selection of the wavelet transform for its loads analysis and identification is discussed. The important hydraulic load of turbine blades is identified and analyzed. It shows that the amplitude of certain frequency load on the plant concrete structure is not invariant with the time varying, and it is not a standard harmonic. The traditional Fourier analysis cannot obtain the real maximum values of loads. So the advanced time-frequency analysis of wavelet transform is very important and significant for investigating the vibrating source and predicting the dynamic response of powerhouse structures.

References

[1] Ma, Zhenyue, Sun,Wanquan 2004 Dynamic identification for underground power house based on genetic algorithm. *J. of Dalian Univ. of Tech.* **44**(2): 292–296

[2] Sun Wanquan, Huang Xiong-hui 2014 Identification of vibration transfer path for the coupled system of water turbine generator set and power house. *J. of vib. and shock*, **33**（6）: 23-28.

[3] S.G. Mallat 1998, *A Wavelet Tour of Signal Processing*, Academic Press, San Diego, USA.

[4] Liao Xiaoyun, Wang G. Gary 2005 Wavelets-based method for variation analysis of non-rigid assemblies. *Inter. J. of Machine Tools & Manufacture*, **45**: 1551–1559.

[5] Sun Wanquan 2004 *Research on Structure Vibration and Identification of Hydropower House*[D]. Dalian University of Technology, China.

Acknowledgments

Authors wishing to acknowledge financial support from the National Natural Science Foundation of China (No.51379030).

IOP Conf. Series: Journal of Physics: Conf. Series **813** (2017) 012051 doi:10.1088/1742-6596/813/1/012051

On random pressure pulses in the turbine draft tube

P A Kuibin, S I Shtork, S G Skripkin, M A Tsoy

Kutateladze Institute of Thermophysics SB RAS, 1 Lavrentyev ave., Novosibirsk, 630090, Russia

E-mail: kuibin@itp.nsc.ru

Abstract. The flow in the conical part of the hydroturbine draft tube undergoes various instabilities due to deceleration and flow swirling at off-design operation points. In particular, the precessing vortex rope develops at part-load regimes in the draft tube. This rope induces periodical low-frequency pressure oscillations in the draft tube. Interaction of rotational (asynchronous) mode of disturbances with the elbow can bring to strong oscillations in the whole hydrodynamical system. Recent researches on flow structure in the discharge cone in a regime of free runner had revealed that helical-like vortex rope can be unstable itself. Some coils of helix close to each other and reconnection appears with generation of a vortex ring. The vortex ring moves toward the draft tube wall and downstream. The present research is focused on interaction of vortex ring with wall and generation of pressure pulses.

1. Introduction

The hydro power stations often play role of regulating instruments in electric grids. Moreover the levels of head water as well as tail water undergo considerable variations due to seasons changes or meteorological conditions etc. Thus, the hydraulic machines need to operate under wide range of technological parameters. Most of the operating points lying away from the best efficiency point are subjected to instability or cavitation developments inducing undesirable unsteady phenomena. In a most extent this refers to the turbines of Francis type. The state of the art in area of different aspects of the hydrodynamic instabilities can be found in recent reviews [1, 2]. In particular, one type of instability, vortex breakdown, leads to generation of helix-like vortex rope. Our recent research [3] on the flow structure behind model of the runner had demonstrated that the helical vortex rope can be unstable itself. During recording with high speed camera we observed that sometimes the neighboring coils of the helix close to each other and process of reconnection starts with further generation of a vortex ring. The vortex ring after the reconnection moves apart from the main vortex rope toward the wall and downstream generating pressure pulses on the wall (some details on interaction of a vortex ring with the wall can be found in [4]). Such event belongs to random (or aperiodical) pulsations which present a particular interest, due to the fact that their nature is much more complex in comparison with precessional pulsations or blade pulsations [5].

The goal of this paper is to make a better understanding of the random pressure pulses phenomenon. We will perform synchronous registration of the vortex ring propagation via high speed camera together with recording the pressure signal from the probe mounted on the wall.

2. Experimental setup

The studies were performed using a closed hydrodynamic loop with a maximum flow rate (Q) of up to 0.067 m^3/s. For simulating the cavitation regime in the test section, the set-up was equipped with a vacuum pump that provided additional evacuation up to 96 kPa. The working section (Figure 1)

IOP Conf. Series: Journal of Physics: Conf. Series **813** (2017) 012051 doi:10.1088/1742-6596/813/1/012051

analogous to one developed at Politehnical University of Timisoara [5], allowed for an optical access for the LDA system and high-speed visualization video camera. The working section presents a model of the draft tube cone with throat diameter $D = 0.1$ m. Another details on the experimental setup see in [6]. In the present research the velocity distributions at the cone inlet cross-section was qualitatively similar to the flow regime observed during part-load operation of Francis turbine [7]. The flow swirl parameter $s = 2\ M/KD = 0.47$ was found through the angular momentum M and momentum K. The second parameter characterizing the flow is Reynolds number, $Re=4Q/\pi D\nu$. Here ν is the kinematic viscosity. The maximal Reynolds number in experiments was about $5\cdot10^5$.

The vortex structures were visualized by the vapor bubbles formed due to cavitation in the zones of low pressure, i.e., at the vortex axis. Visualization was performed using high speed Photron video camera with a maximum resolution of 1 megapixel and frame rate of up to 20,000 fps. To measure pressure pulsations on the draft tube wall, the turbine model was equipped with a piezo resistive pressure transmitter Keller PA-4LC. The pressure fluctuations were measured at 64 mm downstream of the draft tube inlet (figure 1). Standard backscattering two-component LDV system LAD-06i was used for measuring the average velocity profiles (see details in [6]). The measured velocity profiles were used for evaluating the swirl parameter as well as parameters of the helical-like vortex. In particular, intensity of vortex, $\Gamma = 0.13$ m^2/s was found [6].

3. Flow visualization

We used the modern visualization technique which allows investigating in detail the main stages of the vortex reconnection and vortex ring formation. The structure of the basic flow realized in a conical section of the hydraulic model of the turbine is shown in figure 2. Due to vortex breakdown phenomenon the vortex core takes form of a left-handed helix. Localization of the vorticity near the central axis of the vortex leads to the formation of cavitation rope, which is filled with water vapor. Conditions of vortex breakdown is associated with the formation of counter flow along the geometrical axis of the channel. At the same time, the direct downward flow occurs along the working section walls. It should be noted that such a topology corresponds to the velocity field generated by a left-handed precessing vortex structure.

A complete scenario of reconnection process with the formation of a vortex ring was described by Alekseenko et al. [6] and is shown in Figure 3. Due to the unstable state of the vortex system in the expanded cone, at some moment, one of the spiral coils starts moving upward, in particular, due to the

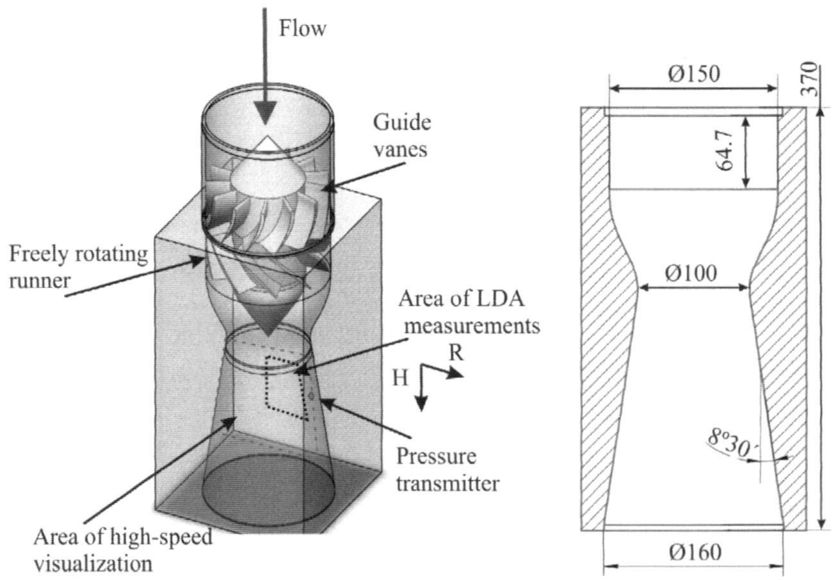

Figure 1. Simplified hydro turbine model

IOP Conf. Series: Journal of Physics: Conf. Series **813** (2017) 012051 doi:10.1088/1742-6596/813/1/012051

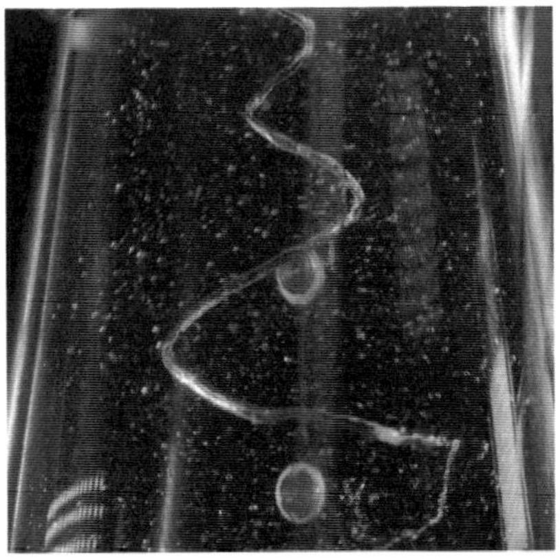

Figure 2. Basic flow regime in a conical draft tube model, $Q = 116\ m^3/h$

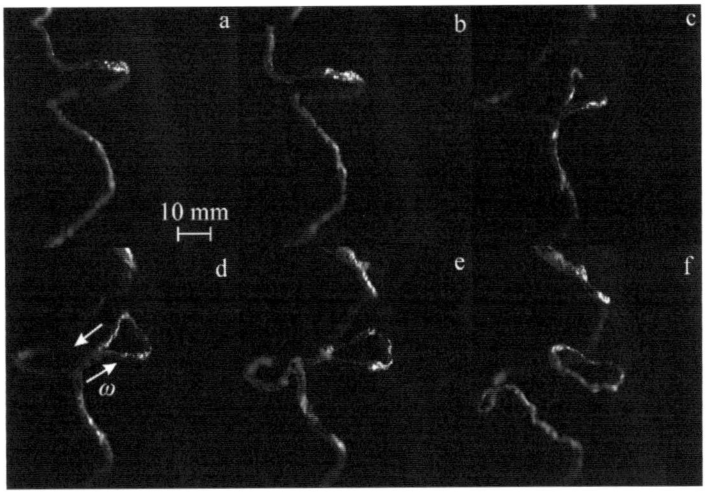

Figure 3. Sequence of the video frames demonstrating formation of isolated vortex ring in a conical part of draft tube model, $Q = 116\ m^3/h$ [6]

reverse flow in the near-axis zone of the channel. Subsequently, we observed an overlap of this coil, and further, the mechanism of convergence of the tube parts with the opposite vorticity vectors turns on (see Figure 3d). The tubes bend at the points of their connections and then the process of reconnection occurs. As a result of the reconnection, vortex ring separates from the spiral tube and moves downstream the cone. The duration of this process is approximately 10 ms from the start of convergence of the spiral parts and ending with full detachment of the ring. After reconnection process, the basic pattern of the vortex has being restored. The reconnection process with the formation of an isolated ring is observed regularly. The size of the vortex ring depends on the separation placement. The upstream reconnection occurs the smaller the vortex ring size. Similar processes of reconnection were investigated numerically [8], but papers on the reconnection in the swirling flow especially in hydro turbine was not found in the literature.

Two different scenarios of vortex ring reconnection were described in [6]: generation of separated vortex ring and generation of ring linked with main helical vortex. Here we are interested in first scenario only which leads to origin of pressure impacts. Based on analysis of video data (statistics up to 1000 revolution of vortex rope precession) we obtained the probability distribution histogram of the events of separations of vortex rings versus time delay between the events (figure 4). The averaged period of vortex rings separations T_{ring} is about 3 times higher than the vortex precession period T_{pr}.

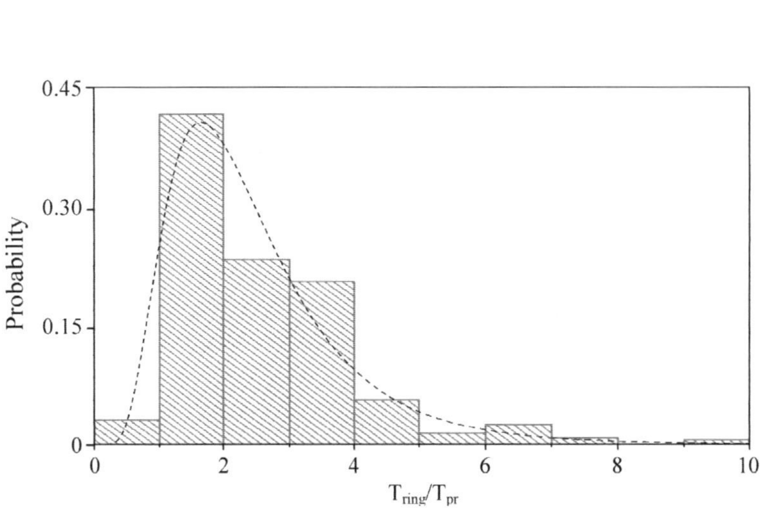

Figure 4. Histogram of probability of the vortex rings separation, $Q = 116$ m^3/h.

4. Pressure measurements

The oscillograms (figure 5) show the variation in time of the pressure on the wall (figure 1) of the draft tube model. Zero on the ordinate axis corresponds to the normal atmospheric pressure. The amplitude of the draft tube wall pressure pulsation in this flow regime caused by precession of the vortex rope is small, approximately 5-7 kPa. In the same time, the pressure pulse caused by the vortex ring is approximately 35-40 kPa, which is 7 times higher than the level of main fluctuations. In some cases, experiments show that the pressure shock caused by the vortex ring may exceed the pulsation induced by PVC by 10 times. After formation of the vortex ring, full collapse of the PVC cavity may occur. Depending on the vortex ring trajectory relative the pressure sensor, the different patterns of pressure pulses can be seen in the oscillogram: pressure pulse (figure 5), pressure pulse with subsequent depressure and vice versa.

5. Conclusion

A phenomenon of vortex ring formation on the precessing vortex rope and pressure pulses induced by vortex ring travelling near probe on the wall have been investigated experimentally in the simplified laboratory model of hydro turbine. Thus, in this experimental work for the first time high-speed visualization synchronized with the wall pressure measurement made it possible to associate random pressure pulses in draft tube with aperiodical process of vortex rings generation due to reconnection between the coils of helical vortex rope. After reconnection the vortex ring can move very close to the wall of the draft tube, causing a sharp local pressure drop. The average period of the vortex rings formation is about 3 times higher than the precession period. The recorded pressure signal depends on the vortex ring trajectory relative to the pressure sensor. Both the sharp rise and fall of the pressure were recorded. The trajectory of the vortex ring and its interaction with the wall of the draft tube are very complex. There are few factors influencing its motion: main swirling flow, PVC induced velocity, self-induced velocity, and velocity induced by interaction with the wall.

Acknowledgments
This work was supported by the Russian Science Foundation, grant No 16-19-00138.

References

[1] Nishi M and Liu S 2013 An outlook on the draft-tube-surge study *Int. J. Fluid Mach. Syst.* **6**(1) 33-48

[2] Nicolet C, Zobeiri A, Maruzewski P and Avellan F 2011 Experimental Investigations on Upper Part Load Vortex Rope Pressure Fluctuations in Francis Turbine Draft Tube *Int. J. Fluid Mach. Syst.* **4**(1) 179-190

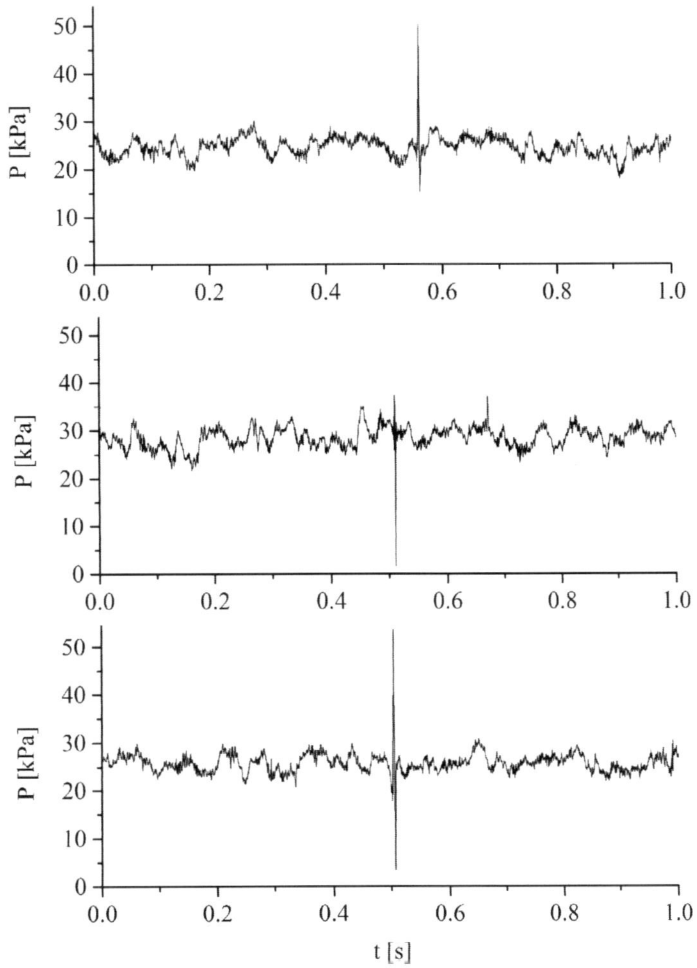

Figure 5. Oscillograms of the pressure pulses associated with vortex rings passing near the sensor.

[3] Alekseenko S V, Kuibin P A, Shtork S I, Skripkin S G and Tsoy M A 2016 Vortex Reconnection in a Swirling Flow *JETP Letters* **103**(7), 455–459

[4] Nicolet C 2007 Hydroacoustic modelling and numerical simulation of unsteady operation of hydroelectric systems, *Phd Thesis, Lausanne, EPFL, 314*

[5] Susan-Resiga R F, Muntean S, Tanasa C and Bosioc A 2008 Hydrodynamic design and analysis of a swirling flow generator. Proc. *of the 4th German – Romanian Workshop on Turbomachinery Hydrodynamics* (GRoWTH), June 12–15, 2008, Stuttgart, Germany http://acad-tim.tm.edu.ro/iSMART-flow/pdf/Resiga_4GRoWTH1.pdf

[6] Alekseenko S V, Kuibin P A, Shtork S I, Skripkin S G, Sonin V I, Tsoy M A and Ustimenko A S 2016 A novel scenario of aperiodical impacts appearance in the turbine draft tube *IOP Conf. Series: Earth and Environmental Science* **49**, 082025

[7] Hosein Foroutan and Savas Yavuzkurt 2014 Flow in the Simplified Draft Tube of a Francis Turbine Operating at Partial Load – Part I: Simulation of the Vortex Rope *J. Appl. Mech* **81**(6), 061010

[8] Hussain F and Duraisamy K 2011 Mechanics of viscous vortex reconnection *Phys. Fluids* **23**(2), 021701

[9] Naguib A, Koochesfahani M 2004 On wall-pressure sources associated with the unsteady separation in a vortex-ring / wall interaction *Phys. Fluids* **16**(7) 2613-2622

Expected load spectra of prototype Francis turbines in low-load operation using numerical simulations and site measurements

M Eichhorn[1], A Taruffi[2] and C Bauer[1]

[1] TU Wien, Institute for Energy Systems and Thermodynamics, Getreidemarkt 9/302, 1060 Vienna, Austria
[2] Andritz Hydro, Rue des Deux Gares 6, 1800 Vevey, Switzerland

E-mail: markus.eichhorn@tuwien.ac.at

Abstract.
The operators of hydropower plants are forced to extend the existing operating ranges of their hydraulic machines to remain competitive on the energy market due to the rising amount of wind and solar power. Faster response times and a higher flexibility towards part- and low-load conditions enable a better electric grid control and assure therefore an economic operation of the power plant. The occurring disadvantage is a higher dynamic excitation of affected machine components, especially Francis turbine runners, due to pressure pulsations induced by unsteady flow phenomena (e.g. draft tube vortex ropes). Therefore, fatigue analysis becomes more important even in the design phase of the hydraulic machines to evaluate the static and dynamic load in different operating conditions and to reduce maintenance costs. An approach including a one-way coupled fluid-structure interaction has been already developed using unsteady CFD simulations and transient FEM computations. This is now applied on two Francis turbines with different specific speeds and power ranges, to obtain the load spectra of both machines. The results are compared to strain gauge measurements on the according Francis turbines to validate the overall procedure.

1. Introduction

As solar and wind power generation is growing steadily, their volatility rises the necessity of large and also flexible energy storage. Even other systems like compressed-air or battery solutions will become more advanced in the future [1], pumped hydro energy storage will remain the most competitive technology in the next years [2]. The pending challenge for manufacturer and operators of pumped storage power plants (PSP) will be an increase of operating flexibility towards off-design operation with a simultaneously prevention of lifetime reduction. Due to unsteady flow phenomena in part-load, low-load and overload conditions of especially Francis and pump turbines, higher pressure fluctuations may appear inducing an increasing dynamic excitation of affected structures like the runner [3]. Therefore, the prediction of critical operating conditions and the fatigue analysis in the design phase of Francis and pump turbines is becoming more significant. The increasing capabilities of numerical computations enable their application for such investigations. To evaluate their accuracy as well, unsteady CFD and transient FEM simulations are performed on two prototype Francis turbines in low-load operating conditions: a high head runner with a specific speed of $n_q \approx 24\,min^{-1}$ (PSP-HH) and a medium head runner with a specific speed of $n_q \approx 52\,min^{-1}$ (PSP-MH). The results are used to compute the expected load spectra with the rainflow counting algorithm. Prototype site measurements should help to validate the whole approach, which has been partially published already (see [4] and [5]). Therefore, several single strain gauges (SG) have been applied on one runner blade at the pressure side (PS) and suction side (SS) close to the trailing edge (see figure 1). The position and orientation of the sensors have been recorded for the comparison with the numerical results.

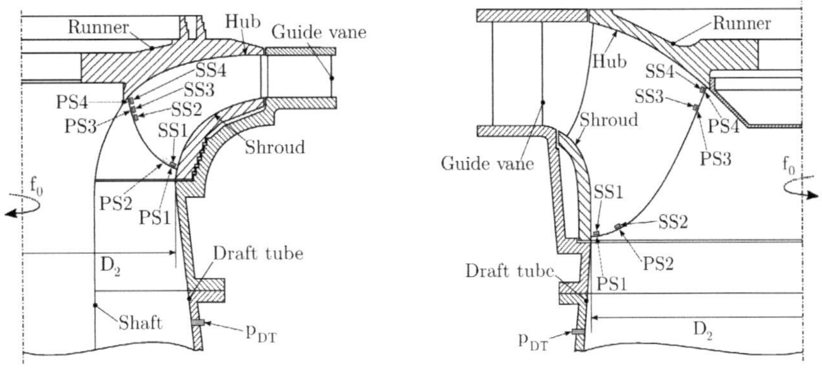

Figure 1: Schematic contour of the meridional section including the positions of the strain gauges on the runner blades and the pressure measurements in the draft tube for PSP-HH (left) and PSP-MH (right).

Further, pressure measurements have been done in the draft tube cone (p_{DT}) of both machines to validate the appearing pressure oscillations in the CFD simulations. One main difference between the two investigated Francis turbines, beside the specific speed, is the shaft going through the draft tube cone at PSP-HH, which has also a major impact on the flow behavior. The measured stress amplitudes σ_a, normalized by the yield strength of the runner material σ_y are displayed in figure 2.

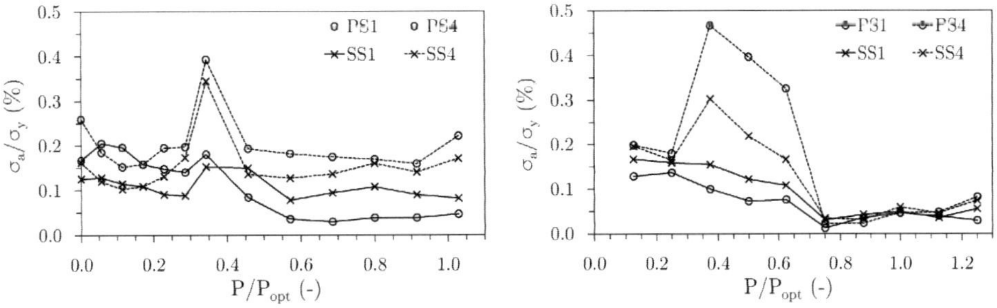

Figure 2: Measured dynamic stress amplitudes at different strain gauge positions and operating points for PSP-HH (left) and PSP-MH (right).

At PSP-HH a relative high stress amplitude level can be observed in all operating points due to the influence of the rotor-stator interaction (RSI). At low-load operation, with a normalized power of $P/P_{opt} \approx 0.34$, the stress amplitudes are obviously increasing with the highest values appearing close to the hub of the runner (SG PS4 and SS4). A similar behavior can be observed for the medium head Francis runner. Due to the higher gap size between the wicket gate and the runner and therefore a lower impact of the RSI, the dynamic stresses are lower in high-

and full-load operation. In part- and low-load operation, the stress amplitudes are dramatically increasing, especially at the runner hub region (SG PS4 and SS4), with the highest values appearing also at $P/P_{opt} \approx 0.37$. Hence, these two low-load operating points are of special interest and will be therefore used for the numerical simulations described in the following sections.

2. Unsteady CFD simulations

As the dynamic excitation of the Francis runners is mainly depending on the pressure oscillation inside the machine, unsteady CFD simulations are performed with OpenFOAM, to investigate the flow phenomena in the critical operating points. Therefore, the relevant components of the two hydraulic machines are considered, including the spiral casing, the stay vanes, the guide vanes the runner and the draft tube. The different domains are discretized separately with hexahedral meshes using ANSYS ICEM CFD and ANSYS Turbogrid (see figure 3). The mesh of PSP-HH consists of about 6 million cells and the one for PSP-MH of about 9 million cells. To accurately predict the pressure fluctuations inside the runner, the discretization is refined in this domain with about 2.6 million cells for PSP-HH and about 3.3 million cells for PSP-MH. The guide vane opening position is adjusted according to the measurements. The unsteady CFD simulations are performed for several runner rotations using a transient rotor-stator analysis. A frozen-rotor simulation is used as initial solution to improve and speed-up the convergence. The discharge is defined as boundary condition at the spiral inlet according to the measurements. At the draft tube outlet the static pressure is set to an averaged value. The time step size is fixed to an according runner rotation of $0.36\,°$ for PSP-HH and $0.6\,°$ for PSP-MH resulting in a maximum Courant number of $Co_{max} = 25$. The CFD simulations are performed using the two-equation turbulence model k-ω SST as well as the Scale Adaptive Simulation (SAS) approach, to evaluate the influence of the turbulence modeling.

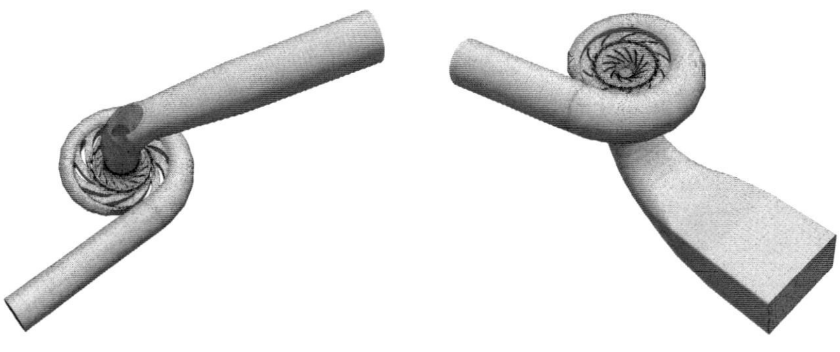

Figure 3: Discretization of the hydraulic components for the CFD simulations of PSP-HH (left) and PSP-MH (right).

In figure 4 the spectral power density of the measured and computed pressure signals in the draft tube cone p_{DT} at low-load operation for PSP-HH ($P/P_{opt} \approx 0.34$) and PSP-MH ($P/P_{opt} \approx 0.37$) are displayed. The pressure fluctuations Δp are normalized by the energy corresponding to the rotational speed $p_E = \rho u_2^2/2$ with the circumferential velocity at the runner outlet $u_2 = D_2 f_0 \pi$ and the rotational frequency of the runner f_0, as proposed in [6]. For a more convincing comparison between the long measurement duration ($120\,s$) and the short CFD simulations, the measured pressure signal is split into several windows according to the computations. Afterward, the mean values (EXP mean) and the standard deviations (EXP range) are evaluated to express the range of the measurements. The measured results of PSP-HH reveal higher dynamic pressure fluctuations in the normalized frequency range of $0 \leq f/f_0 \leq 5$.

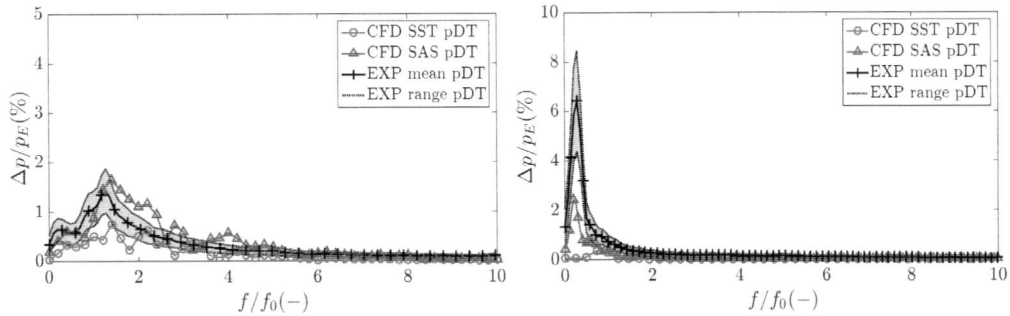

Figure 4: Comparison of the spectral power density between the measured and computed pressure fluctuations in the draft tube cone p_{DT} at low-load conditions for PSP-HH (left) and PSP-MH (right).

The CFD simulations of both turbulence models show also higher values in this range with a slightly underestimation of the SST model. At PSP-MH a more distinct peak is appearing at $f/f_0 \approx 0.8$. The k-ω SST model reveals only a small peak at the rotational frequency $f/f_0 = 1$. The SAS model on the other hand shows more accurate results even the amplitude is underestimated. In figure 5 the numerically computed pressure amplitudes with the SAS model at the runner surfaces of both machines are shown together with appearing draft tube vortices, which are displayed by iso-pressure surfaces.

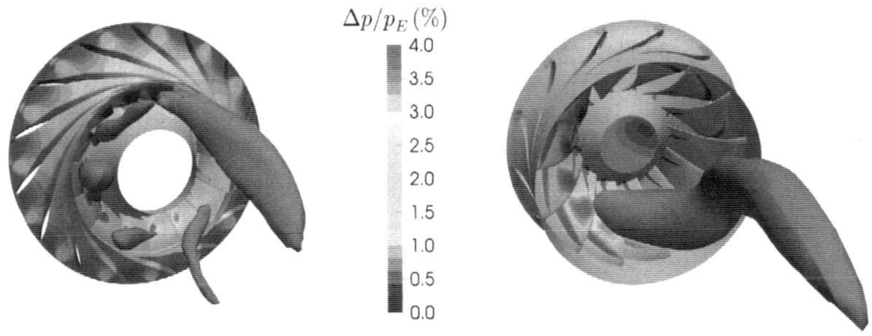

Figure 5: Pressure amplitudes on the Francis runner surfaces computed with the SAS turbulence model and appearing draft tube vortices at PSP-HH (left) and PSP-MH (right).

At PSP-HH high pressure fluctuations are appearing at the runner inlet due to the RSI. At the outlet the rotating draft tube vortices are inducing pressure oscillations as well, especially in the hub region of the runner. One characteristic of PSP-HH is the shaft going through the draft tube. Hence, multiple smaller and bigger vortices are appearing and collapsing again along the time, inducing a broadband excitation as visible in figure 4. At PSP-MH one bigger draft tube vortex rope is appearing and rotating more stable in the runner direction. Therefore, the highest pressure amplitudes are appearing at the runner outlet. Only small excitations are visible at the leading edges due to a less impact of the RSI at PSP-MH.

3. Transient FEM simulations

The results of the unsteady CFD simulations are used to evaluate the structural behavior and the dynamic stresses of the Francis runners by transient FEM simulations with the open-source tool Code_Aster. The model and the discretization of the investigated turbines are displayed in figure 6. The tetrahedral mesh of each runner geometry consists of about 0.5 million nodes. This relative small size is achieved by refining only one runner blade in the critical zones, which is also used for the stress evaluation. A grid convergence study has been performed as well to assure reliable results.

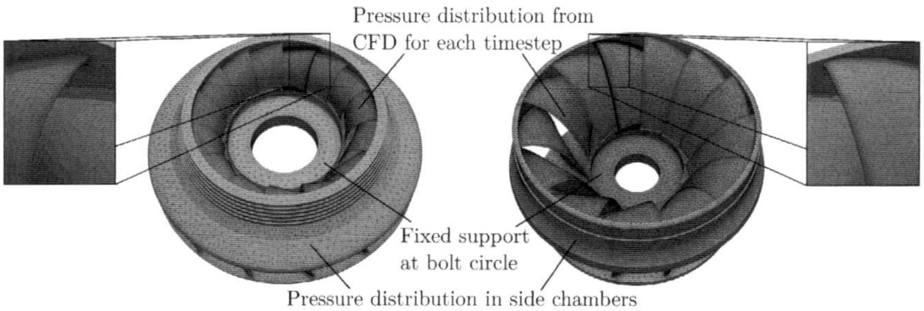

Figure 6: Discretization and boundary conditions for the transient FEM simulations for PSP-HH (left) and PSP-MH (right).

The time-dependent pressure distributions from the CFD computations are applied to the surfaces of the blade channels for each time step. The pressure distributions in the runner side chambers are assumed to be static and are computed by an analytical model as described in [7]. The Francis turbines are fixed at the bolt circles and the circumferential and gravitational forces are considered as well. The transient FEM simulations are performed for several runner rotations with an according time step size of 3.6° for PSP-HH and 3.0° for PSP-MH. The damping of the structure is considered by an equivalent Rayleigh damping as described in [5]. Monitoring points at the refined runner blade are used to compare the stresses with the measured data at the strain gauge positions. The overall dynamic stress amplitudes of the FEM simulations and the measurements at the strain gauge positions for PSP-HH and PSP-MH at the investigated low-load operation points are displayed in figure 7.

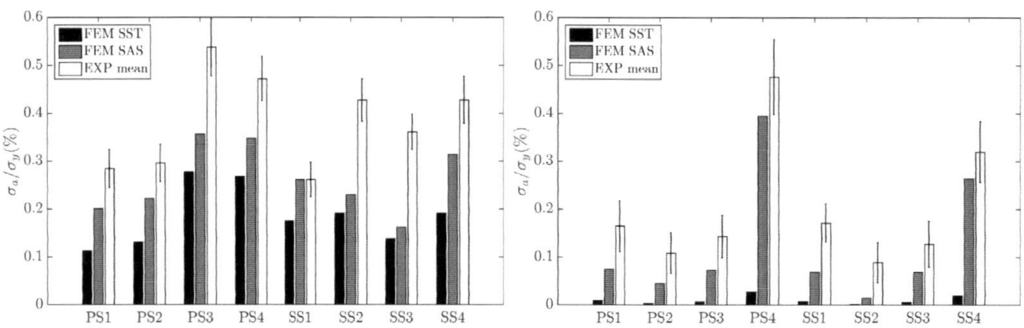

Figure 7: Comparison of the dynamic stress amplitudes calculated by the RMS values between the measured data and the transient FEM simulations at low-load operation for PSP-HH (left) and PSP-MH (right).

The measured stress signal, with a length of 120 s, is separated in smaller windows according to the FEM simulations to calculate the mean values (EXP mean) and the measurement range. To determine the influence of the turbulence model on the structural response, the pressure distributions of the SST and the SAS model are used. The results show an accurate agreement with the measurement tendencies, even the computed values underestimate the measured peaks.

Due to the higher accuracy of the SAS model regarding the pressure fluctuations in the draft tube, the overall stress amplitudes are predicted more precise as well, with the highest values appearing close to the hub of the runner. According to the numerical flow simulations, the results for PSP-MH reveal much bigger differences between the turbulence models, with a total underestimation of the stress amplitudes by the SST model. The SAS model shows the highest amplitudes also close to the hub at SG PS4 and SS4. The spectral power densities of the measured and simulated stress signals at the strain gauge PS4 for PSP-HH reveal higher stress amplitudes in the low frequency range due to the pressure fluctuations in the draft tube (see figure 8). At the first and second harmonic of the blade passing frequency ($f/f_0 = 20$ and $f/f_0 = 40$) higher stress amplitudes are visible, even the SAS turbulence model overestimates the measured values. For PSP-MH a high peak is visible at the normalized frequency $f/f_0 \approx 0.8$ induced by the appearing vortex rope according to figure 4. In contrast to the SST model, the curve of the SAS model is in a good agreement with the measurements.

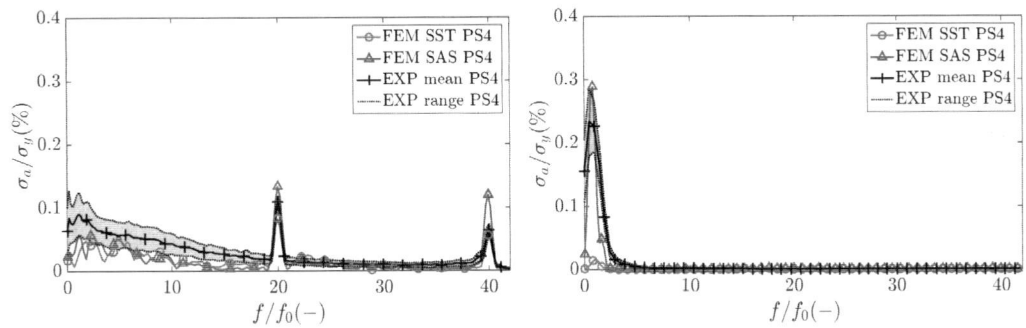

Figure 8: Comparison of the spectral power density between the measurements and the computed stresses at low-load operation for PSP-HH (left) and PSP-MH (right).

4. Expected load spectra

The results of the transient FEM simulations performed with the SAS turbulence model are used for a rainflow counting analysis at the strain gauge positions and at the critical locations of both Francis runners. In figure 9 the dynamic stress amplitudes, calculated by the standard deviation of the time signals, in low-load operation at the refined runner blades are displayed.

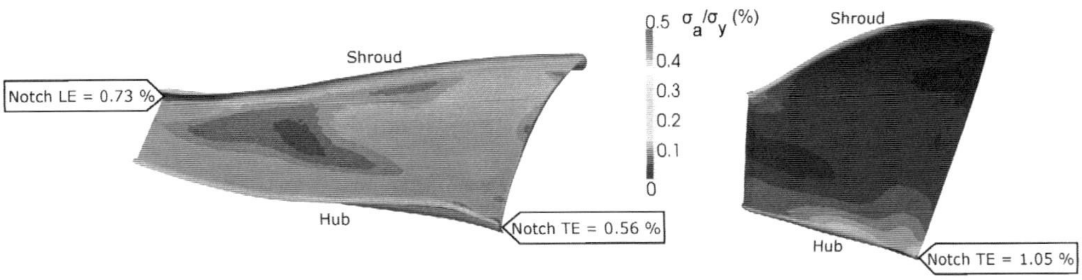

Figure 9: Stress amplitudes at the Francis runner blades at low-load operation for PSP-HH (left) and PSP-MH (right) computed with the SAS turbulence model.

At PSP-HH the highest values are appearing in the notch at the leading edge (LE) of the runner due to the rotor-stator interaction but also at the trailing edge (TE) due to the pressure fluctuations in the draft tube. For PSP-MH the dynamic stresses at the leading edge are lower due to the less impact of the RSI. The critical location with the maximum excitation, induced by the draft tube vortex rope, is appearing at the trailing edge close to the hub of the runner. To determine the impact of the dynamic excitation of the high and medium head runner in low-load operation a rainflow counting analysis is performed. Therefore, the measured and computed uni-axial stress signals at the strain gauge positions and the multi-axial stresses at the critical runner locations, obtained by the FEM simulations, are used. To evaluate the measurement range the signals are again divided into several windows according to the computations. To gain a more convincing comparison between the measurements and the simulations a stress extrapolation, based on the extreme value theory described in [8], is performed. This approach has been already applied for a fatigue analysis of a Francis runner at speed-no-load operation (see [9]). For the evaluation of the damage by the three-dimensional stress state in the notches, the algorithm of Brown and Miller [10] is used, considering the influence of the shear and normal strains. The obtained load spectra of both Francis turbines are displayed in figure 10 including the corresponding stress-cycle (S-N) curve of the runner material for a 99 % survival probability. The measured and computed stress signals are extrapolated up to a time length of 1000 s. The rainflow counting analysis reveals an accurate agreement between the results of the measurements and the FEM simulations at the strain gauge positions for both turbines. The evaluation of the multi-axial stresses at the critical locations shows slightly higher amplitudes compared to the strain gauges as expected. The rainflow curves of both Francis turbines are below the fatigue limit of the runner material, even considering the measurement range.

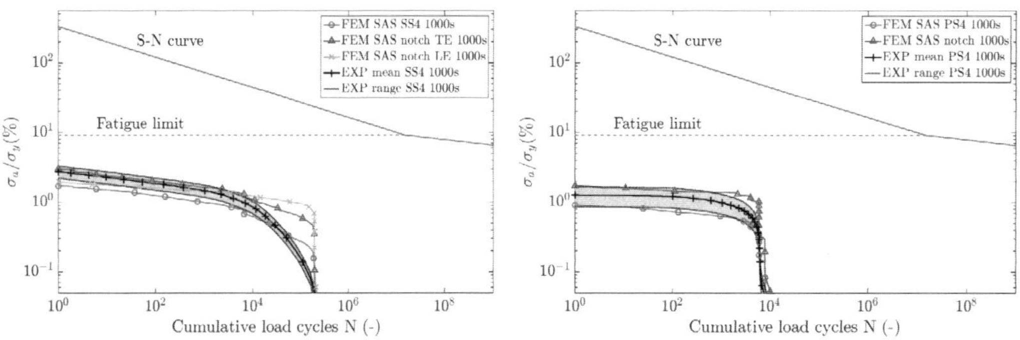

Figure 10: Load spectra of the measured and computed stresses with the SAS turbulence model at low-load operation for PSP-HH (left) and PSP-MH (right).

5. Conclusion

The strain gauge measurements on a high head and a medium head prototype Francis runner revealed increased dynamic excitations in low-load operating conditions at about one third of the rated output power. Additional pressure measurements in the draft tube cone showed higher amplitudes in the lower frequency range for both machines at the same operation points. To investigate the reason for the high stress peaks and to obtain the load spectra of the turbines, numerical computations consisting of unsteady CFD and FEM simulations were performed. Therefore, two different turbulence models - the k-ω SST and the Scale Adaptive Simulation (SAS) model - were used. The flow simulations revealed increased pressure fluctuations at the leading edge of the high head runner due to the major impact of the rotor-stator interaction. At the turbine outlet higher pressure amplitudes occurred as well for both hydraulic machines due to the influence of appearing draft tube vortex ropes. Compared to the pressure measurements, the SAS turbulence model seems to better predict the impact of the vortices on the pressure oscillations. The transient FEM simulations in the low-load operating points revealed an accurate determination of the overall stress amplitudes. The application of the time dependent pressure distributions computed by the SAS turbulence model lead to more precise results, especially for the medium head turbine. The transient FEM simulations exposed different locations with higher dynamic excitations for both turbines: At the leading edge for the high head runner induced by the rotor-stator interaction and at the trailing edge due to the draft tube vortex ropes for both machines. The measured and computed stresses at the strain gauge positions and at the critical notches were further used for a rainflow counting analysis together with a stress extrapolation approach. The results revealed an appropriate agreement between the simulations and the site measurements with stress-cycle curves below the fatigue limit of the runner material. For a valuable lifetime prediction of the Francis runners, additional investigations should be performed considering geometrical deviations, numerical simplifications and further uncertainties.

Acknowledgments

The K-Project GSG-GreenStorageGrid is funded in the framework of COMET - Competence Centers for Excellent Technologies by the Federal Ministry of Transport, Innovation and Technology, the Federal Ministry of Science, Research and Economy, the Vienna Business Agency, the Federal Province of Lower Austria and by the Federal Province of Upper Austria. The program line COMET is administered by the Austrian Research Promotion Agency (FFG). The authors would like to thank the project partners Andritz Hydro, Verbund Hydro Power, ZT Hirtenlehner, TIWAG Tiroler Wasserkraft AG and Vorarlberger Illwerke AG for their contribution in the course of the sub-project PSP-LowLoad. Special thanks goes to Andritz Hydro for their support regarding the application of Code_Aster. The computational results presented have been achieved in part using the Vienna Scientific Cluster (VSC).

References

[1] Teller, O et al 2013 *European Energy Storage Technology Development Roadmap towards 2030.* (Brussels: EASE/EERA Core Working Group)
[2] Guittet M, Capezzali M, Gaudard L, Romerio F, Vuille F and Avellan F 2016 Study of the drivers and asset management of pumped-storage power plants historical and geographical perspective. *Energy* **111** 560–579 ISSN 03605442
[3] Magnoli M V 2014 *Numerical simulation of pressure oscillations in large Francis turbines at partial and full load operating conditions and their effects on the runner structural behaviour and fatigue life.* Ph.D. thesis Technische Universität München
[4] Eichhorn M, Doujak E and Waldner L 2016 Investigation of the fluid-structure interaction of a high head Francis turbine using OpenFOAM and Code_Aster. *28th IAHR symposium on Hydraulic Machinery and Systems*
[5] Eichhorn M and Doujak E 2016 Impact of different operating conditions on the dynamic excitation of a high head francis turbine. *Proceedings of the ASME 2016 International Mechanical Engineering Congress & Exposition*
[6] Dörfler P, Sick M and Coutu A 2013 *Flow-Induced Pulsation and Vibration in Hydroelectric Machinery.* (London, New York: Springer)
[7] Maly A, Eichhorn M and Bauer C 2016 Experimental investigation of transient pressure effects in the side chambers of a reversible pump turbine model. *Proceedings of the 19th International Seminar on Hydropower Plants*
[8] Johannesson P 2006 Extrapolation of load histories and spectra. *Fatigue & Fracture of Engineering Materials Structures* **29** 209–217
[9] Morissette J F, Chamberland-Lauzon J, Nennemann B, Monette C, Giroux A M, Coutu A and Nicolle J 2016 Stress predictions in a francis turbine at no-load operating regime. *28th IAHR Symposium on Hydraulic Machinery and Systems*
[10] Brown M W and Miller K J 1973 A theory of fatigue under multiaxial strain conditions. *Proceedings of the Institution of Mechanical Engineers* **187** 745–755

Effects of Pump-turbine S-shaped Characteristics on Transient Behaviours: Model Setup

Wei Zeng, Jiandong Yang, Jinhong Hu

State Key Laboratory of Water Resources and Hydropower Engineering Science, Wuhan University, Wuhan 430072, China

wzeng@whu.edu.cn

Abstract. Pumped storage stations undergo numerous transition processes, which make the pump turbines go through the unstable S-shaped region. The hydraulic transient in S-shaped region has normally been investigated through numerical simulations, while field experiments generally involve high risks and are difficult to perform. In this research, a pumped storage model composed of a piping system, two model units, two electrical control systems, a measurement system and a collection system was set up to study the transition processes. The model platform can be applied to simulate almost any hydraulic transition process that occurs in real power stations, such as load rejection, startup, frequency control and grid connection.

1. Introduction

To adjust the power system more efficiently and in a timely manner, pumped storage stations undergo numerous transition processes. Many engineering incidents occur in the process, such as turbine lifting accidents as a result of hydraulic transients and unsuccessful grid connections under low head. Therefore, it is important to investigate these transition processes to guarantee the security of the pumped storage station.

The S-shaped characteristics of the pump turbine are a critical factor in determining the transient characteristics of pumped storage stations. Different simulation methods have been applied to study the transition process in the S-shaped region [1-4]. Moreover, field experiments have also been conducted to obtain additional verification. Dörfler et al. partly closed the valve in the field to achieve pump turbine no-load stability [5]. Yang et al. analyzed the pressure pulsation characteristics of a prototype pump turbine during a transition process [6]. The transient experiments of load rejections and power failure must be performed in the commisioning stage to ensure safety and stability [7]. However, owing to the risks involved in field experiments, it is difficult to conduct experiments for specific engineering problems and some extreme conditions.

To compensate for the limitations of field experiments, the corresponding model experiments can be developed. The existing model experiment platforms for hydraulic machineries are generally closed-loop test rigs, which are mainly used to test the turbine characteristics in the constant states [8,9]. To study the transition processes of a Francis turbine, NTNU changed the piping system on the basis of the closed-loop test rig to set up the open-loop test rig. The model is used not only to test the turbine characteristics but also to measure the pressure pulsations at constant states [10,11] or during startup and load rejections [12,13]. However, some complex transition processes in actual pumped storage stations are difficult to achieve.

In summary, to investigate the transient characteristics of pump turbines and pumped storage stations, a more realistic pumped storage system is developed and introduced in this paper.

2. Basic information

The pumped storage system illustrated in Fig. 1 consists of model pipes with circulating water, model units (pump turbines and motor generators), a control system (excitation synchronization protection, frequency and phase conversion, speed control, monitoring system, load system), and a measurement system. The major functions and components of each system are shown in Table 1.

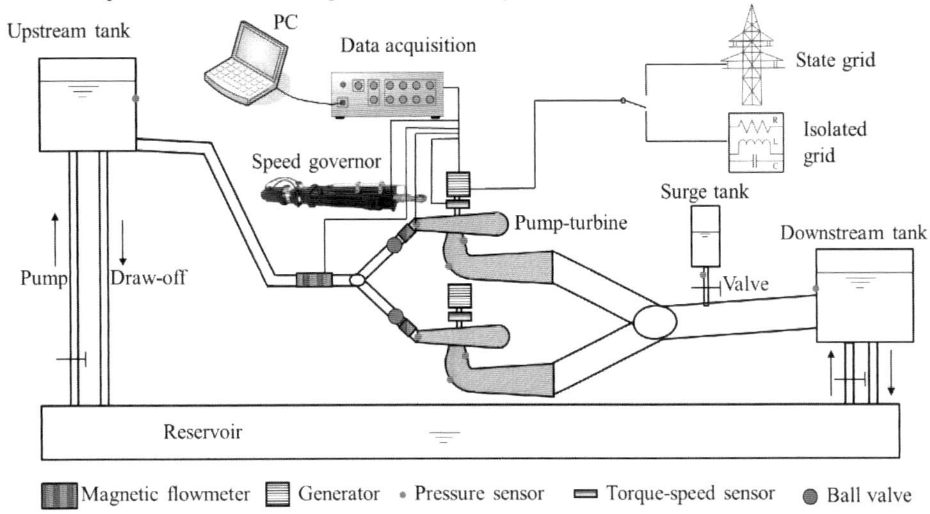

Figure 1. Pumped storage system model

Table 1. Subsystems of the pumped storage system

No.	Subsystem	Main functions
1	Model pipe	Simulates the actual power station pipes and supplies water to the model pump turbines
2	Circulating water system	Supplies power and water circulation for model experiments
3	Model units	Includes two sets of pump turbines and motor generators
4	Excitation and synchronization	Excitation control, synchronization, and protection of the motor generator
5	Frequency and phase conversion	Varies frequency startup in pump mode and conditions transformation in turbine mode
6	Speed control	Controls the speed governor to regulate the guide vane opening and regulates the ball-valve opening degree
7	Local control unit (LCU)	Gives orders to the model units and controls them
8	Load system	Simulates the grid features
9	Measurement system	Collects and processes the parameters

3. Experimental platform setup

3.1 Model pipes with circulating water

The pipe system of the pumped storage system model employs an annular arrangement pattern of two machines sharing the same main pipes. The section area of the downstream impedance surge tank is 0.43 m^2, and the impedance hole can be turned on or off through a valve according to the experimental requirements; the hole is normally closed. The model pipe system is mainly designed based on the pipe layout of the Xianju pumped storage station with the distorted scale including the radial scale

1:17.357 and the axial scale 1:42.366. The pipe system is shown in Fig. 2, and the main dimensions are shown in Table 2.

Figure 2. Model pipe system

Table 2. Pipe system dimensions

Pipe section	Inlet	Upstream main pipe	Upstream branch pipe	Upstream branch pipe	Downstream main pipe	Outlet
Equivalent diameter (m)	0.597	0.357	0.202	0.300	0.426	0.675
Length (m)	2.251	24.154	2.195	4.891	16.285	1.062

The circulating water system primarily provides water circulation for model experiments. Water is pumped from the underground reservoir by centrifugal pumps and supplied to the upstream and downstream tanks through the supply pipe (Fig. 1). The valves on the supply pipe can adjust the flow into the upstream and downstream tanks. Overflow weirs are installed in the tanks to maintain a constant water level, and the overflow out of the weirs flows into the underground reservoir through the draw-off pipes.

3.2 Model units

Two model pump turbines installed in the platform were obtained from the Harbin Electrical Corporation, Harbin, China. They are the pump turbine models at the Xianju pumped storage station (XJ) and the Baoquan pumped storage station (BQ). Both pump turbines basically have similar characteristics in the turbine high efficiency area, but differ significantly in the S-shaped region. The runner geometries are also distinct; these details are shown in Fig. 3. The basic parameters of the two pump turbines are presented in Table 3.

(a) BQ (b) XJ

Figure 3. Model runners

Table 3. Model unit parameters

N_{QE} (m:kW)	Inlet diameter (mm)	Outlet diameter(mm)	Guide vane height (mm)	Number of blades	Number of guide vanes	Rated n (rpm)	Rated H (m)	Rated Q (L/s)	Full rotational inertia (N.m)
37.91	280	146.34	24.44	9	20	1000	10.54	49.1	66.4

3.3 Control system

Each unit has a set of electrical cabinets to control the unit operation, including the local control unit, speed control system, excitation system, and load system. Their basic functions have been presented in Table 1. These systems can make the model units implement almost all the operations and conditions encountered in real power plants. The load system includes resistances, capacitances, and inductances to simulate an isolated grid with different characteristics by varying load combinations. In addition, the units can be also incorporated into the state grid, as shown in Fig. 1.

The PC-III programmable microprocessor speed governor (Fig. 1) is employed to control the pump turbine guide vanes with proportional–integral–derivative control. Its main function is to automatically track the grid frequency after unit startup and to achieve the best process control of startup, fast quasi-synchronization, frequency regulation (isolated network), power regulation, as well as opening degree regulation. In the experiments, the opening degree regulation function can dominate the guide vanes by various means, such as stalling, straight closure, and multi-phase closure.

4. Measurement of experiment data

4.1 Pressure measurement

The PCB 112A pressure sensors and MPM 480 pressure sensors were used to test the transient pressures, and the KYB 335 sensors were used to test the differential pressure of two points at the same section of the spiral case. The Fluke 719 pressure calibrator was used to calibrate the pressure sensors in the field. In the pipe systems, the main pipe, branch pipe, surge chamber, and upstream and downstream tanks were equipped with a certain number of MPM pressure sensors. MPM pressure sensors and PCB pressure sensors were installed on the points shown as Fig. 4 and Table 4. Table 5 presents the basic specifications of the major sensors used in the present tests.

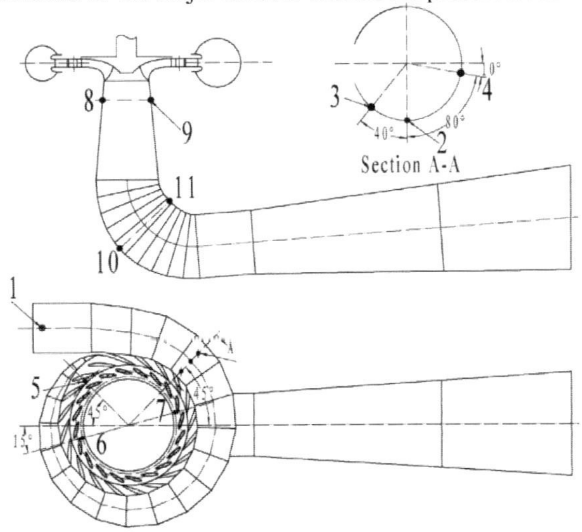

Figure 4. Pressure sensor positions

Table 4. Pressure sensor types and positions

No.	Sensor type	Position
1	MPM 480	Spiral case inlet
2	PCB 112A	The bottom of the spiral case
3 4	KYB 335	Two points at the same section of the spiral case
5	PCB 112A	stay vane
6	PCB 112A	Vaneless space 1
7	PCB 112A	Vaneless space 2
8	MPM 480	Draft tube inlet 1
9	MPM 480	Draft tube inlet 2
10	MPM 480	elbow bend 1
11	MPM 480	elbow bend 2

Table 5. Specifications of the different sensors

Sensor type	Positions	Measurement range	Accuracy degree (%)	Linearity (%)	Uncertainty(%)	Signal
Pressure sensor (Micro MPM 480)	1# Spiral case inlet	0–200 (kPa)	0.25	0.052 F.S	0.109	4–20 mA
	1# Draft tube inlet	−20–50 (kPa)		0.035 F.S	0.083	
	2# Spiral case inlet	0–200 (kPa)		0.051 F.S	0.079	
	2# Draft tube inlet	−20–50 (kPa)		0.053 F.S	0.621	
Flow meter (KROHNE IFM 4110)	Upstream main pipe	−200–200 (L/s)	0.3	0.128 F.S	/	4–20 mA /Frequency
Flow meter (OPTIFLUX 2000)	1# branch pipe	−70–70 (L/s)	0.2	0.157 F.S	/	
	2# branch pipe	−70–70 (L/s)		0.141 F.S		
Torque sensor (HLD09)	Between 1# turbine and 1# generator	−100–100 (N.m)	0.2	0.040% F.S	/	4–20 mA /Frequency
	Between 2# turbine and 2# generator	−100–100 (N.m)		0.020% F.S		

4.2 Flow measurement

Two electromagnetic flowmeters (Table 5) were installed in the model power station to measure the flows. To improve the accuracy of the experiment and to calibrate the electromagnetic flowmeters, a flow measurement device (Fig. 5) was developed by the static weighing method. To measure the forward flow (turbine direction), the supply valve of the downstream tank was closed and the flow passing through the units was drained from the draw-off pipe in the downstream tank. The draw-off pipe connects inlet A of the device and the drained water from outlet B are weighed. The opening and closure times of outlet B are t_1 and t_2, the entire opening duration is t_0, and the weighed water is m. Thus, one can get

$$Q = \frac{m}{\rho(t_0 + t_1/2 + t_2/2)} \tag{1}$$

Here, in the flow calibration test, $t_0 > 60$ s, t_1 and t_2 are less than 0.1 s, and the accuracy of the electronic scale is smaller than 0.05%. The calibration results for the electromagnetic flowmeters using such an apparatus are listed in Table 5.

A more precise measurement of dynamic flow is obtained by minimizing the time constant of the electromagnetic flowmeters to achieve the fastest response. However, there are always some delays in electromagnetic flowmeters. Therefore, the differential pressure sensor was installed in the spiral case to measure the dynamic flow. However, its accuracy is much lesser than that of the electromagnetic flowmeters. Nevertheless, the flow peaks and valleys measured by the differential pressure sensors can be used to calculate the delay time of the electromagnetic flowmeters. Thus, the delay response can be eliminated by adjusting the time coordinate.

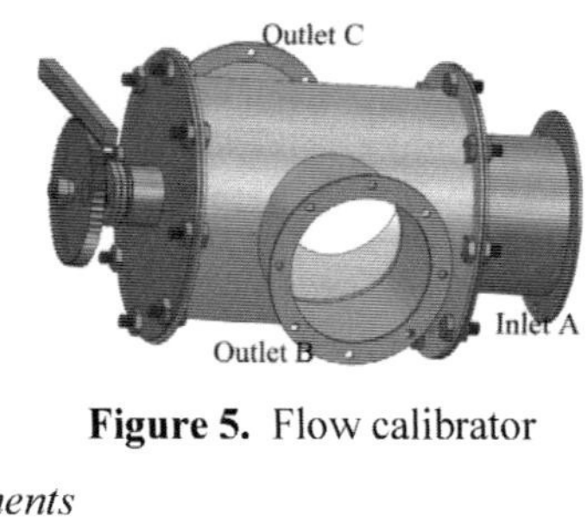

Figure 5. Flow calibrator

4.3 Speed and torque measurements

The HLD09 speed-torque sensor was installed between the turbine and generator, in which the torque applies the strain electrical logging principle: a strain gauge installed in a special shaft deforms with the torque and the resistance changes accordingly. The speed-torque sensors were calibrated as factory default, as shown in Table 5.

4.4 Guide vane movement measurement

The guide vane angles during transients were tested by the angle transducers installed on the top of the guide vane shaft. Using the standard gauge blocks, we obtained the relationship between the guide vane opening and the guide vane angles.

5. Data acquisition and processing

5.1 Data acquisition.

Different analog signals from these sensors were collected by a HBM Gen7i data acquisition system with a 24 bit analog-to-digital resolution, a 250 kHz maximum sampling frequency, and 70 channels. As the experiments in this study focused on the water-hammer pressure, rotational speed, and torque with low frequencies, we chose a 1 kHz sampling frequency because it is adequate for the experiments in this study.

5.2 Data processing.

Based on the previous studies by Trivedi [10-13], the Savitzky-Golay method [14] was used to extract the pressure pulsation in the transition process. The Savitzky-Golay method is a low-pass filter method that performs polynomial fitting of the original data, extracts the timed mean pressure based on the least squares method, and then obtains the pulsating pressure according to the differences between the original data and the timed mean pressure. Fig. 6 shows the measured value of the spiral case pressure and the filtered results by the Savitzky-Golay method under the condition that the one pump turbine with partial rotational inertia rejected the load with the guide vanes closed in 3 s.

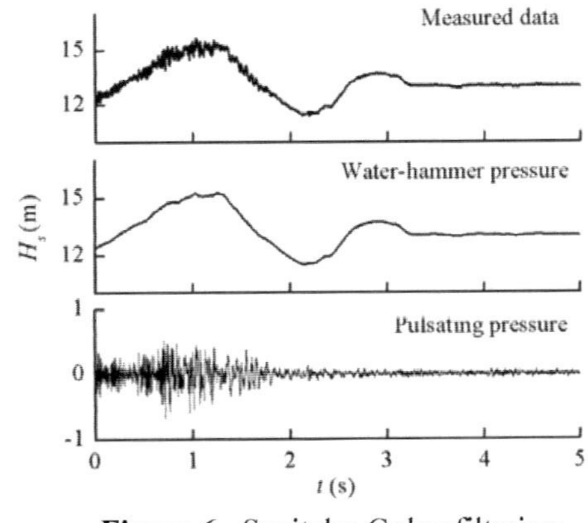

Figure 6. Savitzky-Golay filtering

6. Conclusions

In the present study, a pumped storage model was built to study transition processes of a pumped storage station. The model composed of two pump-turbine scales, two motor generator scales, pipe system and control system is available to conduct a variety of transient experiments happened in the real pumped storage stations. The transient pressures, flow, torque, rotational speed and guide vane movement can be tested with high precisions

In the next paper, the author would introduce the transient experiments conducted on the model to investigate the effects of pump-turbine S-shaped characteristics on hydraulic transients.

References

[1] C. Wang, J.D. Yang, Water Hammer Simulation Using Explicit-Implicit Coupling Methods,

IOP Conf. Series: Journal of Physics: Conf. Series **813** (2017) 012053 doi:10.1088/1742-6596/813/1/012053

ASCE J. Hydraul. Eng., 141(2015) 04014086. DOI: 10.1061/(ASCE)HY. 1943-7900.0000979
[2] W.J. Yang, J.D. Yang, W.C. Guo, W. Zeng, C. Wang, L. Saarinen and P. Norrlund, A mathematical model and its application for hydro power units under different operating conditions,' Energies, 8(2015) 10260-10275. (DOI: 10.3390/en80910260)
[3] C. Nicolet, Hydroacoustic Modeling and Numerical Simulation of Unsteady Operation of Hydroelectric Systems. PhD thesis, EPFL, Lausanne , Switzerland, 2007.
[4] X.X. Zhang, Y.G. Cheng, L.S. Xia and J.D. Yang, Dynamic Characteristics of a Pump-Turbine during Hydraulic Transients of a Model Pumped Storage System: 3D CFD Simulation. 27th IAHR Symposium on Hydraulic Machinery and System, Montreal, Canada, 2014.
[5] P.K. Dörfler, A.J. Engineer, R.N. Pendse, P. Huvet, M.V. Brahme, Stable Operation Achieved on a Single-Stage Reversible Pump-Turbine Showing Instability at no-Load, 19th IAHR Symposium, Section on Hydraulic Machinery and Cavitation, Singapore, 1998.
[6] J.D. Yang, J.H. Hu, W. Zeng, J.B. Yang, Transient Pressure Pulsations of Prototype Francis Pump-turbines. J. Hydraulic Engineering. (In Chinese)
[7] International Electrotechnical Commission, Field Acceptance Tests to Determine The Hydraulic Performance of Hydraulic Turbines, Storage pumps and Pump-Turbines, CEI/IEC 41, Third ed., Switzerland, 1999
[8] http://lmh.epfl.ch/page-57350-en.html
[9] http://lamh.gmc.ulaval.ca/laboratoire/installations/plateforme-dessais/
[10] C. Trivedi, M.J. Cervantes, B.K. Gandhi, O.G. Dahlhaug, Experimental and Numerical Studies for a High Head Francis Turbine at Several Operating Points, ASME J. Fluids Eng., 135 (2013) 111102. DOI: 10.1115/ 1.4024805
[11] C. Trivedi, M.J. Cervantes, O.G. Dahlhaug, B. K. Gandhi, Experimental Investigation of a High Head Francis Turbine During Spin-No-Load Operation, ASME J. Fluids Eng., 137 (2015) 061106. DOI: 10.1115/ 1.4029729
[12] C. Trivedi, M.J. Cervantes, B.K. Gandhi, O.G. Dahlhaug, Transient Pressure Measurements on a High Head Model Francis Turbine During Emergency Shutdown, Total Load Rejection, and Runaway, ASME J. Fluids Eng., 136(2014) 121107. DOI: 10.1115/ 1.4027794
[13] C. Trivedi, M. Cervantes, B. Gandhi, and O. Dahlhaug, 2014, "Experimental Investigations of Transient Pressure Variations in a High Head Model Francis Turbine During Start-up and Shutdown," J Hydrodyn, 26(2), pp. 277-290. doi:10.1016/S1001-6058(14)60031-7.
[14] R.W. Schafer, What is a Savitzky-Golay filter, IEEE Signal Processing Magazine, 7(2011) 111-117. (DOI: 10.1109/MSP.2011.941097)

Experimental investigation of check valve behaviour during the pump trip

D Himr, V Habán and M Hudec

Victor Kaplan Department, Faculty of Mechanical Engineering, Brno University of Technology, Technická 2896/2, 616 69 Brno, Czech Republic

himr@fme.vutbr.cz

Abstract. The paper is focused on the description of an experimental investigation of a nozzle check valve. Attention is paid to the behaviour during transitions when the flow is decelerating. Flow reversal is often accompanied with the check valve slam, because the check valve closes with a delay after the flow changes its direction. The maximal back flow velocity determines the maximal pressure peak when the check valve closes, so this velocity was found for different flow decelerations.

1. Introduction

The check valve is supposed to prevent flow reversal in a system. Typically, check valves are placed in the discharge pipe immediately behind the pump, but they can be placed anywhere, where only one directional flow is desired. Basic demands on check valves are: low resistance in the positive flow direction and infinite resistance in the negative flow direction (no leakage).

There are many different designs of check valves with different properties, but all of them try to fulfill the two basic demands written above. Several examples of design are: ball check valve, swing check valve, tilted disc check valve, dual disc check valve, nozzle check valve, etc. It is not an easy task to choose the best check valve for a specific system and, according to [1], many criteria should be taken into account: The initial cost, the maintenance cost, the head loss, the flowing fluid, the flow characteristic and others including the non-slam characteristic, which describes the potential of the check valve to slam. When the flow decelerates, e.g. due to pump stop, the check valve is supposed to shut precisely at the moment when the flow velocity reaches zero, so that no reverse flow appears. However, the check valve closes with a shorter or a longer delay, when some reversal flow velocity has already developed. Sudden closure immediately stops the water column and causes water hammer. Check valve slam can damage the check valve and cause problems anywhere in the hydraulic system.

One can see that the selection of the most suitable check valve is not an easy task and requires a deep knowledge of the entire system and it is not possible to find a specific design appropriate for all systems.

Regarding safety and reliability, the non-slam characteristic of the check valve is crucial. There are many cases when the piping was damaged by the check valve slam [2], [3] and [4].

Generally, there are two ways to prevent problems with check valve slam. The first way is to secure fast closing, when the flow reverses. According to [5], who cites [6], fast closing requires: low inertia of the disc and low friction, the travel of the disc should be short or the motion should be assisted with springs. Similarly, the role of a spring could be fulfilled by a counterweight, which also helps with closing.

The other way is to slow down closing. In this case, a high reverse flow can evolve and closing has to be slow enough to avoid the water hammer effect and the vortex structures creation [7]. The time period of closing depends on the system and it can be even several minutes. Slow closing is often secured by oil dashpots. Devices slowing down closing have to be robust enough to withstand the hydraulic forces acting on the check valve disc. Authors of this article have already seen systems where dashpots were completely destroyed.

Obviously, the behavior of the check valve during flow reversal is important and the non-slam characteristic gives an indication of the suitability of the specific check valve for the system. This characteristic shows maximal reverse velocity which develops in the system before closing of the check valve. It is, however, necessary to obtain the deceleration value of the system, for example by computation using a special software allowing simulation of the unsteady flow in pipelines.

2. Experiments

MSA, the Czech producer of check valves, is widening their product line and needs to know the dynamic characteristics of the check valves. For the first test, we got a nozzle check valve of an old design to obtain data which will serve for comparison. The first step was building an experimental circuit which allows flow reversal. The circuit scheme is shown in figure 1.

A suction tank (volume 10 m^3) feeds a pump with water, which is pumped through a discharge pipe DN100 into a small tank (volume 2.5 m^3). The water can flow back to the suction tank through a return pipe. The small tank can be pressurized with the compressor and depressurized with an air valve. The check valve is installed in the straight part of the discharge pipe, which has a length of 6 m. The scheme of the check valve is shown in figure 2. The body of the check valve is equipped with a window, so that it is possible to observe the position of the disc. This is the only possibility, because the check valve is compact and no moving parts are outside of the body.

The pump speed is controlled with a dynamometer which also measures torque and speed of the pump. The steady flow rate is measured with an electromagnetic flowmeter (Q in figure 1) and pressure is measured with six pressure sensors – BD sensors DMP 331 (three pieces are mounted on the pipe wall in front of the check valve and thee sensors are behind the check valve).

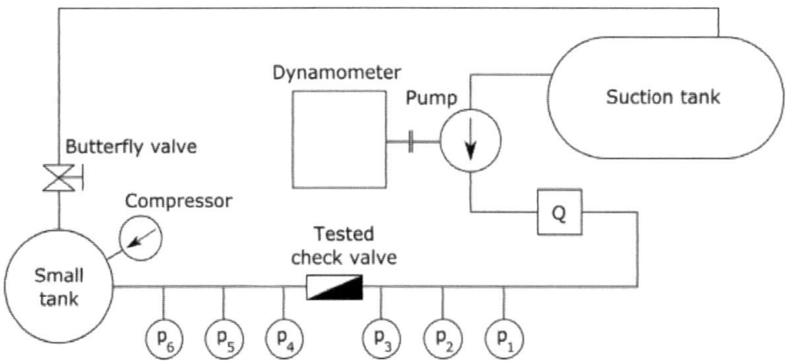

Figure 1. Test circuit.

The steady characteristic of the check valve can be determined using data from the flowmeter and nearest pressure sensors (p_3 and p_4 in figure 1) while the small tank is full of water. The butterfly valve must be open to allow water to circulate in the system.

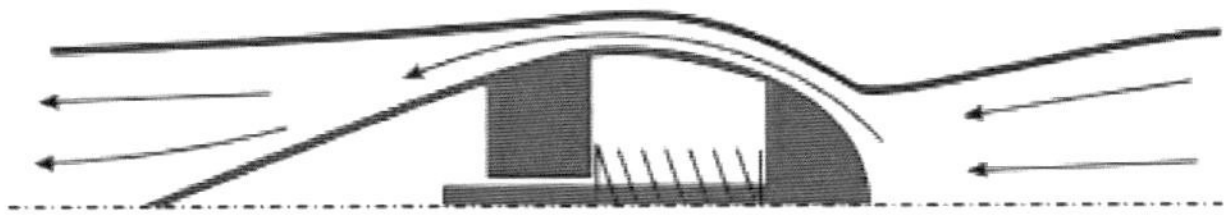

Figure 2. Nozzle check valve.

When dynamic characteristic is measured, the butterfly valve is closed and only 20% of the small tank is filled with water. The experiment starts when the pump runs with steady speed. The flow rate reaches its maximal value and starts going down due to increasing pressure in the small tank (the air in the tank is compressed). Then, the pump is braked so the deceleration becomes more intense. The flow changes its orientation and the check valve closes with a slam.

2.1. Static characteristic

For evaluation of the static characteristic, data from pressure sensors p_3 and p_4 (pressure loss) were used together with data from the flowmeter. Flow coefficient was computed according to equation (1) and the result is shown in figure 3.

$$K_v = 10Q\left(\frac{\rho}{\Delta p}\right)^{1/2} \tag{1}$$

The flow coefficient is changing with the flow rate. The higher flow rate the higher force acts on the spring and the check valve is more open. The stiffness of the spring is 2550 N/m. It was measured in the absence of water. The positon of the disc was determined from pictures taken with a camera during the experiment. The mean position for given flow rate was taken as the disc was not steady but slowly oscillating around the mean position. This is a prevention from getting stuck due to rust or limescale and reduces the risk that the disc does not close when the flow rate goes to zero. Maximal valve opening is 50 mm.

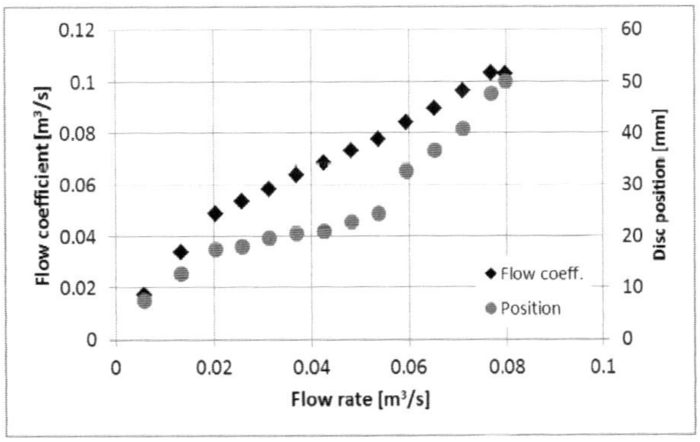

Figure 3. Static characteristic of tested nozzle check valve.

2.2. Dynamic characteristic

Dynamic characteristic shows a relationship between fluid deceleration and maximal reverse flow velocity through the check valve. As the electromagnetic flowmeter is not suitable for the measurement of unsteady flow, Gibson (the so-called time-pressure) method was used to obtain flow rate data, see equation (2).

$$Q(t+\Delta t) = Q(t) - \frac{\Delta t S}{L\rho}\left[\Delta p(t) + RQ(t)|Q(t)|\right] \tag{2}$$

Δp, in equation (2), can be substituted with the pressures in front of the check valve p_1 and p_3 or behind the check valve p_4 and p_6. The results correspond to each other since there is no column separation. Figure 4 shows the evaluated flow rate according to equation (2).

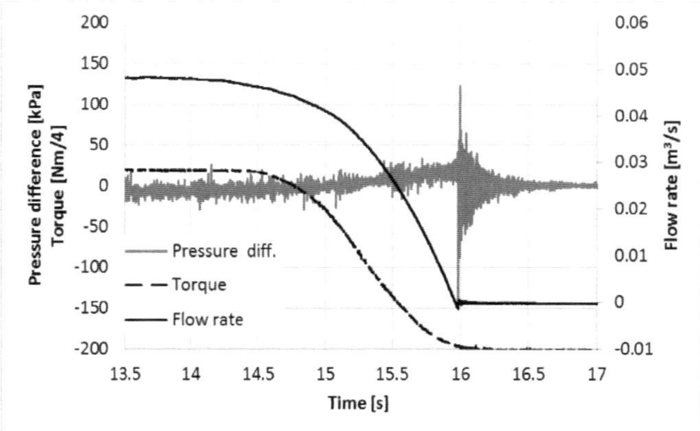

Figure 4. Flow rate change during pump braking.

Figures show pressure difference in front of the check valve. The pump started being braked when the flow rate was 48 l/s in this case. The maximal reverse flow rate was found 1.4 l/s, then the disc hit the seat and caused water hammer.

The limitation of the dynamometer is 800 Nm in either direction. This value defines maximal deceleration that can be reached in this system. In this case, the limit was reached after the check valve shut.

There is a question how to define the flow deceleration dv/dt. [5] uses the time period when the flow rate reaches zero from steady state without any further comments and value of steady flow velocity is used for velocity difference dv. [8] defines the time period from the moment when the downstream pressure starts increasing to the moment when this pressure reaches its maximal value. Velocity difference is the sum of the steady flow velocity and the maximal reverse velocity.

We determined expression dv/dt from the time when the torque at dynamometer is zero until the time when the reverse flow rate reaches maximal value. From figure 4: dt=15.98-14.77 and dv=(0.0448+0.0014)/S. Complete results are plotted in figure 5 and compared with results by [5].

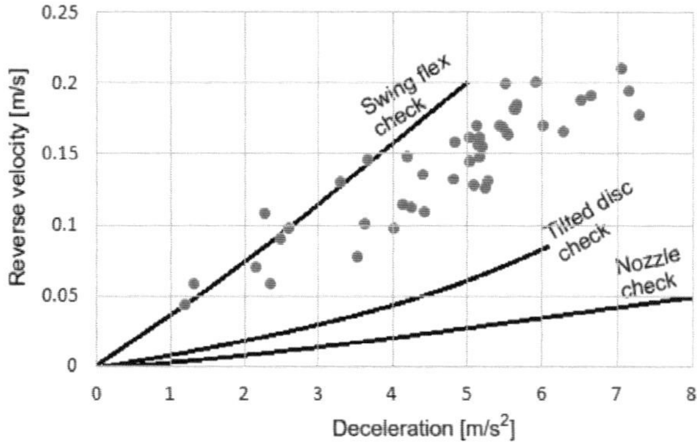

Figure 5. Dynamic characteristic of measured nozzle check valve (spots) and results by [5] (solid lines).

3. Summary and discussion

Static and dynamic characteristics of a nozzle check valve were measured. The check valve was mounted into the test circuit and, at first, the static characteristic was determined. Beside the flow coefficient, the disc position was observed with the camera. For dynamic characteristic determination, the flow deceleration was induced with the dynamometer which braked the pump. Maximal deceleration rate was 7.3m/s^2. The obtained results were compared with results presented in [5].

Unlike the static characteristic, the dynamic characteristic measurement is not treated in the standards. Different authors use different approach to the measurement as well as to the evaluation. Definition of deceleration rate dv/dt is crucial in this case. We decided to use two clearly identifiable points: the moment when the torque of the dynamometer is zero (pump starts braking) and the moment when the reverse flow rate is maximal (disc is hitting the seat). This can be one of the reasons why results by [5] vary noticeably (figure 5). Other reasons can include different construction of the check valve, different mass of moving parts or different spring stiffness, so the comparison is rather informative.

The measured points are scattered because they were obtained from different initial flow rates, so the disc opening varied and closing time depends on the current opening.

4. Acknowledgement

This paper is an output of NETME Centre, regional R&D centre built with the financial support from the Operational Programme Research and Development for Innovations within the project NETME Centre (New Technologies for Mechanical Engineering), Reg. no. CZ.1.05/2.1.00/ 01.0002 and, in the follow-up sustainability stage, supported through NETME CENTRE PLUS (LO1202) by financial means from the Ministry of Education, Youth and Sports under the "National Sustainability Programme I."

Table 1. Nomenclature.

Symbol	Unit	Meaning	Symbol	Unit	Meaning
K_v	m^3/s	flow coefficient	S	m^2	pipe cross-section
L	m	length	t	s	time
p	Pa	pressure	Δp	Pa	pressure difference
Q	m^3/s	flow rate	Δt	s	time difference
R	kg/m^7	resistance	ρ	kg/m^3	density

References

[1] Val-Matic valve and manufacturing corp. 2011 *Design and Selection Criteria of Check Valves* (Elmhurst: Val-Matic valve and manufacturing corp.) p 26.

[2] Chiu C 1987 Lessons learned from the songs unit 1 water hammer event *American Nuclear Society* **55** pp 289–91.

[3] Travis J R, Torrey M D 1985 Thermal hydraulics of a feed-water pipe breakage with a back pressure check-valve *American society of mechanical engineers*.

[4] McElhaney K L 2000 An analysis of check valve performance characteristics based on valve design *Nuclear engineering and design* **197** pp 169–82.

[5] Ballun J V 2007 A methodology for predicting check valve slam *Journal AWWA* **99** 3 pp 60–65.

[6] Thorley A R D 1991 *Fluid Transients in Pipeline Systems* (Hadley Wood: D&L George Ltd.).

[7] Pochyly F, Kozubkova M, Fialova S 2008 *9th International Conference on Flow-Induced Vibration (Prague)* pp 53–58

[8] Rahmeyer W 1996 Dynamic flow testing of check valves *Nuclear industry check valve group* (St. Petersburg, Florida) pp 9.

Effects of Pump-turbine S-shaped Characteristics on Transient Behaviours: Model Setup

Wei Zeng, Jiandong Yang, Jinhong Hu

State Key Laboratory of Water Resources and Hydropower Engineering Science, Wuhan University, Wuhan 430072, China

wzeng@whu.edu.cn

Abstract. Pumped storage stations undergo numerous transition processes, which make the pump turbines go through the unstable S-shaped region. The hydraulic transient in S-shaped region has normally been investigated through numerical simulations, while field experiments generally involve high risks and are difficult to perform. In this research, a pumped storage model composed of a piping system, two model units, two electrical control systems, a measurement system and a collection system was set up to study the transition processes. The model platform can be applied to simulate almost any hydraulic transition process that occurs in real power stations, such as load rejection, startup, frequency control and grid connection.

1. Introduction

To adjust the power system more efficiently and in a timely manner, pumped storage stations undergo numerous transition processes. Many engineering incidents occur in the process, such as turbine lifting accidents as a result of hydraulic transients and unsuccessful grid connections under low head. Therefore, it is important to investigate these transition processes to guarantee the security of the pumped storage station.

The S-shaped characteristics of the pump turbine are a critical factor in determining the transient characteristics of pumped storage stations. Different simulation methods have been applied to study the transition process in the S-shaped region [1-4]. Moreover, field experiments have also been conducted to obtain additional verification. Dörfler et al. partly closed the valve in the field to achieve pump turbine no-load stability [5]. Yang et al. analyzed the pressure pulsation characteristics of a prototype pump turbine during a transition process [6]. The transient experiments of load rejections and power failure must be performed in the commisioning stage to ensure safety and stability [7]. However, owing to the risks involved in field experiments, it is difficult to conduct experiments for specific engineering problems and some extreme conditions.

To compensate for the limitations of field experiments, the corresponding model experiments can be developed. The existing model experiment platforms for hydraulic machineries are generally closed-loop test rigs, which are mainly used to test the turbine characteristics in the constant states [8,9]. To study the transition processes of a Francis turbine, NTNU changed the piping system on the basis of the closed-loop test rig to set up the open-loop test rig. The model is used not only to test the turbine characteristics but also to measure the pressure pulsations at constant states [10,11] or during startup and load rejections [12,13]. However, some complex transition processes in actual pumped storage stations are difficult to achieve.

In summary, to investigate the transient characteristics of pump turbines and pumped storage stations, a more realistic pumped storage system is developed and introduced in this paper.

2. Basic information

The pumped storage system illustrated in Fig. 1 consists of model pipes with circulating water, model units (pump turbines and motor generators), a control system (excitation synchronization protection, frequency and phase conversion, speed control, monitoring system, load system), and a measurement system. The major functions and components of each system are shown in Table 1.

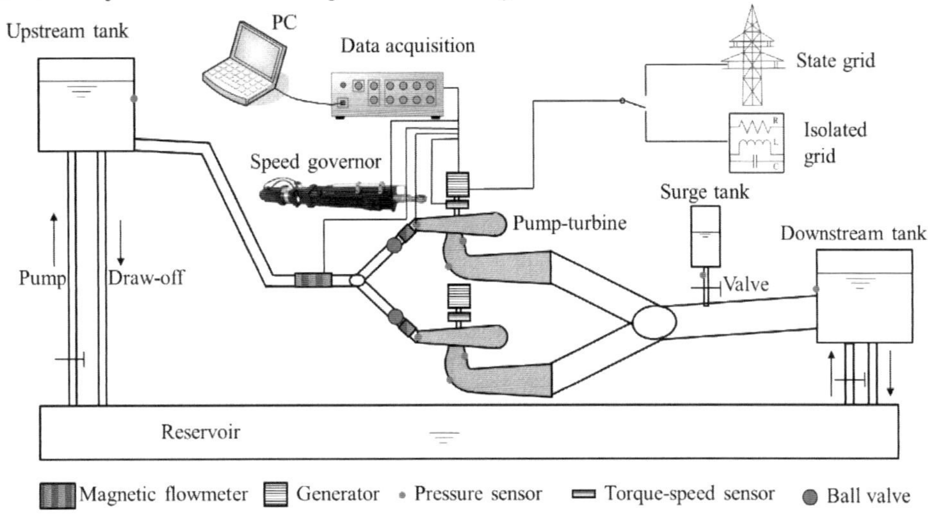

Figure 1. Pumped storage system model

Table 1. Subsystems of the pumped storage system

No.	Subsystem	Main functions
1	Model pipe	Simulates the actual power station pipes and supplies water to the model pump turbines
2	Circulating water system	Supplies power and water circulation for model experiments
3	Model units	Includes two sets of pump turbines and motor generators
4	Excitation and synchronization	Excitation control, synchronization, and protection of the motor generator
5	Frequency and phase conversion	Varies frequency startup in pump mode and conditions transformation in turbine mode
6	Speed control	Controls the speed governor to regulate the guide vane opening and regulates the ball-valve opening degree
7	Local control unit (LCU)	Gives orders to the model units and controls them
8	Load system	Simulates the grid features
9	Measurement system	Collects and processes the parameters

3. Experimental platform setup

3.1 Model pipes with circulating water

The pipe system of the pumped storage system model employs an annular arrangement pattern of two machines sharing the same main pipes. The section area of the downstream impedance surge tank is 0.43 m^2, and the impedance hole can be turned on or off through a valve according to the experimental requirements; the hole is normally closed. The model pipe system is mainly designed based on the pipe layout of the Xianju pumped storage station with the distorted scale including the radial scale

1:17.357 and the axial scale 1:42.366. The pipe system is shown in Fig. 2, and the main dimensions are shown in Table 2.

Figure 2. Model pipe system

Table 2. Pipe system dimensions

Pipe section	Inlet	Upstream main pipe	Upstream branch pipe	Upstream branch pipe	Downstream main pipe	Outlet
Equivalent diameter (m)	0.597	0.357	0.202	0.300	0.426	0.675
Length (m)	2.251	24.154	2.195	4.891	16.285	1.062

The circulating water system primarily provides water circulation for model experiments. Water is pumped from the underground reservoir by centrifugal pumps and supplied to the upstream and downstream tanks through the supply pipe (Fig. 1). The valves on the supply pipe can adjust the flow into the upstream and downstream tanks. Overflow weirs are installed in the tanks to maintain a constant water level, and the overflow out of the weirs flows into the underground reservoir through the draw-off pipes.

3.2 Model units

Two model pump turbines installed in the platform were obtained from the Harbin Electrical Corporation, Harbin, China. They are the pump turbine models at the Xianju pumped storage station (XJ) and the Baoquan pumped storage station (BQ). Both pump turbines basically have similar characteristics in the turbine high efficiency area, but differ significantly in the S-shaped region. The runner geometries are also distinct; these details are shown in Fig. 3. The basic parameters of the two pump turbines are presented in Table 3.

(a) BQ (b) XJ

Figure 3. Model runners

Table 3. Model unit parameters

N_{QE} (m:kW)	Inlet diameter (mm)	Outlet diameter(mm)	Guide vane height (mm)	Number of blades	Number of guide vanes	Rated n (rpm)	Rated H (m)	Rated Q (L/s)	Full rotational inertia (N.m)
37.91	280	146.34	24.44	9	20	1000	10.54	49.1	66.4

3.3 Control system

Each unit has a set of electrical cabinets to control the unit operation, including the local control unit, speed control system, excitation system, and load system. Their basic functions have been presented in Table 1. These systems can make the model units implement almost all the operations and conditions encountered in real power plants. The load system includes resistances, capacitances, and inductances to simulate an isolated grid with different characteristics by varying load combinations. In addition, the units can be also incorporated into the state grid, as shown in Fig. 1.

The PC-III programmable microprocessor speed governor (Fig. 1) is employed to control the pump turbine guide vanes with proportional–integral–derivative control. Its main function is to automatically track the grid frequency after unit startup and to achieve the best process control of startup, fast quasi-synchronization, frequency regulation (isolated network), power regulation, as well as opening degree regulation. In the experiments, the opening degree regulation function can dominate the guide vanes by various means, such as stalling, straight closure, and multi-phase closure.

4. Measurement of experiment data

4.1 Pressure measurement

The PCB 112A pressure sensors and MPM 480 pressure sensors were used to test the transient pressures, and the KYB 335 sensors were used to test the differential pressure of two points at the same section of the spiral case. The Fluke 719 pressure calibrator was used to calibrate the pressure sensors in the field. In the pipe systems, the main pipe, branch pipe, surge chamber, and upstream and downstream tanks were equipped with a certain number of MPM pressure sensors. MPM pressure sensors and PCB pressure sensors were installed on the points shown as Fig. 4 and Table 4. Table 5 presents the basic specifications of the major sensors used in the present tests.

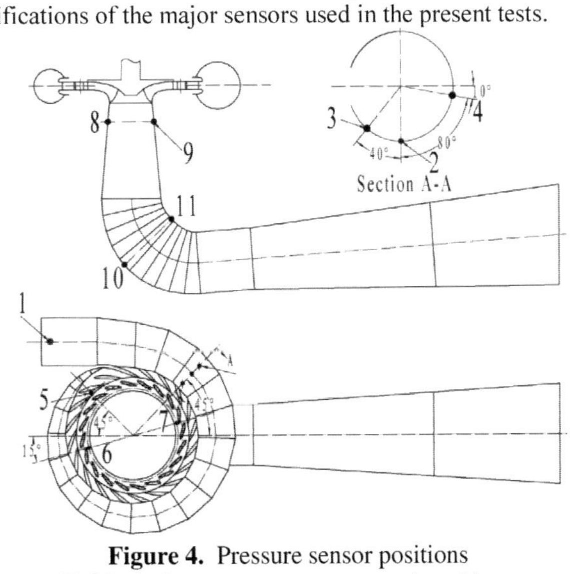

Figure 4. Pressure sensor positions

Table 4. Pressure sensor types and positions

No.	Sensor type	Position
1	MPM 480	Spiral case inlet
2	PCB 112A	The bottom of the spiral case
3 4	KYB 335	Two points at the same section of the spiral case
5	PCB 112A	stay vane
6	PCB 112A	Vaneless space 1
7	PCB 112A	Vaneless space 2
8	MPM 480	Draft tube inlet 1
9	MPM 480	Draft tube inlet 2
10	MPM 480	elbow bend 1
11	MPM 480	elbow bend 2

IOP Conf. Series: Journal of Physics: Conf. Series **813** (2017) 012055 doi:10.1088/1742-6596/813/1/012055

Table 5. Specifications of the different sensors

Sensor type	Positions	Measurement range	Accuracy degree (%)	Linearity (%)	Uncertainty(%)	Signal
Pressure sensor (Micro MPM 480)	1# Spiral case inlet	0–200 (kPa)	0.25	0.052 F.S	0.109	4–20 mA
	1# Draft tube inlet	−20–50 (kPa)		0.035 F.S	0.083	
	2# Spiral case inlet	0–200 (kPa)		0.051 F.S	0.079	
	2# Draft tube inlet	−20–50 (kPa)		0.053 F.S	0.621	
Flow meter (KROHNE IFM 4110)	Upstream main pipe	−200–200 (L/s)	0.3	0.128 F.S	/	4–20 mA /Frequency
Flow meter (OPTIFLUX 2000)	1# branch pipe	−70–70 (L/s)	0.2	0.157 F.S	/	
	2# branch pipe	−70–70 (L/s)		0.141 F.S		
Torque sensor (HLD09)	Between 1# turbine and 1# generator	−100–100 (N.m)	0.2	0.040% F.S	/	4–20 mA /Frequency
	Between 2# turbine and 2# generator	−100–100 (N.m)		0.020% F.S		

4.2 Flow measurement

Two electromagnetic flowmeters (Table 5) were installed in the model power station to measure the flows. To improve the accuracy of the experiment and to calibrate the electromagnetic flowmeters, a flow measurement device (Fig. 5) was developed by the static weighing method. To measure the forward flow (turbine direction), the supply valve of the downstream tank was closed and the flow passing through the units was drained from the draw-off pipe in the downstream tank. The draw-off pipe connects inlet A of the device and the drained water from outlet B are weighed. The opening and closure times of outlet B are t_1 and t_2, the entire opening duration is t_0, and the weighed water is m. Thus, one can get

$$Q = \frac{m}{\rho(t_0 + t_1/2 + t_2/2)} \tag{1}$$

Here, in the flow calibration test, $t_0 > 60$ s, t_1 and t_2 are less than 0.1 s, and the accuracy of the electronic scale is smaller than 0.05%. The calibration results for the electromagnetic flowmeters using such an apparatus are listed in Table 5.

A more precise measurement of dynamic flow is obtained by minimizing the time constant of the electromagnetic flowmeters to achieve the fastest response. However, there are always some delays in electromagnetic flowmeters. Therefore, the differential pressure sensor was installed in the spiral case to measure the dynamic flow. However, its accuracy is much lesser than that of the electromagnetic flowmeters. Nevertheless, the flow peaks and valleys measured by the differential pressure sensors can be used to calculate the delay time of the electromagnetic flowmeters. Thus, the delay response can be eliminated by adjusting the time coordinate.

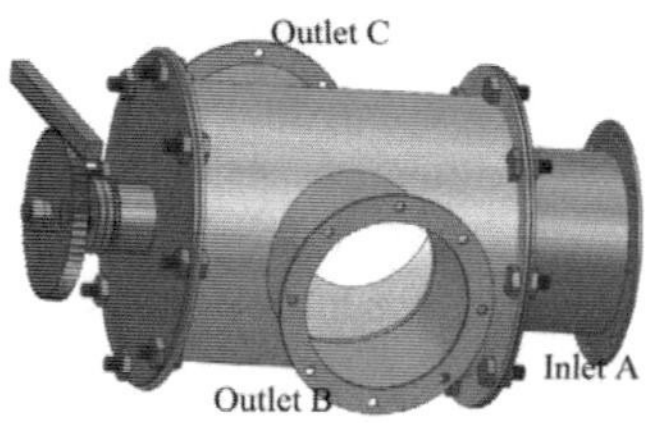

Figure 5. Flow calibrator

4.3 Speed and torque measurements

The HLD09 speed-torque sensor was installed between the turbine and generator, in which the torque applies the strain electrical logging principle: a strain gauge installed in a special shaft deforms with the torque and the resistance changes accordingly. The speed-torque sensors were calibrated as factory default, as shown in Table 5.

4.4 Guide vane movement measurement

The guide vane angles during transients were tested by the angle transducers installed on the top of the guide vane shaft. Using the standard gauge blocks, we obtained the relationship between the guide vane opening and the guide vane angles.

5. Data acquisition and processing

5.1 Data acquisition.

Different analog signals from these sensors were collected by a HBM Gen7i data acquisition system with a 24 bit analog-to-digital resolution, a 250 kHz maximum sampling frequency, and 70 channels. As the experiments in this study focused on the water-hammer pressure, rotational speed, and torque with low frequencies, we chose a 1 kHz sampling frequency because it is adequate for the experiments in this study.

5.2 Data processing.

Based on the previous studies by Trivedi [10-13], the Savitzky-Golay method [14] was used to extract the pressure pulsation in the transition process. The Savitzky-Golay method is a low-pass filter method that performs polynomial fitting of the original data, extracts the timed mean pressure based on the least squares method, and then obtains the pulsating pressure according to the differences between the original data and the timed mean pressure. Fig. 6 shows the measured value of the spiral case pressure and the filtered results by the Savitzky-Golay method under the condition that the one pump turbine with partial rotational inertia rejected the load with the guide vanes closed in 3 s.

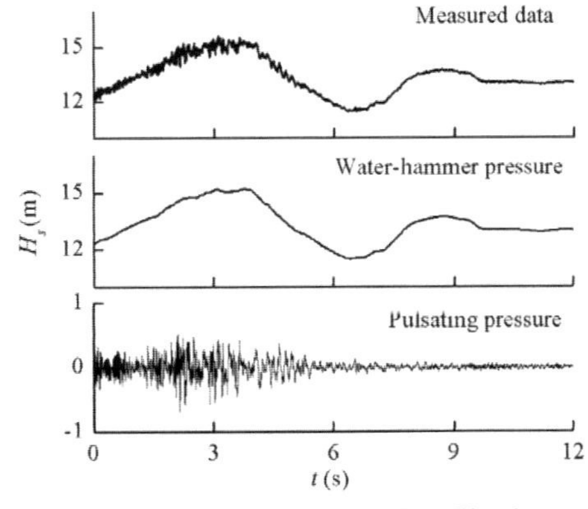

Figure 6. Savitzky-Golay filtering

6. Conclusions

In the present study, a pumped storage model was built to study transition processes of a pumped storage station. The model composed of two pump-turbine scales, two motor generator scales, pipe system and control system is available to conduct a variety of transient experiments happened in the real pumped storage stations. The transient pressures, flow, torque, rotational speed and guide vane movement can be tested with high precisions

In the next paper, the author would introduce the transient experiments conducted on the model to investigate the effects of pump-turbine S-shaped characteristics on hydraulic transients.

References

[1] C. Wang, J.D. Yang, Water Hammer Simulation Using Explicit-Implicit Coupling Methods, ASCE J. Hydraul. Eng., 141(2015) 04014086. DOI: 10.1061/(ASCE)HY. 1943-

7900.0000979

[2] W.J. Yang, J.D. Yang, W.C. Guo, W. Zeng, C. Wang, L. Saarinen and P. Norrlund, A mathematical model and its application for hydro power units under different operating conditions,' Energies, 8(2015) 10260-10275. (DOI: 10.3390/en80910260)

[3] C. Nicolet, Hydroacoustic Modeling and Numerical Simulation of Unsteady Operation of Hydroelectric Systems. PhD thesis, EPFL, Lausanne , Switzerland, 2007.

[4] X.X. Zhang, Y.G. Cheng, L.S. Xia and J.D. Yang, Dynamic Characteristics of a Pump-Turbine during Hydraulic Transients of a Model Pumped Storage System: 3D CFD Simulation. 27th IAHR Symposium on Hydraulic Machinery and System, Montreal, Canada, 2014.

[5] P.K. Dörfler, A.J. Engineer, R.N. Pendse, P. Huvet, M.V. Brahme, Stable Operation Achieved on a Single-Stage Reversible Pump-Turbine Showing Instability at no-Load, 19th IAHR Symposium, Section on Hydraulic Machinery and Cavitation, Singapore, 1998.

[6] J.D. Yang, J.H. Hu, W. Zeng, J.B. Yang, Transient Pressure Pulsations of Prototype Francis Pump-turbines. J. Hydraulic Engineering. (In Chinese)

[7] International Electrotechnical Commission, Field Acceptance Tests to Determine The Hydraulic Performance of Hydraulic Turbines, Storage pumps and Pump-Turbines, CEI/IEC 41, Third ed., Switzerland, 1999

[8] http://lmh.epfl.ch/page-57350-en.html

[9] http://lamh.gmc.ulaval.ca/laboratoire/installations/plateforme-dessais/

[10] C. Trivedi, M.J. Cervantes, B.K. Gandhi, O.G. Dahlhaug, Experimental and Numerical Studies for a High Head Francis Turbine at Several Operating Points, ASME J. Fluids Eng., 135 (2013) 111102. DOI: 10.1115/ 1.4024805

[11] C. Trivedi, M.J. Cervantes, O.G. Dahlhaug, B. K. Gandhi, Experimental Investigation of a High Head Francis Turbine During Spin-No-Load Operation, ASME J. Fluids Eng., 137 (2015) 061106. DOI: 10.1115/ 1.4029729

[12] C. Trivedi, M.J. Cervantes, B.K. Gandhi, O.G. Dahlhaug, Transient Pressure Measurements on a High Head Model Francis Turbine During Emergency Shutdown, Total Load Rejection, and Runaway, ASME J. Fluids Eng., 136(2014) 121107. DOI: 10.1115/ 1.4027794

[13] C. Trivedi, M. Cervantes, B. Gandhi, and O. Dahlhaug, 2014, "Experimental Investigations of Transient Pressure Variations in a High Head Model Francis Turbine During Start-up and Shutdown," J Hydrodyn, 26(2), pp. 277-290. doi:10.1016/S1001-6058(14)60031-7.

[14] R.W. Schafer, What is a Savitzky-Golay filter, IEEE Signal Processing Magazine, 7(2011) 111-117. (DOI: 10.1109/MSP.2011.941097)